Advanced Structured Materials

Volume 226

Series Editors

Common engineering materials are reaching their limits in many applications, and new developments are required to meet the increasing demands on engineering materials. The performance of materials can be improved by combining different materials to achieve better properties than with a single constituent, or by shaping the material or constituents into a specific structure. The interaction between material and structure can occur at different length scales, such as the micro, meso, or macro scale, and offers potential applications in very different fields.

This book series addresses the fundamental relationships between materials and their structure on overall properties (e.g., mechanical, thermal, chemical, electrical, or magnetic properties, etc.). Experimental data and procedures are presented, as well as methods for modeling structures and materials using numerical and analytical approaches. In addition, the series shows how these materials engineering and design processes are implemented and how new technologies can be used to optimize materials and processes.

Advanced Structured Materials is indexed in Google Scholar and Scopus.

Holm Altenbach · Leonhard Hitzler ·
Michael Johlitz · Markus Merkel · Andreas Öchsner
Editors

Lectures Notes on Advanced Structured Materials 3

Springer

Editors
Holm Altenbach
Institut für Werkstoffe, Technologien und Mechanik
Otto-von-Guericke-Universität Magdeburg
Magdeburg, Germany

Michael Johlitz
Institut für Mechanik
Universität der Bundeswehr München
Neubiberg, Germany

Andreas Öchsner
Faculty of Mechanical and Systems Engineering
Esslingen University of Applied Sciences
Esslingen am Neckar, Baden-Württemberg
Germany

Leonhard Hitzler
Department of Materials Engineering
Technische Universität München
Garching, Bayern, Germany

Markus Merkel
Institute for Virtual Product Development
Aalen University of Applied Sciences
Aalen, Baden-Württemberg, Germany

ISSN 1869-8433 ISSN 1869-8441 (electronic)
Advanced Structured Materials
ISBN 978-3-031-77402-7 ISBN 978-3-031-77403-4 (eBook)
https://doi.org/10.1007/978-3-031-77403-4

This Springer imprint is published by the registered company Springer Nature Switzerland AG
The registered company address is: Gewerbestrasse 11, 6330 Cham, Switzerland

Preface

The postgraduate seminar series on advanced structured materials is designed to facilitate teaching and informal discussion in a supportive and friendly environment. The seminar provides a forum for postgraduate students to present their research results and train their presentation and discussion skills. Furthermore, it allows for extensive discussion of current research being conducted in the wider area of advanced structured materials. Doing so, it builds a wider postgraduate community and offers networking opportunities for early career researchers. In addition to focused lectures, the seminar provides specialized teaching/overview lectures from experienced senior academics. The 2024 Postgraduate Seminar entitled "Advanced Structured Materials: Development—Manufacturing—Characterization—Applications" was held from 20th till 24th May 2024 in Porto. The presented postgraduate lectures had a strong focus on polymer mechanics, composite materials and additive manufacturing.

Magdeburg, Germany — Holm Altenbach
Garching, Germany — Leonhard Hitzler
Neubiberg, Germany — Michael Johlitz
Aalen, Germany — Markus Merkel
Esslingen am Neckar, Germany — Andreas Öchsner

Contents

Part I
Overview Lectures

Chapter 1
Creep Mechanics—A Long History and Current Developments

Holm Altenbach

Abstract Creep mechanics is a branch of continuum mechanics that began to develop independently in the nineteenth century. In the 1930s, the first theories allowing a structural analysis considering creep were developed. Later, the one-dimensional approaches were gradually extended to three-dimensional, isotropic, anisotropic, etc. equations. In recent years, there have been many publications on the state of the art of research and numerical solution methods. Some important steps of creep and creep-damage mechanics will be discussed below. Finally, some contributions of the author's group will be highlighted.

1.1 Some Historical Remarks

1.1.1 Starting Point—The Early Period of Creep Mechanics

The starting point for creep mechanics were some accidents in the nineteenth century with tragic outcome. The analysis of the accidents showed that the mechanical structures (for example, steam machines and locomotives) were under relatively low mechanical loads. The failure occurred at stresses that were significantly below the yield stress of the materials used. At the same time, the accidents observed were a result of the elevated operating temperature level and the damage did not occur immediately after the start of the mechanical load, but with a time delay. The question of how to avoid these cases of damage was raised and what countermeasures can be taken. Unfortunately, at that time there was no mechanical theory to describe the time-dependent material behavior, and there was also a lack of understanding in materials science of the microstructural changes in the material at elevated temperatures.

H. Altenbach (✉)
Lehrstuhl für Technische Mechanik, Institut für Werkstoffe,Technologien und Mechanik, Fakultät für Maschinenbau, Otto-von-Guericke-Universität, Magdeburg, Germany
e-mail: holm.altenbach@ovgu.de

H. Altenbach et al. (eds.), *Lectures Notes on Advanced Structured Materials 3*,
Advanced Structured Materials 226, https://doi.org/10.1007/978-3-031-77403-4_1

The creep phenomenon was noticed by Navier[1] (1826) and Coriolis[2] (1830), first experiments concerning the creep process were performed by Vicat (1834) for iron wire. There were publications on creep mechanics already in the nineteenth century. However, the first systematic overview on creep was published by da Costa Andrade[3] (1910, 1914). Since these publications, it is undisputed that creep phenomena can only be described if the time dependence is taken into account and the corresponding constitutive equation is nonlinear. At the same time, it must be accepted that the number of constitutive parameters increases in comparison to the elastic behavior.

1.1.2 Papers on the History of Creep Mechanics and Scientific Events Devoted to Problems in Creep Mechanics

The main stages of the history of creep mechanics until the 1970s are given, for example, by Odqvist[4] and Hult[5] (1962) and Odqvist (1981). Recently, some additional papers on the history and actual tendencies in research were published; see, for example, Hudramovich (2015), Naumenko and Altenbach (2016), Altenbach and Eisentrager (2020), Altenbach and Knape (2020), and Saitova et al. (2024). A rough classification of the main stages of the development of creep mechanics can be given as follows:

- experimental observations,
- first constitutive equations,
- introduction of three-dimensional equations of different levels,
- development of damage mechanics, and
- advanced (for example, mechanism-based) models.

[1] Claude Louis Marie Henri Navier (*10 February 1785, Dijon, Kingdom of France; †21 August 1836, Paris, July Monarchy), French mechanical engineer working for the French government and physicist developing basics in continuum mechanics, for example, the general theory of elasticity (1821) and Navier-Stokes equations in fluid mechanics (1822).

[2] Gaspard-Gustave de Coriolis (*21 May 1792, Paris, France; †19 September 1843, Paris, France), French mathematician, mechanical engineer, and scientist, known for the introduction of supplementary forces in a rotating frame of reference (Coriolis effect); first introduced the term work (travail).

[3] Edward Neville da Costa Andrade (*27 December 1887, London, UK; †6 June 1971, London, UK), English physicist, writer, and poet, known for a creep law for the primary creep stage and the determination of the wavelengths of gamma-rays from radium.

[4] Folke Karl Gustav Odqvist (*29 July 1899, Stockholm, Sweden; †7 May 1984, Danderyd, Sweden), Swedish engineer and professor of strength of materials with significant contributions in creep mechanics.

[5] Jan Alrik Harald Hult (*9 December 1927, Stockholm, Sweden; †20 February 2013, Lerum, Sweden), Swedish civil engineer with outstanding contributions in particular in creep and damage mechanics.

The creep behavior of a material is characterized by three stages: the primary or transient creep (the creep rate is decreasing and the hardening is dominant), secondary or steady-state creep (the creep rate is constant and hardening and softening are in a balance), and tertiary or accelerating creep (the creep rate is increasing and softening and damaging processes are dominant). This type of behavior can be observed for many materials, although the stages may vary.

The first creep constitutive equation was introduced by Norton[6] (1929) and Bailey[7] (1935)—the Norton-Bailey law

$$\dot{\varepsilon} = A\sigma^n \tag{1.1}$$

with the creep rate $\dot{\varepsilon}$, the applied stress σ, and two material parameters—the magnitude A and the creep exponent n. Note that the identification of the parameters requires more experimental effort in comparison to elastic behavior and, in addition, there are more constitutive parameters in comparison with the analog elastic models. Equation (1.1) is a one-dimensional power law (in the sense of mathematics). Later, also other one-dimensional equations were introduced (see, for example, Naumenko and Altenbach (2016)—but all suggested equations were nonlinear). With other words, in creep mechanics there is no physically linear variant like in elasticity. Among others, there is one advantage of these simple constitutive equations for the secondary creep: the dependence on the temperature can be considered in a very easy manner by multiplying an Arrhenius[8]-type term

$$\mathrm{e}^{-\frac{Q}{RT}}$$

with Q as the activation energy for creep, R as the universal gas constant, and T as the absolute temperature.

In the 1930s, the number of reports on several applications to energetic systems in theoretical journals like the Zeitschrift für Angewandte Mathematik und Mechanik increased even more. As an example we present here the publication of Stodola[9] (1933), who reported about gas turbine applications. At the same time he presented a nice definition of creep (see Fig. 1.1).

Mechanical loads are multiaxial. Equation (1.1) can be extended to this case in a simple manner. The first step is the substitution of the stress and the strain rate in

[6] Frederick Harwood Norton (*23 October 1896, Manchester-by-the-Sea, Essex, Massachusetts, U.S.A.; †27 November 1986, Gloucester, Essex, Massachusetts, U.S.A.), US-American physicist, professor at MIT.

[7] Richard William Bailey (*6 January 1885, Romford, Essex, UK; †4 September 1957, Hinckley, Leicestershire, UK), Research Department of the Metropolitan-Vickers Electrical Company, Ltd., Manchester, F.R.S. since 1949.

[8] Svante August Arrhenius (*19 February 1859, Wik, Sweden-Norway; †2 October 1927, Stockholm, Sweden), Swedish scientist, founder of physical chemistry, Nobel prize in Chemistry 1903.

[9] Aurel Boreslav Stodola (*10 May 1859, Liptószentmiklós/Liptovský Mikuláš, Kingdom of Hungary; †25 December 1942, Zurich, Switzerland), Slovak engineer, physicist, and inventor, a pioneer in the area of technical thermodynamics and its applications.

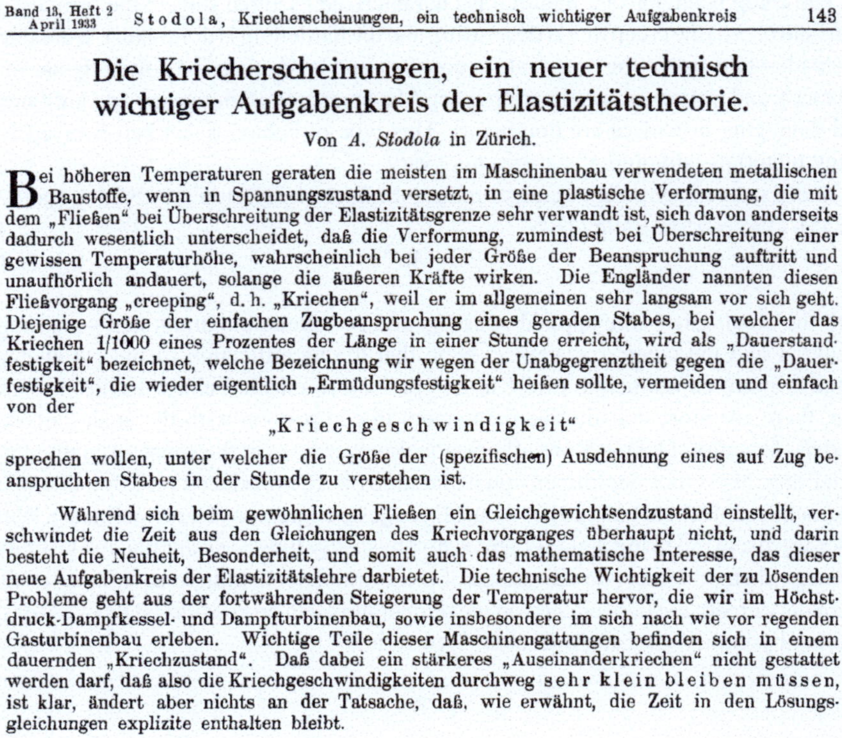

Band 13, Heft 2 April 1933 Stodola, Kriecherscheinungen, ein technisch wichtiger Aufgabenkreis 143

Die Kriecherscheinungen, ein neuer technisch wichtiger Aufgabenkreis der Elastizitätstheorie.

Von *A. Stodola* in Zürich.

Bei höheren Temperaturen geraten die meisten im Maschinenbau verwendeten metallischen Baustoffe, wenn in Spannungszustand versetzt, in eine plastische Verformung, die mit dem „Fließen" bei Überschreitung der Elastizitätsgrenze sehr verwandt ist, sich davon anderseits dadurch wesentlich unterscheidet, daß die Verformung, zumindest bei Überschreitung einer gewissen Temperaturhöhe, wahrscheinlich bei jeder Größe der Beanspruchung auftritt und unaufhörlich andauert, solange die äußeren Kräfte wirken. Die Engländer nannten diesen Fließvorgang „creeping", d. h. „Kriechen", weil er im allgemeinen sehr langsam vor sich geht. Diejenige Größe der einfachen Zugbeanspruchung eines geraden Stabes, bei welcher das Kriechen 1/1000 eines Prozentes der Länge in einer Stunde erreicht, wird als „Dauerstand·festigkeit" bezeichnet, welche Bezeichnung wir wegen der Unabgegrenztheit gegen die „Dauer·festigkeit", die wieder eigentlich „Ermüdungsfestigkeit" heißen sollte, vermeiden und einfach von der

„Kriechgeschwindigkeit"

sprechen wollen, unter welcher die Größe der (spezifischen) Ausdehnung eines auf Zug be·anspruchten Stabes in der Stunde zu verstehen ist.

Während sich beim gewöhnlichen Fließen ein Gleichgewichtsendzustand einstellt, ver·schwindet die Zeit aus den Gleichungen des Kriechvorganges überhaupt nicht, und darin besteht die Neuheit, Besonderheit, und somit auch das mathematische Interesse, das dieser neue Aufgabenkreis der Elastizitätslehre darbietet. Die technische Wichtigkeit der zu lösenden Probleme geht aus der fortwährenden Steigerung der Temperatur hervor, die wir im Höchst·druck·Dampfkessel· und Dampfturbinenbau, sowie insbesondere im sich nach wie vor regenden Gasturbinenbau erleben. Wichtige Teile dieser Maschinengattungen befinden sich in einem dauernden „Kriechzustand". Daß dabei ein stärkeres „Auseinanderkriechen" nicht gestattet werden darf, daß also die Kriechgeschwindigkeiten durchweg sehr klein bleiben müssen, ist klar, ändert aber nichts an der Tatsache, daß, wie erwähnt, die Zeit in den Lösungs·gleichungen explizite enthalten bleibt.

Fig. 1.1 Copy of the first part of Stodola's paper (Stodola 1933)

the one-dimensional equation by their tensorial counterparts and the introduction of an equivalence hypotheses between the one- and three-dimensional states. Such an extension was presented by Odqvist (1962) and Bailey (1935) in the 1930s. They introduced a corresponding theory for isotropic material behavior using invariants of the stress and the strain rate tensors. The complete and consistent tensorial description was provided by Prager[10] (1945) and Reiner[11] (1945), which includes, in addition, anisotropy. The starting point of these developments was the theory of invariants of isotropic second-rank tensors. In Betten (2008) an excellent summary of the creep equations is given for the primary, the secondary, and the tertiary creep behavior of isotropic and anisotropic materials. Since that time there are some open questions:

[10] William/Willy Prager (*23 May 1903, Karlsruhe, German Empire; †17 March 1980, Savognin/Zurich, Switzerland), German-born American applied mathematician, in mechanics well-known for his contributions to the theory of plasticity, for example, the Drucker-Prager yield criterion.

[11] Markus Reiner (*5 January 1886, Czernowitz, Bukovina, Austria-Hungary; †25 April 1976, Haifa, Israel), Austrian-Israeli scientist and a founder of rheology.

- The split into deviatoric part and hydrostatic (volumetric) part of the stress and the strain tensors and the constitutive equations, based on the axioms of rheology (Reiner 1960), is essential for isotropic material behavior. What should we do in the case of anisotropy? The split can be used, but we cannot consider such a split for the constitutive equations.
- The hydrostatic behavior is assumed to be purely elastic. But this is not always correct—see Bridgeman's[12] experiments (Bridgeman 1952).

Further modifications were established considering existing mismatches with some experimental results. In this case, modified creep equations were suggested, e.g. the strain hardening theory presented by Nadai[13] (1938) and Soderberg[14] (1938), and which were discussed, for example, by Rimrott[15] (1968). Note that the suggested theories are very "sophisticated"—no rational ideas like in continuum mechanics or theory of viscoelasticity are included.

In the 1950s, new structural-mechanical problems also emerged in connection with creep problems. Stability problems considering creep were presented by Hoff (1953, 1957). Elements of the geometrically nonlinear theory had to be developed (von Kármán[16]-type theories). Generally large strain problems were not considered because in applications time-dependent large strain are not allowed.

The increasing use of polymer materials resulted in the development of new theories, partly based on analogies between viscoelasticity and creep. The viscoelastic behavior is often mathematically described with the help of integral equations based on the hereditary theory of elasticity developed by Rabotnov[17] (1980) among others.

Hereditary mechanics is a branch of continuum mechanics in which such processes are studied when the state of a mechanical system depends on the history of actions performed on the system. The mathematical apparatus of hereditary mechanics is the theory of integral equations and fractional differential equations. The main object of

[12] Percy Williams Bridgman (*21 April 1882, Cambridge, Massachusetts, U.S.A.; †20 August 1961, Randolph, New Hampshire, U.S.A.), American physicist who received 1946 the Nobel Prize in Physics.

[13] Arpád Ludwig Nádai (*3 April 1883, Budapest, Hungary; †18 July 1963, Budapest, Hungary), Hungarian-born academic, professor of mechanics.

[14] Carl Richard Söderberg/Soderberg (*3 February 1895, Ulvohamm, Sweden; †17 October 1979, Cambridge, Massachusetts, U.S.A.), Swedish-born academic, professor of mechanical engineering at MIT.

[15] Friedrich Paul Johannes Rimrott (*4 August 1926, Halle, Germany; †25 August 2003, Toronto, Canada), German-Canadian mechanical engineer, professor University Toronto, 1992–1997 Otto-von-Guericke-Universität Magdeburg, Germany.

[16] Theodore von Kármán/Tódor Kármán (*11 May 1881, Budapest, Austria-Hungary; †6 May 1963, Aachen, West Germany), Hungarian-American mathematician, aerospace engineer, and physicist, considering linear in-plane and quadratic deflection approximations in the kinematics of thin structures.

[17] Yuri Nikolaevich Rabotnov (*11[Julian calendar](24[Gregorian calender]) February 1914, Nizhny Novgorod, Russian Empire; †13 May 1985, Moscow, Soviet Union), Soviet scientist-mechanic, full member of the USSR Academy of Sciences since 1958, Head of the Department of Plasticity Theory of Moscow State University (1953–1985), Deputy Director of the Institute of Hydrodynamics of the Siberian Branch of the USSR Academy of Sciences (1958–1965).

the study of hereditary mechanics is viscoelastic materials. Considering this another application field should be mentioned—creep processes in concrete which were studied, for example, by Arutunyan[18] (1966). The last one was the starting point for the development of a new branch of mechanics of deformable bodies—mechanics of growing solids developed by Arutunyan and his student Manzhirov[19] (Arutiunian 1977; Manzhirov and Lychev 2012).

From the end of 1950s special attention was paid to tertiary creep as the final stage of creep in materials and structures. This was the starting point of a new branch of creep mechanics which was called creep-damage mechanics or only damage mechanics. The development of this research field started with pioneering works of Kachanov[20] (1958) and Rabotnov (1959). Both introduced a new variable characterizing the damage—the first one named this variable continuity ($1 \leq \psi \leq 0$), the second one—damage ($0 \leq \omega \leq 1$). The ranges of these parameters are bounded; however, the bounds are theoretical values. As was mentioned by Lemaitre (1996), there is no perfect material that means $\psi = 1$ or $\omega = 0$ are not possible and fracture occurs far from the final value $\psi = 0$ or $\omega = 1$. The concepts of Kachanov or Rabotnov have an analogy to the ratio number relating the cycle to the number of cycles at failure introduced by Palmgren (1924), Langer (1937), and Miner (1945). They are valid for isotropic creep-damage. The pure phenomenological approach was later more and more founded upon knowledge of material scientists. From the beginning of 1980s, mechanism-based approaches started to develop (see, for example, Trampczynski et al. 1981; Kowalewski et al. 1994; Othman et al. 1994; McLean and Dyson 2000). During the last years, more and more models for anisotropic damage were suggested (Chow and Wang 1987; Chow et al. 2000; Skrzypek and Ganczarski 2003; Lokoshchenko and Platonov 2004; Hyde and Sun 2005; Desmorat and Cantournet 2008; Nosal and Ganczarski 2023).

[18] Nagush Khachaturovich Harutyunyan/Arutunyan (*10 Julian calendar (23 Gregorian calender) November 1912, Yerevan, Russian Empire; †18 January 1993, Moscow, Russian Federation), Armenian-Soviet scientist in the field of mechanics and statesman, academician of the National Academy of Sciences of the Republic of Armenia.

[19] Alexander Vladimirovich Manzhirov (*24 May 1957, Rostov-on-Don, Soviet Union; †3 September 2018, Moscow, Russian Federation), Soviet and Russian scientist-mechanic, foreign member of the National Academy of Sciences of the Republic of Armenia.

[20] Lazar Markovich Kachanov (*26 June Julian calendar (9 Gregorian calender July) 1914, Surazh, Chernihiv Governorate, Russian Empire; †10 December 1993, Boston, U.S.A.), Russian-American scientist in the field of mechanics with fundamental contributions to theory of plasticity, fracture mechanics, and damage mechanics.

1.1.3 Scientific Events Devoted to Creep Mechanics

In 1960, a symposium "Creep in Structures" among the IUTAM[21]-Symposia was organized for the first time. The complete list of all symposia in this series is given in Table 1.1.

There are other conferences devoted to selected creep and creep-damage problems, for example Skrzypek and Ganczarski (2003), Ohno and Uehara (2007), and Huh et al. (2009). The International Association for Applied Mathematics and Mechanics (GAMM) offered plenary lectures during their annual conferences, and in 1998 and 2001 the topic of plenary lectures were related to creep mechanics (Betten 1998; Altenbach 2002). In 1987 and 1999, the International Centre for Mechanical Sciences organized a special courses on creep problems. The first one was coordinated by Kraijcinovic[22] and Lemaitre (1987) and the second one by Altenbach and

Table 1.1 IUTAM symposia "Creep in Structures"

Year	Location	Organizer(s)	Proceedings
1960	Stanford (U.S.A.)	Nicolas J. Hoff[a]	Hoff (1962)
1970	Gothenburg (Sweden)	Jan Hult	Hult (1972)
1980	Leicester (U.K.)	Alan Ponter and David Hayhurst	Ponter (1981)
1990	Kraków (Poland)	Michaił Życzkowski[b]	Zyczkowski (1991)
2000	Nagoya (Japan)	Sumio Murakami[c] and Nobotada Ohno	Murakami and Ohno (2001)
2012	Paris (France)	Serge Kruch	Altenbach and Kruch (2013)
2023	Magdeburg (Germany)	Holm Altenbach and Konstantin Naumenko	Altenbach and Naumenko (2023)

Comments The IUTAM symposium, held in *Paris in 2012*, was a continuation, but with a new title "Advanced Materials Modelling for Structures" with the focus on new materials and on generalized and unified models of inelastic deformation. For the IUTAM symposium, held in *Magdeburg in 2023*, the traditional name was used again, but there was a delay with respect to corona disaster

[a] Nicolas/Miklós John Hoff (*3 January 1906 in Magyarovar, Austro-Hungarian Empire; †4 August 1997 in Stanford, Kalifornien, U.S.A.), Hungarian-born American engineer specializing in aeronautics and astronautics, contributions to structural stability analysis

[b] Michaił Życzkowski (*12 April 1930 in Kraków, Poland; †24 May 2006 in Kraków, Poland), Polish mechanical engineer with focus on analysis of the inelastic behavior of structures

[c] Sumio Murakami (*8 January 1937, Nagoya, Japan; †9 October 2015, Nagoya, Japan), Japanese mechanical engineer with contributions to creep-damage mechanics

[21] International Union of Theoretical and Applied Mechanics.

[22] Dusan Krajcinovic (*28 March 1935, Zagreb, Kingdom of Yugoslavia; †10 August 2007, Madison, Wisconsin, U.S.A.), specialist in mechanics of materials, and structural analysis and design.

Skrzypek[23] (1999). Apart from the above-mentioned conferences on creep in structures, conferences on damage were held since 1979 (EUROMECH Colloquia 115 Boehler 1982, 147, and 251 and Symposium on Continuum Damage and Fracture Benallal 2000).

1.2 Contributions from the Author's Research Group at the Universities in Halle and Magdeburg

Research work began in the author's working group at the end of the 1980s, initially at the University of Magdeburg[24] until 1995, later at the University of Halle[25] until 2011, and then again in Magdeburg. Selected research topics are briefly presented below and important references are provided.

The starting point for our research work was some fundamental textbooks and monographs on creep and damage mechanics (for example, Finnie and Heller 1959; Kachanov 1960; Odqvist and Hult 1962; Nowacki 1965; Hult 1966; Odqvist 1974; Rzhanitsyn 1968; Rabotnov 1969; Penny and Marriott 1971; Odqvist 1974; Malinin 1975; Kraus 1980; Malinin 1981; Boyle and Spence 1983; Kachanov 1986; Čadek 1988; Findley et al. 1989; Lemaitre and Chaboche 1990; Skrzypek 1993; Haddad 1995; Penny and Mariott 1995; Krajcinovic 1996; Lemaitre 1996; Voyiadjis et al. 1998; Lemaitre and Desmorat 2005; Betten 2008; Murakami 2012; Hyde et al. 2013; Lokoshchenko 2017 among others). Most of our results are summarized in Naumenko and Altenbach (2007, 2016, 2020).

1.2.1 Kachanov-Rabotnov Approach and Mechanism-Based Models

Since the mid of the 1990s, classical creep mechanics considering damage and applied to thin-walled structures was in the focus (Altenbach and Naumenko 1995; Altenbach et al. 1997a, b, c; Altenbach 1999). All these papers were based on classical creep mechanics and the damage approach of Rabotnov. The analysis of thin-walled structural elements was based on material independent and dependent equations. The first one can be divided into geometrical relations (e.g. strain-displacement equations) considering the geometrical linear approach and equations of motion or equations of equilibrium. The second one is a set of equations describing the constitutive behavior.

[23] Jacek Jan Skrzypek (*6 July 1940, Sułkowice, Poland (occupied by Germany); †3 October 2018, Kraków, Poland), full professor Politechnika Krakowska (Cracow University of Technology) with contributions to the theory of plasticity and creep mechanics.

[24] Technische Universität Magdeburg (since 1987), now Otto-von-Guericke-Universität Magdeburg (since 1993).

[25] Martin-Luther-Universität Halle-Wittenberg.

The behavior of plates and shells operating at elevated temperatures is characterized by time-dependent equations reflecting creep. This behavior can be presented in different manners (depending on the effects which are taken into account). In the mentioned papers, traditional equations for creep (like Norton's law) and damage (in Rabotnov's sense) are introduced; however, also more complicated equations for different behavior at tension and compression are presented. Applications were related to plate and shell problems.

In Altenbach et al. (1999, 2000), results of finite element solutions using the ANSYS code are presented. With the help of the numerical analysis, the following questions were discussed:

- Special solutions for beams and plates under creep-damage conditions are presented applying the Ritz method. The solutions allowing the verification of the ANSYS user defined creep-damage subroutine taking into account the damage evolution.
- The mesh sensitivity of long-term solutions for creep-damage bending of beams and plates was studied.
- Finally, the results for time-dependent deflections and stresses based on different plane stress, shell, and solid elements available in the ANSYS code for the plasticity and creep analysis were compared. For the simplicity of solutions steady state loads and temperatures were assumed and the problems of damage localization were ignored.

The main conclusions can be summarized as

- The approximations or meshes justified for the elastic solutions using displacement-based variational methods cannot be used for the creep damage analysis which was in contrast to the commercial finite element code manuals of that time.
- The mesh established basing on convergent steady state creep solutions can be used for the continuum damage mechanics analysis of thin-walled structures in bending.

1.2.2 Damage Activation, Shear Correction, and Edge Effects

The damage activation and deactivation were considered, e.g. in Altenbach et al. (2001). Based on the experimental results of copper at 250 °C (Murakami and Sanomura 1985), the isotropic and anisotropic damage models as well as the mechanism of the damage activation or deactivation (Hansen and Schreyer 1995) are established and coupled in one constitutive equation. With the help of the finite element method, the creep-damage behavior of copper under different stress states is simulated. In the cases of the spontaneous reversal of the shear stress, the rotation of the principal directions of the stress tensor leads to a delayed rotation of the principal strain directions; therefore, a change of the damage state (close or reopen of the micro-cracks) is induced. The predictions were compared with those based on the isotropic model as well as the anisotropic model without the activation mechanism.

The shear correction factors in improved theories of beams, plates, and shells are sometimes essential. In the case of creep-damage analysis of beams, plates, and shells, this is not well studied and in Altenbach (2002) a shear correction factor was presented. The applicability of classical engineering models for beams, plates, and shells to the creep-damage analysis was discussed, and it was shown that a first improvement of the classical approach can be given within the assumptions of the first order shear deformation theory first presented by Reissner[26] (1944, 1945) and Mindlin[27] (1951). Based on the beam equations, it was demonstrated that the shear correction factors have to be modified within the time-step analysis.

Edge effects were investigated in Altenbach et al. (2004). Finite element solutions for moderately thick plates under creep-damage conditions based on shell and solid type finite elements were discussed. The results illustrate the time-dependent stress redistributions in the edge zones. It was shown that the transverse normal and shear stresses may have a significant influence on the damage evolution and must be considered in numerical lifetime estimations which means that "more accurate" cross-section approximations for the transverse normal and shear stresses have to be used in the case of creep.

1.2.3 *Applications and Non-classical Effects*

An interesting application case—creep in a multipass weld—was discussed in Naumenko and Altenbach (2005). Based on the contributions of T. Hyde and co-workers (Hyde et al. 1997, 2003a, b), an engineering creep theory using the creep potential hypothesis, the flow rule, and assumption of transverse isotropy was developed. The main result was a coordinate-free equation for secondary creep formulated in terms of the Norton-Bailey-Odqvist creep potential and transversally isotropic invariants of the stress tensor. The material constants were identified according to the experimental data presented in the literature.

In Altenbach et al. (1999) non-classical effects of creep and creep-damage coupling are considered. Here were suggested traditional equations for creep (like Norton's law) and damage (in Rabotnov's sense), but also more complicated equations

[26] Max Erich (Eric) Reissner (*5 January 1913, Aachen, German Empire; †1 November 1996, La Jolla, CA, U.S.A.), German-American civil engineer and mathematician, Professor of Mathematics at the Massachusetts Institute of Technology. He was recipient of the Theodore von Kármán Medal in 1964, and the ASME Medal in 1988. He was a honorary member of the American Society of Mechanical Engineers and the German Society for Applied Mathematics and Mechanics (Gesellschaft für Angewandte Mathematik und Mechanik).

[27] Raymond David Mindlin (*17 September 1906, New York City, U.S.A.; †22 November 1987, Hanover, New Hampshire, U.S.A.), American mechanical engineer, Professor of Applied Science at Columbia University. He is known as mechanician, who made seminal contributions to many branches of applied mechanics, applied physics, and engineering sciences.

for different behavior in tension and compression. Some mathematical aspects of the material parameter's ranges were discussed in Altenbach et al. (1999).

The geometrical linear and nonlinear applications and the wide stress ranges were in the focus since the end of the 1990s. Additionally, considering engineering applications, problems of thin-walled structures were analyzed using both geometrical linear and nonlinear formulations. Since 1997, instead of a purely phenomenological approach, mechanism-based models were applied (Altenbach et al. 1997a). In addition, the classical creep theory was extended to creep behavior considering wide ranges of stress levels (Altenbach et al. 2008; Naumenko et al. 2009) introducing a double power law (Figs. 1.2 and 1.3).

The creep for materials exhibiting non-classical creep behavior (for example, different tension and compression behavior, the equivalent behavior under tension, compression and torsion is different, Poynting-Swift effect or other second order effects) was in the spotlight of research activities from 1989 until 1996. A detailed discussion of non-classical behavior is presented, e.g. in Altenbach et al. (1995). The development of this branch of creep mechanics was connected with the use of some light alloys, gray cast irons, polymers, ceramics, composites, and other materials whose creep behavior depends on the type of loading. Most of the approaches in this field are related to tensorial-nonlinear creep equations and the generalization of the Rabotnov concept. In Altenbach et al. (1995), a systematic development of isotropic and anisotropic creep equations considering the influence of the loading type is suggested. The suggested equations can be applied to metals and alloys, but

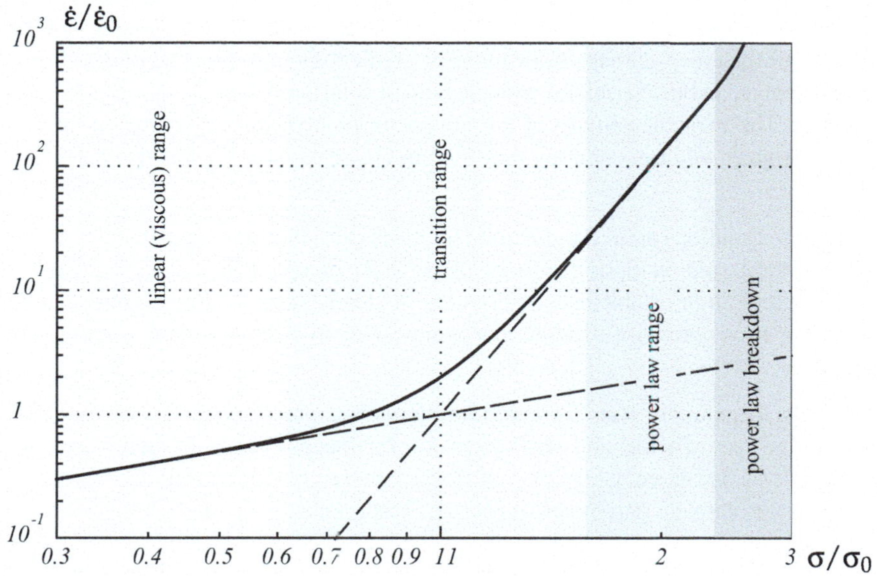

Fig. 1.2 Minimum creep rate versus stress dependence—double power law (Naumenko et al. 2009, reprinted with permission of Springer)

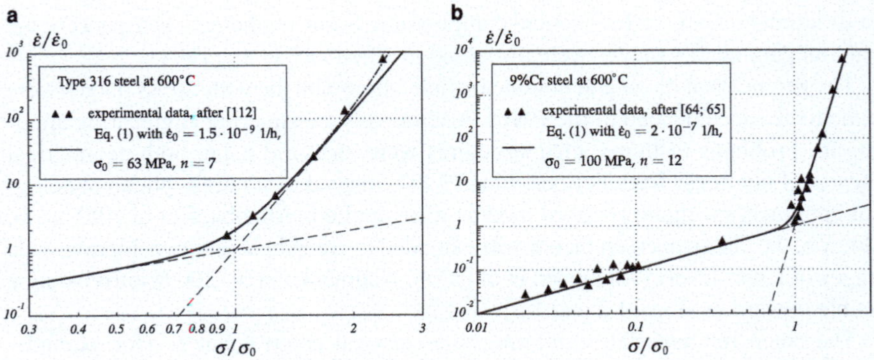

Fig. 1.3 Minimum creep rate versus stress. **a** Type 316 steel at 600°C, experimental data (Rieth et al. 2004), **b** 9%Cr steel at 600°C, experimental data (Kloc and Sklenička 2004; Kloc et al. 2001) (Naumenko et al. 2009, reprinted with permission of Springer)

also to polymers (Altenbach et al. 1995). In Altenbach and Zolochevsky (1992), the energetic variant of creep constitutive equations based on the dissipation power was discussed and applied to the titanium alloy OT-4 at 748 °K. Application to shells of revolution are presented in Altenbach and Zolochevsky (1991).

Numerous commercial numerical software packages are currently available. When using them, the problem arises that one has to prove the correctness of the solutions. So-called benchmark tests are used for this purpose. In the author's team benchmark tests for creep problems were established. The accuracy of creep-damage predictions was investigated, e.g. in Altenbach et al. (2000). It was shown that the accuracy of the creep-damage finite element predictions in beams and plates (bending problems) can be determined with the help of solutions based on the Ritz method (Fig. 1.4). The mesh sensitivity of the long-term predictions for beams and plates was investigated while accounting for different element types. The main conclusions were

- the approximations or meshes justified for the elastic solutions using displacement-based variational methods cannot be used for the creep damage analysis, and
- the discretization established on convergent steady state creep solutions can be used for the continuum damage mechanics analysis of thin-walled structures in bending.

Another benchmark tests for creep problems was presented in Altenbach et al. (2001). The use of solid and shell elements for creep-damage problems in thin-structures was discussed. Conventional creep-damage constitutive equations considering a scalar damage parameter were used. The applicability of shell theory (here a first order shear deformation theory) was studied for the case that the stress state has an influence on the constitutive behavior. As a result it was established that

> If the damage evolution differs significantly at tension and compression, the classical transverse shear stress parabolic distribution is no more valid.

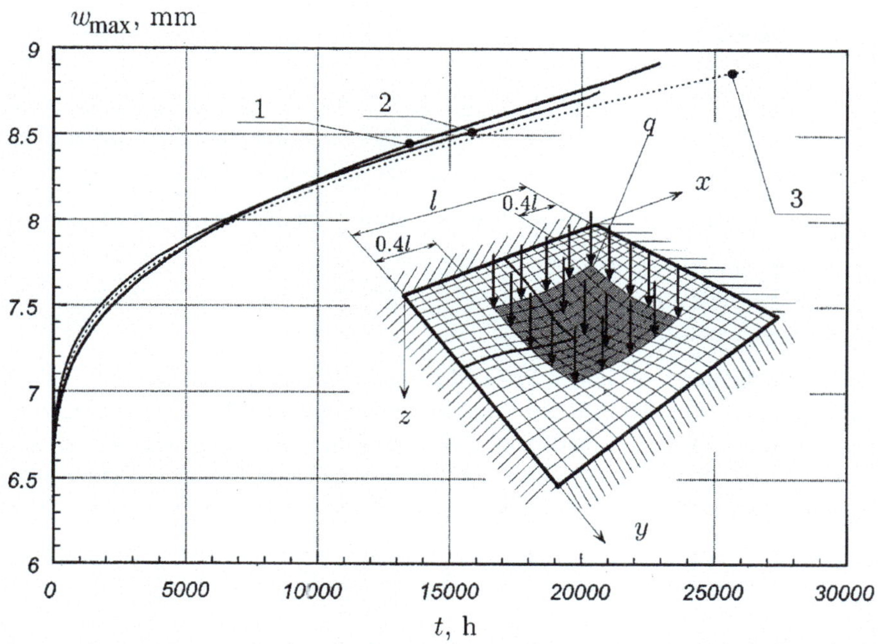

Fig. 1.4 Time-dependent maximum deflection of a clamped square plate: 1—SHELL 43, 2—SOLID 95, 3—solution based on the Ritz method and assuming Kirchhoff's theory. (Altenbach et al. 2000, reprinted with permission from Springer)

1.2.4 Rheological Models

The use of rheological models in solid mechanics as a tool for material modeling and simulation is widely used (Altenbach and Eisenträger 2021; Altenbach 2023 among others). The starting point of rheology as a science was in the twenties of the last century, but the first rheological models were developed in the seventeenth century by Robert Hooke[28] and other scientists. With the increasing use of plastics in the 1950s, the rheological modeling became very popular; however the application was limited due to the lack of computers. The main focus was on plastics and their time-dependent material behavior. However, in the last years, rheological models were used also in other cases with applications to metals, concrete, geomaterials, etc. Here a creep mechanics example is briefly presented.

Starting with Naumenko et al. (2011) the creep behavior of an alloy containing a hard and soft phase was described using a rheological model. Developments of this approach are presented in Eisenträger et al. (2017, 2018a, b) and Eisenträger (2019). In Altenbach et al. (2015) this methodology was applied to a polymer.

[28] Robert Hooke (*18 July 1635, Freshwater, Isle of Wight, England; †3 March 1703, London, England), English polymath who was active as a physicist ("natural philosopher"), astronomer, geologist, meteorologist, and architect.

1.3 Outlook

Due to the fact that creep mechanics still lacks a closed and rational theory, there is a need for further fundamental research. At the same time, new application tasks demand further investigations. Let us summarize some of these tasks:

- Further impulses for creep research are mainly expected from the following areas:
  - power plant constructions,
  - aerospace constructions, and
  - microsystem technology.

  Note that the first two items are related to traditional application fields.
- The performance and efficiency improvement leads partly to an increase of exploitation temperature, and creep and creep-damage became more important.
- As the component size decreases, the temperature increases and failure can be avoided only after intensive investigation of this problem.

References

Altenbach H (1999) Creep-damage behaviour of plates and shells. Mech Time-Depend Mater 3(2):103–123. https://doi.org/10.1023/A:1009834707257

Altenbach H (2002) Creep analysis of thin-walled structures. ZAMM - J Appl Math Mech/Zeitschrift für Angewandte Mathematik und Mechanik 82(8):507–533. https://doi.org/10.1002/1521-4001(200208)82:8<507::AID-ZAMM507>3.0.CO;2-Y

Altenbach H (2023) Rheological modeling - historical remarks and actual trends in solid mechanics. In: Altenbach H, Kaplunov J, Lu H, Nakada M (eds) Advances in mechanics of time-dependent materials. Advanced structured materials, vol 188, pp 1–16. Springer Nature, Cham. https://doi.org/10.1007/978-3-031-22401-0_1

Altenbach H, Altenbach J, Naumenko K (1997a) On the prediction of creep damage by bending of thin-walled structures. Mech Time-Depend Mater 1(2):181–193. https://doi.org/10.1023/A:1009794001209

Altenbach H, Altenbach J, Zolochevsky A (1995) Erweiterte Deformationsmodelle und Versagenskriterien der Werkstoffmechanik. Deutscher Verlag für Grundstoffindustrie

Altenbach H, Altenbach J, Zolochevsky A (1995) A generalized constitutive equation for creep of polymers at multiaxial loading. Mech Compos Mater 31(6):511–518. https://doi.org/10.1007/BF00634897

Altenbach H, Deuring P, Naumenko K (1999) A system of ordinary and partial differential equations describing creep behaviour of thin-walled shells. Zeitschrift für Analysis und ihre Anwendungen 18(4):1003–1030. https://doi.org/10.4171/ZAA/925

Altenbach H, Eisenträger J (2020) Introduction to creep mechanics - introduction to creep mechanics. In: Altenbach H, Öchsner A (eds) Encyclopedia of continuum mechanics, pp 1337–1344. Springer, Berlin, Heidelberg. https://doi.org/10.1007/978-3-662-55771-6_155

Altenbach H, Eisenträger J (2021) Rheological modeling in solid mechanics from the beginning up to now. Lecture Notes of TICMI, vol 22, pp 13–29

Altenbach H, Girchenko A, Kutschke A, Naumenko K (2015) Creep behavior modeling of polyoxymethylene (pom) applying rheological models. In: Altenbach H, Brünig M (eds) Inelastic

behavior of materials and structures under monotonic and cyclic loading, Advanced structured materials, vol 57, pp 1–15. Springer. https://doi.org/10.1007/978-3-319-14660-7_1
Altenbach H, Gorash Y, Naumenko K (2008) Steady-state creep of a pressurized thick cylinder in both the linear and the power law ranges. Acta Mech 195(1–4):263–274. https://doi.org/10.1007/s00707-007-0546-5
Altenbach H, Huang C, Naumenko K (2001) Modelling of the creep-damage under reversed stress states by considering damage activation and deactivation. Technische Mechanik 21(4):273–282
Altenbach H, Knape K (2020) On the main directions in creep mechanics of metallic materials. Proc Natl Acad Sci Armenia: Mech 73(3), 24–43. https://doi.org/10.33018/73.3.2
Altenbach H, Kolarov G, Morachkovsky OK, Naumenko K (2000) On the accuracy of creep-damage predictions in thinwalled structures using the finite element method. Comput Mech 25(1):87–98. https://doi.org/10.1007/s004660050018
Altenbach H, Kolarov G, Naumenko K (1999) Solution of creep-damage problems for beams and rectangular plates using the ritz and the finite element method. Technische Mechanik 19(4):249–258
Altenbach H, Kruch S (eds) (2013) Advanced materials modelling for structures, Advanced structured materials, vol 19. Springer, Berlin, Heidelberg. https://doi.org/10.1007/978-3-642-35167-9
Altenbach H, Kushnevsky V, Naumenko K (2001) On the use of solid- and shell-type finite elements in creep-damage predictions of thinwalled structures. Arch Appl Mech 71(2):164–181. https://doi.org/10.1007/s004190000132
Altenbach H, Morachkovsky O, Naumenko K, Sychov A (1997b) Geometrically nonlinear bending of thin-walled shells and plates under creep-damage conditions. Arch Appl Mech 67(5):339–352. https://doi.org/10.1007/s004190050122
Altenbach H, Naumenko K (1995) Geometrically nonlinear small strain creep theory for thin plates including isotropic damage. Math Modell Sci Comput 5(2):89–99
Altenbach H, Naumenko K (eds) (2023) Creep in structures VI - IUTAM symposium proceedings, Advanced structured materials, vol 194. Springer Nature, Cham. https://doi.org/10.1007/978-3-031-39070-8
Altenbach H, Skrzypek JJ (eds) (1999) Creep and damage in materials and structures, CISM international centre for mechanical sciences, vol 399. Springer. https://doi.org/10.1007/978-3-7091-2506-9
Altenbach H, Zolochevsky A (1991) Kriechen dünner schalen aus anisotropen werkstoffen mit unterschiedlichem zug-druckverhalten. Forsch Ingenieurwes 57(6):172–179. https://doi.org/10.1007/BF02575157
Altenbach H, Zolochevsky A (1992) Energy version of creep and stress-rupture strength theory for anisotropic and isotropic materials which differ in resistance to tension and compression. J Appl Mech Tech Phys 33(1):101–106. https://doi.org/10.1007/BF00864514
Altenbach J, Altenbach H, Naumenko K (1997c) Lebensdauerabschätzung dünnwandiger Flächentragwerke auf der Grundlage phänomenologischer Materialmodelle für Kriechen und Schädigung. Tech Mech 17(4):353–364
Altenbach J, Altenbach H, Naumenko K (1999) A phenomenological creep theory including nonclassical effects and creep-damage coupling. Appl Mech Eng 4:11–16. Proceedings of the international conference on engineering rheology ICER '99
Altenbach J, Altenbach H, Naumenko K (2004) Edge effects in moderately thick plates under creep-damage conditions. Technische Mechanik 24(3):254–263
Arutiunian NK (1977) Boundary value problem of the theory of creep for a body with accretion. J Appl Math Mech 41(5):804–810. https://doi.org/10.1016/0021-8928(77)90161-7
Arutyunyan NK (1966) Some problems in the theory of creep. Pergamon Press, Oxford
Bailey RW (1935) The utilization of creep test data in engineering design. Proc Instit Mech Eng 131(1):131–349. https://doi.org/10.1243/PIME_PROC_1935_131_012_02

Benallal A (ed) (2000) Continuous Damage and fracture. Elsevier, Paris (2000). The symposium on continuous damage and fracture was held in Cachan to celebrate the 66th birthday of Pr. Jean Lemaitre from 23rd to 27th October 2000

Betten J (1998) Anwendungen von Tensorfunktionen in der Kontinuumsmechanik anisotroper Materialien. ZAMM - J Appl Math Mech/Zeitschrift für Angewandte Mathematik und Mechanik 78(8):507–521. https://doi.org/10.1002/(SICI)1521-4001(199808)78:8<507::AID-ZAMM507>3.0.CO;2-F

Betten J (2008) Creep mechanics, 3rd edn. Springer, Berlin, Heidelberg (2008). https://doi.org/10.1007/978-3-540-85051-9

Boehler JP (ed) (1982) Mechanical behavior of anisotropic solids/Comportment Méchanique des Solides Anisotropes. Martinus Nijhoff Publishers and Editions du CNRS, Den Haag. https://doi.org/10.1007/978-94-009-6827-1. Proceedings of the Euromech Colloquium 115 Villard-de-Lans, June 19–22, 1979/Colloque Euromech 115 Villard-de-Lans, 19-22 Juin 1979

Boyle JT, Spence J (1983) Stress analysis for creep. Butterworth, London

Bridgeman PW (1952) Studies in large plastic flow and fracture with special emphasis on the effects of hydrostatic pressure. Metallurgy and metallurgical engineering series. McGraw-Hill, New York, Toronto, London

Chow C, Wang J (1987) An anisotropic theory of continuum damage mechanics for ductile fracture. Eng Fract Mech 27(5):547–558. https://doi.org/10.1016/0013-7944(87)90108-1

Chow CL, Yang XJ, Chu E (2000) Viscoplastic constitutive modeling of anisotropic damage under nonproportional loading. J Eng Mater Technol 123(4):403–408. https://doi.org/10.1115/1.1395575

da Costa Andrade EN (1910) On the viscous flow in metals, and allied phenomena. Proc R Soc Lond. Ser A, Containing Papers of a Mathematical and Physical Character 84(567):1–12. https://doi.org/10.1098/rspa.1910.0050

da Costa Andrade EN (1914) The flow in metals under large constant stresses. Proc R Soc Lond. Ser A, Containing Papers of a Mathematical and Physical Character 90(619):329–342. https://doi.org/10.1098/rspa.1914.0056

Desmorat R, Cantournet S (2008) Modeling microdefects closure effect with isotropic/anisotropic damage. Int J Damage Mech 17(1):65–96. https://doi.org/10.1177/1056789507069541

Eisenträger J, Naumenko K, Altenbach H (2018a) Calibration of a phase mixture model for hardening and softening regimes in tempered martensitic steel over wide stress and temperature ranges. J Strain Anal Eng Design 53(3):156–177. https://doi.org/10.1177/0309324718755956

Eisenträger J, Naumenko K, Altenbach H (2018b) Numerical implementation of a phase mixture model for rate-dependent inelasticity of tempered martensitic steels. Acta Mech 229(7):3051–3068. https://doi.org/10.1007/s00707-018-2151-1

Eisenträger J, Naumenko K, Altenbach H (2019) Numerical analysis of a steam turbine rotor subjected to thermo-mechanical cyclic loads. Technische Mechnik 19(3):261–281. https://doi.org/10.24352/UB.OVGU-2019-024

Eisenträger J, Naumenko K, Altenbach H, Gariboldi E (2017) Analysis of temperature and strain rate dependencies of softening regime for tempered martensitic steel. J Strain Anal Eng Design 52(4):226–238. https://doi.org/10.1177/0309324717699746

Findley WN, Lai JS, Onaran K (1989) Creep and relaxation of nonlinear viscoelastic materials. Dover, New York

Finnie I, Heller WR (1959) Creep of engineering materials. McGraw-Hill

Haddad YM (1995) Viscoelasticity of engineering materials. Chapman & Hall, London

Hansen NR, Schreyer HL (1995) Damage deactivation. J Appl Mech 62(2):450–458. https://doi.org/10.1115/1.2895951

Hoff N (ed) (1962) Creep in structures. Springer, Berlin

Hoff NJ (1953) The necking and the rupture of rods subjected to constant tensile loads. J Appl Mech 20(1):105–108. https://doi.org/10.1115/1.4010601

Hoff NJ (1957) Buckling at high temperature. J R Aeronaut Soc 61(563):756–774. https://doi.org/10.1017/S0368393100133735

Hudramovich V (2015) Review of studies in creep theory conducted at institute of technical mechanics of the National academy of sciences of Ukraine and State space agency of Ukraine. Technical Mechanics 4(4):85–91. [В.С. Гудрамович: Онаучныхисследованияхвобластитеорииползучестивинститутетехническоймеханики Национальнойакадемиинаук Украины и Государственномкосмическомагентстве Украины. ТехническаяМеханика4(NO.4), с. 85-91 (2015).]

Huh H, Park CG, Lee CS, Keum YT (eds) (2009) Engineering plasticity and its applications: from nanoscale to macroscale. World Scientific. https://doi.org/10.1142/7225

Hult J (ed) (1972) Creep in structures. Springer, Berlin, Heidelberg, New York

Hult JA (1966) Creep in engineering structures. Blaisdell Publishing Company, Waltham

Hyde T, Sun W, Hyde C (2013) Applied creep mechanics. McGraw-Hill Education

Hyde TH, Sun W (2005) A study of anisotropic creep behaviour of a 9CrMoNbV weld metal using damage analyses with a unit cell model. Proc Inst Mech Eng Part L: J Mater Design Appl 219(4):193–206. https://doi.org/10.1243/146442005X34539

Hyde TH, Sun W, Agyakwa PA, Shipway PH, Williams JA (2003) Anisotropic creep and fracture behaviour of a 9CrMoNbV weld metal at 650°C. In: Skrzypek JJ, Ganczarski AW (eds.) Anisotropic behaviour of damaged materials, pp 295–316. Springer, Berlin, Heidelberg. https://doi.org/10.1007/978-3-540-36418-4_9

Hyde TH, Sun W, Becker AA, Williams JA (1997) Creep continuum damage constitutive equations for the base, weld and heat-affected zone materials of a service-aged 1/2Cr1/2Mo1/4V:2 1/4Cr1Mo multipass weld at 640°C. J Strain Anal Eng Des 32(4):273–285. https://doi.org/10.1243/0309324971513409

Hyde TH, Sun W, Williams JA (2003b) Creep analysis of pressurized circumferential pipe weldments - a review. J Strain Anal Eng Design 38(1):1–27. https://doi.org/10.1243/030932403762671854

Kachanov LM (1958) On the time to rupture under creep conditions (in Russ.). Izvestiya AN SSSR Otdelenie Tekhnicheskikh Nauk (8):26–31. [Качанов, Л.М.: Овремениразрушениявусловияхползучести. Изв. АНСССР. Отд. Техн. Наук. No. 8, c. 26-31 (1958).]

Kachanov LM (1960) Theory of Creep (in Russ. [КачановЛ.М.: Теорияползучести. М.: физматгиз1960.]). FizMatGiz, Moscow (1960)

Kachanov LM (1986) Introduction to continuum damage mechanics. Martinus Nijhoff, Dordrecht

Kloc L, Sklenička V, Ventruba J (2001) Comparison of low stress creep properties of ferritic and austenitic creep resistant steels. Mater Sci Eng A 319:774–778. https://doi.org/10.1016/S0921-5093(01)00943-1

Kloc L, Sklenička V (2004) Confirmation of low stress creep regime in 9% chromium steel by stress change creep experiments. Mater Sci Eng A 387–389:633–638. https://doi.org/10.1016/j.msea.2003.12.078. 13th International Conference on the Strength of Materials

Kowalewski ZL, Hayhurst DR, Dyson BF (1994) Mechanisms-based creep constitutive equations for an aluminium alloy. J Strain Anal Eng Design 29(4):309–316. https://doi.org/10.1243/03093247V294309

Krajcinovic D (1996) Damage mechanics. Applied mathematics and mechanics, vol 41. North-Holland, Amsterdam

Krajcinovic D, Lemaitre J (eds) (1987) Continuum damage mechanics - theory and application. CISM international centre for mechanical sciences, vol 295. Springer. https://doi.org/10.1007/978-3-7091-2806-0

Kraus H (1980) Creep analysis. Wiley, New York

Langer BF (1937) Fatigue failure from stress cycles of varying amplitude. J Appl Mech 4(4):A160–A162. https://doi.org/10.1115/1.4008807

Lemaitre J (1996) A course on damage mechanics, 2nd edn. Springer, Berlin

Lemaitre J, Chaboche JL (1990) Mechanics of solid materials. Cambridge University Press, Cambridge

Lemaitre J, Desmorat R (2005) Engineering damage mechanics: ductile. Fatigue and Brittle Failures Springer, Creep. https://doi.org/10.1007/b138882

Lokoshchenko A (2017) Creep and long-term strength of metals. Routledge & CRC Press, Boca Raton. https://doi.org/10.1201/b22242

Lokoshchenko A, Platonov D (2004) Creep rupture of anisotropic tubes under complex stress state. In: Proceedings of the 7th international conference "Biaxial/multiaxial fatigue and fracture", pp 567–571. DVM, Berlin

Malinin NN (1975) Applied theory of plasticity and creep (in Russ. [Малинин, Н.Н.: Прикладнаятеорияпластичностииползучести. М.: Машиностроение(1975).]). Mashinostroenie, Moscow

Malinin NN (1981) Creep calculations of structural elements (in Russ. [Малинин, Н.Н.: Расчетыnaползучестьзлементовмашиностроительныхконструкций. М: Машиностроение(1981).]). Mashinostroenie, Moscow

Manzhirov AV, Lychev SA (2012) Doklady physics. J Appl Math Mech 57(4):160–163. https://doi.org/10.1134/S1028335812040015

McLean M, Dyson BF (2000) Modeling the effects of damage and microstructural evolution on the creep behavior of engineering alloys. J Eng Mater Technol 122(3):273–278. https://doi.org/10.1115/1.482798

Mindlin RD (1951) Influence of rotatory inertia and shear on flexural motions of isotropic, elastic plates. J Appl Mech 18(2):31–38. https://doi.org/10.1115/1.4010217

Miner MA (1945) Cumulative damage in fatigue. J Appl Mech 12(3):A159–A164. https://doi.org/10.1115/1.4009458

Murakami S (2012) Continuum damage mechanics: a continuum mechanics approach to the analysis of damage and fracture. Solid mechanics and its applications. Springer. https://doi.org/10.1007/978-94-007-2666-6

Murakami S, Ohno N (eds) (2001) IUTAM symposium on creep in structures. Kluwer, Dordrecht

Murakami S, Sanomura Y (1985) Creep and creep damage of copper under multiaxial states of stress. In: Sawczuk A, Bianci G (eds) Plasticity today: modelling, methods and applications. Elsevier Applied Science, London, pp 535–551

Nadai A (1938) The influence of time upon creep. the hyperbolic sine creep law. In: Contributions to the mechanics of solids dedicated to Stephen Timoshenko by his friends on the occasion of his sixties birthday anniversary, pp 155–170. Macmillian

Naumenko K, Altenbach H (2005) A phenomenological model for anisotropic creep in a multipass weld metal. Arch Appl Mech 74(11):808–819. https://doi.org/10.1007/s00419-005-0409-2

Naumenko K, Altenbach H (2007) Modeling of creep for structural analysis. Foundations of engineering mechanics. Springer. https://doi.org/10.1007/978-3-540-70839-1

Naumenko K, Altenbach H (2016) Modeling high temperature materials behavior for structural analysis - part I: Continuum mechanics foundations and constitutive models, Advanced structured materials, vol 28. Springer. https://doi.org/10.1007/978-3-319-31629-1

Naumenko K, Altenbach H (2020) Modeling high temperature materials behavior for structural analysis. Advanced structured materials, vol 112. Springer. https://doi.org/10.1007/978-3-030-20381-8

Naumenko K, Altenbach H, Gorash Y (2009) Creep analysis with a stress range dependent constitutive model. Arch Appl Mech 79(6–7):619–630. https://doi.org/10.1007/s00419-008-0287-5

Naumenko K, Altenbach H, Kutschke A (2011) A combined model for hardening, softening, and damage processes in advanced heat resistant steels at elevated temperature. Int J Damage Mech 20(4):578–597. https://doi.org/10.1177/1056789510386851

Norton FH (1929) Creep of steel at high temperatures. McGraw-Hill, New York

Nosal P, Ganczarski A (2023) Modeling solid materials in dem using the micropolar theory. In: Altenbach H, Naumenko K (eds) Creep in structures VI - IUTAM symposium proceedings, Advanced structured materials, vol 194, pp 233–248. Springer Nature Switzerland, Cham (2023). https://doi.org/10.1007/978-3-031-39070-8_14

Nowacki W (1965) Theorie des Kriechens. Verlag H, Deutsch, Wien

Odqvist FKG (1974) Mathematical Theory of Creep and Creep Rupture. 2nd edn. Oxford University Press, Oxford, Oxford Mathematical Monographs

Odqvist FKG (1981) Historical survey of the development of creep mechanics from its beginnings in the last century to 1970. In: Ponter ARS, Hayhurst DR (eds) Creep in structures, pp 1–12. Springer, Berlin, Heidelberg. https://doi.org/10.1007/978-3-642-81598-0_1

Odqvist FKG, Hult J (1962) Kriechfestigkeit metallischer Werkstoffe. Springer, Berlin u.a

Ohno N, Uehara T (eds) (2007) Engineering plasticity and its applications from nanoscale to macroscale, vol 9. Lecture notes in applied and computational mechanics. Trans. Tech. Publications Ltd., Stafa-Zürich

Othman AM, Dyson BF, Hayhurst DR, Lin J (1994) Continuum damage mechanics modelling of circumferentially notched tension bars undergoing tertiary creep with physically-based constitutive equations. Acta Metall Mater 42(3):597–611. https://doi.org/10.1016/0956-7151(94)90256-9

Palmgren AG (1924) Die Lebensdauer von Kugellagern. Zeitschrift des Vereines Deutscher Ingenieure (ZVDI) 68(14):339–341

Penny RK, Mariott DL (1995) Design for creep. Chapman & Hall, London. https://doi.org/10.1007/978-94-011-0561-3

Penny RK, Marriott D (1971) Design for creep. McGraw-Hil, London

Ponter ARS, Hayhurst DR (eds) (1981) Creep in structures. Springer, Berlin

Prager W (1945) Strain hardening under combined stresses. J Appl Phys 16(12):837–840. https://doi.org/10.1063/1.1707548

Rabotnov YN (1959) A mechanism of the long term fracture (in Russ.). In: Voprosy prochnosti materialov i konstruktsii, pp 5–7. Izdatel'stvo AN SSSR, Moscow (1959). [Работнов, Ю.Н.: Омеханизмедлительногоразрушения. Вопросыпрочностиматериаловиконструкций. М.: Изд-воАНСССР, с. 5-7 (1959).]

Rabotnov YN (1969) Creep Problems in Structural Members. North-Holland, Amsterdam

Rabotnov YN (1980) Elements of hereditary solid mechanics. Mir, Moscow

Reiner M (1945) A mathematical theory of dilatancy. Am J Math 67(3):350–362

Reiner M (1960) Deformation, Strain and Flow: an Elementary Introduction to Rheology. Lewis, London, H.K

Reissner E (1944) On the theory of bending of elastic plates. J Math Phys 23:184–191. https://doi.org/10.1002/sapm1944231184

Reissner E (1945) The effect of transverse shear deformation on the bending of elastic plates. J Appl Mech 12(11):A69–A77. https://doi.org/10.1115/1.4009435

Rieth M, Falkenstein A, Graf P, Heger S, Jäntsch U, Klimiankou M, Materna-Morris E, Zimmermann H (2004) Creep of the austenitic steel AISI 316 L(N). experiments and models. Technical Report, Forschungszentrum Karlsruhe. https://doi.org/10.5445/IR/270059325.31.40.03; LK 01; Wissenschaftliche Berichte, FZKA-7065 (November 2004)

Rimrott FPJ (1968) The hyperbolic sine creep law in engineering practice. In: Rimrott FJP, Schwaighofer J (eds) Mechanics of the solid state. University of Toronto Press, Toronto, pp 168–180

Rzhanitsyn AR (1968) Theory of creep (in Russ. [РжаницынА.Р.: Теорияползучести. М.: Стройиздат1968.]). Stroyizdat, Moscow

Saitova R, Arutyunyan A, Altenbach H (2024) High temperature creep and embrittlement in metals and alloys under conditions of the long-term usage. Acta Mech (online first). https://doi.org/10.1007/s00707-024-04026-6

Skrzypek JJ (1993) Plasticity and creep. Examples, and problems. CRC Press, Boca Raton, Theory

Skrzypek JJ, Ganczarski A (eds) (2003) Anisotropic behaviour of damaged materials, Lecture notes in applied and computational mechanics, vol 9. Springer, Berlin, Heidelberg. https://doi.org/10.1007/978-3-540-36418-4

Soderberg CR (1938) Plasticity and creep in machine design. In: Contributions to the mechanics of solids dedicated to Stephen Timoshenko by his friends on the occasion of his sixties birthday anniversary, pp 97–110. Macmillian

Stodola A (1933) Die Kriecherscheinungen, ein neuer technisch wichtiger Aufgabenkreis der Elastizitätstheorie. ZAMM - J Appl Math Mech/Zeitschrift für Angewandte Mathematik und Mechanik 13(2):143–146. https://doi.org/10.1002/zamm.19330130221

Trampczynski W, Hayhurst D, Leckie F (1981) Creep rupture of copper and aluminium under non-proportional loading. J Mech Phys Solids 29(5):353–374. https://doi.org/10.1016/0022-5096(81)90034-X

Čadek J (1988) Creep in metallic materials, Materials science monographs, vol 48. Elsevier, Amsterdam et al

Vicat (1834) Ueber die allmähliche Verlängerung des Eisendrahtes bei verschiedener Strekung. Polytechnisches J 51(XCVI):434–438

Voyiadjis GZ, Ju JW, Chaboche JL (eds) (1998) Damage mechanics in engineering materials. Studies in applied mechanics, vol 46. Elsevier, Amsterdam

Życzkowski M (ed) (1991) Creep in structures. Springer, Berlin, Heidelberg

Chapter 2
Yield Conditions in the Invariant Space: Theory and Experimental Investigations

Andreas Öchsner

Abstract This chapter presents the general concept of stress in the three-dimensional space and its representation by a stress matrix. The physical content of such a stress matrix can be represented by so-called stress invariants and a common set of these quantities is introduced in detail. The second part is devoted to the graphical representation of yield conditions and different experimental approaches to investigate their corresponding shape.

2.1 Fundamentals

2.1.1 Comments on the Stress Matrix

Let us consider a three-dimensional body which is sufficiently supported and loaded (i.e., by any point or distributed loads) as schematically shown in Fig. 2.1a. Considering the symmetry of the stress matrix, six independent stress components, i.e., three normal and three shear stresses, can be identified; see Fig. 2.1b.

The stress components acting on a differential volume element may have, for example, the values as shown in Eq. (2.1) for the given (x, y, z) coordinate system. A coordinate transformation from the original (x, y, z) to the (x', y', z') coordinate system results in a stress matrix with different stress components, while a principal axis transformation (PAT) calculates the principal stresses σ_i, $(i = 1, 2, 3)$.

A. Öchsner (✉)
Faculty of Mechanical and Systems Engineering, Esslingen University of Applied Sciences, Kanalstrasse 33, 73728 Esslingen, Germany
e-mail: andreas.oechsner@hs-esslingen.de

H. Altenbach et al. (eds.), *Lectures Notes on Advanced Structured Materials 3*,
Advanced Structured Materials 226, https://doi.org/10.1007/978-3-031-77403-4_2

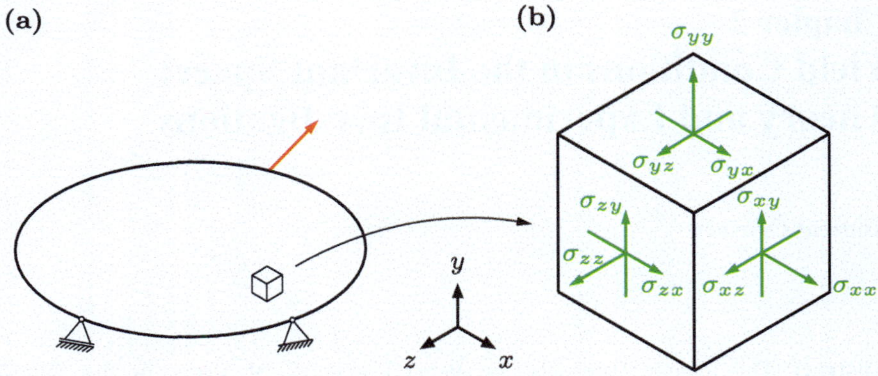

Fig. 2.1 **a** Three-dimensional body under arbitrary load and boundary conditions; **b** infinitesimal volume element with acting normal and shear stress components (see Altenbach and Öchsner 2020)

$$\sigma_{ij} = \begin{bmatrix} \sigma_{xx} & 0 & \sigma_{xz} \\ 0 & \sigma_{yy} & \sigma_{yz} \\ \sigma_{zx} & \sigma_{zy} & \sigma_{zz} \end{bmatrix}_{(x,y,z)} \overset{\text{rotation}}{\Rightarrow} \begin{bmatrix} \sigma_{x'x'} & \sigma_{x'y'} & \sigma_{x'z'} \\ \sigma_{y'x'} & \sigma_{y'y'} & 0 \\ \sigma_{z'x'} & 0 & \sigma_{z'z'} \end{bmatrix}_{(x',y',z')}$$

$$\overset{\text{PAT}}{\Rightarrow} \begin{bmatrix} \sigma_1 & 0 & 0 \\ 0 & \sigma_2 & 0 \\ 0 & 0 & \sigma_3 \end{bmatrix}_{(1,2,3)} . \tag{2.1}$$

Looking at this simple example, the following characteristics of the stress matrix can be summarized:

- The components of the stress matrix depend on the orientation of the user-defined coordinate system.
- There is a specific coordinate system $(1, 2, 3)$ where the shear stresses vanish and only normal stresses remain on the main diagonal, i.e., the so-called principal stresses σ_i $(i = 1, 2, 3)$.
- The six or three stress components cannot easily be compared to experimental values from uniaxial tests (e.g., the initial yield stress in tension k_t^{init}).

In addition to the above facts, it can be postulated that:

- A graphical representation of any surface is much easier in a principal stress space $(1, 2, 3)$ with three coordinates than in a space with six coordinates (σ_{ij}).

Further information on continuum mechanics and plasticity theory can be taken from the classical textbooks by Backhaus (1983), Chen and Han (1988), Altenbach (2018), Altenbach et al. (1995), Betten (1987), Itskov (2009), and Lurie (2005).

Let us review in the following the determination of the principal stresses and the axes directions of the corresponding $(1, 2, 3)$ coordinate system. From a mathematical point of view, this question can be answered by determining the eigenvalues of the stress matrix (principal stresses) and the corresponding eigenvectors (principal directions). The solution of the so-called *characteristic equation* gives the three principal stresses σ_i $(i = 1, 2, 3)$, i.e.,

$$\det\left(\sigma_{ij} - \sigma_i \delta_{ij}\right) = 0, \tag{2.2}$$

where Kronecker delta (δ_{ij} is 1 if $i = j$, and 0 if $i \neq j$) was used (see Chen and Han 1988). Equation (2.2) can be written in components as follows:

$$\det\left(\begin{bmatrix} \sigma_{xx} & \sigma_{xy} & \sigma_{xz} \\ \sigma_{yx} & \sigma_{yy} & \sigma_{yz} \\ \sigma_{zx} & \sigma_{zy} & \sigma_{zz} \end{bmatrix} - \sigma_i \begin{bmatrix} 1 & 0 & 0 \\ 0 & 1 & 0 \\ 0 & 0 & 1 \end{bmatrix}\right) = \det\left(\begin{bmatrix} \sigma_{xx} - \sigma_i & \sigma_{xy} & \sigma_{xz} \\ \sigma_{yx} & \sigma_{yy} - \sigma_i & \sigma_{yz} \\ \sigma_{zx} & \sigma_{zy} & \sigma_{zz} - \sigma_i \end{bmatrix}\right) = 0. \tag{2.3}$$

The calculation of the determinant ('det') results in the following cubic equation in σ_i:

$$\begin{aligned} &\sigma_i^3 - \left(\sigma_{xx} + \sigma_{yy} + \sigma_{zz}\right)\sigma_i^2 + \left(\sigma_{xx}\sigma_{yy} + \sigma_{xx}\sigma_{zz} + \sigma_{yy}\sigma_{zz} - \sigma_{xy}^2 - \sigma_{xz}^2 - \sigma_{yz}^2\right)\sigma_i \\ &\quad - \left(\sigma_{xx}\sigma_{yy}\sigma_{zz} - \sigma_{xx}\sigma_{yz}^2 - \sigma_{yy}\sigma_{xz}^2 - \sigma_{zz}\sigma_{xy}^2 + 2\sigma_{xy}\sigma_{xz}\sigma_{yz}\right) = 0, \end{aligned} \tag{2.4}$$

where the three roots $(\sigma_1, \sigma_2, \sigma_3)$ of Eq. (2.4) are the principal stresses.

The coordinates of the ith eigenvector (x_i, y_i, z_i)—which correspond to the direction of one of the new $(1, 2, 3)$ coordinate axes—result from the following system of three equations:

$$\begin{bmatrix} \sigma_{xx} - \sigma_i & \sigma_{xy} & \sigma_{xz} \\ \sigma_{yx} & \sigma_{yy} - \sigma_i & \sigma_{yz} \\ \sigma_{zx} & \sigma_{zy} & \sigma_{zz} - \sigma_i \end{bmatrix} \begin{bmatrix} x_i \\ y_i \\ z_i \end{bmatrix} = \begin{bmatrix} 0 \\ 0 \\ 0 \end{bmatrix}. \tag{2.5}$$

To represent the physical content of the stress matrix, the set of the so-called *basic invariants* can be used (see Backhaus 1983). The definition of this set of invariants is given in Table 2.1.

Table 2.1 Definition of the three basic (J_i) stress invariants of the stress matrix

First Basic Stress Invariant of σ_{ij}
$J_1 = \sigma_{xx} + \sigma_{yy} + \sigma_{zz}$
Second Basic Stress Invariant of σ_{ij}
$J_2 = \frac{1}{2}\left(\sigma_{xx}^2 + \sigma_{yy}^2 + \sigma_{zz}^2\right) + \sigma_{xy}^2 + \sigma_{xz}^2 + \sigma_{yz}^2$
Third Basic Stress Invariant of σ_{ij}
$J_3 = \frac{1}{3}(\sigma_{xx}^3 + \sigma_{yy}^3 + \sigma_{zz}^3 + 3\sigma_{xy}^2\sigma_{xx} + 3\sigma_{xy}^2\sigma_{yy} + 3\sigma_{xz}^2\sigma_{xx} + 3\sigma_{xz}^2\sigma_{zz}$ $+3\sigma_{yz}^2\sigma_{yy} + 3\sigma_{yz}^2\sigma_{zz} + 6\sigma_{xy}\sigma_{xz}\sigma_{yz})$

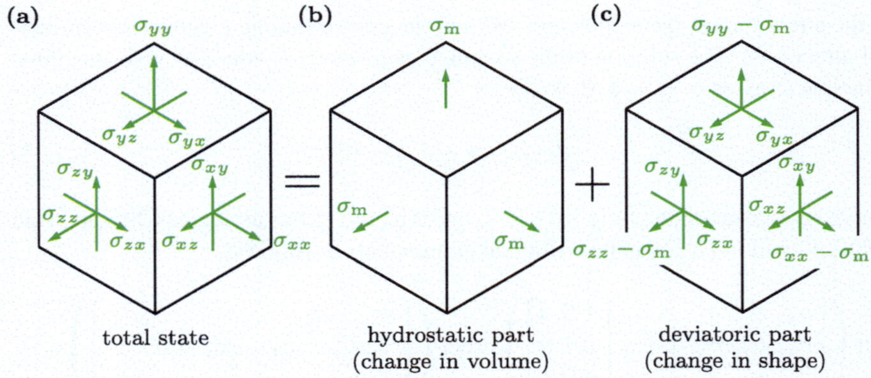

Fig. 2.2 Decomposition of the stress matrix (**a**) into its spherical (**b**) and the deviatoric (**c**) part

It is common in the framework of the plasticity theory of isotropic materials to decompose the stress matrix σ_{ij} into a pure volume changing (spherical or hydrostatic) matrix σ^{o}_{ij} and a pure shape changing (deviatoric) stress matrix s_{ij} (cf. Fig. 2.2)[1]:

$$\sigma_{ij} = \sigma^{\text{o}}_{ij} + s_{ij} = \sigma_{\text{m}}\delta_{ij} + s_{ij}. \tag{2.6}$$

In Eq. (2.6), $\sigma_{\text{m}} = \frac{1}{3}(\sigma_{xx} + \sigma_{yy} + \sigma_{zz})$ denotes the mean normal stress.[2]

Equation (2.6) can be written in components as

$$\underbrace{\begin{bmatrix} \sigma_{xx} & \sigma_{xy} & \sigma_{xz} \\ \sigma_{xy} & \sigma_{yy} & \sigma_{yz} \\ \sigma_{xz} & \sigma_{yz} & \sigma_{zz} \end{bmatrix}}_{\text{stress matrix } \sigma_{ij}} = \underbrace{\begin{bmatrix} \sigma_{\text{m}} & 0 & 0 \\ 0 & \sigma_{\text{m}} & 0 \\ 0 & 0 & \sigma_{\text{m}} \end{bmatrix}}_{\text{hydrostatic matrix } \sigma^{\text{o}}_{ij}} + \underbrace{\begin{bmatrix} s_{xx} & s_{xy} & s_{xz} \\ s_{xy} & s_{yy} & s_{yz} \\ s_{xz} & s_{yz} & s_{zz} \end{bmatrix}}_{\text{deviatoric matrix } s_{ij}}. \tag{2.7}$$

It can be seen that the elements outside the diagonal terms, i.e., the shear stresses, are the same for the stress and the deviatoric stress matrix

$$s_{ij} = \sigma_{ij} \quad \text{for} \quad i \neq j, \tag{2.8}$$

$$s_{ij} = \sigma_{ij} - \sigma_{\text{m}} \quad \text{for} \quad i = j, \tag{2.9}$$

and it can be shown that the first invariant of the deviator is (i.e., the so-called deviator equation):

$$s_{xx} + s_{yy} + s_{zz} = 0. \tag{2.10}$$

[1] It should be noted that in the case of anisotropic materials, a hydrostatic stress state may result in a shape change (Betten 2001).

[2] Also called the hydrostatic stress; in the context of soil mechanics, the pressure $p = -\sigma_{\text{m}}$ is also used.

The following list summarizes the calculation of the stress deviator components:

$$s_{xx} = \sigma_{xx} - \sigma_{\text{m}} = \frac{2}{3}\sigma_{xx} - \frac{1}{3}(\sigma_{yy} + \sigma_{zz}), \tag{2.11}$$

$$s_{yy} = \sigma_{yy} - \sigma_{\text{m}} = \frac{2}{3}\sigma_{yy} - \frac{1}{3}(\sigma_{xx} + \sigma_{zz}), \tag{2.12}$$

$$s_{zz} = \sigma_{zz} - \sigma_{\text{m}} = \frac{2}{3}\sigma_{zz} - \frac{1}{3}(\sigma_{xx} + \sigma_{yy}), \tag{2.13}$$

$$s_{xy} = \sigma_{xy}, s_{yz} = \sigma_{yz}, s_{xz} = \sigma_{xz}. \tag{2.14}$$

The definitions of the basic invariants can be applied directly to the hydrostatic and deviatoric part of the stress matrix to obtain a similar representation as provided in Table 2.1; see summary in Table 2.2 for the hydrostatic matrix and the summary in Table 2.3 for the deviatoric matrix.

The hydrostatic part of σ_{ij} has in the case of metallic materials (full dense materials) for temperatures approximately under $0.3T_{\text{mt}}$ (T_{mt}: melting temperature) nearly no influence on the occurrence of inelastic strains since dislocations slip only under the influence of shear stresses (for higher temperatures from 0.3 till $0.5T_{\text{mt}}$ also non-conservative climbing is possible) (Suzuki et al. 1985). On the other hand, the hydrostatic stress has a considerable influence on the yielding behavior in the case of soil mechanics, cellular materials, or in damage mechanics (formation of pores, e.g., Gurson 1977).

Table 2.2 Definition of the three basic (J_i°) stress invariants of the hydrostatic stress matrix σ_{ij}°

First Stress Invariant of σ_{ij}°
$J_1^{\circ} = 3\sigma_{\text{m}}$
Second Stress Invariant of σ_{ij}°
$J_2^{\circ} = \frac{3}{2}\sigma_{\text{m}}^2$
Third Stress Invariant of σ_{ij}°
$J_3^{\circ} = \sigma_{\text{m}}^3$

Table 2.3 Definition of the three basic (J_i') stress invariants of the deviatoric stress matrix s_{ij}

First Stress Invariant of s_{ij}
$J_1' = 0$
Second Stress Invariant of s_{ij}
$J_2' = -I_2'$
Third Stress Invariant of s_{ij}
$J_3' = I_3'$

Table 2.4 Basic invariants in terms of σ_{ij} and principal values

Invariants	Components of σ_{ij}	Principal stresses $\sigma_1, \sigma_2, \sigma_3$
	Stress matrix	
J_1	$\sigma_{xx} + \sigma_{yy} + \sigma_{zz}$	$\sigma_1 + \sigma_2 + \sigma_3$
J_2	$\frac{1}{2}\left(\sigma_{xx}^2 + \sigma_{yy}^2 + \sigma_{zz}^2\right) +$ $+\sigma_{xy}^2 + \sigma_{xz}^2 + \sigma_{yz}^2$	$\frac{1}{2}\left(\sigma_1^2 + \sigma_2^2 + \sigma_3^2\right)$
J_3	$\frac{1}{3}(\sigma_{xx}^3 + \sigma_{yy}^3 + \sigma_{zz}^3 + 3\sigma_{xy}^2\sigma_{xx} +$ $+3\sigma_{xy}^2\sigma_{yy} + 3\sigma_{xz}^2\sigma_{xx} + 3\sigma_{xz}^2\sigma_{zz} +$ $+3\sigma_{yz}^2\sigma_{yy} + 3\sigma_{yz}^2\sigma_{zz} + 6\sigma_{xy}\sigma_{xz}\sigma_{yz})$	$\frac{1}{3}\left(\sigma_1^3 + \sigma_2^3 + \sigma_3^3\right)$
	Spherical matrix	
J_1^{o}	$\sigma_{xx} + \sigma_{yy} + \sigma_{zz}$	$\sigma_1 + \sigma_2 + \sigma_3$
J_2^{o}	$\frac{1}{6}\left(\sigma_{xx} + \sigma_{yy} + \sigma_{zz}\right)^2$	$\frac{1}{6}\left(\sigma_1 + \sigma_2 + \sigma_3\right)^2$
J_3^{o}	$\frac{1}{27}\left(\sigma_{xx} + \sigma_{yy} + \sigma_{zz}\right)^3$	$\frac{1}{27}\left(\sigma_1 + \sigma_2 + \sigma_3\right)^3$
	Stress deviator matrix	
J_1'	0	0
J_2'	$\frac{1}{6}\left[(\sigma_{xx} - \sigma_{yy})^2 + (\sigma_{yy} - \sigma_{zz})^2\right.$ $\left.+(\sigma_{zz} - \sigma_{xx})^2\right] + \sigma_{xy}^2 + \sigma_{yz}^2 + \sigma_{zx}^2$	$\frac{1}{6}\left[(\sigma_1 - \sigma_2)^2 + (\sigma_2 - \sigma_3)^2\right.$ $\left.+(\sigma_3 - \sigma_1)^2\right]$ or $-(s_1 s_2 + s_2 s_3 + s_3 s_1)$
J_3'	$s_{xx}s_{yy}s_{zz} + 2\sigma_{xy}\sigma_{yz}\sigma_{zx}$ $-s_{xx}\sigma_{yz}^2 - s_{yy}\sigma_{zx}^2 - s_{zz}\sigma_{xy}^2$	$s_1 s_2 s_3$
with	$s_{xx} = \frac{1}{3}(2\sigma_{xx} - \sigma_{yy} - \sigma_{zz})$ $s_{yy} = \frac{1}{3}(-\sigma_{xx} + 2\sigma_{yy} - \sigma_{zz})$ $s_{zz} = \frac{1}{3}(-\sigma_{xx} - \sigma_{yy} + 2\sigma_{zz})$ $s_{xy} = \sigma_{xy}, s_{xz} = \sigma_{xz}, s_{yz} = \sigma_{yz}$	$s_1 = \frac{1}{3}(2\sigma_1 - \sigma_2 - \sigma_3)$ $s_2 = \frac{1}{3}(-\sigma_1 + 2\sigma_2 - \sigma_3)$ $s_3 = \frac{1}{3}(-\sigma_1 - \sigma_2 + 2\sigma_3)$

The evaluation of the basic invariants for the stress matrix, as well as the hydrostatic and deviatoric part, is presented in Table 2.4, expressed in components of σ_{ij} and the principal stresses $\sigma_1, \sigma_2, \sigma_3$.

It can be seen in Table 2.4 that the spherical matrix is completely characterized by its first invariant because the second and third invariants are powers of it. The stress deviator matrix is completely characterized by its second and third invariants. Therefore, the physical contents of the stress state σ_{ij} can be described either by the three basic stress invariants J_i or if we use the decomposition in its spherical and deviatoric part by the first invariant of the spherical matrix and the second and third invariant of the stress deviator matrix. In the following, we will only use these three basic invariants to describe yield and failure conditions. Thus, the physical content of a state of stress will be described by the following set of invariants:

$$\sigma_{ij} \rightarrow J_1^{\mathrm{o}}, J_2', J_3'. \tag{2.15}$$

The representation of a stress state in terms of invariants is also very useful in the context of the implementation of yield conditions into commercial finite element codes. This significantly facilities the calculation of derivatives (see Öchsner 2003; Trapp and Öchsner 2018).

2.1.2 Graphical Representation of Yield Conditions

Plastic flow starts in a uniaxial tensile test as soon as the acting tensile stress σ reaches the initial yield stress k^{init}; see Öchsner (2016). In the case of a multiaxial stress state, this simple comparison is replaced by the yield condition. To this end, a scalar value is calculated from the acting six stress components and compared to an experimental scalar value. The yield condition in stress space can be expressed in its most general form ($\mathbb{R}^6 \times \mathbb{R}^{\text{dim}(\boldsymbol{q})} \rightarrow \mathbb{R}$) as

$$F = F(\boldsymbol{\sigma}, \boldsymbol{q}). \tag{2.16}$$

For further characterization, we assume in the following the special case of ideal plastic material behavior (vector of hardening variables $\boldsymbol{q} = \boldsymbol{0}$) so that for ($\mathbb{R}^6 \rightarrow \mathbb{R}$)

$$F = F(\boldsymbol{\sigma}) \tag{2.17}$$

depends now only on the stress state. The values of F have—as in the uniaxial case—the following mechanical meaning:

$$F(\boldsymbol{\sigma}) = 0 \rightarrow \text{plastic material behavior}, \tag{2.18}$$

$$F(\boldsymbol{\sigma}) < 0 \rightarrow \text{elastic material behavior}, \tag{2.19}$$

$$F(\boldsymbol{\sigma}) > 0 \rightarrow \text{invalid}. \tag{2.20}$$

A[3] further simplification is obtained under the assumption that the yield condition can be split in a pure stress part $f(\boldsymbol{\sigma})$ and an experimental material parameter k:

$$F(\boldsymbol{\sigma}) = f(\boldsymbol{\sigma}) - k. \tag{2.21}$$

The yield condition $F = 0$ represents in a n-dimensional space a hyper-surface that is also called the yield surface or the yield loci. The number n is equal to the independent stress matrix components. A direct graphical representation of the yield surface is not possible due to its dimensionality, i.e., six variables. However, a reduction of the dimensionality is possible to achieve if a principle axis transformation (see Eq. (2.2)) is applied to the argument σ_{ij}. The components of the stress matrix reduce to the

[3] Under the restriction of rate-independent plasticity.

Fig. 2.3 Particular octahedral plane with normal vector $n = 1/\sqrt{3}[1\ 1\ 1]^{\mathrm{T}}$

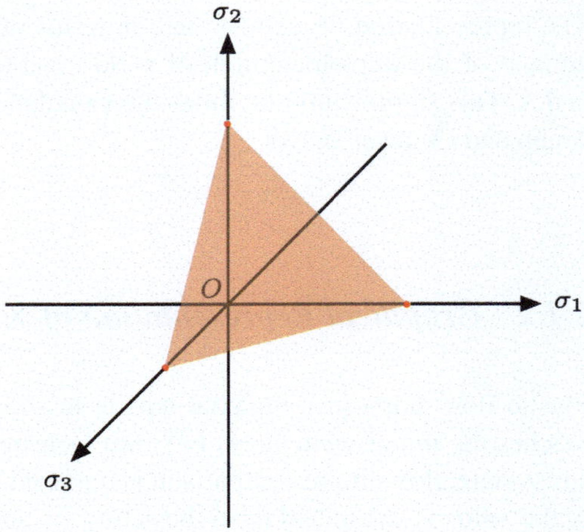

principal stresses σ_1, σ_2, and σ_3 on the principal diagonal of the stress matrix and the non-diagonal elements are equal to zero. In such a principal stress space, it is possible to graphically represent the yield condition as a three-dimensional surface. This space is also called the Haigh–Westergaard stress space (see Chen and Zhang 1991). A hydrostatic stress state lies in such a principal stress system on the space diagonal (hydrostatic axis). Planes with normal vectors $\boldsymbol{n} = 1/\sqrt{3}\left[\pm 1\ \pm 1\ \pm 1\right]^{\mathrm{T}}$ are called octahedral planes.

In the following, we will focus only on octahedral planes perpendicular to the hydrostatic axis, i.e., with a normal vector $\boldsymbol{n} = 1/\sqrt{3}\left[+1\ +1\ +1\right]^{\mathrm{T}}$; see Fig. 2.3. The particular octahedral plane passing through the origin of the coordinate system is called the deviatoric plane or π-plane (see Chen and Han 1988). Because $\sigma_1 + \sigma_2 + \sigma_3 = 0$, it follows from Eq. (2.6) that $\sigma_{ij} = s_{ij}$, i.e., any stress state on the π-plane is pure deviatoric.

The possibility of a representation of a yield condition based on a set of independent stress invariants (e.g., according to Eq. (2.15)) is the characteristic of any isotropic yield condition, regardless of the choice of coordinate system. Therefore, Eq. (2.17) can also be written as

$$F = F(J_1^{\mathrm{o}}, J_2', J_3'). \tag{2.22}$$

On the basis of the dependency of the yield condition on the invariants, a descriptive classification can be performed. Yield conditions independent of the hydrostatic stress (J_1^{o}) can be represented by the invariants J_2' and J_3'. Stress states with $J_2' =$ const. lie on a circle around the hydrostatic axis in an octahedral plane. A dependency of the yield condition on J_3' results in a deviation from the circular shape. The yield surface forms a prismatic body whose longitudinal axis is represented by the hydrostatic axis. A dependency on J_1^{o} denotes a size change of the cross-section of the yield surface along the hydrostatic axis. However, the shape of the cross-section

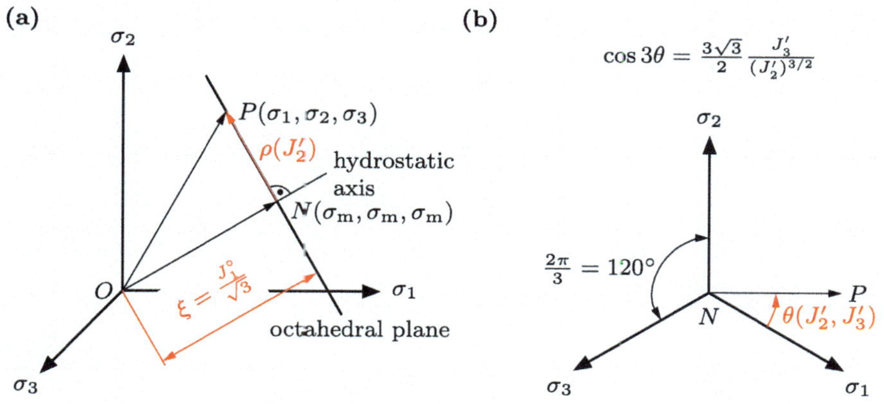

Fig. 2.4 Geometrical interpretation of basic stress invariants: **a** principal stress space; **b** octahedral plane

remains similar in the mathematical sense. Therefore, a dependency on J_1° can be represented by sectional views through planes along the hydrostatic axis.

The geometrical interpretation of stress invariants (see Chen and Han 1988) is given in Fig. 2.4:

It can be seen that an arbitrary stress state P can be expressed by its position along the hydrostatic axis $\xi = \frac{1}{\sqrt{3}} J_1^{\circ}$ and its polar coordinates ($\rho = \sqrt{2J_2'}$, $\theta(J_2', J_3')$) in the octahedral plane through P. For the set of polar coordinates, the so-called stress Lode angle θ is defined in the range $0 \leq \theta \leq 60°$ as (see Nayak and Zienkiewicz 1972 and Bruhns 2024 for more details),

$$\cos(3\theta) = \frac{3\sqrt{3}}{2} \cdot \frac{J_3'}{(J_2')^{3/2}}. \tag{2.23}$$

It can be concluded from Eq. (2.23) that

$$3\theta = 3\theta(J_2', J_3') \quad \text{or} \quad \theta = \theta(J_2', J_3'). \tag{2.24}$$

The set of coordinates ($\xi, \rho, \cos(3\theta)$) is known in the literature as the Haigh–Westergaard coordinates (Chen and Han 1988). The expressions for the coordinates ξ and ρ can be obtained as follows: The location of the stress state P (see point P in Fig. 2.4a) can be represented by the following vector:

$$\boldsymbol{OP} = \begin{bmatrix} \sigma_1 \\ \sigma_2 \\ \sigma_3 \end{bmatrix}. \tag{2.25}$$

Projecting this vector in direction of the normal vector ($\boldsymbol{n}$) of the octahedral plane gives

$$\boldsymbol{\xi} = (\boldsymbol{OP}:\boldsymbol{n})\,\boldsymbol{n} = \left(\begin{bmatrix}\sigma_1\\ \sigma_2\\ \sigma_3\end{bmatrix} : \frac{1}{\sqrt{3}}\begin{bmatrix}1\\1\\1\end{bmatrix}\right)\frac{1}{\sqrt{3}}\begin{bmatrix}1\\1\\1\end{bmatrix} = \frac{1}{3}(\sigma_1+\sigma_2+\sigma_3)\begin{bmatrix}1\\1\\1\end{bmatrix}. \tag{2.26}$$

The absolute value of the projection in the direction of the octahedral normal vector is obtained as follows:

$$\xi = |\boldsymbol{\xi}_{\parallel}| \tag{2.27}$$

$$= \frac{1}{3}\sqrt{(\sigma_1+\sigma_2+\sigma_3)^2 + (\sigma_1+\sigma_2+\sigma_3)^2 + (\sigma_1+\sigma_2+\sigma_3)^2} \tag{2.28}$$

$$= \frac{\sqrt{3}}{3}(\sigma_1+\sigma_2+\sigma_3) = \frac{1}{\sqrt{3}}(\sigma_1+\sigma_2+\sigma_3) = \frac{1}{\sqrt{3}}J_1^{\text{o}}. \tag{2.29}$$

The projection of the vector $\boldsymbol{OP}$ perpendicular to the normal vector of the octahedral plane gives

$$\boldsymbol{\rho} = \boldsymbol{OP} - \boldsymbol{\xi} = \begin{bmatrix}\sigma_1\\ \sigma_2\\ \sigma_3\end{bmatrix} - \frac{1}{3}(\sigma_1+\sigma_2+\sigma_3)\begin{bmatrix}1\\1\\1\end{bmatrix} \tag{2.30}$$

$$= \begin{bmatrix}\sigma_1 - \frac{1}{3}(\sigma_1+\sigma_2+\sigma_3)\\ \sigma_2 - \frac{1}{3}(\sigma_1+\sigma_2+\sigma_3)\\ \sigma_3 - \frac{1}{3}(\sigma_1+\sigma_2+\sigma_3)\end{bmatrix}. \tag{2.31}$$

The absolute value of this perpendicular component is obtained as

$$\rho = |\boldsymbol{\rho}| = \sqrt{\sum_{i=1}^{3}(\sigma_i - \tfrac{1}{3}(\sigma_1+\sigma_2+\sigma_3))^2}, \tag{2.32}$$

or after a short calculation as

$$\rho = \sqrt{\frac{1}{3}\left((\sigma_1-\sigma_2)^2 + (\sigma_2-\sigma_3)^2 + (\sigma_1-\sigma_3)\right)^2} = \sqrt{2J_2'}. \tag{2.33}$$

To derive the expression for the Lode angle θ according to Eq. (2.23), it is helpful to recall the geometrical definition of this angle (see Fig. 2.5), i.e., the smallest angle between the projection of the σ_1 axis and the projection of the vector $\boldsymbol{OP}$ on the octahedral plane.

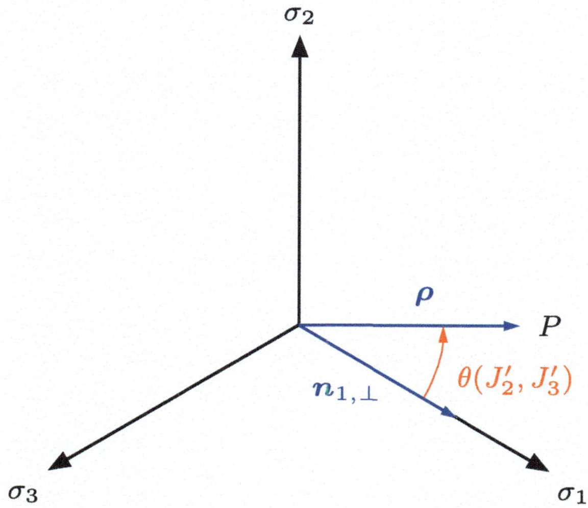

Fig. 2.5 On the definition of the Lode angle θ in an octahedral plane

The projection of the vector $\boldsymbol{OP}$ on the octahedral plane is equal to the vector $\boldsymbol{\rho}$ given in Eq. (2.31). On the other hand, the projection of the σ_1 axis on the octahedral plane can be obtained as follows. The normal vector in direction of the σ_1 axis, i.e., $\boldsymbol{n}_1$, can be split in its parallel ($\boldsymbol{n}_{1,\parallel}$) and perpendicular projection ($\boldsymbol{n}_{1,\perp}$) in regards to the normal vector ($\boldsymbol{n}$) of the octahedral plane:

$$\boldsymbol{n}_{1,\parallel} = (\boldsymbol{n}_1 : \boldsymbol{n})\,\boldsymbol{n} = \left(\begin{bmatrix}1\\0\\0\end{bmatrix} : \frac{1}{\sqrt{3}}\begin{bmatrix}1\\1\\1\end{bmatrix}\right)\frac{1}{\sqrt{3}}\begin{bmatrix}1\\1\\1\end{bmatrix} = \frac{1}{3}\begin{bmatrix}1\\1\\1\end{bmatrix}. \tag{2.34}$$

$$\boldsymbol{n}_{1,\perp} = \boldsymbol{n}_1 - \boldsymbol{n}_{1,\parallel} = \begin{bmatrix}1\\1\\1\end{bmatrix} - \frac{1}{3}\begin{bmatrix}1\\1\\1\end{bmatrix} = \frac{1}{3}\begin{bmatrix}2\\-1\\-1\end{bmatrix}, \tag{2.35}$$

whereas the absolute value of $\boldsymbol{n}_{1,\perp}$ is obtained as

$$n_{1,\perp} = |\boldsymbol{n}_{1,\perp}| = \sqrt{\frac{4}{9} + \frac{1}{9} + \frac{1}{9}} = \sqrt{\frac{2}{3}}. \tag{2.36}$$

Thus, the Lode angle θ can be obtained based on the definition of the scalar product between two vectors as

$$\theta = \frac{\boldsymbol{\rho} : \boldsymbol{n}_{1,\perp}}{|\boldsymbol{\rho}||\boldsymbol{n}_{1,\perp}|} = \frac{\begin{bmatrix}\sigma_1 - \frac{1}{3}(\sigma_1+\sigma_2+\sigma_3)\\ \sigma_2 - \frac{1}{3}(\sigma_1+\sigma_2+\sigma_3)\\ \sigma_3 - \frac{1}{3}(\sigma_1+\sigma_2+\sigma_3)\end{bmatrix} : \frac{1}{3}\begin{bmatrix}2\\-1\\-1\end{bmatrix}}{\rho n_{1,\perp}} \tag{2.37}$$

$$= \frac{\begin{bmatrix} s_1 \\ s_2 \\ s_3 \end{bmatrix} : \frac{1}{3} \begin{bmatrix} 2 \\ -1 \\ -1 \end{bmatrix}}{\rho n_{1,\perp}} = \frac{\frac{1}{3}(2s_1 - s_2 - s_3)}{\sqrt{2J_2'}\sqrt{\frac{2}{3}}}. \tag{2.38}$$

With the trigonometric identity, i.e., $\cos(3\theta) = 4\cos^3(\theta) - 3\cos\theta$ and the relationship $s_1 + s_2 + s_3 = 0$, one gets

$$\cos(3\theta) = 4\left(\frac{\sqrt{3}}{2}\frac{s_1}{\sqrt{J_2'}}\right)^3 - 3\left(\frac{\sqrt{3}}{2}\frac{s_1}{\sqrt{J_2'}}\right) = \frac{3\sqrt{3}}{2(J_2')^{3/2}}\left(s_1^3 - s_1 J_2'\right). \tag{2.39}$$

With $J_2' = -(s_1 s_2 + s_2 s_3 + s_3 s_1)$ and $J_3' = s_1 s_2 s_3$, one gets finally (Chen and Han 1988):

$$\cos(3\theta) = \frac{3\sqrt{3}}{2(J_2')^{3/2}}\left(s_1^3 + s_1^2(s_2 + s_3) + s_1 s_2 s_3\right) \tag{2.40}$$

$$= \frac{3\sqrt{3}}{2(J_2')^{3/2}}(s_1 s_2 s_3) = \frac{3\sqrt{3}}{2} \cdot \frac{J_3'}{(J_2')^{3/2}}. \tag{2.41}$$

To investigate the shape of the yield surface, particular experiments, including multiaxial stress states, must be realized and the initial yield points marked and approximated in the Haigh–Westergaard space, the $J_1^{\circ} - \sqrt{3J_2'}$ invariant space (this representation is also known as the Burzyński-plane Kolupaev 2018), and octahedral planes, see Figs. 2.6 and 2.7.

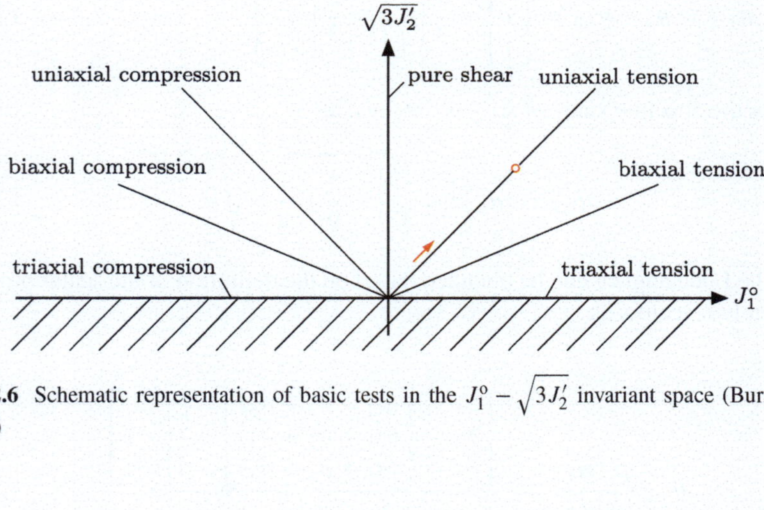

Fig. 2.6 Schematic representation of basic tests in the $J_1^{\circ} - \sqrt{3J_2'}$ invariant space (Burzyński-plane)

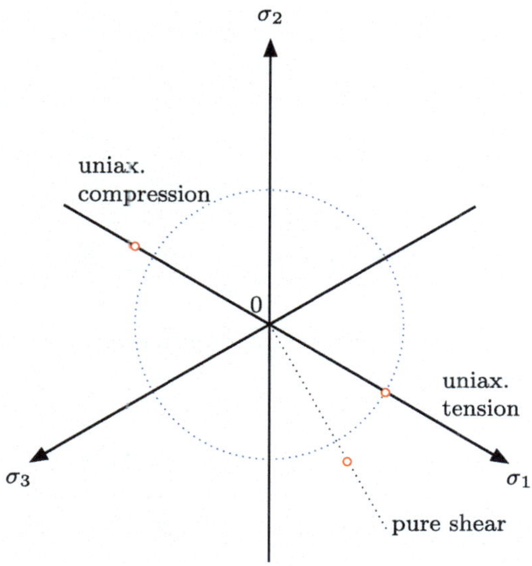

Fig. 2.7 Identification of the shape of a yield condition in an octahedral plane (σ_m = const.)

Table 2.5 Definition of basic tests in the $J_1^o - \sqrt{3J_2'}$ invariant space

Case	J_1^o	$\sqrt{3J_2'}$	Comment
uniax. tension (σ)	σ	σ	slope: 1
uniax. compression ($-\sigma$)	$-\sigma$	σ	slope: -1
biax. tension (σ)	2σ	σ	slope: 0.5
biax. compression ($-\sigma$)	-2σ	σ	slope: -0.5
triax. tension (σ)	3σ	0	horiz. axis
triax. compression ($-\sigma$)	-3σ	0	horiz. axis
pure shear (τ)	0	$\sqrt{3}\,\tau$	vertical axis

The loading path, for example, for the biaxial tension case in the $J_1^o - \sqrt{3J_2'}$ invariant space (see Fig. 2.6) is obtained as follows:

$$\sqrt{3J_2'} = \sigma = \frac{1}{2} J_1^o. \tag{2.42}$$

Further load paths for basic experiments in the $J_1^o - \sqrt{3J_2'}$ invariant space are summarized in Table 2.5.

The *pure* shear test mentioned in Fig. 2.6 and Table 2.5 can be realized with cylindrical specimens or tubes which are subjected to a torsional load. It should be mentioned here that scientific literature also suggests the so-called *simple* shear test (Walrath and

Adams 2024; Stojcevski et al. 2018). The corresponding stress state is characterized by a shear stress which is superimposed by a bending load, i.e., normal stress (Heyder et al. 2002). A possible experimental setup for the simple shear test is reported in Öchsner (2003).

2.2 Experimental Realization of Multiaxial Stress States

2.2.1 Thin-Walled Tubes Under Internal/External Pressure and Additional Loads

Many different possibilities are available to experimentally realize multiaxial stress states. The particular approach depends on many factors such as the material itself, the type of specimen that is possible to manufacture, and many other factors. In addition, depending on the specimen type, effects such as bucking, wrinkling, and stress concentrations must be many times considered and excluded. The scientific literature offers plenty of resources and some overview on the different approaches can be found, for example, in Evans (1972) and Bardenheier (1982).

The stress state in a thin-walled tube[4] under an internal pressure p_i is characterized by three orthogonal normal stress components, which are acting in the tangential ('t'), axial ('a'), and radial ('r') direction:

$$\sigma_\mathrm{t} = p_\mathrm{i} \times \frac{d_\mathrm{i}}{2s}, \tag{2.43}$$

$$\sigma_\mathrm{a} = p_\mathrm{i} \times \frac{d_\mathrm{i}}{4s} = \frac{\sigma_\mathrm{t}}{2}, \tag{2.44}$$

$$\sigma_\mathrm{r} = -\frac{p_\mathrm{i}}{2}. \tag{2.45}$$

The basic course of these principal stresses is illustrated in Fig. 2.8. It can be seen that all stress components are assumed constant across the wall thickness ($s = (d_\mathrm{o} - d_\mathrm{i})/2$) and that the radial component (σ_r) is significantly smaller than the tangential (σ_t) and axial (σ_a) stress contribution. Thus, we may neglect in a first attempt the radial stress and assume a two-dimensional normal stress state.

[4] 'Thin' means that the ratio between the outer and inner diameter is below a certain threshold: $d_\mathrm{o}/d_\mathrm{i} \leq 1.2$ (Altenbach 2022). Alternatively, this condition can be expressed based on the wall thickness s as $s/d_\mathrm{i} \leq 0.1$.

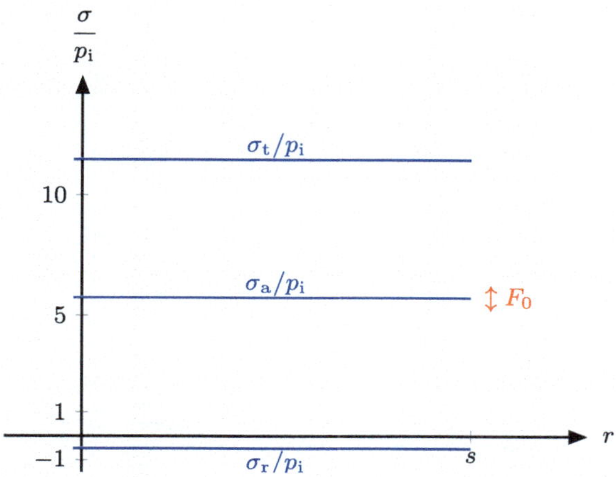

Fig. 2.8 Stress distribution in a thin-walled tube under internal pressure p_i ($d_o = 25$ mm, $d_i = 23$ mm, $p_i = 50$ MPa)

Based on the stress components provided in Eqs. (2.43) and (2.44) and under the simplification that the radial stress component can be neglected, the slope (m) of the load path in the $J_1^o - \sqrt{3J_2'}$ invariant space is obtained as

$$m = \frac{\sqrt{3J_2'}}{J_1^o} = \frac{\sqrt{\sigma_a^2 + \sigma_t^2 - \sigma_a\sigma_t}}{\sigma_a + \sigma_t} = \frac{\sqrt{3}}{3} \approx 0.577, \tag{2.46}$$

which corresponds to an angle of $\alpha = 30°$ (see Fig. 2.9b). This means that an experiment using a thin-walled tube under increasing internal pressure can deliver only a single point of the yield surface (indicated in Fig. 2.9b by the circle marker) in the $J_1^o - \sqrt{3J_2'}$ invariant space. To further explore the shape of the yield surface, additional external loads can be imposed to the pressured tube.

Let us start to investigate the case that the pressured tube is additionally loaded by an external tensile force (F_0) (Stout and Petrovic 1984; Ripley and Korkolis 2016); see Fig. 2.10a. This tensile force results in an additional axial stress component and the total stress state can be characterized as

$$\sigma_t = p_i \times \frac{d_i}{2s}, \tag{2.47}$$

$$\sigma_a = p_i \times \frac{d_i}{4s} + \frac{F_0}{A}, \tag{2.48}$$

where the cross-sectional area is obtained as $A = \frac{\pi}{4}(d_o^2 - d_i^2) \approx \pi s d_i$. This means that the tangential stress component is not affected by the external tensile force and the total axial stress component can be influenced; see Fig. 2.8.

Assuming that a particular internal pressure p_i is acting, one may gradually increase the external tensile load F_0. This means that we have first a state with σ_t and $\sigma_a = \sigma_t/2$, which corresponds to a point on the line through the origin with

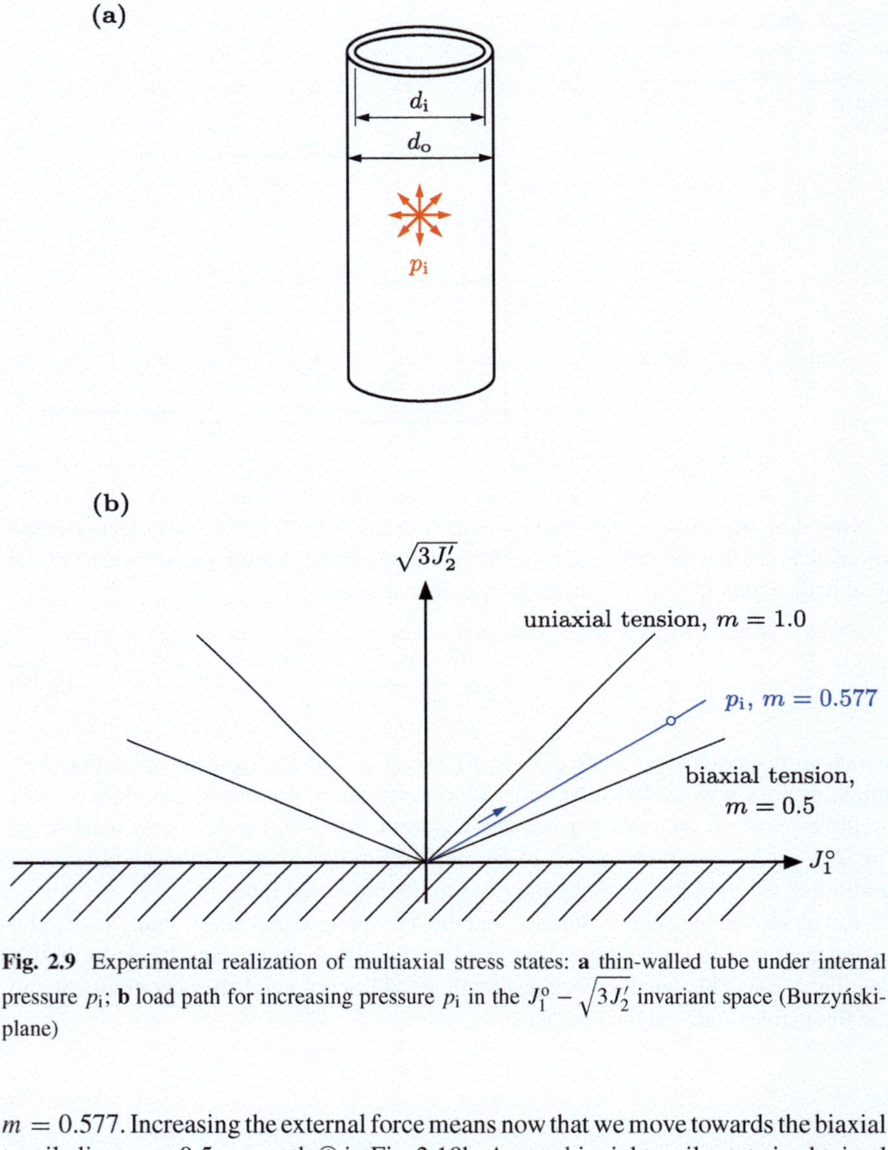

Fig. 2.9 Experimental realization of multiaxial stress states: **a** thin-walled tube under internal pressure p_i; **b** load path for increasing pressure p_i in the $J_1^o - \sqrt{3J_2'}$ invariant space (Burzyński-plane)

$m = 0.577$. Increasing the external force means now that we move towards the biaxial tensile line $m = 0.5$; see path ① in Fig. 2.10b. A pure biaxial tensile state is obtained for $\sigma_t = \sigma_a$, or

$$F_0|_{\text{biax. tension}} = \frac{p_i d_i A}{4s}. \tag{2.49}$$

A further increase of F_0 beyond this value results in a state $\sigma_a > \sigma_t$ or finally $\sigma_a \gg \sigma_t$, which means in a more or less uniaxial stress state; see path ② in Fig. 2.10b.

A different range in the $J_1^o - \sqrt{3J_2'}$ invariant space can be covered by inverting the external load F_0, i.e., by applying an external compressive force (Stout and Petrovic

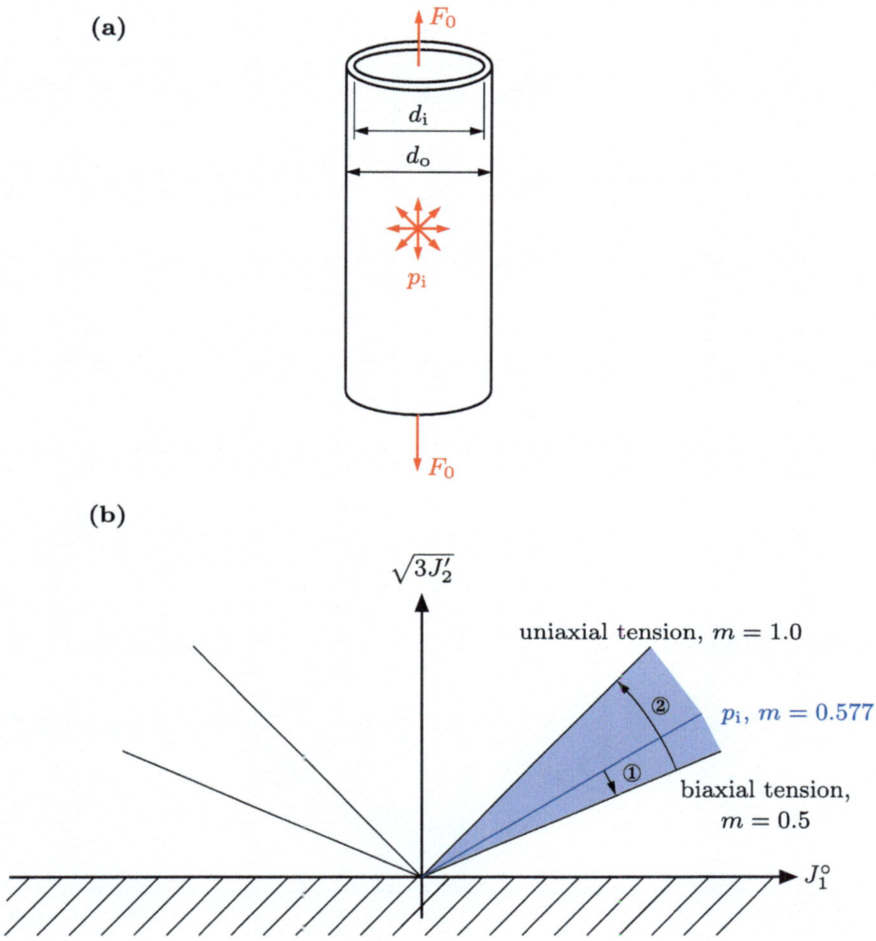

Fig. 2.10 Experimental realization of multiaxial stress states: **a** thin-walled tube under internal pressure p_i and external tensile force F_0; **b** representation in the $J_1^o - \sqrt{3J_2'}$ invariant space (Burzyński-plane)

1984); see Fig. 2.11a. This compressive force results in a reduction of the axial stress component and the total stress state can be characterized as

$$\sigma_t = p_i \times \frac{d_i}{2s}, \tag{2.50}$$

$$\sigma_a = p_i \times \frac{d_i}{4s} - \frac{F_0}{A}, \tag{2.51}$$

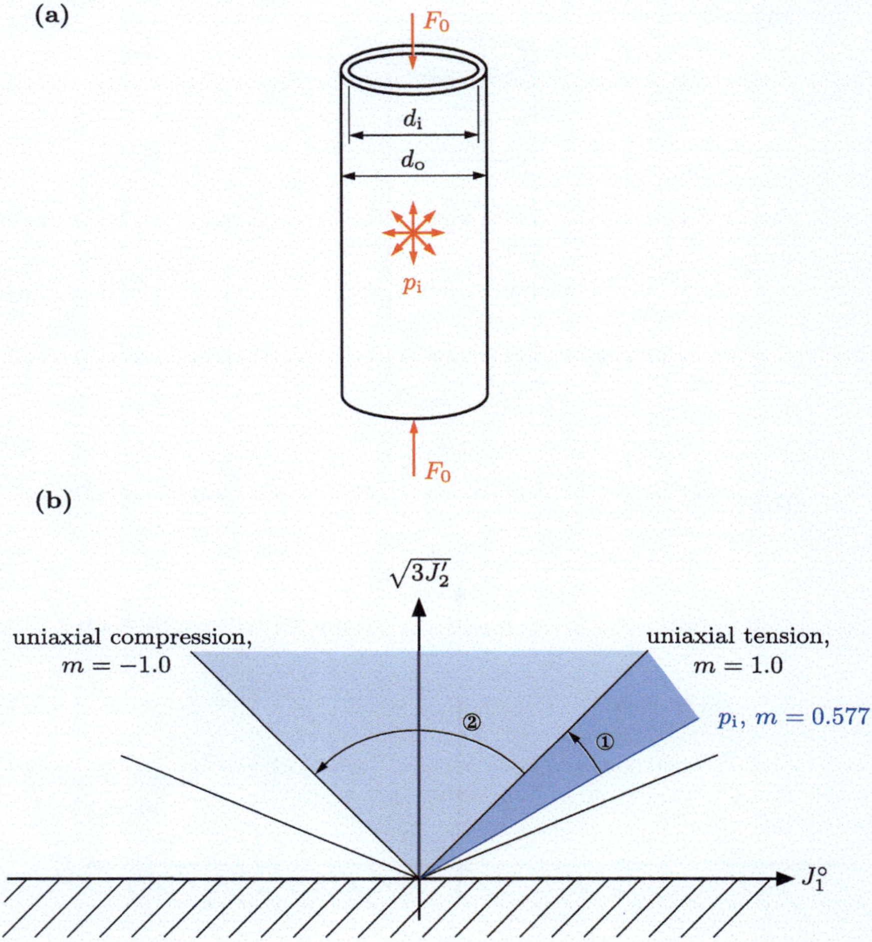

Fig. 2.11 Experimental realization of multiaxial stress states: **a** thin-walled tube under internal pressure p_i and external compressive force F_0; **b** representation in the $J_1^{\circ} - \sqrt{3J_2'}$ invariant space (Burzyński-plane)

where the cross-sectional area is again obtained as $A = \frac{\pi}{4}(d_o^2 - d_i^2) \approx \pi s d_i$. As in the tensile case, the tangential stress component is not affected by the external compressive force and the total axial stress component can be influenced; see Fig. 2.8.

Assuming that a particular internal pressure p_i is acting, one may gradually increase the external compressive force F_0. This means that we have first a state with σ_t and $\sigma_a = \sigma_t/2$, which corresponds to a point on the line through the origin with $m = 0.577$. Increasing the magnitude of the external compressive force means now that we move towards the uniaxial tensile line $m = 1.0$; see path ① in Fig. 2.11b.

A pure uniaxial tensile state is obtained for $\sigma_a = 0$, or

$$|F_0|_{\text{uniax. tension}}| = \frac{p_i d_i A}{4s}. \tag{2.52}$$

A further increase of $|F_0|$ beyond this value results in a state $|\sigma_a| > |\sigma_t|$ or finally $|\sigma_a| \gg |\sigma_t|$, which means in more or less uniaxial compression; see path ② in Fig. 2.11b.

A different experimental setup is based on the pressured thin tube with a superimposed torsional load (Stout and Petrovic 1984); see Fig. 2.12a. This external torsional moment results in a linear shear stress distribution across the wall thickness, which can be approximated for thin tubes as

$$\overline{\tau} \approx \frac{M_0}{I_p} \times \overline{r}, \tag{2.53}$$

where $I_p = \frac{\pi}{32}(d_o^4 - d_i^4) \approx \frac{\pi}{4} s d_i^3$ is the polar second moment of area and $\overline{r} = (d_i + d_o)/4$ is the average radius. Thus, a biaxial stress state with σ_a, σ_t, and $\overline{\tau}$ is acting in the tube wall. Based on these stress components, the slope (m) of a load path in the $J_1^o - \sqrt{3J_2'}$ invariant space is obtained as

$$m = \frac{\sqrt{3J_2'}}{J_1^o} = \frac{\sqrt{\sigma_a^2 + \sigma_t^2 - \sigma_a\sigma_t + 3\overline{\tau}^2}}{\sigma_a + \sigma_t}. \tag{2.54}$$

Assuming that a particular internal pressure p_i is acting, one may gradually increase the external torsional moment M_0. This means that we have first a state with σ_t and $\sigma_a = \sigma_t/2$, which corresponds to a point on the line through the origin with $m = 0.577$. Increasing the external torsional moment means now that we move towards the pure shear line ($m \to \infty$); see path ① in Fig. 2.12b. An approximate pure shear state is obtained for $\overline{\tau} \gg \sigma_t$.

An experimental setup for the realization of a thin-walled tube under internal pressure p_i and external torsional moment M_0 is shown in Fig. 2.13.

Let us look now at a different experimental setup. The stress state in a thin-walled tube under an *external* pressure p_o is characterized by three orthogonal normal stress components, which are acting in the tangential, axial, and radial direction:

$$\sigma_t = -p_o \times \frac{d_o}{2s}, \tag{2.55}$$

$$\sigma_a = -p_o \times \frac{d_o}{4s} = \frac{\sigma_t}{2}, \tag{2.56}$$

$$\sigma_r = -\frac{p_o}{2}. \tag{2.57}$$

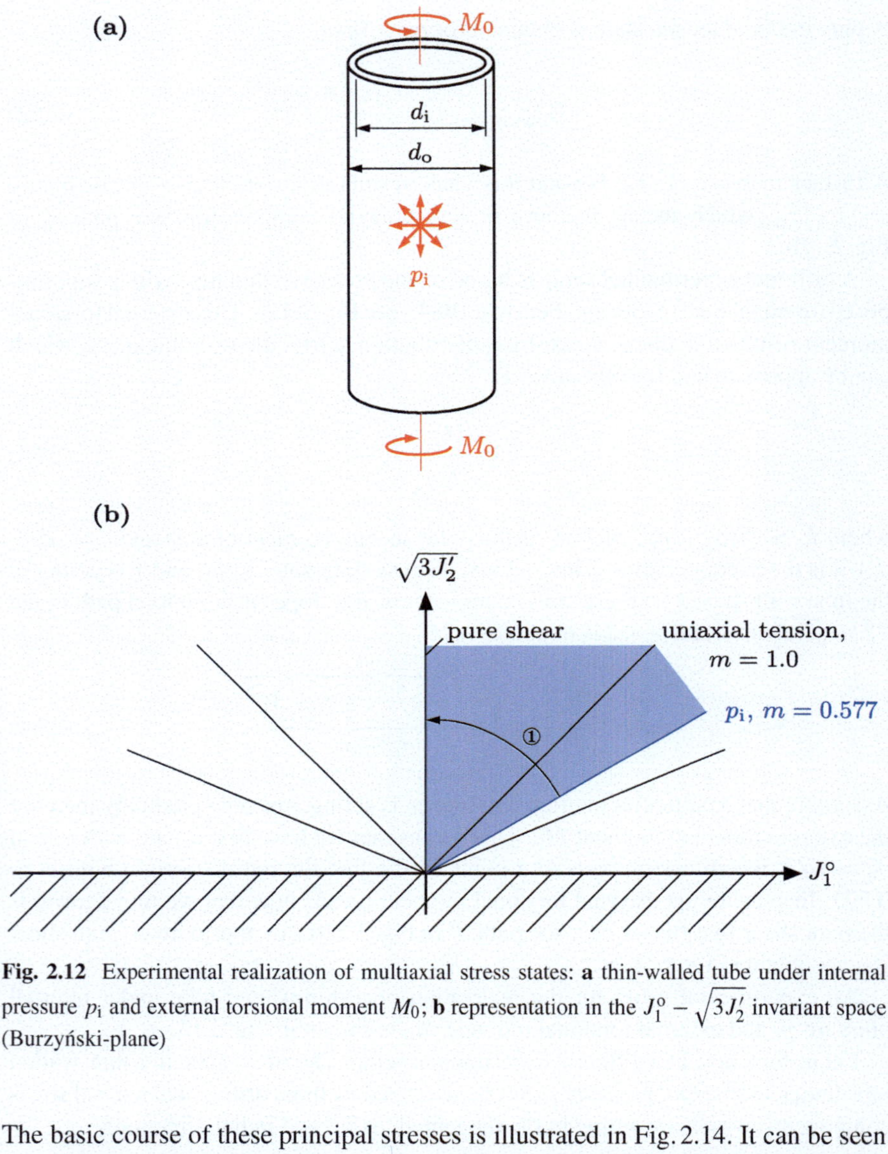

Fig. 2.12 Experimental realization of multiaxial stress states: **a** thin-walled tube under internal pressure p_i and external torsional moment M_0; **b** representation in the $J_1^o - \sqrt{3J_2'}$ invariant space (Burzyński-plane)

The basic course of these principal stresses is illustrated in Fig. 2.14. It can be seen that all stress components are assumed constant across the wall thickness (s) and that the magnitude of the radial component (σ_r) is significantly smaller than the tangential (σ_t) and axial (σ_a) stress contribution. Thus, we may neglect in a first attempt the radial stress and also assume here a two-dimensional normal stress state.

Based on the stress components provided in Eqs. (2.55) and (2.56) and under the simplification that the radial stress component can be neglected, the slope (m) of the load path in the $J_1^o - \sqrt{3J_2'}$ invariant space is obtained as

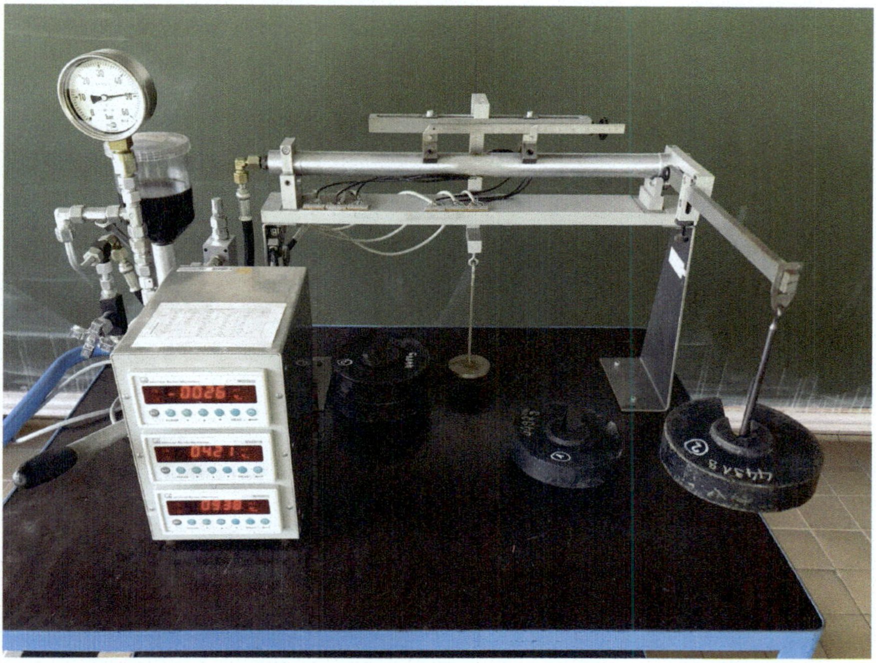

Fig. 2.13 Experimental realization of multiaxial stress states based on a thin-walled tube under internal pressure p_i and external torsional moment M_0. In addition, there is a possibility to superimpose a bending load (HS Esslingen)

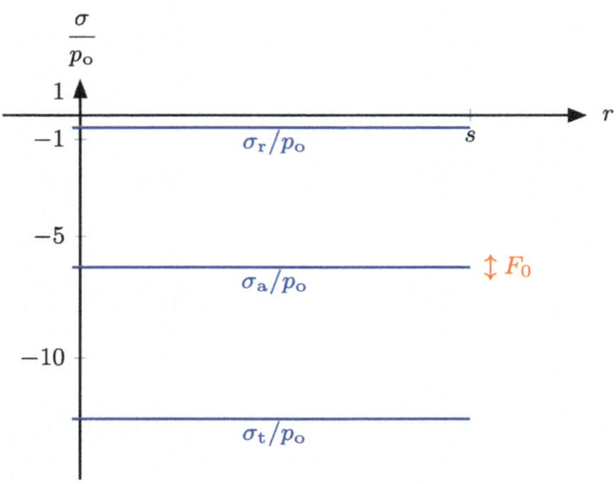

Fig. 2.14 Stress distribution in a thin-walled tube under external pressure p_o ($d_o = 25$ mm, $d_i = 23$ mm, $p_o = 50$ MPa)

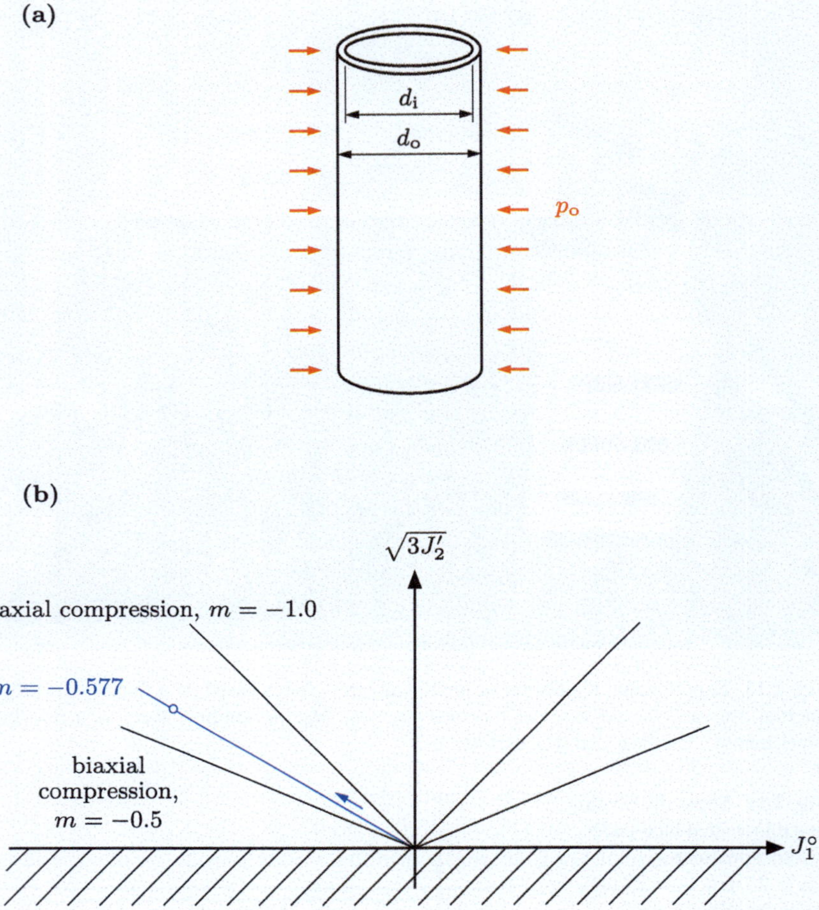

Fig. 2.15 Experimental realization of multiaxial stress states: **a** thin-walled tube under external pressure p_o; **b** load path for increasing pressure p_o in the $J_1^o - \sqrt{3J_2'}$ invariant space (Burzyński-plane)

$$m = \frac{\sqrt{3J_2'}}{J_1^o} = \frac{\sqrt{\sigma_a^2 + \sigma_t^2 - \sigma_a\sigma_t}}{\sigma_a + \sigma_t} = -\frac{\sqrt{3}}{3} \approx -0.577, \tag{2.58}$$

which corresponds to an angle of $\alpha = 150°$ (see Fig. 2.15b). This means that an experiment using a thin-walled tube under increasing external pressure can deliver only a single point of the yield surface (indicated in Fig. 2.15b by the circle marker) in the $J_1^o - \sqrt{3J_2'}$ invariant space. To further explore the shape of the yield surface, additional external loads can be imposed to the pressured tube.

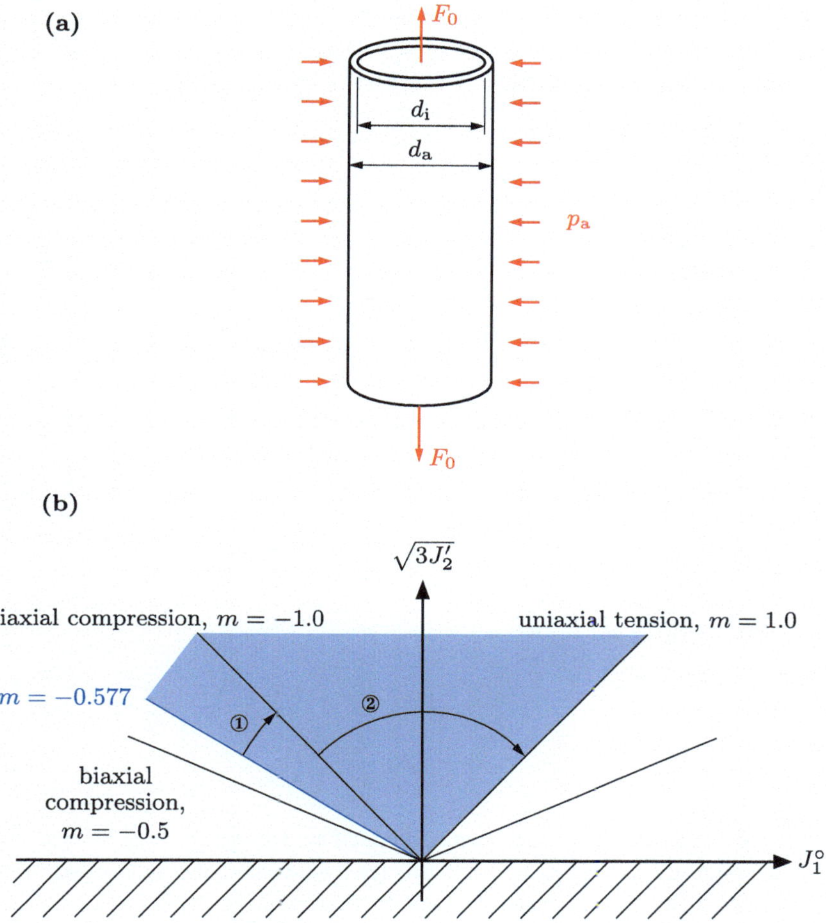

Fig. 2.16 Experimental realization of multiaxial stress states: **a** thin-walled tube under external pressure p_a and external tensile force F_0; **b** representation in the $J_1^\circ - \sqrt{3J_2'}$ invariant space (Burzyński-plane)

Let us start to investigate the case that the from the outside pressured tube is additionally loaded by an external tensile force (F_0); see Fig. 2.16a. This tensile force results in an additional axial stress component and the total stress state can be characterized as

$$\sigma_t = -p_o \times \frac{d_o}{2s}, \tag{2.59}$$

$$\sigma_z = -p_o \times \frac{d_o}{4s} + \frac{F_0}{A}, \tag{2.60}$$

where the cross-sectional area is again obtained as $A = \frac{\pi}{4}(d_o^2 - d_i^2) \approx \pi s d_i$. This means that the tangential stress component is not affected by the external tensile force and the total axial stress component can be influenced; see Fig. 2.14.

Assuming that a particular external pressure p_o is acting, one may gradually increase the external tensile load F_0. This means that we have first a state with σ_t and $\sigma_a = \sigma_t/2$, which corresponds to a point on the line through the origin with $m = -0.577$. Increasing the external force means now that we move towards the uniaxial compression line $m = -1.0$, see path ① in Fig. 2.16b. A pure uniaxial compression state is obtained for $\sigma_a = 0$, or

$$\left| F_0|_{\text{uniax. compression}} \right| = \frac{p_o d_o A}{4s}. \tag{2.61}$$

A further increase of F_0 beyond this value results in a state $\sigma_a > \sigma_t$ or finally $\sigma_a \gg \sigma_t$, which means in a more or less uniaxial tensile stress state; see path ② in Fig. 2.16b.

A different range in the $J_1^o - \sqrt{3J_2'}$ invariant space can be covered by inverting the external load F_0, i.e., by applying an external compressive force, see Fig. 2.17a. This compressive force results in an increase in the magnitude of the axial stress component and the total stress state can be characterized as

$$\sigma_t = -p_o \times \frac{d_o}{2s}, \tag{2.62}$$

$$\sigma_a = -p_o \times \frac{d_o}{4s} - \frac{F_0}{A}, \tag{2.63}$$

where the cross-sectional area is again obtained as $A = \frac{\pi}{4}(d_o^2 - d_i^2) \approx \pi s d_i$. Again, the tangential stress component is not affected by the external compressive force and the total axial stress component can be influenced; see Fig. 2.14.

Assuming that a particular external pressure p_o is acting, one may gradually increase the external compressive force F_0. This means that we have first a state with σ_t and $\sigma_a = \sigma_t/2$, which corresponds to a point on the line through the origin with $m = -0.577$. Increasing the magnitude of the external compressive force means now that we move towards the biaxial compression line $m = -0.5$; see path ① in Fig. 2.17b. A pure biaxial compression state is obtained for $\sigma_a = \sigma_r$, or

$$\left| F_0|_{\text{biax. compression}} \right| = \frac{p_o d_o A}{4s}. \tag{2.64}$$

A further increase of $|F_0|$ beyond this value results in a state $|\sigma_a| > |\sigma_t|$ or finally $|\sigma_a| \gg |\sigma_t|$, which means in a more or less uniaxial compression; see path ② in Fig. 2.17b.

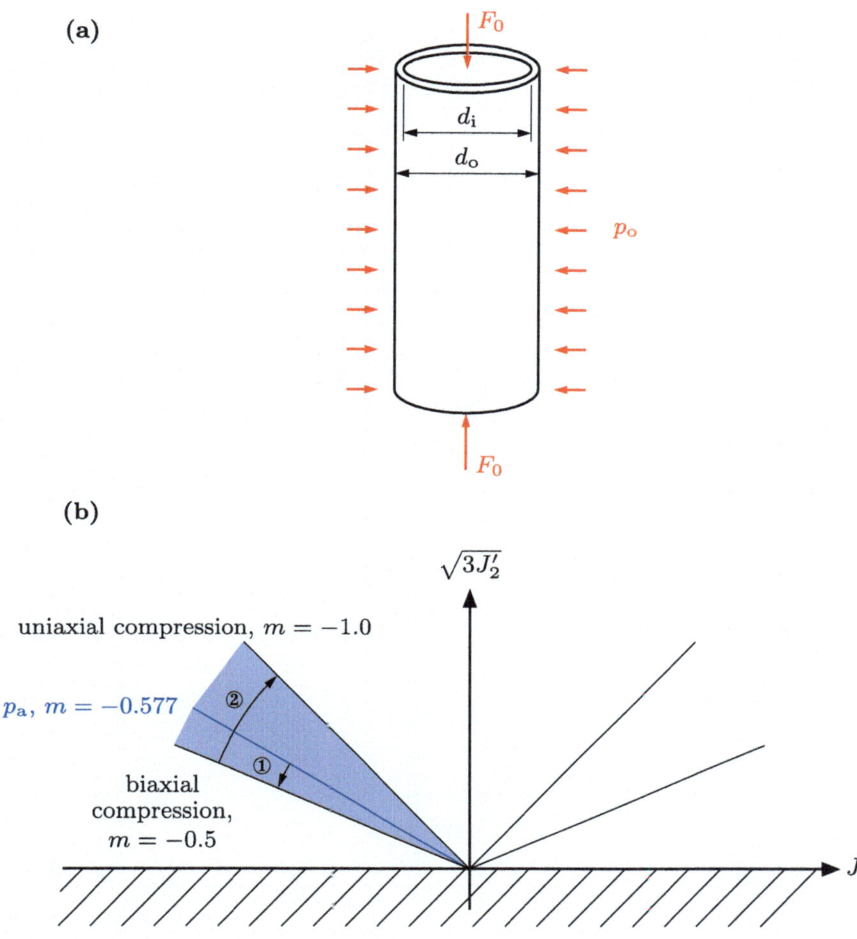

Fig. 2.17 Experimental realization of multiaxial stress states: **a** thin-walled tube under external pressure p_o and external compressive force F_0; **b** representation in the $J_1^o - \sqrt{3J_2'}$ invariant space (Burzyński-plane)

A different experimental setup is based from the outside pressured thin tube with a superimposed torsional load; see Fig. 2.18a. This external torsional moment results again in a linear shear stress distribution across the wall thickness, which can be approximated for thin tubes as

$$\overline{\tau} \approx \frac{M_0}{I_p} \times \overline{r}, \tag{2.65}$$

where $I_p = \frac{\pi}{32}(d_o^4 - d_i^4) \approx \frac{\pi}{4} s d_i^3$ is the polar second moment of area and $\overline{r} = (d_i + d_o)/4$ is the average radius. Thus, a biaxial stress state with σ_a, σ_t, and $\overline{\tau}$ is acting in

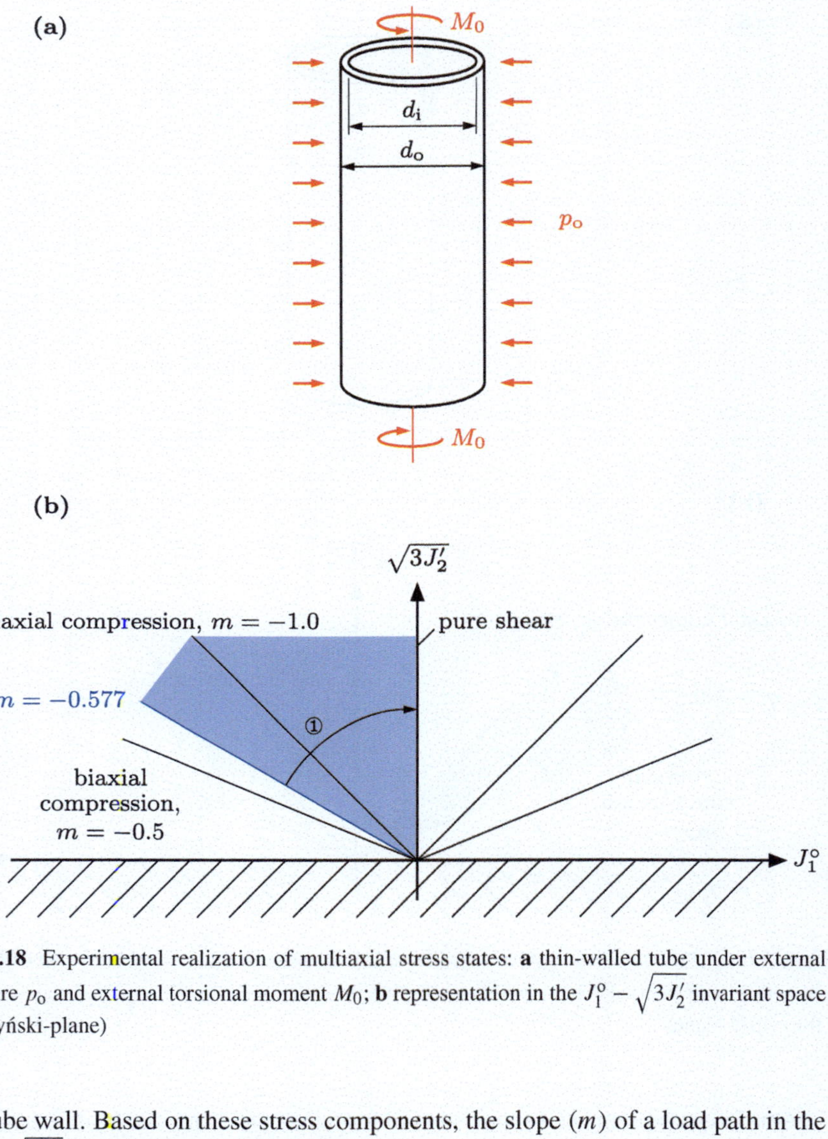

Fig. 2.18 Experimental realization of multiaxial stress states: **a** thin-walled tube under external pressure p_o and external torsional moment M_0; **b** representation in the $J_1^o - \sqrt{3J_2'}$ invariant space (Burzyński-plane)

the tube wall. Based on these stress components, the slope (m) of a load path in the $J_1^o - \sqrt{3J_2'}$ invariant space is obtained as

$$m = \frac{\sqrt{3J_2'}}{J_1^o} = \frac{\sqrt{\sigma_a^2 + \sigma_t^2 - \sigma_a\sigma_t + 3\tau^2}}{\sigma_a + \sigma_t}. \tag{2.66}$$

Assuming that a particular external pressure p_o is acting, one may gradually increase the external torsional moment M_0. This means that we have first a state with σ_t and $\sigma_a = \sigma_t/2$, which corresponds to a point on the line through the origin with $m = -0.577$. Increasing the external torsional moment means now that we move

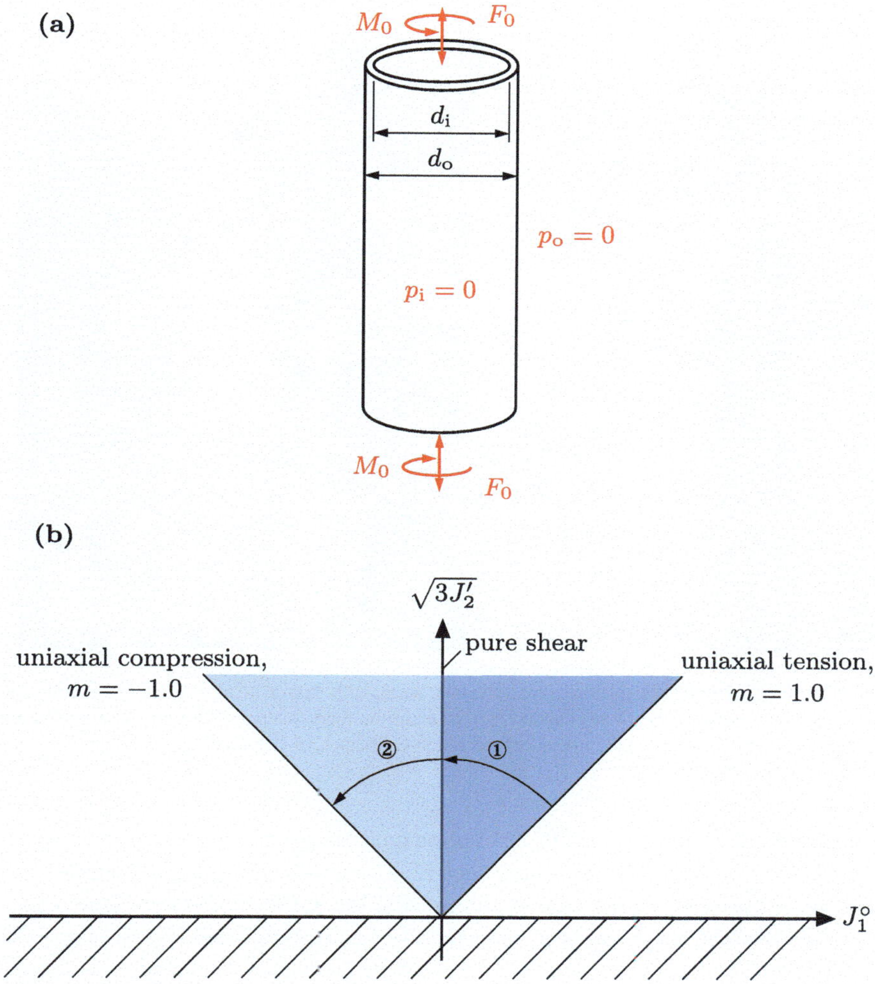

Fig. 2.19 Experimental realization of multiaxial stress states: **a** thin-walled tube under external tensile/compressive force F_0 and external torsional moment M_0 (Bauwens 1970); **b** representation in the $J_1^o - \sqrt{3J_2'}$ invariant space (Burzyński-plane)

towards the pure shear line ($m \to \infty$); see path ① in Fig. 2.18b. An approximate pure shear state is obtained for $|\overline{\tau}| \gg |\sigma_t|$.

In certain circumstances, the following experimental approach might be easier to realize. Combining a classical tensile/compression test with a torsional load (see Fig. 2.19a), i.e., avoiding the application of an internal or external pressure, it might be possible to cover the range between uniaxial compression and uniaxial tension in the $J_1^o - \sqrt{3J_2'}$ invariant space; see Fig. 2.19b. Corresponding experimental setups and/or experimental results are reported, for example, in Bauwens (1970), Mehrabi

Fig. 2.20 Experimental realization of multiaxial stress states based on an axial force F_0 (± 200 kN) and external torsional moment M_0 (2000 Nm). In addition, there is a possibility to superimpose an internal pressure (400–500 bar) in case of tubular specimens (HS Esslingen)

et al. (2015), and Chang et al. (2022). Caution must be taken in the design of the specimen and the experimental procedure because a thin-walled tube under compressive load (the corresponding region is highlighted in Fig. 2.19b by the light blue color) might be subjected to global or local instability, i.e., buckling and/or wrinkling.

A possible experimental setup for the combined tensile/torsion loading is shown in Fig. 2.20.

2.2.2 *Loading Devices with Two or Three Axes*

A different way to realize multiaxial stress states can be based on a biaxial testing machine; see Fig. 2.21a (Shiratori and Ikegami 1968; Öchsner 2003).

A cruciform specimen can be loaded in tension and/or compression in two perpendicular directions (let us call in the following the horizontal direction '1' and the vertical direction '2'). Let us assume that we start from a pure biaxial stress state ($\sigma_1 = \sigma_2$), this means that we are on the line through the origin with $m = 0.5$. Gradually reducing, for example, the horizontal stress component ($\sigma_1 < \sigma_2$) leads us finally to a uniaxial stress state ($\sigma_2 > 0$) with $m = 1.0$; see path ①. Introducing now

(a)

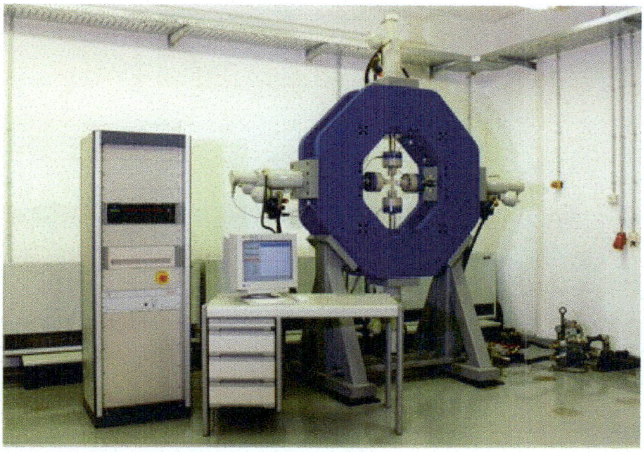

(b)

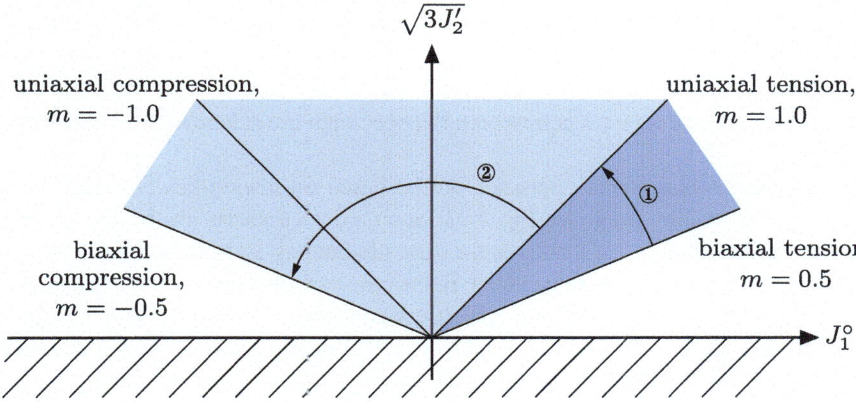

Fig. 2.21 Experimental realization of multiaxial stress states: **a** biaxial testing machine (Öchsner 2003; Öchsner et al. 2005); **b** representation in the $J_1^\circ - \sqrt{3J_2'}$ invariant space (Burzyński-plane)

in the horizontal direction an increasing compressive load leads us in a counter clockwise path (see ②), which finally ends in a pure shear stress state with ($\sigma_1 = -\sigma_2$) or $J_1^\circ = 0$. Gradually reducing the tensile stress leads us finally to a uniaxial compressive state ($\sigma_1 < 0$) with $m = -1.0$. Activating now in the vertical direction as well as a compressive load leads us finally to a biaxial compressive state ($-\sigma_1 = -\sigma_2$) with $m = -0.5$. If the specimen is thin, a compressive load may result in instability failure (buckling). This region is highlighted in Fig. 2.21b by the light blue color.

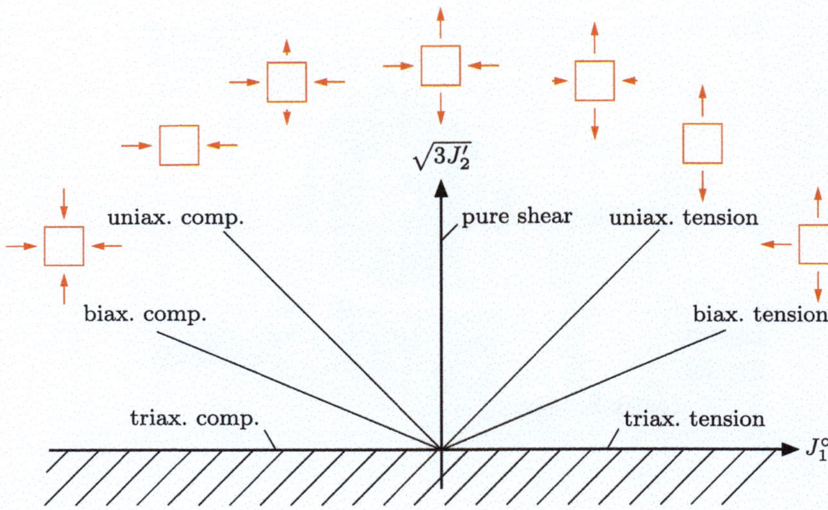

Fig. 2.22 Experimental realization of multiaxial stress states due to biaxial testing: representation in the $J_1^\circ - \sqrt{3J_2'}$ invariant space (Burzyński-plane) and corresponding stress states acting on the specimen

The different stress states acting on a flat specimen are schematically illustrated in Fig. 2.22.

A different experimental approach can be based on testing devices with three perpendicular loading axes; see Fig. 2.23. Such a device exerts on the specimen a triaxial tensile/compressive load (see for example Launay and Gachon 1971 for the triaxial compression of concrete cubes). In the case of multiaxial *tensile* loading, the clamping of the specimen and potential stress concentrations require some sophisticated specimen design. Thus, at least from a theoretical point of view, the entire range in the $J_1^\circ - \sqrt{3J_2'}$ invariant space can be experimentally investigated.

2.2.3 Hydrostatic Pressure Due to a Pressure Chamber and Additional Loads

A completely different experimental approach is based on the idea to apply a hydrostatic pressure to the specimen and to superimpose, for example, a macroscopic tensile/compressive load (Sardar et al. 1968; Mears et al. 1969; Sauer et al. 1970; Christiansen et al. 1971; Pae 1977; Spitzig and Richmond 1979; Sfer et al. 2002); see Fig. 2.24a. Such an experiment can be realized by placing a tensile/compressive machine in a pressure chamber where the hydrostatic pressure is applied by an hydraulic oil.

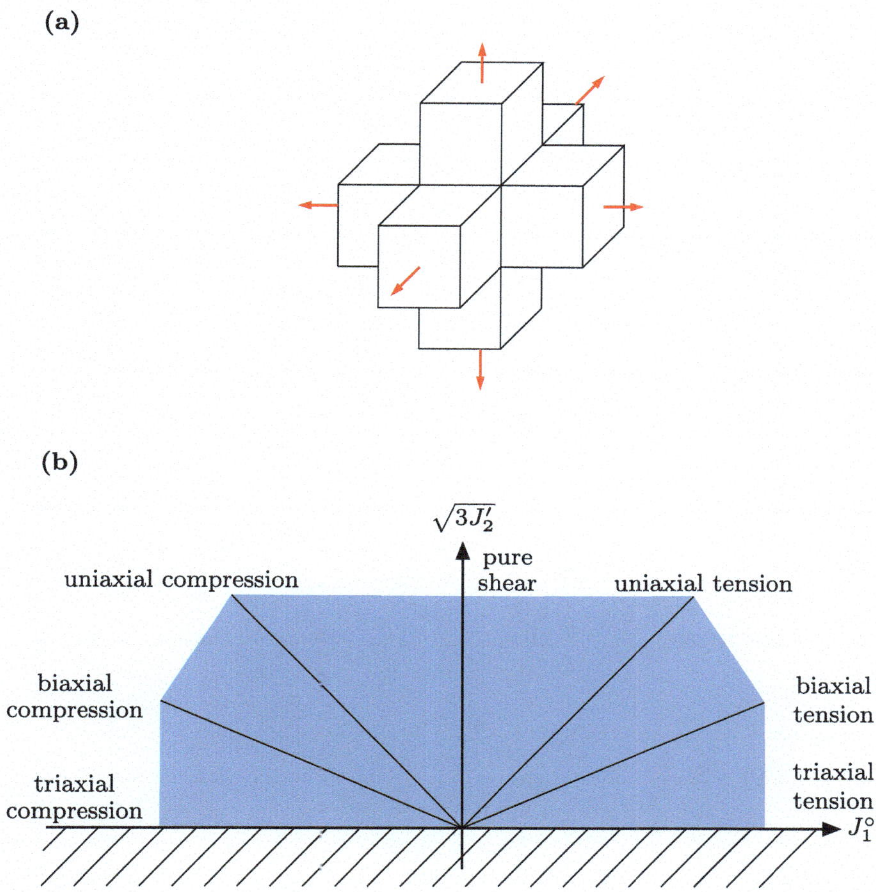

Fig. 2.23 Experimental realization of multiaxial stress states: **a** specimen under triaxial tensile/compressive load; **b** representation in the $J_1^\circ - \sqrt{3J_2'}$ invariant space (Burzyński-plane)

The different stress states acting on a, in general, cylindrical specimen are schematically illustrated in Fig. 2.25. Looking at this figure and Table 2.6, it can be seen that the classical traces of the fundamental tests (e.g., uniaxial compression with $\alpha = -45°$) can be alternatively investigated by particular multiaxial stress states.

Similar to the idea of having a tensile/compression devise in a pressure chamber (see Fig. 2.24), one may also locate a torsional device in a hydrostatic pressure chamber (Ishihara and Li 1972; Bažant et al. 1986); see Fig. 2.26.

Looking at Table 2.7, it can be seen that the classical traces of the fundamental tests (e.g., uniaxial compression with $\alpha = -45°$) can be alternatively investigated by particular multiaxial stress states.

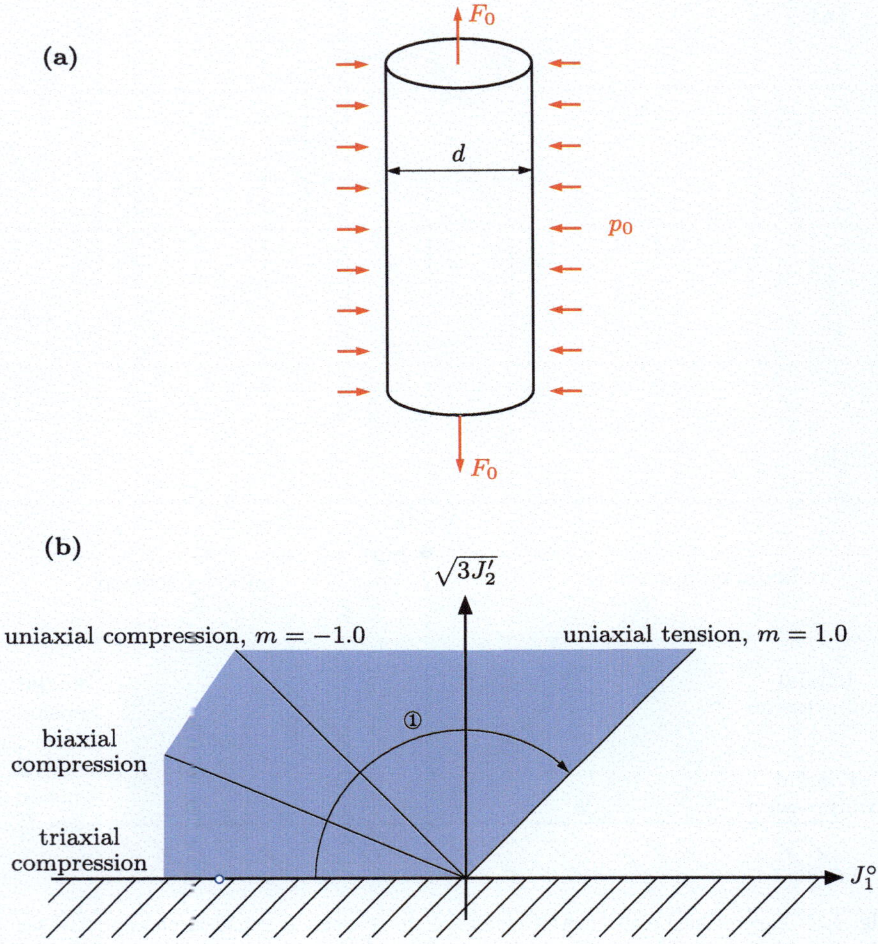

Fig. 2.24 Experimental realization of multiaxial stress states: **a** specimen under hydrostatic pressure p_0 and external tensile/compressive force F_0 (Sauer et al. 1970); **b** representation in the $J_1^\circ - \sqrt{3J_2'}$ invariant space (Burzyński-plane)

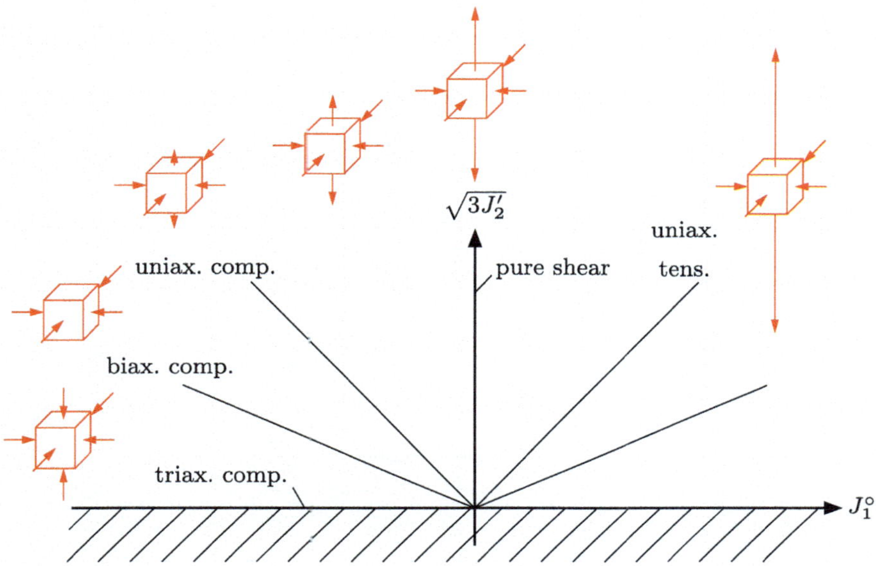

Fig. 2.25 Experimental realization of multiaxial stress states due to hydrostatic pressure and a tensile/compressive force: representation in the $J_1^\circ - \sqrt{3J_2'}$ invariant space (Burzyński-plane) and corresponding normal stress states acting on the specimen

Table 2.6 Experimental realization of multiaxial stress states due to hydrostatic pressure p_0 and external force F_0: stress states, slopes and angles with reference to Fig. 2.24

Stress state	Slope m	Angle α
$\sigma_1 = -\sigma, \sigma_2 = -\sigma, \sigma_3 = -\sigma$	0	0°
$\sigma_1 = -\frac{1}{2}\sigma, \sigma_2 = -\sigma, \sigma_3 = -\sigma$	−0.2	−11.310°
$\sigma_1 = 0, \sigma_2 = -\sigma, \sigma_3 = -\sigma$	−0.5	−26.565°
$\sigma_1 = \frac{1}{2}\sigma, \sigma_2 = -\sigma, \sigma_3 = -\sigma$	−1.0	−45.000°
$\sigma_1 = \sigma, \sigma_2 = -\sigma, \sigma_3 = -\sigma$	−2.0	−63.435°
$\sigma_1 = 2\sigma, \sigma_2 = -\sigma, \sigma_3 = -\sigma$	$-\infty$	−90.000°
$\sigma_1 = 3\sigma, \sigma_2 = -\sigma, \sigma_3 = -\sigma$	4.0	75.964°
$\sigma_1 = 4\sigma, \sigma_2 = -\sigma, \sigma_3 = -\sigma$	2.5	68.199°
$\sigma_1 = 5\sigma, \sigma_2 = -\sigma, \sigma_3 = -\sigma$	2.0	63.435°
$\sigma_1 = 10\sigma, \sigma_2 = -\sigma, \sigma_3 = -\sigma$	1.375	53.973°
$\sigma_1 = 100\sigma, \sigma_2 = -\sigma, \sigma_3 = -\sigma$	1.031	45.864°

(a)

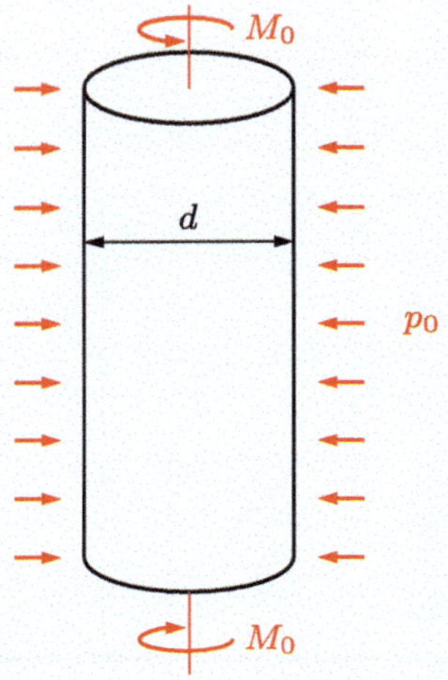

(b)

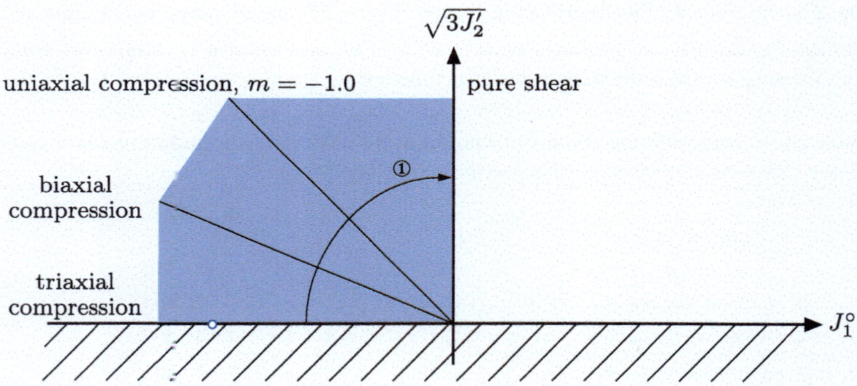

Fig. 2.26 Experimental realization of multiaxial stress states: **a** specimen under hydrostatic pressure p_0 and external torsional moment M_0; **b** representation in the $J_1^\circ - \sqrt{3J_2'}$ invariant space (Burzyński-plane)

Table 2.7 Experimental realization of multiaxial stress states due to hydrostatic pressure p_0 and external moment M_0: stress states, slopes, and angles with reference to Fig. 2.26

Stress state	Slope m	Angle α
$\sigma_1 = \sigma_2 = \sigma_3 = -\sigma; \tau = 0$	0	0°
$\sigma_1 = \sigma_2 = \sigma_3 = -\sigma; \tau = -0.5\sigma$	−0.289	−16.102°
$\sigma_1 = \sigma_2 = \sigma_3 = -\sigma; \tau = -1.0\sigma$	−0.577	−30°
$\sigma_1 = \sigma_2 = \sigma_3 = -\sigma; \tau = -2.0\sigma$	−1.155	−49.107°
$\sigma_1 = \sigma_2 = \sigma_3 = -\sigma; \tau = -5.0\sigma$	−2.887	−70.893°
$\sigma_1 = \sigma_2 = \sigma_3 = -\sigma; \tau = -10.0\sigma$	−5.774	−80.174°
$\sigma_1 = \sigma_2 = \sigma_3 = -\sigma; \tau = -50.0\sigma$	−28.868	−88.016°

References

Altenbach H, Altenbach J, Zolochevsky A (1995) Erweiterte Deformationsmodelle und Versagenskriterien der Werkstoffmechanik. Deutscher Verlag für Grundstoffindustrie, Stuttgart

Altenbach H (2018) Kontinuumsmechanik: Einführung in die materialunabhängigen und materialabhängigen Gleichungen. Springer Vieweg, Berlin

Altenbach H, Öchsner A (eds) (2020) Encyclopedia of continuum mechanics. Springer, Berlin

Altenbach H (2022) Holzmann/Meyer/Schumpich Technische Mechanik Festigkeitslehre. Springer Vieweg, Wiesbaden

Backhaus G (1983) Deformationsgesetze. Akademie-Verlag, Berlin

Bardenheier R (1982) Mechanisches Versagen von Polymerwerkstoffen. Carl Hanser Verlag, München

Bauwens JC (1970) Yield condition and propagation of Lüders' lines in tension-torsion experiments on poly(vinyl chloride). J Polym Sci A2 8(6):893–901. https://doi.org/10.1002/pol.1970.160080606

Bažant ZP, Prasannan S, Hagen M, Meiri S, Vaitys R, Klima R, Hess JD (1986) Large triaxial-torsional testing machine with hygrothermal control. Mater Struct 19:285–294. https://doi.org/10.1007/BF02472112

Betten J (1987) Tensorrechnung für Ingenieure. Teubner Verlag, Stuttgart

Betten J (2001) Kontinuumsmechanik. Springer, Berlin

Bruhns OT, Some remarks on Lode angle and Lode parameter. Proc Appl Math Mech (PAMM), in press

Chang Y, Wen W, Xu Y, Cui H, Xu Y (2022) Quasi-static mechanical behavior of filament wound composite thin-walled tubes: tension, torsion, and multi-axial loading. Thin Wall Struct 177:109361. https://doi.org/10.1016/j.tws.2022.109361

Chen WF, Han DJ (1988) Plasticity for structural engineers. Springer, New York

Chen WF, Zhang H (1991) Structural plasticity: theory, problems, and CAE software. Springer, New York

Christiansen AW, Baer E, Radcliffe SV (1971) The mechanical behaviour of polymersunder high pressure. Philos Mag 24(188):451–467. http://dx.doi.org/10.1080/14786437108227400

Evans WJ (1972) Deformation and failure under multiaxial stresses - a survey of laboratory techniques and experimental data. C.P. No. 1306. https://reports.aerade.cranfield.ac.uk/bitstream/handle/1826.2/1319/arc-cp-1306.pdf?sequence=1&isAllowed=y. Accessed 21 Jan 2024

Gurson AL (1977) Continuum theory of ductile rupture by void nucleation and growth: part I — yield criteria and flow rules for ductile media. J Eng Mater-T ASME 99(1):2–15. https://doi.org/10.1115/1.3443401

Heyder M, Öchsner A, Ströhla S (2002) Evaluation of stress states for the experimental determination of the shear modulus of cellular materials. Materialwiss Werkst 33:85–89. https://doi.org/10.1002/1521-4052(200202)33:2<85::AID-MAWE85>3.0.CO;2-%23

Ishihara K, Li S-I (1972) Liquefaction of saturated sand in triaxial torsion shear test. Soils Found 12(2):19–39. https://doi.org/10.3208/sandf1972.12.19

Itskov M (2009) Tensor algebra and tensor analysis for engineers: with applications to continuum mechanics. Springer, Berlin

Kolupaev VA (2018) Equivalent stress concept for limit state analysis. Springer, Cham

Launay P, Gachon H (1971) Strain and ultimate strength of concrete under triaxial stress. In: Proceedings of the first international conference on structural mechanics in reactor technology, Berlin, vol 4, Part H, Paper H1/3

Lurie AI (2005) Theory of elasticity. Springer, Berlin

Mears DR, Pae KD, Sauer JA (1969) Effects of hydrostatic pressure on the mechanical behavior of polyethylene and polypropylene. J Appl Phys 40:4229–4237. https://doi.org/10.1063/1.1657180

Mehrabi R, Andani MT, Kadkhodaei M, Elahinia M (2015) Experimental study of NiTi thin-walled tubes under uniaxial tension, torsion, proportional and non-proportional loadings. Exp Mech 55:1151–1164. https://doi.org/10.1007/s11340-015-0016-2

Nayak GC, Zienkiewicz OC (1972) Convenient form of stress invariants for plasticity. J Struct Div-ASCE 98(4):949–953. https://doi.org/10.1061/JSDEAG.0003219

Öchsner A (2003) Experimentelle und numerische Untersuchung des elastoplastischen Verhaltens zellularer Modellwerkstoffe. VDI-Verlag, Düsseldorf

Öchsner A, Kuhn G, Grácio J (2005) Investigation of cellular solids under biaxial stress states. Exp Mech 45(4):325–330. https://doi.org/10.1007/BF02428162

Öchsner A (2016) Continuum damage and fracture mechanics. Springer, Singapore

Pae KD (1977) The macroscopic yielding behaviour of polymers in multiaxial stress fields. J Mater Sci 12:1209–1214. https://doi.org/10.1007/BF02426859

Ripley PW, Korkolis YP (2016) Multiaxial deformation apparatus for testing of microtubes under combined axial-force and internal-pressure. Exp Mech 56:273–286. https://doi.org/10.1007/s11340-015-0097-y

Sardar D, Radcliffe SV, Baer E (1968) Effects of high hydrostatic pressure on the mechanical behavior of a crystalline polymer-polyoxymethylene. Polym Eng Sci 8(4):290–301. https://doi.org/10.1002/pen.760080409

Sauer JA, Mears DR, Pae KD (1970) Effects of hydrostatic pressure on the mechanical behaviour of polytetrafluoroethylene and polycarbonate. Eur Polym J 6:1015–1032. https://doi.org/10.1016/0014-3057(70)90034-0

Sfer D, Carol I, Gettu R, Etse G (2002) Study of the behavior of concrete under triaxial compression. J Eng Mech 128(2):156–163. https://doi.org/10.1061/(ASCE)0733-9399(2002)128:2(156)

Shiratori E, Ikegami K (1968) Experimental study of the subsequent yield surface by using cross-shaped specimens. J Mech Phys Solids 16(6):373–394. https://doi.org/10.1016/0022-5096(68)90002-1

Spitzig WA, Richmond O (1979) Effect of hydrostatic pressure on the deformation behavior of polyethylene and polycarbonate in tension and in compression. Polym Eng Sci 19(16):1129–1139. https://doi.org/10.1002/pen.760191602

Stojcevski F, Hilditch T, Henderson LC (2018) A modern account of Iosipescu testing. Compos Part A- Appl S 107:545–554. https://doi.org/10.1016/j.compositesa.2018.02.011

Stout MG, Petrovic JJ (1984) Multiaxial loading fracture of Al_2O_3 tubes: I, experiments. J Am Ceram Soc 67(1):14–18. https://doi.org/10.1111/j.1151-2916.1984.tb19138.x

Suzuki H, Ninomiya T, Sumino K, Takeuchi S (1985) Dislocation in solids: proceedings of the IX Yamada. VNU Science Press, Utrecht

Trapp M, Öchsner A (2018) Computational plasticity for finite elements: a fortran-based introduction. Springer, Cham

Walrath DE, Adams DF (2024) Verification and application of the Iosipescu shear test method. Department Report, Department of Mechanical Engineering, University of Wyoming (UWME-DR-401-103-1). https://ntrs.nasa.gov/api/citations/19850010067/downloads/19850010067.pdf. Accessed 11 Jan 2024

Chapter 3
Fine-Grained Microstructures in Single-Step Metal Additive Manufacturing: Comparison of AlSiMg and AlMgScZr Alloys

Leonhard Hitzler

Abstract Single-step additive manufacturing (AM) techniques, such as powder-bed fusion (PBF-LB/M), paved the way for the direct fabrication of ready-to-use metal parts. The direct fabrication of aluminum parts became accessible with the rise of high-power laser setups in AM machines. Especially aluminum–silicon (AlSi) cast alloys became a popular choice and due to the fast quenching in PBF-LB/M, the achieved static mechanical properties surpassed those of the as-cast and heat-treated state. However, certain limitations are present, the most prevalent being the reoccurring weakening in the heat-affected zone surrounding every single scan track, which reduces the overall strength and especially reduces the fatigue performance. Heat treatments can improve fatigue performance, but thereby noticeably reduce the static mechanical strength. Since this limiting factor is tied to the chemical composition of the AlSi alloys, efforts were made to design aluminum alloys for the PBF-LB/M process, which overcome those limitations. This chapter serves to explain the reason the reoccurring weakness is induced in the AlSi alloy family and to explain in which manner this weakness was overcome in aluminum–magnesium–scandium–zirconium alloys.

3.1 Introduction

Aluminium (Al) alloys, in general, are frequently used in engineering and are appreciated for their combined properties of low cost, ease of manufacturing, attractive strength-to-weight ratio, and good corrosion resistance (Hitzler et al. 2023). Their wide acceptance in the industry is related to their long history in casting and, as far as additive manufacturing (AM) is concerned, they slowly became popular with the availability of highly focused high-energy laser setups around the year 2014 (Hitzler

L. Hitzler (✉)
Institute of Materials Science, Technical University of Munich, Boltzmannstraße 15, 85748 Garching, Germany
e-mail: leonhard.hitzler@tum.de

H. Altenbach et al. (eds.), *Lectures Notes on Advanced Structured Materials 3*, Advanced Structured Materials 226, https://doi.org/10.1007/978-3-031-77403-4_3

et al. 2018a; Jerrard et al. 2010). The availability of high local energy densities and thus, the option to enable full melting[1] at reasonable scanning speeds, was crucial for their successful fabrication and required to overcome the hurdles of high thermal conductivity, high reflectivity and effectively breaking up the aluminium oxide layer (Tang, M.: Inclusions 2017).

AM of Al alloys began with known cast alloys, predominantly aluminium–silicon (AlSi) and aluminium–silicon–magnesium (AlSiMg) alloys, due to their good weldability, favourable melt flow behaviour, eutectic solidification, and low shrinkage (Hitzler et al. 2018a; Jerrard et al. 2010). High cooling rates from the molten state resulted in very fine-grained microstructures and superb hardness and static mechanical strength, especially for the age-hardenable AlSiMg alloys. This further sparked interest to utilise AM technologies, predominantly the laser powder-bed fusion (PBF-LB/M) process, to fabricate ready-to-use parts. However, noteworthy fluctuations in mechanical strength and hardness, as well as a macroscopic anisotropy were discovered, which rendered the qualification processes more tedious and also fostered the general perception that AM processes are unreliable (Hitzler et al. 2017a). Today, we know better and can attribute the scatter and fluctuations to microstructural features linked to the process and the alloy behaviour and understand it through material science. For a comprehensive overview, the reader is referred to Hitzler et al. (2023).

Already in the early days of AM of Al alloys, researchers pointed out that none of the current Al alloys were designed for the unique process conditions, thereby predominantly hinting at the PBF-LB/M process called selective laser sintering (SLS) or selective laser melting (SLM) at the time (Hitzler et al. 2018a; Jerrard et al. 2010). As such, it was recommended to develop Al alloys better suited for PBF-LB/M to achieve even better material properties. Alloy development, of course, is a lengthy procedure and requires dedication, funding, and a very crucial, interest of the industry to adopt it and thereby, to achieve some demand to enable the fabrication of the alloy outside the laboratory scale. This initial hurdle often is not met and progress is halted. For Al alloys and their processing via PBF-LB/M, however, there is the example of Scalmalloy, an aluminium–magnesium–scandium–zirconium (AlMgScZr) alloy, which showcases that there are significant gains to be had.

In this chapter, we will explore the main weaknesses of the AlSiMg alloy family, when fabricated via PBF-LB/M and understand how most of these weaknesses were overcome through AlMgScZr alloy development.

[1] Prior to this AM of metals often was restricted to sintering and during that time period, AM of Al alloys was rare (Olakanmi et al. 2015).

3.2 Microstructural Development

3.2.1 Background

Additive manufacturing can be split up into single-step and multi-step processes. For metal processing, single-step refers to melting the feedstock (welding process), which results in a usable metal part being achieved directly after the AM process (Davids et al. 2022; DebRoy et al. 2018). Multi-step processes utilise a binding media to shape the part and employ a debinding and sintering, or infiltration process thereafter, to obtain decent material strength. For ductile metal alloys with good or decent weldability, single-step processes are preferred and return better material properties and induce less defects. In the following, only single-step processes and full melting of the feedstock are considered.

Melting of the feedstock directly implies that solidification is the driving factor for microstructural development, coupled with remelting and in-situ heat treatments caused by repetitive heat input and/or elevated temperatures in the built environment. For an in-depth explanation of the solidification microstructure, the reader is referred to Hitzler (2024).

Solidification of the melt starts either by existing grains with renewed access to the melt growing further (epitaxial grain growth) or through nucleation, followed by growth of the nuclei thereafter. These two solidification options are depicted in Fig. 3.1 on the examples of austenitic stainless steel (AISI 316L, EN 1.4404) and hypoeutectic AlSi alloy (AlSi10Mg).

3.2.2 Aluminium–Silicon Alloys

Nucleation on the melt pool boundary and the solid–liquid phase boundary occurs when the primary solidifying phase is incompatible with the surrounding solid matter (DebRoy et al. 2018; Hitzler 2024). For common AlSi and ASiMg alloys, ranging from 7 wt.-% to 12 wt.-%, the following constellation is present: The as-fabricated microstructure consists of aluminium solid solution (α-Al) and a eutectic network comprised of α-Al and silicon phase (Si) (Fig. 3.1) (Hitzler et al. 2023; Aboulkhair et al. 2019). With the maximum solubility[2] of 1.6 wt.-% (equilibrium condition at solidus temperature), the majority of the silicon content can be assumed to be present in the eutectic network. With the Si being present in the fine eutectic, it can form an effective phase boundary during partial remelting.[3] The α-Al is molten, while the Si with its much higher melting point (1414 °C) becomes the solid phase in contact to

[2] The maximum solubility of Si in the α-Al based on the thermodynamic equilibrium is 1.6 wt.-%. However, fast quenching rates increase the solubility (Hitzler et al. 2023; Li et al. 2015). Due to the fast quenching from the molten state, the dissolved Si in the α-Al it does not follow the equilibrium and exceeds this maximum (Prashanth et al. 2014).

[3] Due to necessary overlaps, no matter how small or large these may be.

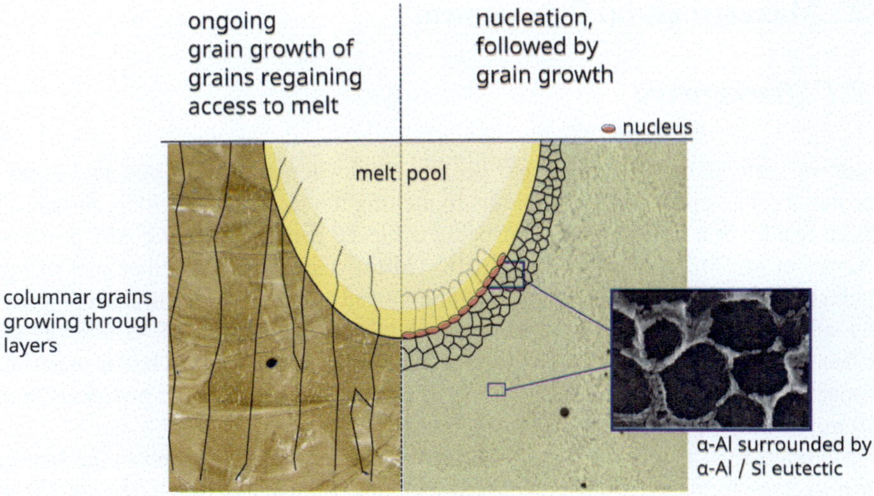

Fig. 3.1 Example of the two solidification options: (left-hand side) the surrounding solid phase equals the primary solidifying phase, leading to existing grains with renewed access to the melt growing further; and (right-hand side) a mismatch of primary solidifying phase (here α-Al) solidifying on a secondary phase (here Si). This schematic was generated with adopted images from Hitzler et al. 2017a; Hitzler et al. 2017b; Heilgeist 2018) and is reprinted with permission from Hitzler (2024)

the melt. This holds true for hypoeutectic AlSi alloys above 4 wt.-% of Si, below the phase fraction of Si is insufficient to form an effective boundary (Kimura et al. 2017).

In short, multiple criteria must be met to achieve the typical microstructure of hypoeutectic AlSi and AlSiMg alloys fabricated with PBF-LB/M:

- A heterogenous microstructure with two or more phases is required. For ease of explanation, these are called phase A (primary) and phase B (secondary).
- Phase B (Si) must effectively shield the melt pool from the solid grains of phase A (α-Al). This necessitates a sufficient quantity and distribution of phase B to form an effective boundary. In addition, phase B requires a noteworthy higher melting point than phase A, otherwise, it is unlikely for the melt pool penetration to stop at phase B.
- Primary solidification must be phase A, so a layer of phase A is initially formed on top of phase B, thereby, in turn, cutting off phase B from the melt.
- With no single phase being able to grow unhindered, large columnar grains through layers are absent.

The governing weakness is the heat-affected zone (HAZ) with a lower base strength and increased Si phase fraction (Tang, M.: Inclusions 2017; Prashanth et al. 2014). The main reason can be summarised as follows (Hitzler et al. 2023, 2020a; Hafenstein et al. 2020):

- α-Al solid solution is metastable and has a high amount of Si dissolved directly after initial quenching from the molten state.
- Reheating cycles lead to the ejection of Si from the α-Al, whereby Si precipitates in the α-Al are formed. However, the ejection of Si and diffusion of Si predominantly lead to the growth of Si grains in the HAZ.
- This, in turn, leads to a higher phase fraction of Si in the HAZ, causing an embrittlement, coupled with enlarged and weakened α-Al grains in close proximity.

Tensile samples consistently show the rupture to follow the HAZ (Aboulkhair et al. 2016). This weakening affects tensile and compressive strength in a similar manner, thus, both show a qualitative equal dependency on the direction of stress relative to the layering (Hitzler et al. 2018b; Sert et al. 2020). In fracture toughness tests a similar behaviour is seen, whereby the crack path follows the weakened HAZ, even when this means the crack path becomes longer (Hitzler et al. 2020b). Within successive layers, the scanning strategy and in particular, the rotation angle increment going from layer *m*-1 to layer *m* (Fig. 3.2) can be used to minimise the impact of the weakening in plane (in *xy*) (Hitzler et al. 2017a, 2019). As analogy, this can be compared to a layered composite in which the fibre-reinforcement is directional and through the stacking schemata great strength can be achieved in the confinement of the *xy*-plane. However, for 2D layering, no fibre-reinforcement is present in *z*-direction.

In summary, this poses a dilemma. On one hand, the eutectic network with its Si grains enables fine-grained microstructures and enables to reach static mechanical strength results far beyond cast counterparts. Thereby showcasing that AlSi and

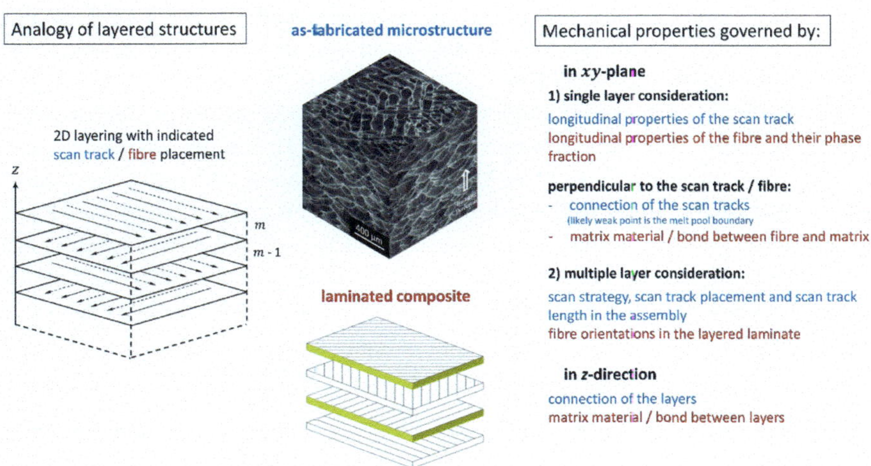

Fig. 3.2 Analogy of layered structures, here on the example of AlSi10Mg and a laminated composite. This schematic was generated with images adopted from Rafieazad et al. (2019); Yibre et al. 2020; Krakhmalev et al. 2018)

ASiMg alloys are well suited for PBF-LB/M, whilst being a limiting factor at the same time. This poses the challenge of how to hold on to the fine-grained microstructure, effectively sticking to a solidification starting with nucleation, but without the eutectic network and thus, without Si.

3.2.3 Aluminium–Magnesium–Scandium–Zirconium Alloys

For solidification to begin via nucleation the driving factor is the incompatibility of the primary solidifying phase with the surrounding solid phase. As such, there are the possibilities to have a second phase to act as a shield (for example AlSi) or to alter the primary solidifying phase. The latter is the key used in AlMgScZr. The base alloy is an aluminium–magnesium–manganese (AlMgMn) alloy, to which Scandium (Sc) and Zirconium (Zr) are added (Table 3.1). Both form trialuminides (intermetallic phases: Al_3Sc and Al_3Zr) on their own and, when both are simultaneously present $Al_3Sc_{1\text{-}x}Zr_x$ trialuminide (Hallem et al. 2006; Zakharov 1997). These trialuminides solidify at temperatures much higher than the α-Al. They are low in phase fraction compared to the Si phase fraction in AlSi alloys and form small dispersoids in the melt. Their presence facilitates nucleation sides for the α-Al. Hence, the solidification of the majority of the melt happens in a superimposed manner and is comprised of ongoing grain growth of existing α-Al grains with renewed access to the melt and nucleation of α-Al on the trialuminide dispersoids (Spierings et al. 2017).

The as-built microstructure of AlMgScZr is described as a bi-modal microstructure and showcases distinct zones (Fig. 3.3). The prior melt pool boundaries exhibit an ultra-fine-grained microstructure and towards the melt pool centre columnar grains are evident. These columnar grains only span the height of a single scan track and their ongoing growth is hindered by the dispersoids and the α-Al grains emerging on their free surfaces the very next layer.

High strength is achieved through the ultra-fine-grained microstructure, coupled with precipitation hardening through the dispersoids. Hence, the dispersoids are dual use, enabling nucleation of α-Al during solidification and acting as precipitates to increase strength in the solid. Comparing their phase fraction to Mg_2Si precipitates in AlSiMg alloys, their phase fraction is high and their strengthening impact is substantial. Moreover, unlike the AlSiMg alloys, the AlMgScZr alloys can be strengthened through post-heat treatments and can reach significantly higher strength values (Table 3.2).

Table 3.1 Chemical composition of AlMgScZr (Scalmalloy), main alloy elements in mass percentage

Element	Mg	Sc	Zr	Mn	Al
Min	4.00	0.60	0.20	0.30	Bal
Max	4.90	0.80	0.50	0.80	Bal

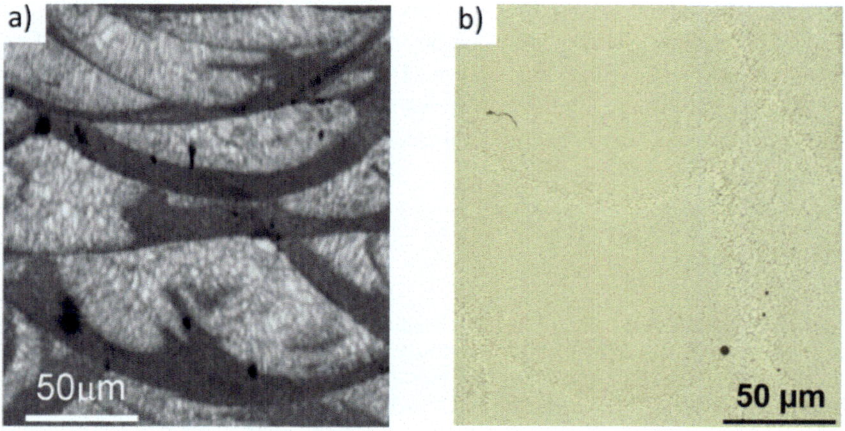

Fig. 3.3 As-built microstructure of **a** AlMgScZr (Spierings et al. 2018) and **b** AlSi10Mg. Image **a** reprinted with permission, original source: Spierings, A.B., Dawson, K., Dumitraschkewitz, P., Pogatscher, S., Wegener, K.: Microstructure characterisation of SLM-processed AlMgScZr alloy in the heat treated and HIPed condition. Addit. Manuf. 20, 173–181 (2018)

Table 3.2 Overview of reported tensile properties (yield strength (YS), ultimate tensile strength (UTS), and elongation at fracture (A_t) for various conditions (solution annealed (SA), artificially aged (AG), hot isostatic pressed (HIP), direct aged (DA))

Alloy	Condition	YS [MPa]	UTS [MPa]	A_t [%]
AlSi10Mg Hitzler et al. (2018a); Hitzler et al. (2020a); Hafenstein et al. (2020)	As-built	250	420	7
		200	320	4
	SA + AG	240	300	5
	HIP + DA	250	330	12
Scalmalloy Deillon et al. (2022)	As-built	280	415	15
	SA + AG	450	515	10
	HIP + AG	480	520	10

3.3 Summary

The AlMgScZr alloy (Scalmalloy) developed for the PBF-LB/M process showcases the potential of dedicated alloy development and the substantial gains that can be achieved. It also sparked further alloy developments of Al alloys for AM. The primary reason is that Scandium (Sc) is rather expensive, and as such, the AlMgScZr alloys are only suitable for aerospace and military applications. Therefore, replacements for Sc are highly sought after. An up-to-date overview of new Al alloys is provided in the current review from Rometsch et al. (Rometsch et al. 2022). Additionally, the AlMgScZr alloy itself is being improved as well. It still shares one issue with the

AlSiMg family, which is evaporation losses of Mg during PBF-LB/M fabrication, introducing fluctuations in the alloy composition. The addition of Calcium (Ca) was reported to drastically reduce evaporation (Deillon et al. 2022). In summary, while the AlMgScZr alloys are a huge leap forward, there still is a great need for further improvements, especially concerning affordable Al alloys for AM.

References

Aboulkhair NT, Maskery I, Tuck C, Ashcroft I, Everitt NM (2016) The microstructure and mechanical properties of selectively laser melted AlSi10Mg: the effect of a conventional T6-like heat treatment. Mater Sci Eng A 667:139–146. https://doi.org/10.1016/j.msea.2016.04.092

Aboulkhair NT, Simonelli M, Parry L, Ashcroft I, Tuck C, Hague R (2019) 3D printing of aluminium alloys: additive manufacturing of aluminium alloys using selective laser melting. Prog Mater Sci 106. https://doi.org/10.1016/j.pmatsci.2019.100578

Davids A, Apfelbacher L, Hitzler L, Krempaszky C (2022) Multi-step additive manufacturing technologies utilizing the powder metallurgical manufacturing route. In: Altenbach H, Johlitz M, Merkel M, Öchsner A (eds) Lectures notes on advanced structured materials. Advanced structured materials, pp 99–117

DebRoy T, Wei HL, Zuback JS, Mukherjee T, Elmer JW, Milewski JO, Beese AM, Wilson-Heid A, De A, Zhang W (2018) Additive manufacturing of metallic components—process, structure and properties. Prog Mater Sci 92:112–224. https://doi.org/10.1016/j.pmatsci.2017.10.001

Deillon L, Jensch F, Palm F, Bambach M (2022) A new high strength Al–Mg–Sc alloy for laser powder bed fusion with calcium addition to effectively prevent magnesium evaporation. J Mater Process Technol 300. https://doi.org/10.1016/j.jmatprotec.2021.117416

Hafenstein S, Hitzler L, Sert E, Öchsner A, Merkel M, Werner E (2020) Hot isostatic pressing of Aluminum-silicon alloys fabricated by laser powder-bed fusion. Technologies 8(3):48. https://doi.org/10.3390/technologies8030048

Hallem H, Lefebvre W, Forbord B, Danoix F, Marthinsen K (2006) The formation of Al3(ScxZryHf1−x−y)-dispersoids in aluminium alloys. Mater Sci Eng A 421(1–2):154–160. https://doi.org/10.1016/j.msea.2005.11.063

Heilgeist S (2018) The influence of post-heat treatments on the microstructure, tensile strength and surface hardness of selectively laser-melted AlSi10Mg. Master thesis, University of applied sciences esslingen/University of applied sciences aalen

Hitzler L, Janousch C, Schanz J, Merkel M, Heine B, Mack F, Hall W, Öchsner A (2017a) Direction and location dependency of selective laser melted AlSi10Mg specimens. J Mater Process Technol 243:48–61. https://doi.org/10.1016/j.jmatprotec.2016.11.029

Hitzler L, Hirsch J, Heine B, Merkel M, Hall W, Öchsner A (2017b) On the anisotropic mechanical properties of selective laser melted stainless steel. Materials 10(10):1136. https://doi.org/10.3390/ma10101136

Hitzler L, Merkel M, Hall W, Öchsner A (2018a) A review of metal fabricated with laser- and powder-bed based additive manufacturing techniques: process, nomenclature, materials, achievable properties, and its utilization in the medical sector. Adv Eng Mater 20(5):1700658. https://doi.org/10.1002/adem.201700658

Hitzler L, Schoch N, Heine B, Merkel M, Hall W, Öchsner A (2018b) Compressive behaviour of additively manufactured AlSi10Mg. Mat-Wiss U Werkstofftech 49(5):683–688. https://doi.org/10.1002/mawe.201700239

Hitzler L, Hirsch J, Tomas J, Merkel M, Hall W, Öchsner A (2019) In-plane anisotropy of selective laser melted stainless steel: the importance of the rotation angle increment and the limitation window. P I Mech Eng l: J Mat 233(7):1419–1428. https://doi.org/10.1177/1464420718757068

Hitzler L, Hafenstein S, Mendez-Martin F, Clemens H, Sert E, Öchsner A, Merkel M, Werner E (2020a) Heat treatments and critical quenching rates in additively manufactured aluminium-silicon alloys. Materials 13(3):720. https://doi.org/10.3390/ma13030720

Hitzler L, Sert E, Schuch E, Öchsner A, Merkel M, Heine B, Werner E (2020b) Fracture toughness of L-PBF fabricated aluminium-silicon: a quantitative study on the role of crack growth direction with respect to layering. Prog Addit Manuf 5(3):259–266. https://doi.org/10.1007/s40964-020-00113-x

Hitzler L (2024) Classical solidification structures in single-step metal additive manufacturing. In: Altenbach H, Hitzler L, Johlitz M, Merkel M, Öchsner A (eds) Lectures notes on advanced structured materials 2. Advanced structured materials. Springer Nature, pp 57–70

Hitzler L, Sert E, Öchsner A, Werner E (2023) Microstructure and mechanical property correlation for additively manufactured aluminum-silicon alloys. In: Kadkhodapour J, Schmauder S, Sajadi F (eds) Quality analysis of additively manufactured metals. Elsevier, pp 355–387

Jerrard P, Hao L, Dadbakhsh S, Evans J (2010) Consolidation behaviour and mircrostructure characteristics of pure aluminium and alloy following selective laser melting processing. In: Hinduja S, Li L (eds) Proceedings of the 36th international MATADOR conference. Springer, London, United Kingdom, pp 487–490

Kimura T, Nakamoto T, Mizuno M, Araki H (2017) Effect of silicon content on densification, mechanical and thermal properties of Al-xSi binary alloys fabricated using selective laser melting. Mater Sci Eng A 682:593–602. https://doi.org/10.1016/j.msea.2016.11.059

Krakhmalev P, Fredriksson G, Svensson K, Yadroitsev I, Yadroitsava I, Thuvander M, Peng R (2018) Microstructure, solidification texture, and thermal stability of 316 L stainless steel manufactured by laser powder bed fusion. Metals 8(8). https://doi.org/10.3390/met8080643

Li XP, Wang XJ, Saunders M, Suvorova A, Zhang LC, Liu YJ, Fang MH, Huang ZH, Sercombe TB (2015) A selective laser melting and solution heat treatment refined Al–12Si alloy with a controllable ultrafine eutectic microstructure and 25% tensile ductility. Acta Mater 95:74–82. https://doi.org/10.1016/j.actamat.2015.05.017

Olakanmi EO, Cochrane RF, Dalgarno KW (2015) A review on selective laser sintering/melting (SLS/SLM) of aluminium alloy powders: processing, microstructure, and properties. Prog Mater Sci 74:401–477. https://doi.org/10.1016/j.pmatsci.2015.03.002

Prashanth KG, Scudino S, Klauss HJ, Surreddi KB, Löber L, Wang Z, Chaubey AK, Kühn U, Eckert J (2014) Microstructure and mechanical properties of Al–12Si produced by selective laser melting: effect of heat treatment. Mater Sci Eng A 590:153–160. https://doi.org/10.1016/j.msea.2013.10.023

Rafieazad M, Mohammadi M, Nasiri AM (2019) On microstructure and early stage corrosion performance of heat treated direct metal laser sintered AlSi10Mg. Addit Manuf 28:107–119. https://doi.org/10.1016/j.addma.2019.04.023

Rometsch PA, Zhu Y, Wu X, Huang A (2022) Review of high-strength aluminium alloys for additive manufacturing by laser powder bed fusion. Mater Des 219. https://doi.org/10.1016/j.matdes.2022.110779

Sert E, Hitzler L, Hafenstein S, Merkel M, Werner E, Öchsner A (2020) Tensile and compressive behaviour of additively manufactured AlSi10Mg samples. Prog Addit Manuf 5:305–313. https://doi.org/10.1007/s40964-020-00131-9

Spierings AB, Dawson K, Heeling T, Uggowitzer PJ, Schäublin R, Palm F, Wegener K (2017) Microstructural features of Sc- and Zr-modified Al-Mg alloys processed by selective laser melting. Mater des 115:52–63. https://doi.org/10.1016/j.matdes.2016.11.040

Spierings AB, Dawson K, Dumitraschkewitz P, Pogatscher S, Wegener K (2018) Microstructure characterization of SLM-processed Al-Mg-Sc-Zr alloy in the heat treated and HIPed condition. Addit Manuf 20:173–181. https://doi.org/10.1016/j.addma.2017.12.011

Tang M (2017) Inclusions, porosity, and fatigue of AlSi10Mg parts produced by selective laser melting. PhD thesis, Carnegie Mellon University

Yibre AM, Koçer B, Esleman EA, Önal G (2020) Weight optimization of hybrid composite laminate using learning-oriented artificial algae algorithm. SN Appl Sci 2(8). https://doi.org/10.1007/s42452-020-3126-0

Zakharov VV (1997) Stability of the solid solution of scandium in aluminum. Met Sci Heat Treat 39(2):61–66. https://doi.org/10.1007/bf02467664

Chapter 4
Heat Transfer—Modeling, Simulation, Measurement

Jens Kortsch-Banzhaf, Josef Tomas, and Markus Merkel

Abstract In numerous engineering applications, heat plays a critical role in characterizing the properties and performance of products and processes. As the demand for increased energy efficiency grows, the importance of understanding and controlling heat transfer and temperature profiles has become even more significant within the product development process. Engineers increasingly rely on advanced numerical simulation and precise measurement techniques to analyze, optimize, and enhance the thermal properties and overall behavior of products and manufacturing processes. This focus on thermal management is essential for ensuring that products not only meet general performance criteria but also adhere to energy efficiency standards. Additive Manufacturing has emerged as a powerful tool in this context, enabling engineers to implement innovative concepts that prioritize mass reduction, efficient cooling, and effective heating. Through additive manufacturing, parts can be designed with intricate geometries and complex shapes. Additionally, the use of material composites in additive manufacturing allows for the tailoring of thermal properties to meet specific application requirements. By carefully considering the interplay between shape, geometry, and material selection, engineers can create highly optimized components that meet the rigorous demands of modern engineering challenges, particularly in terms of energy efficiency and thermal management.

Keywords Heat transfer · Fourier's law · Thermal conductivity · Transient plane source method · Additive manufacturing

---

J. Kortsch-Banzhaf (✉) · J. Tomas · M. Merkel
Institute for Virtual Product Development, Aalen University of Applied Sciences, Beethovenstraße 1, 73430 Aalen, Germany
e-mail: jens.kortsch-banzhaf@hs-aalen.de

J. Tomas
e-mail: josef.tomas@hs-aalen.de

M. Merkel
e-mail: markus.merkel@hs-aalen.de

H. Altenbach et al. (eds.), *Lectures Notes on Advanced Structured Materials 3*, Advanced Structured Materials 226, https://doi.org/10.1007/978-3-031-77403-4_4

4.1 Understanding Heat: A Comprehensive Overview

Heat is a form of energy that is transferred from one system to another due to a temperature difference. In physics, heat is defined as the energy that flows between two systems due to their different temperatures. Thermal energy is one of the most fundamental forms of energy and plays a central role in many natural and technical processes (Tipler and Mosca 2008).

Heat is observed when particles of a fluid (e.g., molecules in a liquid or gas) are in motion. The faster these particles move, the higher the temperature of the substance and the more heat energy is present. It is important to understand that heat is not the same as temperature, although the two are closely related. Temperature is a measure of the average kinetic energy content of the particles in a material, while heat is the energy that is transferred when two materials with different temperatures are exposed to each other (Halliday et al. 2013).

Heat transport describes the transfer of thermal energy from one place to another. Conduction, convection, and radiation are three basic mechanisms by which temperature transport can take place (Cengel and Ghajar 2015).

Conduction is the process by which thermal energy is transferred by the direct contact of particles in a solid or between solids. When a hot object is exposed to a colder object, within a fluid, the faster (hotter) molecules transfer their energy to the slower (colder) molecules (Incropera and DeWitt 2007). An example of this can be seen when a fluid within a tube is heated or cooled from one side, while the opposite side is insulated and is not exposed to external sources (Lienhard et al. 2012).

The efficiency of heat conduction depends on the material. Metals, such as copper and aluminum, are good conductors of heat, while materials such as wood or plastic are poor conductors of heat (they are nearly insulators) (Bergman et al. 2011).

Convection occurs when heat is transported from one place to another by the movement of a fluid (liquid or gas) (Bejan 2013). When part of a fluid is heated, it expands and its density decreases, causing it to rise while the cooler, denser fluid sinks. This movement of the fluid creates a convection current that transports the heat. An example of convection is the heating of water in a pot, where the hot water rises and the cold water sinks, resulting in uniform heating (White 2016).

Thermal radiation is the transfer of thermal energy by electromagnetic waves, in particular infrared radiation. In contrast to heat conduction and convection, thermal radiation does not require a medium; it can also take place in a vacuum (Siegel and Howell 2002). A classic example of thermal radiation is the heating of the earth by the sun. The sun's rays transport thermal energy through space and heat the earth's surface (Modest 2013).

The ability of the object to conduct heat is known as its thermal conductivity and is denoted by the parameter λ. As per the law of heat conduction, commonly referred to as Fourier's law, the rate of heat transfer through a material is directly proportional to the negative temperature gradient and the area, located at right angles to that gradient, through which the heat flows. It is possible to express this law in two equivalent forms: the integral form, wherein we examine the quantity of energy flowing into or out of a body as a whole, and the differential form, wherein we examine the flow rates or fluxes of energy locally.

The differential form of Fourier's law of thermal conduction indicates that the local heat flux density $\dot{\boldsymbol{q}}$ is proportional to the product of thermal conductivity $\boldsymbol{\lambda}$ and the local temperature gradient $\nabla \boldsymbol{T}$. The "–" indicates that heat always flows from the higher level of temperature to the lower level. The heat flux density is the quantity of energy that flows through a unit area in a unit time.

$$\dot{\boldsymbol{q}} = -\boldsymbol{\lambda} \nabla \boldsymbol{T} \tag{4.1}$$

The thermal conductivity $\boldsymbol{\lambda}$ is mostly regarded as a constant, although this may not always be the case. Some common materials have a small variation in thermal conductivity over a significant range of temperatures. The thermal conductivity of anisotropic materials typically varies with orientation, which is represented by a second-order tensor. In materials that are not homogenous, it varies with spatial location (Incropera et al. 2011; COMSOL 2020; Çengel and Ghajar 2015).

In an isotropic medium, Fourier's law leads to a heat equation

$$\frac{\partial T}{\partial t} = \alpha \left(\frac{\partial^2 T}{\partial x^2} + \frac{\partial^2 T}{\partial y^2} + \frac{\partial^2 T}{\partial z^2} \right) \tag{4.2}$$

with a fundamental solution famously known as heat kernel.

For many simplified applications, Fourier's law is used in its one-dimensional form, for example, in the z-direction

$$\dot{q} = -\lambda \frac{\mathrm{d}T}{\mathrm{d}z}. \tag{4.3}$$

By integrating the differential form over the materials total surface $\boldsymbol{S}$, we obtain the integral form of Fourier's law

$$\oiint_S \dot{\boldsymbol{q}} \cdot d\boldsymbol{S} = -\lambda \oiint_S \nabla \boldsymbol{T} \cdot d\boldsymbol{S}^{\circ}. \tag{4.4}$$

The heat flow rate is given by the differential equation when applied to a homogeneous material with a one-dimensional geometry between two endpoints at constant temperature

$$\frac{Q}{\Delta t} = -\lambda \frac{A}{L} \Delta T \tag{4.5}$$

If the temperature gradient between adjacent particles remains constant over time, stationary heat conduction, also known as steady-state heat conduction, takes place. The heat flow $\mathrm{d}\dot{Q}$ in the x-direction normal to the surface element $\mathrm{d}A$

$$d\dot{Q} = -\lambda \frac{d\vartheta}{dx} dA \tag{4.6}$$

with proportionality factor λ as thermal conductivity and $d\vartheta/dx$ as temperature gradient. The negative sign is necessary because the temperature gradient points in the direction of higher temperature.

4.2 Simulation

Heat simulation is a computational technique used to model the transfer and distribution of heat within a system or material over time. This process involves several steps, including model building, defining boundary conditions, and specifying the material properties.

In heat transfer simulation, the first step is to create a mathematical or computational model of the system or structure under study. This model represents the geometry of the object or environment in which heat transfer will be analyzed. The model can range from simple shapes (like rods, plates, or cubes) to complex structures (like automotive engines, electronic devices, or building materials).

Boundary conditions are crucial in defining how heat interacts with the surroundings of the model. These conditions specify how heat enters, exits, or behaves at the boundaries of the material or system being studied. Common boundary conditions in heat simulation include:

- **Fixed Temperature (Dirichlet Condition)**: The temperature at the boundary is held constant. For example, a wall in a building might be maintained at a specific temperature.
- **Heat Flux (Neumann Condition)**: The rate of heat transfer per unit area across the boundary is specified. This could represent, for instance, the heat being radiated from a surface.
- **Convective Boundary Condition**: This considers heat transfer between a surface and a surrounding fluid, where the rate of heat transfer depends on the temperature difference and the properties of the fluid.
- **Adiabatic Boundary Condition**: This assumes no heat flow across the boundary, meaning the surface is perfectly insulated.

The material properties significantly influence how heat is conducted, convected, or radiated within the model. Key material properties required for heat simulation include:

- **Thermal Conductivity**: This measures how well a material conducts heat. High thermal conductivity materials, like metals, transfer heat efficiently, while low conductivity materials, like insulators, do not.

- **Specific Heat Capacity**: This is the amount of heat required to change the temperature of a unit mass of the material by one degree Celsius. Materials with high specific heat capacity can absorb more heat without a significant temperature change.
- **Density**: This affects how much heat a material can store. Denser materials generally store more heat.
- **Emissivity**: In simulations involving radiation, emissivity defines how effectively a material emits thermal radiation.

Once the model is built, boundary conditions are set, and material properties are assigned, the heat simulation can be run using appropriate software, such as finite element analysis (FEA) tools. The simulation calculates temperature distribution, heat flux, and other thermal properties over time, allowing engineers and scientists to analyze how heat will behave in the real system.

This process is widely used in engineering fields such as mechanical design, electronics cooling, thermal management in buildings, and energy systems to optimize designs, ensure safety, and improve energy efficiency.

4.3 Measurement of Thermal Properties

In the study and analysis of heat transfer, several key measurements are critical for understanding the thermal behavior of materials and systems. These measurements include temperature, heat flux, and heat. These measurements are essential in various fields, including engineering, materials science, and environmental studies, to analyze and optimize thermal processes.

4.3.1 Temperature Measurement

Temperature is a measure of the average kinetic energy of the particles in a substance and is one of the most fundamental thermal properties.

Thermometers are traditional liquid-in-glass thermometers (mercury or alcohol) that are used for measuring temperature by the expansion of the liquid with temperature changes. Thermocouples are devices consisting of two dissimilar metal wires joined at one end, creating a voltage that correlates with temperature. Thermocouples are widely used due to their broad temperature range and rapid response. Resistance temperature detectors (RTDs) measure temperature based on the change in electrical resistance of a metal (usually platinum) with temperature. They provide high accuracy and stability. Thermistors are temperature-sensitive resistors that change resistance with temperature. They are typically used in narrow temperature ranges.

Infrared (IR) Thermometers measure the thermal radiation emitted by an object to determine its surface temperature without contact. They are useful for measuring high temperatures or the temperature of moving or inaccessible objects. Thermographic cameras provide a visual map of temperature distribution on a surface by detecting infrared radiation, commonly used for identifying hot spots or thermal anomalies.

4.3.2 Heat Flux Measurement

Heat flux indicates how much heat is moving through a material or across a surface over time. Heat flux sensors (heat flux transducers) consist of a thin material with known thermal properties, placed in contact with the surface where heat flux is to be measured. The temperature difference across the sensor gives a direct measurement of heat flux. They are often used in building materials, electronics, and aerospace applications.

In some cases, heat flux can be inferred using calorimetric methods, where the heat absorbed or released by a material is measured, and this data is used to calculate the heat flux. Thermal gradient methods involve measuring the temperature difference across a known distance within a material, combined with its thermal conductivity, to calculate heat flux. Transient source plane methods belong to this category.

4.3.3 Heat Measurement

Unlike temperature, heat depends on the amount of material and the specific conditions of heat transfer. Calorimeters are devices used to measure the heat exchanged in a process. There are different types of calorimeters: bomb calorimeter measures the heat of combustion of a substance in a constant volume. A simpler device used for measuring heat changes at constant pressure is a coffee cup calorimeter, typically for reactions in aqueous solutions.

Differential Scanning Calorimetry (DSC) measures the amount of heat required to increase the temperature of a sample relative to a reference. It is widely used in materials science to study phase transitions and heat capacity. In chemical processes, the change in enthalpy (ΔH) can be measured using calorimetric techniques, which directly relates to the heat absorbed or released during a reaction. In larger systems, such as HVAC (heating, ventilation, and air conditioning) systems, energy meters are used to measure the total heat energy delivered or consumed over time.

4.3.4 Transient Plane Source Method

The thermal constants analyzer Hot Disk 2500 S is a measurement system for thermal properties, Fig. 4.1. This system allows an absolute, dynamic, and direct measurement of thermal parameters of materials with only one measurement. The complex

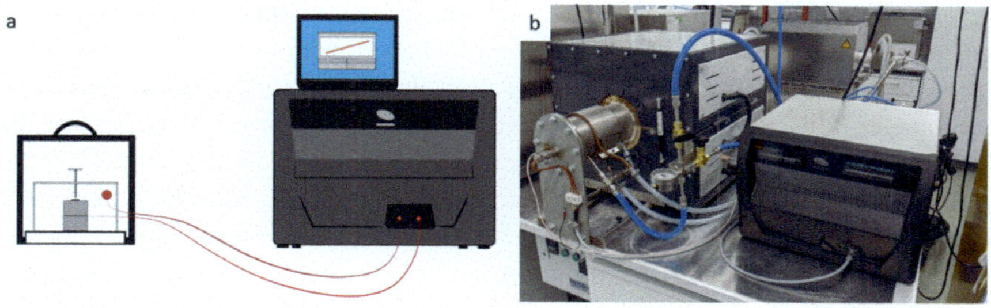

Fig. 4.1 Thermal constants analyzer Hot Disk 2500S **a** schematic measurement setup **b** measurement system with oven on the left

mathematical correlation is given by Gustafsson (1991)

$$R(t) = R_0\{1 + \alpha \cdot (\Delta \mathrm{T}_i + \Delta \mathrm{T}_{ave}(\tau))\}. \tag{4.7}$$

R_0 is the resistance of the sensor before the current is applied, α is the Temperature coefficient of resistance, ΔT_i is the constant temperature difference over the thin insulating layers which are covering the sensor, and $\Delta \mathrm{T}_{ave}(\tau)$ is the temperature increase of the sample surface. More information can be found in the literature (ISO 2017, 1987; Gustafsson 1991; Hot Disk 2015; Daniel 2500).

It covers an extensive range of materials of various geometries and meets ISO Standard 22,007-2 (Hot Disk 2500; ISO 2017). The device is compatible with different modules. This allows the thermal properties of isotropic as well as anisotropic materials to be measured. In addition, one-dimensional measurements can be made on rods, and it is capable of doing measurements on the surface of thin film test specimens. The great advantages of this method over other measurement methods are the simplicity with which the experiments can be carried out and the versatility of its use.

Depending on the selected sensor, different temperature ranges can be covered, and how thermal conductivity can be measured at these. The hot disk covers a measuring range from −235 °C up to 1000 °C. For measurements in the range from 750 °C to 1000 °C, however, the system must be kept free of oxygen. In addition to thermal conductivity, the Hot Disk offers the possibility of measuring the specific heat capacity up to 5 MJ/m^3K. Three insulation materials are available for the sensors: Kapton, Mica, and Teflon.

In the year 1990, Silas E. Gustafsson proposed the transient plane source method as a novel way to describe the thermal transport paths of arbitrary materials (Gustafsson 1991). Also referred to as the "Hot Disk method", this measurement method is standardized in the ISO 22007-2 (Hot Disk 2015). It is based on the "hot wire method" (ISO 1987) and proves to be an easy-to-use transient procedure, which allows testing almost any type of material over a broad spectrum of temperatures (30K-1273 K) (Daniel 2500). The basic principle of this method is to induce a precisely known amount of power in the form of a heat pulse into a sample of unknown properties and meanwhile record the temperature response of the sample through voltage change over the sensor. To realize this, a thin, spiral-shaped Kapton-insulated nickel foil with a thickness of 10 μm, functions as both the heat source and the resistance thermometer (Gustafsson 1991). The sensor data can then be used to calculate the TD and conductivity of the sample. The sensor is to be placed between two identical parts of a sample with as much contact area as possible, Fig. 4.2 (Hot Disk 2500). This means solids need a flat surface in the contact area of the sensor. Other than that the shape, size, and physical condition of the test medium are not relevant, which is a great advantage of the TPSM. Solids, powders, pastes, liquids, and gases can be investigated, making it a convenient tool for determining the thermal transport properties of materials.

Fig. 4.2 Hot disk method scheme (Hot Disk 2500)

4.4 Additive Manufacturing

In research and development, temperature and heat play a crucial role in understanding material properties and optimizing processes. Scientists use controlled heating and cooling to study phase transitions, such as melting, crystallization, and chemical reactions, which are essential for developing new materials. Thermal analysis techniques, like differential scanning calorimetry (DSC), provide insights into the thermal stability and heat capacity of materials, guiding the design of products with enhanced performance. In the mechanical application industry, heat treatment processes, such as annealing, quenching, and tempering, are used to alter the microstructure of metals, improving their strength, hardness, and ductility. Temperature control is critical in machining and manufacturing processes, where excessive heat can lead to tool wear and material deformation. Thermal management systems are designed to dissipate heat in engines, ensuring efficient operation and preventing overheating. In automotive and aerospace industries, temperature regulation is vital for the reliability and safety of components exposed to extreme conditions. Advanced cooling technologies, such as heat exchangers and radiators, are developed to maintain optimal operating temperatures. These applications highlight the importance of temperature and heat in both advancing scientific knowledge and enhancing the durability and performance of mechanical systems.

Additive manufacturing, particularly 3D metal printing, has significantly impacted various industries by enabling the production of complex, custom, and lightweight metal components. In the aerospace industry, it allows for the creation of strong yet lightweight parts, such as turbine blades and structural brackets, enhancing fuel efficiency and overall performance. The automotive sector benefits from rapid prototyping and the ability to manufacture custom, high-performance engine components. In the medical field, 3D metal printing is revolutionizing the production of patient-specific implants and surgical tools, tailored to individual needs. The energy sector utilizes this technology for creating complex turbine components and for repairing worn-out parts, thereby extending their lifespan. Industrial tooling has also seen advancements, with 3D printing enabling the production of intricate molds and custom fixtures, improving manufacturing efficiency. Additionally, the robotics industry leverages 3D metal printing for creating custom end-effectors and integrated sensor components. Consumer electronics benefit from the production of customized heat sinks and enclosures, while the construction industry uses it for innovative architectural designs. Finally, artists and designers are exploring new creative possibilities with custom metal artwork, demonstrating the broad and transformative potential of 3D metal printing across diverse fields.

Additive manufacturing has many advantages, particularly in the degrees of freedom of design. These are used for efficient heat transport, among other things. Cooling channels can be placed close to the contours of the zones to be tempered, and the orientation of the three-dimensional paths and the cross-sections can be selected as required.

The result is increased and efficient cooling and increased heating. Material is saved, including no local structural transformation due to the production heat. Environmentally harmful coolants/lubricants can be dispensed with in additive manufacturing. This results in shorter cycle times in production, improved component quality thanks to a more homogeneous temperature profile, and a reduction in weight or component size.

The optimization of the cooling channels with regard to their dimensioning, taking into account the fluid mechanics to ensure an optimal fluid flow, is a task of product design. Structural mechanics, in particular strength and rigidity, must not suffer as a result. The cross sections in additive manufacturing must be selected in such a way that there are no internal support structures; subsequent removal is only possible with a great deal of technical effort and time.

The production of such structures or components is possible and economical on commercially available systems using powder bed-based laser beam melting (Kortsch-Banzhaf et al. 2024; Sykora et al. 2023; Schuhmann et al. 2022). Temperature and heat play critical roles in the components and processes of additive manufacturing, especially in 3D metal printing. The control of temperature during the printing process is essential for ensuring the proper melting and solidification of metal powders, which directly affects the quality and mechanical properties of the printed parts (Vafadar et al. 2021; Renken et al. 2019). Inconsistent temperature control can lead to defects such as porosity, warping, and residual stresses, compromising the integrity and precision of the final product (Vafadar et al. 2021; Ning et al. 2019; Marchese et al. 2017). Heat also influences the bonding between layers in additive manufacturing, where insufficient heat may result in weak inter-layer adhesion, while excessive heat can cause deformation or overheating of the material (Craeghs et al. 2010; Majeed et al. 2019; Tomas et al. 2022). The thermal properties of the material being printed, such as thermal conductivity and expansion, must be carefully managed to prevent cracking and distortion as the material cools (Ning et al. 2019; Renken et al. 2018). Additionally, the temperature of the build chamber and surrounding environment needs to be controlled to avoid thermal gradients that can lead to uneven cooling and internal stresses (Renken et al. 2019; Majeed et al. 2019). Post-processing heat treatments are often required to relieve residual stresses and enhance the mechanical properties of the printed components (Silva et al. 2011; Hussain et al. 2021). Overall, precise temperature and heat management are crucial for achieving high-quality, reliable, and durable products in additive manufacturing (Mani et al. 2015).

Understanding the complex relationship between temperature, heat, and material behavior requires a deep knowledge of the thermal properties of materials and components. These properties, such as thermal conductivity, heat capacity, and expansion, are critical for predicting how materials will respond to thermal stresses during manufacturing and operation. Accurate measurement of these thermal properties is essential to optimize processes like heat treatment and additive manufacturing. Without this knowledge, it becomes challenging to ensure the quality, durability, and performance of the final products. Therefore, comprehensive thermal analysis is key to advancing material science and engineering applications.

4.5 Conclusion and Outlook

Temperature and heat are fundamental factors in various fields, playing a crucial role in research, mechanical industries, and additive manufacturing. Controlled heating and cooling are essential for studying material properties and optimizing processes, leading to the development of advanced products with improved performance. In mechanical applications, precise temperature management ensures the durability and reliability of metal components through heat treatment and machining processes. Additive manufacturing, particularly 3D metal printing, benefits significantly from effective thermal management, which is critical for achieving high-quality, defect-free parts. The integration of optimized heat sinks and cavities in 3D printing enhances production efficiency and product quality, while the ability to tailor thermal properties highlights the potential for innovation in this field. Overall, the careful control, measurement, and optimization of temperature and heat is key to advancing both manufacturing techniques and the performance of the resulting products.

The future of additive manufacturing holds significant promise with the integration of thermal and mechanical anisotropy, enabling the creation of load-adaptive structures that are precisely tailored to their operational environments. By leveraging anisotropic properties, engineers can design components that respond optimally to specific mechanical loads and thermal conditions, enhancing performance and durability. As the understanding of thermal properties in 3D printing materials improves, new possibilities emerge for fine-tuning the printing process, leading to components with highly controlled thermal behavior. This advancement will allow for the development of more efficient heat exchangers, thermal barriers, and other critical components in industries like aerospace and automotive. Ultimately, the combination of improved thermal knowledge and anisotropic design capabilities will revolutionize the way complex, multifunctional structures are conceived and produced through additive manufacturing.

References

22007-2: 2015. Plastics–Determination of thermal conductivity and diffusivity–Part 2: Transient plane source (Hot Disk) method (2015).

Bejan A (2013) Convection heat transfer, 4th ed. Wiley

Bergman TL, Lavine AS, Incropera FP, DeWitt DP (2011) Fundamentals of heat and mass transfer, 7th ed. Wiley

Cengel YA, Ghajar AJ (2015) Heat and mass transfer: fundamentals & applications, 5th ed. McGraw-Hill Education

Çengel YA, Ghajar AJ (2015) Heat and mass transfer: fundamentals & applications, 5th ed. New York, NY, USA: McGraw-Hill Education

COMSOL, Heat Transfer in Solids. COMSOL Documentation (2020). https://doc.comsol.com/5.5/doc/com.comsol.help.comsol/comsol_ref_heattransfer.21.25.html. Accessed: 12 Aug 2024

Craeghs T, Bechmann F, Berumen S, Kruth JP (2010) Feedback control of layerwise laser melting using optical sensors. Phys Procedia 5:505–514

Daniel, Instruction Manual of Hot Disk thermal constants analyser tps 2500 s

Hot Disk, TPS 2500 S - Instruments - Hot Disk, 4 June 2023. https://www.hotdiskinstruments.com/products-services/instruments/tps-2500-s/

Gustafsson SE (1991) Transient plane source techniques for thermal conductivity and thermal diffusivity measurements of solid materials. Rev Sci Instrum 62(3):797–804

Halliday D, Resnick R, Walker J (2013) Halliday, Resnick, Walker - Physik (9 Aufl). Wiley-VCH

He Y (2005) Rapid thermal conductivity measurement with a hot disk sensor: part 1 theoretical considerations. Thermo Chimica Acta 436(1–2):122–129

Hussain SZ, Kausar Z, Koreshi ZU, Sheikh SR, Rehman HZU, Yaqoob H, Shah MF, Abdullah A, Sher F (2021) Feedback control of melt pool area in selective laser melting additive manufacturing process. Processes 9:1547

Incropera FP, DeWitt DP, Bergman TL, Lavine AS (2011) Fundamentals of heat and mass transfer, 7th edn. Wiley, Hoboken, NJ, USA

Incropera FP, DeWitt DP (2007) Fundamentals of heat and mass transfer, 6th ed. Wiley

ISO, ISO 8894-1:1987, 10 Nov 2023. https://www.iso.org/standard/16420.html

ISO, ISO 22007-4:2017, 10 Oct 2023. https://www.iso.org/standard/65085.html

Kortsch-Banzhaf J, Schultheis G, Merkel M (2024) Investigation of the thermophysical properties of porous steel components made by selective laser melting, lecture notes on advanced structured materials 2, vol 203. Springer Nature

Lienhard IV JH, Lienhard V JH (2012) A heat transfer textbook, 4th ed. Cambridge, MA, USA: Phlogiston Press

Majeed M, Khan HM, Rasheed I (2019) Finite element analysis of melt pool thermal characteristics with passing laser in SLM process. Optik 194:163068

Mani M, Feng S, Lane B, Donmez A, Moylan S, Fesperman R (2015) Measurement science needs for real-time control of additive manufacturing powder bed fusion processes. NIST 2015

Marchese G, Garmendia Colera X, Calignano F, Lorusso M, Biamino S, Minetola P, Manfredi D (2017) Characterization and comparison of Inconel 625 processed by selective laser melting and laser metal deposition. Adv Eng Mater 19(4)

Modest MF (2013) Radiative heat transfer, 3rd ed. Academic Press

Ning J, Sievers D, Garmestani H, Liang S (2019) Analytical modeling of in-process temperature in powder bed additive manufacturing considering laser power absorption, latent heat, scanning strategy, and powder packing. Materials 12(5)

Renken V, Lübbert L, Blom H, von Freyberg A, Fischer A (2018) Model assisted closed-loop control strategy for selective laser melting. Procedia CIRP 74:659–663

Renken V, von Freyberg A, Schünemann K, Pastors F, Fischer A (2019) In-process closed-loop control for stabilizing the melt pool temperature in selective laser melting. Prog Addit Manuf 4(4):411–421

Schuhmann D, Pinto G, Merkel M, Harrison AK (2022) A study on additive manufacturing of metal components for mobility in the area of after-sales with spare and performance parts. Vehicles 4:957–977

Siegel R, Howell JR (2002) Thermal radiation heat transfer, 4th ed. Taylor & Francis

Silva BL, Araujo IJC, Silva WS, Goulart PR, Garcia A, Spinelli JE (2011) Correlation between dendrite arm spacing and microhardness during unsteady-state directional solidification of Al-Ni alloys. Philos Mag Lett 91(5):337–343

Sykora J, Sedlmajer M, Schubert T, Merkel M, Kroft L, Kucerova L, Rehor J (2023) Additive manufacturing of WC-Co specimens with internal channels. Materials 16:3907

Tipler PA, Mosca G (2008) Physik (6 Aufl). Spektrum Akademischer Verlag

Tomas J, Merkel M, Altenbach H (2022) On the optical thermography in selective laser melting process. Mech Solids 57:597–603

Vafadar A, Guzzomi F, Rassau A, Hayward K (2021) Advances in metal additive manufacturing: a review of common processes, industrial applications, and current challenges. Appl Sci 11(3):1213

White FM (2016) Fluid mechanics, 8th ed. McGraw-Hill Education

Chapter 5
A Constitutive Approach to Represent the Influence of Ageing on the Dynamic Mechanical Properties of Polymers

Alexander Lion and Michael Johlitz

Abstract Polymers, ubiquitous in modern life, undergo static and dynamic ageing processes that significantly impact their mechanical, thermal and chemical properties over time. This contribution provides a comprehensive overview of the static and dynamic ageing phenomena observed in polymers, encompassing mechanisms, influencing factors and implications across various industries. The ageing of polymers involves intricate chemical reactions such as oxidation, hydrolysis and photodegradation, leading to molecular rearrangements, chain scission and the formation of degradation products. Factors influencing polymer ageing include environmental conditions (e.g. temperature, humidity and UV radiation), mechanical stress, chemical exposure and material composition. Understanding the statics and dynamics of polymer ageing is crucial for predicting material behaviour, optimizing product performance, and ensuring safety and reliability in diverse applications. In the automotive sector, for instance, ageing-related degradation can affect the structural integrity of polymer components, impacting vehicle safety and longevity. Similarly, in the packaging industry, knowledge of polymer ageing kinetics is essential for designing sustainable packaging solutions with extended shelf life and reduced environmental impact. Therefore, the computer-based simulation and prediction of polymer ageing with adequate calibrated models is essential. This is the point where this article applies. A one-dimensional thermomechanical consistent model is developed and presented which is able to capture dynamic ageing processes in the frequency and time domains. Meaningful simulations demonstrate the model's ability to reasonably depict various ageing effects. A brief summary and a comprehensive outlook on the upcoming work complete this contribution.

A. Lion (✉) · M. Johlitz
Universität der Bundeswehr, Werner-Heisenberg-Weg 39, 85577 Neubiberg, Germany
e-mail: alexander.lion@unibw.de

M. Johlitz
e-mail: michael.johlitz@unibw.de

H. Altenbach et al. (eds.), *Lectures Notes on Advanced Structured Materials 3*,
Advanced Structured Materials 226, https://doi.org/10.1007/978-3-031-77403-4_5

5.1 Introduction and Motivation

Polymers are utilized in a myriad of applications in modern times due to their versatility, durability and cost-effectiveness. One common sector is the packaging industry where polymers are ubiquitous in packaging due to their ability to provide lightweight, durable and customizable solutions. They help in preserving food, reducing transportation costs and extending shelf life. From plastic bottles and containers to flexible packaging films, polymers dominate this industry.

Another major sector is the automotive industry. Polymers play a vital role in modern vehicles, offering lightweight alternatives to metal components. They contribute to fuel efficiency, design flexibility and improved safety. Polymers are found in interior trims, dashboards, bumpers, engine components and various other parts.

They are also increasingly used in construction and building materials for their durability, resistance to corrosion and ease of installation, e.g. they are utilized in pipes, insulation materials, roofing, windows, doors and composite materials. Polymers help to enhance the energy efficiency and reduce maintenance costs in buildings.

Overall, polymers are indispensable in modern society, offering versatile solutions across numerous industries due to their unique combination of properties and cost-effectiveness. However, the environmental impact of polymer or so-called plastic usage remains a significant concern, driving efforts towards sustainable alternatives and recycling initiatives.

Researching the ageing and lifespan of polymers is crucial for several reasons and has been investigated by a lot of authors, e.g. Budzien et al. (2008), Duarte and Achenbach (2007) and Perera (2003). Polymers are integral to our daily lives, but they often have a long lifespan and can accumulate in the environment for centuries. Understanding their ageing and lifespan allows us to develop methods to extend their durability and optimize their disposal, ultimately helping to reduce environmental pollution from plastic waste. By studying ageing, we can comprehend how polymers behave over time, thereby enhancing product performance by selecting materials that last longer and are less susceptible to wear and tear. In some applications, such as aerospace or medical technology, safety is paramount. Understanding the ageing of polymers allows us to ensure that these materials remain stable under extreme conditions or over long periods, avoiding unexpected failures or risks. Overall, researching the ageing and lifespan of polymers is crucial for reducing environmental pollution, improving product performance, ensuring safety and mitigating economic impacts.

Polymer ageing is a multifaceted field encompassing diverse research endeavours aimed at understanding the complex processes that occur as polymers undergo changes over time. This literature review provides an overview of key works in the realm of polymer ageing, highlighting major themes, methodologies and advancements.

Studies on ageing of polymers typically investigate changes in the mechanical properties over time under the influence of ageing factors such as temperature, humidity and stress.

There is an extensive literature with numerous experiments and modelling approaches for the influence of ageing on the quasi-static behaviour of polymers, e.g. Musil et al. (2018, 2020). However, if one considers the influence of ageing on the dynamic behaviour, the corresponding literature is sparse. Petríková et al. (2011) have studied the influence of thermal ageing with respect to the static and dynamic ageing mechanisms in specific polymer systems and published many interesting experimental data which motivate the development of a model. Also Karekar et al. (2023) have shown that the dynamic behaviour in the glassy regime is weakly influenced by ageing since the storage and loss moduli change with the ageing duration. Regarding the application in the automotive industry, the frequency-dependent dynamic stiffness of bearings is of particular importance, cf. Kurian et al. (2021), Kruse and Chauvet (2020) and Lion and Johlitz (2020). It depends on the dynamic mechanical material properties and the density of the elastomers used as well as the geometry of the bearing. In Lion and Johlitz (2020), a simplified model was used to estimate that the first natural frequency of typical aggregate bearings is a few hundred hertz. If such a bearing is excited at this or higher natural frequencies, pronounced resonances and anti-resonances and thus increases and decreases in its dynamic stiffness occur. Thus, the influence of the frequency-dependent transmission behaviour of aggregate bearings must be taken into account in the design.

Pioneering work on the investigation, interpretation and modelling of the ageing of elastomers under constant load, oxygen and temperature influence goes back to Tobolsky et al. (1944) and Andrews et al. (1946). Ageing manifests itself in a degeneration of the current polymer network in combination with the formation of a second network due to oxygen and temperature. Depending on how pronounced these effects are, the material stiffens or softens. For details on the kinetics and chemical principles of ageing, please refer to the work by Ehrenstein and Pongratz (2007), the habilitation of Johlitz (2015) and the review article of Tayefi et al. (2023). For continuum mechanical modelling of this type of ageing, scalar internal variables are usually introduced into material models of finite elasticity or viscoelasticity, whose evolution depends on the temperature and, if applicable, the concentration of diffusing oxygen.

Works by Johlitz (2012); Johlitz and Lion (2013), Lion and Johlitz (2012) and Musil et al. (2018, 2020) have contributed significantly to our understanding of polymer ageing, particularly in the context of material characterization and environmental degradation mechanisms. Their research has provided valuable insights into the molecular-level changes and degradation pathways occurring in polymers subjected to ageing conditions, further advancing the knowledge base in this field. But all these models are not able to capture the influence of ageing on the dynamic properties of the investigated materials.

In conclusion, research on polymer ageing encompasses a wide range of investigations into mechanisms, environmental influences, characterization techniques, modelling approaches, applications and mitigation strategies. Continued advancements in this field are essential for developing durable, sustainable and resilient polymer materials to meet the evolving needs of modern society.

This study concentrates on the effects of ageing on the viscoelastic properties of polymers, particularly the fundamental elasticity and relaxation characteristics. It analyses how long-term ageing processes affect the frequency-dependent dynamic properties of polymers.

5.2 Influence of Ageing on the Material Behaviour

The ageing of polymers significantly influences their material behaviour, undergoing various physical, chemical and mechanical changes. These alterations can impact both the macroscopic properties of the material and its microscopic structure.

With ageing, the mechanical properties of polymers may experience a reduction in strength and elasticity. This can lead to increased brittleness, deformation or cracking, especially under mechanical stress. Ageing can also impair the ductility and toughness of the material, diminishing its ability to absorb energy and deformation. In addition, ageing can alter the thermal properties of polymers, including their thermal conductivity, coefficient of thermal expansion and temperature resistance. This can result in heightened sensitivity to temperature fluctuations and restrict the applications of the polymers. Furthermore, the chemical resistance of polymers may lessen due to ageing, leading to deterioration in their resistance to chemical substances, solvents or environmental influences. This can result in material degradation, surface discoloration or corrosion, compromising the integrity of the material. At a microscopic level, ageing can induce changes in the polymer structure, such as chain degradation, cross-linking, or loss of crystallinity. These alterations can directly impact the mechanical properties of the material and reduce its long-term stability. Another important point is that ageing can also influence the environmental compatibility of polymers by increasing the release of harmful chemicals or microplastic particles. This can cause ecological damage and impair the recyclability of the material. Overall, ageing can affect the material behaviour of polymers in various ways, presenting both technical and environmental challenges. A better understanding of ageing mechanisms and appropriate measures for ageing resistance are therefore crucial for the development of durable and sustainable polymer products.

Simulation and modelling can help to understand the influence of ageing to the changes in the behaviour of the technical system and to reduce costs but meaningful models need a sufficient database in order to make meaningful predictions of the life time of the investigated materials.

This article is intended as a first step and deals exclusively with the thermo-oxidative ageing of rubber, whose static and mechanical properties change as a result of the ageing process in atmospheric oxygen and at elevated temperatures.

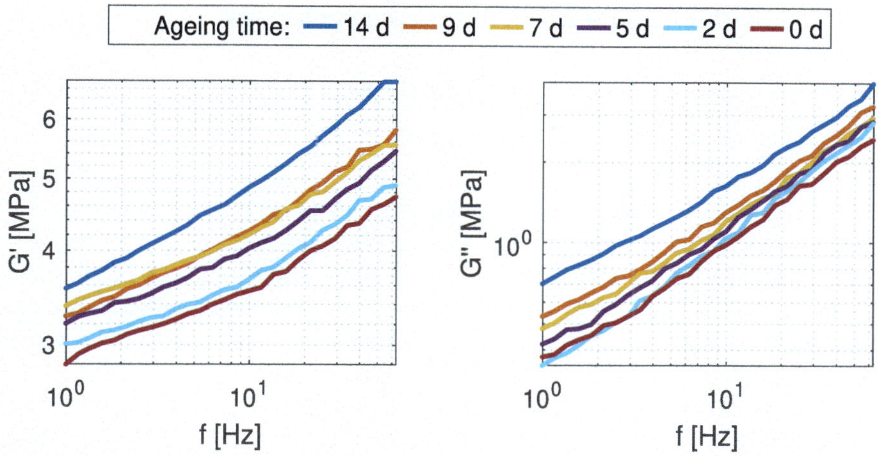

Fig. 5.1 Ageing dependence of storage modulus (left) and of loss modulus (right)

The preliminary test of Fig. 5.1 shows both the dependence of the storage modulus and the loss modulus on ageing. The material was aged under atmospheric oxygen at 100 °C for a period of 14 days. A frequency sweep at room temperature was carried out both on the virgin material and at different ageing times.

The results of this pretest show a significant stiffening of the storage modulus and loss modulus with ageing time. This indicates that the chemical structure of the polymer is changing. A reason could be the well-known degradation or creation of chemical bonds under thermo-oxidative conditions. These changes are irreversible and affect both the quasi-static stress–strain behaviour as well as the dynamic mechanical parameters in form of the storage and loss moduli in the frequency domain and of the relaxation and creep responses in the time domain.

As an intermediate summary, chemical ageing induces changes in the material behaviour as follows:

- Quasi-static stress–strain responses become stiffer.
- Equilibrium stress increases.
- Rate of stress relaxation lessens.
- Storage modulus increases with ageing duration.
- Loss modulus increases with ageing duration.

Hence, the challenge is the development of a constitutive model to represent these phenomena.

5.3 One-Dimensional Modelling Approach

5.3.1 *Analysis of a One-Dimensional Model of Linear Viscoelasticity Without Ageing*

First, the equation set of the standard three-parameter model of linear viscoelasticity is to be repeated and provided (cf. e.g. Lion (2000)). This is based on a spring with Young's modulus E, to which a Maxwell element, consisting of a series connection of a spring element with Young's modulus $\hat{E}$ and a damper element with the viscosity η, is connected in parallel. The total stress σ of the model results from the additive split of the equilibrium stress σ_{eq} and the overstress σ_{ov} to

$$\sigma = \sigma_{\mathrm{eq}} + \sigma_{\mathrm{ov}} \tag{5.1}$$

while the constitutive equation of the equilibrium stress reads

$$\sigma_{\mathrm{eq}} = E\,\varepsilon \tag{5.2}$$

with Young's modulus E and the strain ε. The overstress is given by the following differential equation:

$$\dot{\sigma}_{\mathrm{ov}} = \hat{E}\,\dot{\varepsilon} - \frac{1}{\tau}\,\sigma_{\mathrm{ov}}\,. \tag{5.3}$$

Herein the parameter τ represents the relaxation time and $\hat{E}$ represents Young's modulus of the spring of the Maxwell element. The differential equation also contains the rates of the overstress and of the strain.

Now the following question arises: Which material parameters have to depend in which manner on the ageing process to describe the experimental observations?

It is well known that quasi-static stress–strain curves change with ageing. Hence, the equilibrium modulus E in Eq. (5.2) should depend on the ageing variable(s).

Let us now have a look at the dynamic properties of the materials in the frequency domain. Likewise the frequency-dependent loss modulus $E''(\omega)$ has to depend on ageing. As a consequence, the overstress modulus $\hat{E}$ or the relaxation time τ should also depend on the ageing variable(s):

$$E''(\omega) = \hat{E}\,\frac{\omega\,\tau}{1 + (\omega\,\tau)^2}\,. \tag{5.4}$$

In the time domain, the stress relaxation behaviour has to be considered in dependence of ageing, i.e. the equilibrium modulus E, the overstress modulus $\hat{E}$ and the relaxation time τ should depend on the ageing variable(s):

$$\sigma(t) = \varepsilon_0 \left(E + \hat{E}\, e^{-\frac{t}{\tau}} \right) . \tag{5.5}$$

Last but not least, also the storage modulus $E'(\omega)$ changes with ageing. It is obvious that equilibrium modulus E, overstress modulus $\hat{E}$ and the relaxation time τ should depend on the ageing variable(s) as formulated before for the loss modulus:

$$E'(\omega) = E + \hat{E}\, \frac{(\omega\,\tau)^2}{1 + (\omega\,\tau)^2} . \tag{5.6}$$

Finally, the consequence of this observations is the following: In order to describe ageing of rubber, each material parameter of the viscoelasticity model should depend on ageing variable(s).

5.3.2 *Definition of Internal State Variables to Describe Irreversible Ageing*

Let us start with the introduction of an internal variable q which acts as a process variable in order to describe the above-mentioned material behaviour with respect to ageing. This variable has to be formulated in that manner, that in the unaged state it equals $q(0) = 0$. If the material is completely aged, the variable reads $q(\infty) = 1$.

$$0 \le q(t) \le 1 \tag{5.7}$$

Taking into account the irreversibility of ageing, the evolution of the ageing variable q has to follow the rule $\dot{q}(t) \ge 0$. Figure 5.2 shows some examples of the development of such a variable.

As a consequence, the experimentally based formulation of evolution laws to model ageing can be based on the variable q itself, on the strain ε and the absolute temperature θ and several other process variables if needed

$$\dot{q} = f(q, \varepsilon, \theta, ...) . \tag{5.8}$$

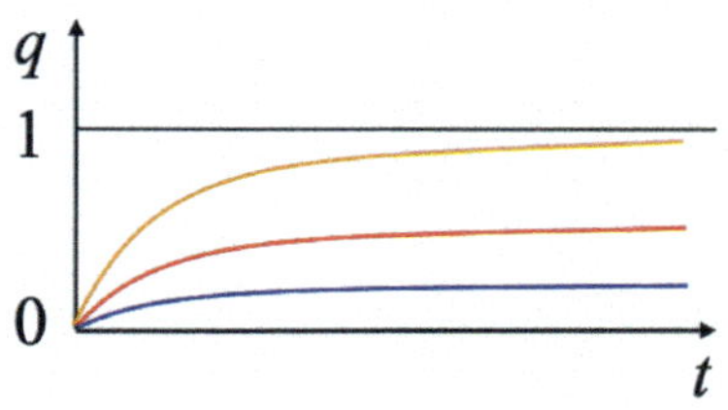

Fig. 5.2 Temporal evolution of the internal variable q

5.3.3 Formulation of a Specific Helmholtz Free Energy to Model Ageing of Viscoelastic Materials

The next step is the formulation of the Helmholtz free-energy function ψ. This function is used to evaluate the 2nd law of thermodynamics in order to guarantee the thermomechanical consistence of the model. The starting point is the additive decomposition of the free energy $\rho\,\psi$ into an equilibrium part and an overstress part

$$\rho\,\psi = \rho\,\psi_{\mathrm{eq}} + \rho\,\psi_{\mathrm{ov}}\,. \tag{5.9}$$

The equilibrium part has to consider both softening and stiffening effects. Therefore, the following approach with two contributions is chosen to

$$\rho\,\psi_{\mathrm{eq}} = \frac{E_-}{2}\,(1 - q_{\mathrm{eq}-})\,\varepsilon^2 + \frac{E_+}{2}\int_0^t \dot{q}_{\mathrm{eq}+}(s)\,(\varepsilon(t) - \varepsilon(s))^2\,\mathrm{ds}\,. \tag{5.10}$$

The first part with Young's modulus E_- and the internal ageing variable $q_{\mathrm{eq}-}$ takes the degeneration of the polymer network into account while the second hypoelastic part with Young's modulus E_+ and the internal ageing variable $q_{\mathrm{eq}+}$ considers the stiffening effects. For more details, the reader is referred to the papers of Johlitz (2012) or Lion and Johlitz (2012).

This idea is now transferred for the viscoelastic part of the model such that its strain energy function reads

$$\begin{aligned}\rho\,\psi_{\mathrm{ov}} &= \frac{\hat{E}_-}{2}\,(1 - q_{\mathrm{ov}-})\,(\varepsilon - \varepsilon_{\mathrm{in}-})^2 \\ &+ \frac{\hat{E}_+}{2}\int_0^t \dot{q}_{ov+}(s)\,(\varepsilon(t) - \varepsilon_{\mathrm{in}+}(t) - (\varepsilon(s) - \varepsilon_{\mathrm{in}+}(s)))^2\,\mathrm{ds}\,.\end{aligned} \tag{5.11}$$

Again, the first part with Young's modulus $\hat{E}_-$ and the internal ageing variable $q_{\mathrm{ov}-}$ take the degeneration of the viscoelastic part of the polymer network into account while the second hypoelastic part with Young's modulus $\hat{E}_+$ and the internal ageing variable $q_{\mathrm{ov}+}$ consider the stiffening effects. In addition to the ageing variables, the equation also contains the internal variables $\varepsilon_{\mathrm{in}-}$ and $\varepsilon_{\mathrm{in}+}$ of strain type from the classical linear viscoelasticity theory.

5.3.4 Evaluation of the Isothermal Clausius–Duhem Inequality and Derivation of the Mechanical Model

In order to derive the mechanical model, the isothermal Clausius–Duhem inequality

$$\sigma\,\dot{\varepsilon} - \rho\,\dot{\psi} \geq 0 \tag{5.12}$$

has to be evaluated. It considers the total stress power and the total time derivative of the free energy function with respect to the process variables. For arbitrary strain histories, the stress power supplied to a material volume element has to be equal to or larger than the temporal change in the free energy:

$$\left(\sigma - \frac{\partial\rho\psi}{\partial\varepsilon}\right)\dot{\varepsilon} + \left(\frac{\partial\rho\psi}{\partial\varepsilon_{\text{in}-}}\right)\dot{\varepsilon}_{\text{in}-} + \left(\frac{\partial\rho\psi}{\partial\varepsilon_{\text{in}+}}\right)\dot{\varepsilon}_{\text{in}+} + \frac{E_-}{2}\,\varepsilon^2\,\dot{q}_{\text{eq}-} + \frac{\hat{E}_-}{2}\,(\varepsilon - \varepsilon_{\text{in}-})^2\,\dot{q}_{\text{ov}-} \geq 0\,. \tag{5.13}$$

Taking into account the argumentation of Coleman and Noll (1961, 1963), the total stress can be derived to

$$\sigma = \sigma_{\text{eq}} + \sigma_{\text{ov}}\,. \tag{5.14}$$

Moreover, the ageing-related equilibrium stress can be splitted into

$$\sigma_{\text{eq}} = \sigma_{\text{eq}-} + \sigma_{\text{eq}+} \tag{5.15}$$

while the constitutive equation for the softening part is

$$\sigma_{\text{eq}-} = E_-\left(1 - q_{\text{eq}-}\right)\varepsilon \tag{5.16}$$

and the hypoelastic rate-dependent constitutive equation for the stiffening part can be developed to

$$\dot{\sigma}_{\text{eq}+} = E_+\,q_{\text{eq}+}\dot{\varepsilon}\,. \tag{5.17}$$

The internal ageing variables must likewise satisfy the inequality (5.13). Both ageing variables are formulated in the same manner and depend on the absolute temperature.

$$\dot{q}_{\text{eq}-} = f_{\text{eq}-}\left(q_{\text{eq}-}, \theta, \ldots\right) \tag{5.18}$$

$$\dot{q}_{\text{eq}+} = f_{\text{eq}+}\left(q_{\text{eq}+}, \theta, \ldots\right). \tag{5.19}$$

This is common practice and already published in literature for many years, cf. Johlitz (2012) or Johlitz and Lion (2013).

The new part of the modelling in this article concerns the ageing of viscoelasticity. There are hardly any approaches and studies on this yet. Taking into account the ageing behaviour of the rubber which is related to the overstress, the evaluation of the Clausius–Duhem inequality provides the additive decomposition of the overstress into a softening and a stiffening part

$$\sigma_{ov} = \sigma_{ov-} + \sigma_{ov+} \, . \tag{5.20}$$

The stress-softening part reads

$$\sigma_{ov-} = \hat{E}_- \, (1 - q_{ov-}) \, (\varepsilon - \varepsilon_{in-}) \tag{5.21}$$

with the corresponding evolution equation of the internal variable

$$\dot{\varepsilon}_{in-} = \frac{1}{\eta_-(p_-)} \sigma_{ov-} \, . \tag{5.22}$$

In this evolution equation, another internal ageing variable p_- is introduced which has to be formulated in the same manner like Eqs. (5.18) and (5.19). For describing the stiffening, another hypoelastic approach of the stress–strain relationship can be achieved to

$$\dot{\sigma}_{ov+} = \hat{E}_+ \, q_{ov+} \, (\dot{\varepsilon} - \dot{\varepsilon}_{in+}) \tag{5.23}$$

with the corresponding evolution equation:

$$\dot{\varepsilon}_{in+} = \frac{1}{\eta_+(p_+)} \sigma_{ov+} \, . \tag{5.24}$$

Again, another internal variable p_+ was introduced. In summary, the evolution equations, which describe both the equilibrium part and the overstress part of the model as a function of ageing, can be recapitulated as follows:

$$\dot{q}_{ov-} = f_{ov-} \, (q_{ov-}, \theta, \ldots) \tag{5.25}$$

$$\dot{q}_{ov+} = f_{ov+} \, (q_{ov+}, \theta, \ldots) \tag{5.26}$$

$$\dot{p}_- = g_- \, (p_-, \theta, \ldots) \tag{5.27}$$

$$\dot{p}_+ = g_+ \, (p_+, \theta, \ldots) \, . \tag{5.28}$$

5.3.5 *Calculation of the Model Responses Regarding Quasi-Static and Harmonic Excitations as Well as Relaxation After Ageing*

The validity of the derived model will now be demonstrated in this subchapter. In order to be able to recalculate the short-term model responses analytically, the

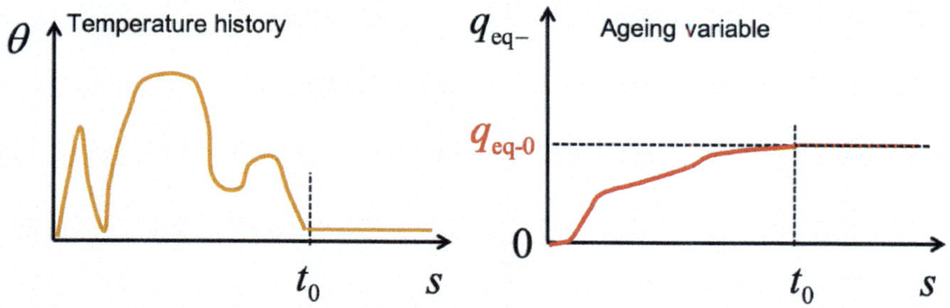

Fig. 5.3 Schematical overview of the temperature history (left) and ageing variable (right)

assumption and simplification is made that all ageing variables are assumed to be constant during such a test. The ageing variables develop depending on the temperature history up to a point in time t_0 as shown in Fig. 5.3. At this time t_0 an intermittent test is performed and during this test the ageing variables are then assumed to be constant. To make this clearer in the equations, they will be given a zero as index.

$$q_{eq-0} = \underset{s \leq t_0}{\Im_{eq-}} (\theta(s)) = \text{const.} \tag{5.29}$$

The same is done for all other ageing variables.

Now, the total stress is divided additively into four parts

$$\sigma = \sigma_{eq-} + \sigma_{eq+} + \sigma_{ov-} + \sigma_{ov+} \tag{5.30}$$

with the equilibrium stresses

$$\sigma_{eq-} = E_- \left(1 - q_{eq-0}\right) \varepsilon \tag{5.31}$$

and

$$\dot{\sigma}_{eq+} = E_+ \, q_{eq+0} \dot{\varepsilon}, \tag{5.32}$$

respectively. The degenerative part of the overstress is given by

$$\sigma_{ov-} = \hat{E}_- \left(1 - q_{ov-0}\right) \left(\varepsilon - \varepsilon_{in-}\right) \tag{5.33}$$

with the corresponding evolution equation

$$\dot{\varepsilon}_{in-} = \frac{1}{\eta_-(p_{-0})} \sigma_{ov-} \tag{5.34}$$

and the stiffening part of the overstress by

$$\dot{\sigma}_{\text{ov}+} = \hat{E}_+ \, q_{\text{ov}+0} \, (\dot{\varepsilon} - \dot{\varepsilon}_{\text{in}+}) \tag{5.35}$$

with the corresponding evolution equation

$$\dot{\varepsilon}_{\text{in}+} = \frac{1}{\eta_+(p_{+0})} \, \sigma_{\text{ov}+} \, . \tag{5.36}$$

Consequence: Under the assumption of a constant state of ageing, the short-term model responses can be calculated in closed form. Hence, the quasi-static stress–strain curve changes with ageing to

$$\sigma = \left(E_- \left(1 - q_{\text{eq}-0}\right) + E_+ \, q_{\text{eq}+0}\right) \varepsilon \, . \tag{5.37}$$

Even the loss modulus depends on ageing

$$E''(\omega) = \hat{E}_+ \, q_{\text{ov}+0} \frac{\omega \, \tau_{+0}}{1 + (\omega \, \tau_{+0})^2} + \hat{E}_-(1 - q_{\text{ov}-0}) \frac{\omega \, \tau_{-0}}{1 + (\omega \, \tau_{-0})^2} \tag{5.38}$$

with the relationships

$$\tau_{+0} = \frac{\eta_+(p_{+0})}{\hat{E}_+ \, q_{\text{ov}+0}} \tag{5.39}$$

and

$$\tau_{-0} = \frac{\eta_-(p_{-0})}{\hat{E}_-(1 - q_{\text{ov}-0})} \, . \tag{5.40}$$

The ageing-dependent stress relaxation solution is derived to

$$\sigma(t) = \varepsilon_0 \left(E_-(1 - q_{\text{eq}-0}) + E_+ \, q_{\text{eq}+0} + \hat{E}_+ \, q_{\text{ov}+0} \, e^{-\frac{t}{\tau_{+0}}} + \hat{E}_-(1 - q_{\text{ov}-0}) \, e^{-\frac{t}{\tau_{-0}}} \right) \tag{5.41}$$

and at least the ageing-dependent storage modulus reads as

$$\begin{aligned} E'(\omega) = {} & E_-(1 - q_{\text{eq}-0}) + E_+ \, q_{\text{eq}+0} \\ & + \hat{E}_+ \, q_{\text{ov}+0} \frac{(\omega \, \tau_{+0})^2}{1 + (\omega \, \tau_{+0})^2} + \hat{E}_-(1 - q_{\text{ov}-0}) \frac{(\omega \, \tau_{-0})^2}{1 + (\omega \, \tau_{-0})^2} \, . \end{aligned} \tag{5.42}$$

5.4 Simulations

The equations summarized in the last subsection are now used to simulate a few standard tests for various ageing times.

5.4.1 *Traditional Ageing Approach: Ageing Only Influences the Equilibrium Stress*

First, the traditional and already known influence of ageing is simulated with regard to the basic elasticity, i.e. only the ageing variables $q_{\mathrm{eq}-0}$ and $q_{\mathrm{eq}+0}$ have values greater than zero. Depending on the material, one or the other of the two variables can dominate the process. Ageing times from 0 d to 54 d are taken into account.

Figure 5.4 is showing the dependence of the storage modulus E' on ageing. In this case, the parameter set was chosen in this manner that the stiffening effect is dominating the process. One can see that ageing results into an ageing time-dependent vertical shift of the storage modulus (left). The loss modulus is not affected by ageing (right) since it contains only the viscoelastic part of the model.

5.4.2 *Ageing-Induced Increase in the Relaxation Times*

In this subsection, the relaxation times are formulated in dependence of the ageing variables. Again, the stiffening effect is assumed to be the dominating process. Figure 5.5 shows the corresponding results of the storage and loss modulus. Now, ageing-dependent horizontal shifts of both dynamic moduli can be seen.

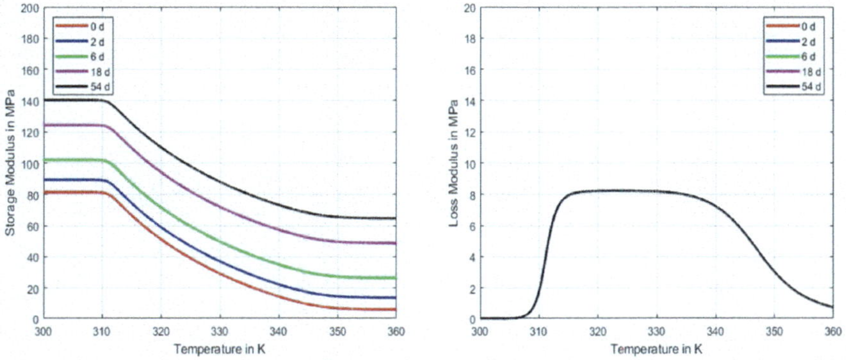

Fig. 5.4 Temperature dependence of storage modulus (left) and temperature dependence of loss modulus (right)

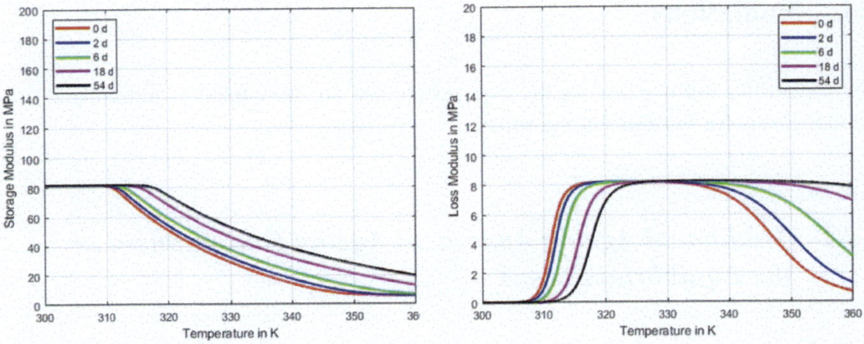

Fig. 5.5 Temperature dependence of storage modulus (left) and temperature dependence of loss modulus (right)

The glass-transition temperature automatically shifts to lower frequencies or higher temperatures and there are no vertical shifts of the dynamic moduli.

5.4.3 *Ageing-Induced Increase in the Overstress Moduli*

In a last step, the ageing-induced increase of the overstress moduli is investigated. Again, the same simulation with respect to different ageing times was done. Figure 5.6 shows the outcoming results of these simulations. It can be clearly seen that there are ageing-dependent horizontal shifts of both dynamic moduli. The glass-transition temperature again shifts to lower frequencies or higher temperatures as seen in the results of the last subsection and there are no vertical shifts of the dynamic moduli.

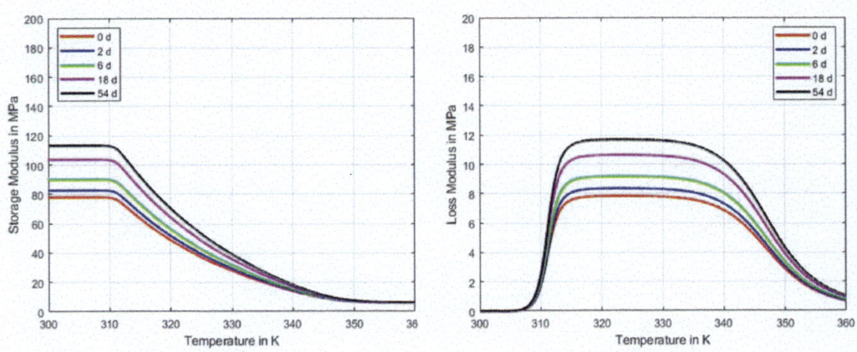

Fig. 5.6 Temperature dependence of storage modulus (left) and temperature dependence of loss modulus (right)

5.5 Conclusion and Outlook

In this contribution, a thermodynamical consistent model to represent the impact of chemical ageing on static and dynamic material properties of polymers was developed. Both the parameters of the equilibrium stress and the elasticity parameters of the overstress and the relaxation times or viscosities can be formulated in dependence of ageing parameters. This model was motivated by a number of preliminary tests and showed qualitatively that it can describe the experimentally observed relationships. Nevertheless, a lot of work has to be done. At first, a meaningful database has to be derived in order to guarantee the identification of the model parameters and a model refinement if necessary. Both things are not done yet. Taking rubber into account, it is obvious that it undergoes large deformations. Hence, the model has to be transferred to finite deformations. Nevertheless, a computer-based simulation by using the Finite-Element Method (FEM) is necessary, since there is no chance to solve three-dimensional nonlinear and coupled equations in a closed analytical form. Advanced computational techniques can enable accelerated ageing simulations, allowing researchers to predict long-term degradation behaviour based on short-term experimental data. By extrapolating degradation kinetics and ageing mechanisms, these simulations can facilitate the rapid assessment of material durability and reliability.

Nevertheless, this one-dimensional formulation is very valuable for gaining a basic understanding of the relationships involved in ageing.

Another idea could be that future models are likely to integrate multi-scale approaches, incorporating molecular-level interactions, mesoscale structure and macroscale behaviour. By capturing phenomena across multiple length and time scales, these models can provide more comprehensive predictions of polymer degradation and lifespan and can help to interpret the physical meaning of internal variables.

Also machine learning techniques may offer the potential to extract patterns and relationships from vast datasets of experimental and simulation data. These data-driven models can enhance predictive capabilities by identifying complex correlations between material properties, environmental conditions and ageing mechanisms.

A challenge is the integration of various physical phenomena such as mechanical loading, chemical degradation and environmental exposure into coupled models. On the one side, this can offer a more realistic representation of polymer ageing processes since coupled physical-based models enable the simulation of complex interactions and feedback mechanisms, leading to more accurate predictions of lifespan under real-world conditions. But on the other side, it is a very stony path to determine such a model through experiments and to determine the model parameters.

In essence, the future of computationally predicting polymer lifespan holds immense potential for advancing our understanding of ageing mechanisms, optimizing material performance and accelerating the development of durable and sustainable polymer-based products. As computational capabilities continue to evolve, these predictive models are poised to play a pivotal role in shaping the future of materials science and engineering.

References

Andrews R, Tobolsky A, Hanson E (1946) The theory of permanent set at elevated temperatures in natural and synthetic rubber vulcanizates. J Appl Phys 17:352–361

Budzien J, Rottach D, Curro J, Lo C, Thompson A (2008) A new constitutive model for the chemical ageing of rubber networks in deformed states. Macromolecules 41:9896–9903

Coleman BD, Noll W (1961) Foundations of linear viscoelasticity. Rev Mod Phys 33:239–249

Coleman BD, Noll W (1963) The thermodynamics of elastic materials with heat conduction and viscosity. Arch Rat Mech Anal 13:167–178

Duarte J, Achenbach M (2007) On the modelling of rubber ageing and performance changes in rubbery components. Kaut Gummi Kunstst 60:172–175

Ehrenstein G, Pongratz S (2007) Beständigkeit von Kunststoffen. Carl Hanser Verlag

Johlitz M (2012) On the representation of ageing phenomena. J Adhesion 88(7):620–648

Johlitz M (2015) Zum alterungsverhalten von polymeren: Experimentell gestützte, thermo-chemomechanische modellbildung und numerische simulation. Universität der Bundeswehr München, Habilitation

Johlitz M, Lion A (2013) Chemo-thermomechanical ageing of elastomers based on multiphase continuum mechanics. Continuum Mech Thermodyn 25(5):605–624

Karekar A, Schicktanz C, Tariq M, Oßwald K, Reincke K, V Cepus B, Saalwächter K (2023) Effects of artificial weathering in nr/sbr elastomer blends. Polymer Degrad Stab 208

Kruse E, Chauvet L (2020) Lagerlösungen für elektromotoren. ATZ - Automobiltechn Zeitschrift 122:50–55

Kurian A, Thakur S, Kunde S, Sankpal K, Wagh S (2021) Challenges in electric vehicle nvh. In: 2021 IEEE transportation electrification conference (ITEC-India)

Lion A (2000) Thermomechanik von Elastomeren. Berichte des Instituts für Mechanik der Universität Kassel (Bericht 1/2000)

Lion A, Johlitz M (2012) On the representation of chemical ageing of rubber in continuum mechanics. Int J Solids Struct 49(10):1227–1240

Lion A, Johlitz M (2020) A mechanical model to describe the vibroacoustic behaviour of elastomeric engine mounts for electric vehicles. Mech Syst Signal Proc 144

Musil B, Johlitz M, Lion A (2018) On the ageing behaviour of NBR: chemomechanical experiments, modelling and simulation of tension set. Continuum Mech Thermodyn. https://doi.org/10.1007/s00,161-018-0728-5

Musil B, Johlitz M, Lion A (2020) On the ageing behaviour of NBR: chemomechanical experiments, modelling and simulation of tension set. Contin Mech Thermodyn 369–385. https://doi.org/10.1007/s00161-018-0728-5

Perera D (2003) Physical ageing of organic coatings. Prog Org Coat 47:61–76

Petríková I, Bohdana M, Duc NP (2011) Influence of thermal ageing on mechanical properties of styrene-butadiene rubber. In: Proceedings of the 7th European conference on constitutive models for rubber

Tayefi M, Eesaee M, Hassanipour M, Elkoun S, David E, Nguyen-Tri P (2023) Recent progress in the accelerated aging and lifetime prediction of elastomers: a review. Polymer Degr Stab 214

Tobolsky AV, Prettyman IB, Dillon JH (1944) Stress relaxation of natural and synthetic rubber stocks. J Appl Phys 15:380–395

Part II
Postgraduate Lectures

Chapter 6
Fabrication and Comparison of Recycled PLA with Recycled PLA/Hemp Composite Using the FFF Process

Anusha Arulalan, Christopher Rockinger, and Markus Merkel

Abstract Bio-composites were previously created with conventional techniques, but recent developments in additive manufacturing have spurred a great deal of study and use in this area. Additive manufacturing layers materials methodically to create three-dimensional objects. In this study, injection molding and fused deposition modeling were utilized to create tensile specimens from hemp and recycled polylactic acid (PLA), and their mechanical properties were assessed. Furthermore, injection-molded tensile specimens composed entirely of recycled polylactic acid were used to assess their mechanical characteristics and contrast them with those of the new natural fiber-reinforced composites (NFRC). These specimens underwent tensile tests to determine their mechanical characteristics. The results imply that these bio-composites could be used as effective, sustainable substitutes for traditional synthetic fibers and plastics, which are highly hazardous to the environment.

Keywords Natural fiber-reinforced composite · Additive manufacturing · Polylactic acid · Hemp fiber · Fused filament fabrication

6.1 Introduction

Hague et al. (2003) highlight that designing a consumer product with biocomposites is a great approach to make it more sustainable and environmentally friendly. Biocomposites blend a biodegradable polymer matrix with natural fibers such as flax, hemp, or bamboo. These materials have a number of advantages, including

A. Arulalan (✉) · C. Rockinger · M. Merkel
Institute for Virtual Product Development, Aalen University of Applied Sciences, Beethovenstraße 1, 73430 Aalen, Germany
e-mail: anusha.arulalan@hs-aalen.de

C. Rockinger
e-mail: christopher.rockinger@hs-aalen.de

M. Merkel
e-mail: markus.merkel@hs-aalen.de

H. Altenbach et al. (eds.), *Lectures Notes on Advanced Structured Materials 3*, Advanced Structured Materials 226 https://doi.org/10.1007/978-3-031-77403-4_6

improved biodegradability, reduced environmental impact, and a particular visual attractiveness. Additive manufacturing (AM) decreases production costs, material waste, the design-manufacturing cycle, and so on. Galantucci et al. (2010), Brischetto et al. (2018) discuss how Fused Filament Fabrication (FFF) enables the production of geometrically complex parts for a variety of applications such as automotive, aerospace, and construction. FFF material extrusion is a commonly used additive manufacturing technology in which filaments are heated and extruded via nozzle tips and deposited layer by layer on the platform to produce the parts (Brooks and Molony 2016; Masood 2014). Joshi et al. (2004) found that natural fiber composites are likely to be more environmentally friendly than glass fiber composites in most situations. Their production has a lesser impact on the environment than glass fiber production. They have greater amounts of fiber for comparable performance, reducing their harmful base polymer content. Also, the lightweight nature of natural fiber composites increases fuel efficiency and lowers emissions during the use phase of the product. Du Maire et al. (2022) emphasize that the manufacture of intricate and customized goods, additive manufacturing has revolutionary potential. With the help of this technology, materials like polyamide 12 powder, which can be recycled several times, can be used more effectively and sustainably in production processes.

Even though there are difficulties related to the mechanical qualities deteriorating with repeated reuse, research on optimizing powder reuse is still advanced to save waste and maintain high production standards. Xiao et al. (2019) studied hemp hurd (HH)/polylactide (PLA) biocomposites by FFF and found that, in comparison to the commercially available PLA filament, FFF printed items with HH loading under 30 phr displayed higher impact toughness. With increasing HH loading, the FFF-printed samples also showed improved dimensional accuracy. Behera et al. (2022) treated hemp fibers with sodium hydrogen carbonate and coated them with PHB and PLA, which in turn showed a visible change in fiber surface, surface roughness and enhanced the interfacial adhesion between the coated fibers and epoxy matrix. Wang et al. (2022) successfully created a PLA/hemp bio-composite with a blending ratio of 30 wt.% hemp fiber achieving comparable tensile strength to virgin PLA. All of the aforementioned publications demonstrate that natural fiber-reinforced composite can be used for improved material characteristics, and additive manufacturing will also be a great choice for applications where PLA with better strength and less plastic is necessary. This paper investigates the use of NFRC in an FFF machine to create consumer products.

6.2 Materials

The material selection and creation are one of the crucial steps determining the strength of the consumer product. Incorporating sustainable materials can significantly lessen the environmental impact of manufacturing processes. Sustainable materials are produced from renewable resources and leave little environmental impact (Narayanan and Gunasekera 2019). PLA is chosen because they are biodegradable, renewable, nontoxic, has antimicrobial properties, oxygen barrier properties, good mechanical strength, light transmission, rigidity, biocompatibility, low cost, high stiffness, and flexibility. Thermoplastic materials that are bio-degradable like polylactic acid (PLA) are a good choice for a consumer product.

Table 6.1 compares fibers like abaca, bamboo, banana, coir, cotton, flax, hemp, and jute by density, diameter, moisture, tensile strength, and elongation. A fiber's properties dictate its application suitability, emphasizing the need for materials that blend environmental sustainability with mechanical needs. Hemp is one of the strongest natural fibers accessible for long-lasting uses like textiles and building materials because of its remarkably high tensile strength, which may reach up to 690 MPa compared to bamboo, coir, cotton, and even abaca. Its sustainable growing qualities, which include low pesticide requirements and a small environmental impact, also make it a great option for environmentally friendly initiatives. In the realm of materials for biocomposites, the combination of natural hemp fibers and recycled polylactic acid stands as a significant step towards sustainability. Hemp grows without the need for pesticides or herbicides, aligning perfectly with environmentally responsible practices (Atmakuri et al. 2022). Its inherent stiffness and tensile strength make it an excellent choice for durable biocomposites, and its low water affinity improves water resistance (Sawpan et al. 2011). Hemp's versatility and widespread cultivation across different climates enhance its economic feasibility. Moreover, its mechanical properties surpass those of many other natural fibers, reducing reliance on petroleum-derived materials (Cigasova et al. 2015). Complementing hemp's attributes, recycled PLA offers a sustainable biopolymer option. Derived from renewable sources, PLA is biodegradable, recyclable, and compostable, replacing petroleum-based polymers (Ilyas et al. 2021). When combined with natural fibers, PLA-based composites gain acceptance across various applications due to their biodegradability, recyclability, high mechanical strength, low toxicity, and impressive barrier properties (Ilyas et al. 2022). PLA's renewability and biodegradability align with environmentally responsible practices, and its recyclability ensures long-term sustainability (Li et al. 2023). These composites find utility in diverse industries, including packaging, textiles, biomedicine, structural applications, and automotive sectors, thanks to their robust mechanical strength and barrier properties (Ilyas and Sapuan 2020). PLA's adaptability to various processing techniques, such as thermoforming and injection molding, adds to its versatility (Trivedi et al. 2023). This underscores its status as a sustainable material choice for biocomposite production. In summary, the synergy between hemp fibers and recycled PLA epitomizes sustainable materials sourcing for biocomposites.

Table 6.1 Different fibers with their properties

Fiber	Density (g/cm^3)	Diameter (μm)	Moisture content (%)	Tensile strength (MPa)	Elongation at break (%)	Reference
Abaca	1.5	10–30	5–10	400	3–10	Faruk et al. (2012), Yan et al. (2016), Dittenber and GangaRao (2012)
Bamboo	0.6–1.1	240–330	9.16	140–230	1.40	Faruk et al. (2012), Yan et al. (2016)
Banana	1.35	50–250	10.71	529–914	3	Reddy and Yang (2005), Badrinath and Senthilvelan (2014), Cherian et al. (2010),
Coir	1.2–1.5	100–450	8–11.36	175–180	30	Yan et al. (2016), Dittenber and GangaRao (2012),
Cotton	1.5–1.6	12–35	7.85–8.5	287–597	7–8	Yan et al. (2016), Dittenber and GangaRao (2012), Ramesh et al. (2017),
Flax	1.5	5–38	1.2–8	345–1036	2.7–3.2	Faruk et al. (2012), Yan et al. (2016),
Hemp	1.48	–	6.2–12	690	1.6	Faruk et al. (2012), Yan et al. (2016),
Jute	1.3–1.5	20–200	12.5–13.7	200–773	1.5–1.8	Faruk et al. (2012), Yan et al. (2016)

6.3 Manufacturing and Testing Methodologies

Sourcing

Figure 6.1a shows the raw carded hemp fiber with an average length of 2–3 mm from the suppliers "Nina Heitmann Felde Fibres". Figure 6.1b shows the previously used PLA filaments that are re-made as pellets.

Hemp/PLA Filament

PLA/hemp filament with a ratio of 9–11% hemp is extruded with a diameter of 1.75 mm with a sprue speed of 95 rpm. The extruded filament is wounded into a spool.

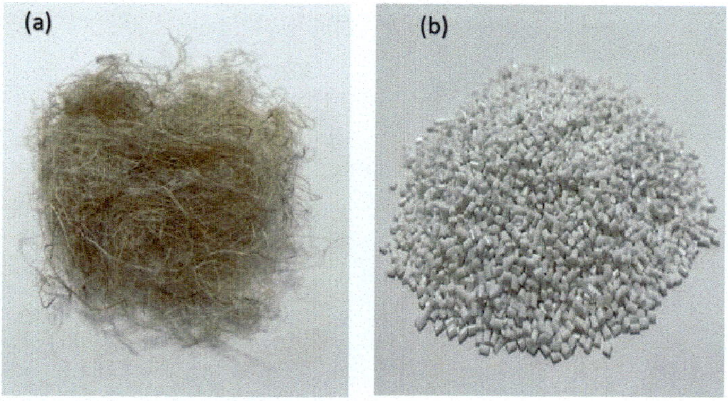

Fig. 6.1 **a** Photograph of raw hemp fiber and **b** sample of recycled PLA

Preparation of the Materials for Extrusion

The methodology for this research entails a systematic approach to create a biocomposite material using hemp fiber and recycled polylactic acid. Prior to the extrusion process, the hemp fiber is dried for 8 h at 50 °C to reduce its moisture content, which enhances processability. The recycled PLA is also subjected to a drying process for 8 h at 60 °C.

Extruding the Materials

The two materials are then combined during the extrusion process; to do this, the Leistritz extrusion machine is adjusted to 211 °C by trial and error, with a sprue speed of 95 rpm and a pressure of 83 bar. The mass of fiber used is 480.5 g and the mass of plastic used is 4590 g. PLA pellets are introduced into the extruder, and hemp fiber is gradually incorporated as the PLA flows through the machine. During this process, the materials merge and are subsequently cooled by flowing water, followed by alignment. The wet filament is fed into a machine that measures filament diameter and optimizes it to around 1.80 mm, which is the ideal diameter for FFF machines.

Preparing the Filament for Further Production

The filament is then wound onto a spool as it dries as shown in Fig. 6.2, creating the basis for further applications. Understanding the complex makeup of the filament requires accuracy in calculating the fiber-to-plastic ratio. A ratio of around 10.47% is obtained by dividing the mass of the plastic by the overall weight of the fiber. This indicates that the mass of the fiber makes up approximately 10.47% of the total mass of the plastic in the filament.

SEM Analysis

Ten different samples were painstakingly made using additive manufacturing, namely, the method of producing biocomposite samples with dimensions of Ø10 × 10 mm each, to enable a thorough examination of porosity properties. This project

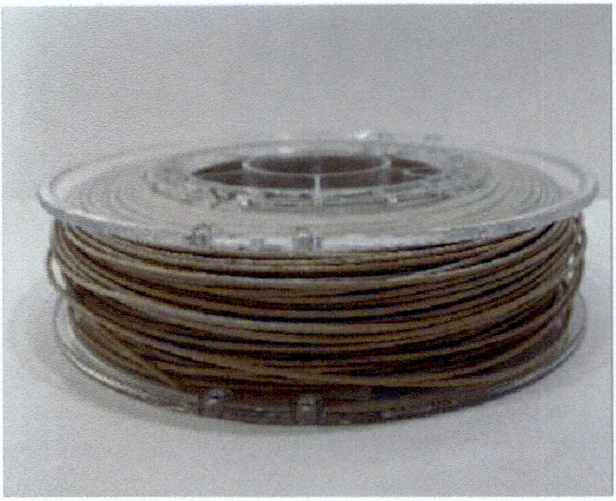

Fig. 6.2 Picture of created hemp and PLA filament

was to examine the details of porosity in the artificial specimens. The samples were created using the same parameters as those used for printing tensile specimens. After production, the samples were cut into cross-sections perpendicular to the printing direction, enhancing the visibility of the internal structure of the printed layers for detailed analysis. The specimens were fabricated using the newly developed NFRC material. In examining the two SEM images, it is noted that both samples were made from the same material and under identical production conditions. However, Fig. 6.3a exhibits noticeable porosity, while Fig. 6.3b does not show any discernible porosity in the cross-sectional plane.

The difference in porosity could be attributed to intrinsic variations in the filament production process, which emphasizes the need for improved filament extrusion methods to enhance material consistency. This consistency is critical as it directly impacts the overall quality of additively manufactured components. This focus on material homogeneity is an important factor to consider in the continuous effort to optimize the performance and quality of additively manufactured components.

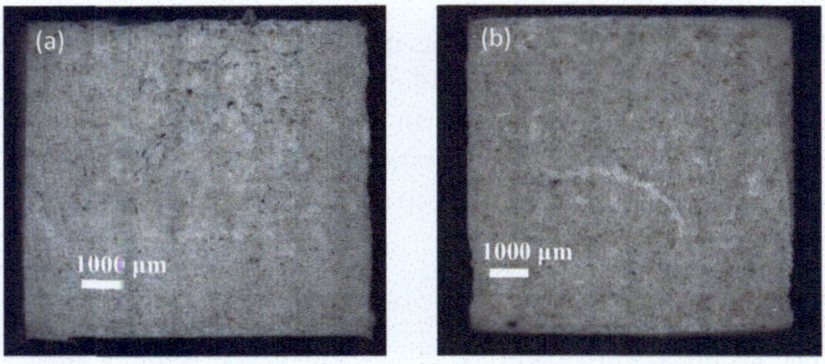

Fig. 6.3 Image at 10 times magnification: high porosity test specimen (**a**), low porosity test specimen (**b**)

Creating Tensile Specimens with the Composite by Injection Molding (IM)

The filament is further processed for injection molding, where specific mold dimensions according to DIN EN ISO 527-2 are used and process parameters are fine-tuned to achieve the desired molding outcomes. For injection molding, the filament was broken down into pellets and fed into the hopper which makes the fibers to be mostly linearly oriented along the direction of flow under high pressure. Through a process of trial and error, the ideal settings were determined to be injection pressure of 1500 bar, reprint pressure of 650 bar, reprint time of 15 s, and cooling time of 10 s. Additionally, recycled PLA was also injection molded for tensile testing to determine the improvements of the filament generated. Ten samples each were molded for both recycled PLA and the hemp/PLA composite.

Creating Tensile Specimens with the Filament by Additive Manufacturing

Finally, additive manufacturing takes place through FFF, utilizing a Prusa i3 3D printer with defined nozzle and bed temperatures as shown in Table 6.2 after a set of trial and error processes for the newly created material. In FFF, the material is deposited layer over layer which can change the orientation of the fibers along the path of the extruder. The schematic representation of the orientation with the polar angle and azimuth angle is given in Fig. 6.4. Specimens are manufactured from the bio-composite material as shown in Fig. 6.5, adhering to a specified orientation (90°/0°).

This method offers a sustainable and environmentally friendly solution for creating biodegradable components, with potential applications spanning various industries, showcasing the potential for more eco-conscious material production.

Table 6.2 shows the parameters that were effective for the composite filament that was created.

Tensile Testing

Figure 6.6 depicts an image of a tensile test setup for all the tested samples.

The bio-composite material and recycled PLA were injection molded/additively manufactured and evaluated using 10 samples each. The tests were conducted using a standard tensile test machine with the testing standards according to DIN EN ISO

Table 6.2 Printing parameters for 3D printing

Speed	50 mm/s
Nozzle temperature	220 °C
Bed temperature	65 °C
Fan speed	150%
Nozzle diameter	0.8 mm
Layer thickness	0.2 mm
Shell thickness	0.5 mm
Infill density	100%

Description	Polar Angle (θ)	Azimuth Angle (φ)
Specimen (Along x-axis)	90°	0°
Nozzle (From z-axis)	0°	-

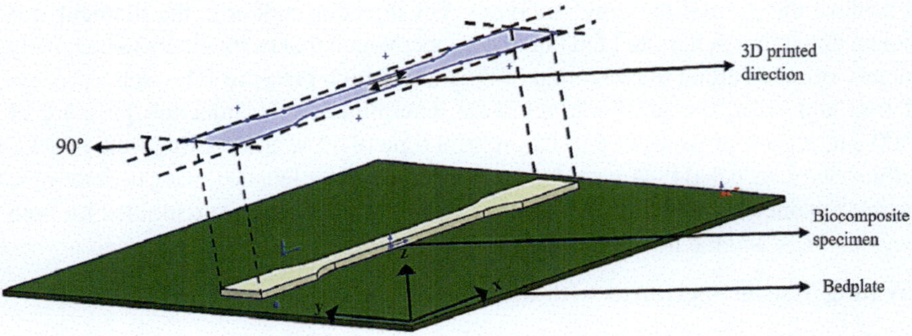

Fig. 6.4 Schematic representation of orientation of the tensile specimen

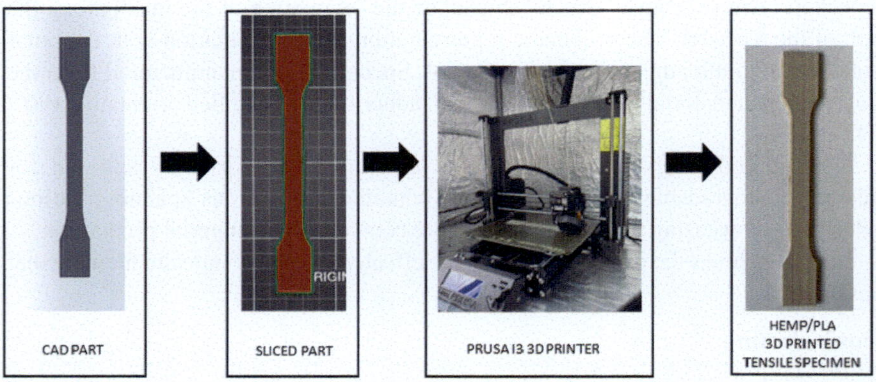

Fig. 6.5 Process of 3D printing

527-1, involving global strain measurement. The tests were performed without the use of a thermo-chamber, and the specimens used were flat samples.

6.4 Results and Discussion

Table 6.3 shows the results obtained from the mechanical tests of the created composite by IM and AM and will be discussed.

A comparative analysis of the mechanical properties of bio-composites produced by injection molding (IM) and additive manufacturing (AM) reveals that the method of manufacture has a major impact on the final product's performance attributes.

Fig. 6.6 Tensile test setup for biocomposite

Because of the nature of the high-pressure material injection into a mold that produces high degrees of fiber orientation along the flow direction, injection molding typically yields composites with better mechanical qualities. According to the provided data, the injection-molded bio-composite has an elastic modulus of 3641.18 MPa and a yield stress of 51.2169 MPa, compared to the additively manufactured composite's 3589.35 and 38.2233 MPa, respectively. This alignment improves the material's ability to bear load.

This discrepancy might be explained by the fact that fibers are oriented during injection molding to optimize their mechanical contact with the polymeric matrix. The greater yield strain (1.77%) in the IM bio-composite, as opposed to the AM bio-composite (1.72%), indicates that this alignment maximizes the transfer of stress between the matrix and the fibers, which not only increases the strength and stiffness but also improves the composite's capacity to withstand external loads before yielding. Furthermore, the uniform fiber distribution obtained by injection molding facilitates improved stress distribution throughout the material, leading to higher ultimate tensile strength and elongation capabilities prior to failure; values of 2.12% for ultimate strain and 49.0686 MPa for ultimate tensile strength in IM composite

Table 6.3 Tensile test results

Property	Biocomposite-IM	Recycled PLA	Biocomposite-AM
Elastic modulus (MPa)	3641.18 ± 659.66	3005.46 ± 554.64	3589.35 ± 774.21
Yield stress (σ_y) (MPa)	51.22 ± 0.65	32.84 ± 0.00	38.22 ± 0.00
Maximum stress (σ_max) (MPa)	51.06 ± 0.66	36.87 ± 3.40	39.28 ± 1.71
Ultimate tensile strength (σ_b) (MPa)	49.07 ± 1.36	36.83 ± 3.40	39.18 ± 1.77
Yield strain (ε_y) (%)	1.77 ± 0.05	1.28 ± 0.00	1.72 ± 0.18
Maximum strain (ε_max) (%)	1.69 ± 0.14	1.15 ± 0.14	1.58 ± 0.28
Ultimate strain (ε_b) (%)	2.12 ± 0.43	1.15 ± 0.14	1.62 ± 0.22
Maximum force (F_max) (N)	1106.21 ± 14.21	798.86 ± 73.72	851.07 ± 37.00
Breaking force (F_b) (N)	1063.06 ± 29.53	797.94 ± 73.62	848.80 ± 38.31

highlight this benefit. Though AM is comparatively lesser, it gives better results compared to recycled PLA. This aspect alone makes the manufactured biocomposite to be better for many applications.

6.5 Result Analysis

Comparing the results of injection molded biocomposite and PLA, it has been found that, when compared to PLA samples, the created composite samples consistently show good mechanical properties, such as higher yield stress, breaking force, maximum stress, yield strain, maximum strain, and ultimate tensile strength as discussed in Table 6.3. The results of this study highlight the composite material's superior deformation capabilities and strength under particular testing conditions. Figure 6.7 highlights the stress–strain points of each sample across all materials, demonstrating that NFRC exhibits superior properties compared to R-PLA. This is particularly evident as the stress and strain values for both NFRC IM and NFRC Am are consistently higher than those of R-PLA, underscoring the enhanced mechanical performance of NFRC. This provides the significance of taking these differences into account when choosing materials for various applications where mechanical performance plays a crucial role in determining suitability.

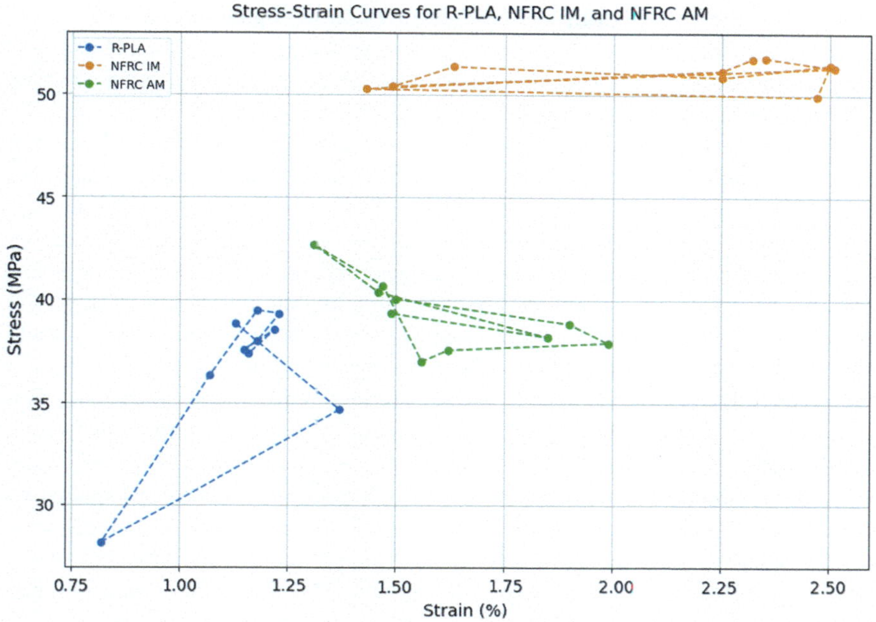

Fig. 6.7 Stress–strain points of samples for R-PLA, NFRC IM, NFRC AM

In summary, the decision between 3D-printed biocomposites and injection-molded biocomposites is based on the particular mechanical needs of the intended use. The IM biocomposite shows promise for applications where high mechanical performance is essential due to its generally superior mechanical qualities, especially in yield stress, maximum stress, ultimate tensile strength, maximum force, and breaking force. The elastic modulus, yield strain, maximum strain, and ultimate strain, however, are similar for both approaches. As a result, while choosing a fabrication technique, one should carefully analyze the target application's unique requirements as well as the desired mechanical qualities.

Young's Modulus

The materials were subjected to controlled tensile loading following the fabrication of standardized specimens and strict adherence to testing protocols. The applied force and subsequent elongation were then meticulously quantified. With careful examination, several stress–strain curve segments associated with elastic deformation were found. These sections show the linear region in which the material exhibits elastic behavior or deformation proportionate to applied stress. Young's modulus, a measure of a material's stiffness or resistance to deformation, can be found by calculating the slope of this linear region. This modulus is a basic feature of materials that sheds light on how they will behave mechanically when subjected to tensile loads. Through precise characterization of Young's modulus, our work advances our knowledge of material behavior and helps develop a range of scientific and engineering applications. Recycled PLA in the injection molding process exhibited

Young's modulus of approximately 3005 MPa, highlighting its mechanical properties in a sustainable context. Moving to composite materials, both the IM and AM processes demonstrated enhanced stiffness, with IM composite yielding a modulus of 3641.18 MPa and AM composite registering at 3589.35 MPa.

A greater Young's modulus is frequently required in engineering and material science to guarantee structural integrity and performance in a variety of applications. This characteristic is important when stability and minimal deflection are important since it represents a material's stiffness and resistance to deformation under stress. When creating buildings, bridges, and other infrastructure components, for example, materials with larger Young's moduli are chosen in structural engineering because they can endure large loads and environmental stresses while retaining their shape and usefulness over time. Similar to this, materials with high Young's moduli are preferred in the aerospace and automotive industries because they can minimize weight while providing strength and rigidity, two important factors. Higher Young's modulus materials also improve dimensional stability and precision during the production process, guaranteeing constant performance and quality of machined parts and components. In comparison, both the IM and AM biocomposites exhibit superior properties over recycled PLA, particularly in terms of Young's modulus, yield stress, and ultimate tensile strength. This makes them more suitable for applications where higher stiffness, strength, and durability are required.

6.6 Conclusions

According to the experimental results, the biocomposite that combines PLA and hemp fibers works better under the specified conditions than only PLA. This implies that adding hemp fibers improves the composite material's functional or mechanical qualities. The biocomposite's enhanced performance suggests that it may find use in situations where mechanical properties like strength and durability are essential. These findings highlight the feasibility and potential of using natural fibers, such as hemp, in composite materials, providing a more environmentally friendly substitute for traditional polymers like PLA. Future developments in this field of study and research may result in the widespread use of biocomposites across a range of industries, thereby promoting environmental sustainability.

References

Atmakuri A, Palevicius A, Janusas G, Eimontas J (2022) Investigation of hemp and flax fiber-reinforced EcoPoxy matrix biocomposites: morphological, mechanical, and hydrophilic properties. Polymers-Basel 14(21). https://doi.org/10.3390/polym14214530

Brischetto S, Ferro CG, Torre R, Maggiore P (2018) 3D FDM production and mechanical behavior of polymeric sandwich specimens embedding classical and honeycomb cores. Curved Layer Struct 5(1):80–94. https://doi.org/10.1515/cls-2018-0007

Brooks H, Molony S (2016) Design and evaluation of additively manufactured parts with three dimensional continuous fibre reinforcement. Mater des 90:276–283. https://doi.org/10.1016/j.matdes.2015.10.123
Behera S, Gautam RK, Mohan S (2022) Polylactic acid and polyhydroxybutyrate coating on hemp fiber: Its effect on hemp fiber reinforced epoxy composites performance. J Compos Mater 56(6):929–939. https://doi.org/10.1177/00219983211066991
Badrinath R, Senthilvelan T (2014) Comparative investigation on mechanical properties of banana and sisal reinforced polymer based composites. Proc Mat Sci 5:2263–2272. https://doi.org/10.1016/j.mspro.2014.07.444
Cherian BM, Leão AL, de Souza SF, Thomas S, Pothan LA, Kottaisamy M (2010) Isolation of nanocellulose from pineapple leaf fibres by steam explosion. Carbohyd Polym 81(3):720–725. https://doi.org/10.1016/j.carbpol.2010.03.046
Cigasova J, Stevulova N, Sicakova A (2015) New biocomposites based on hemp hurds. ZNPRzBiS XXXII(3/II/2015):75–81. https://doi.org/10.7862/rb.2015.141
Dittenber DB, GangaRao HVS (2012) Critical review of recent publications on use of natural composites in infrastructure. Compos Part A-Appl Sci 43(8):1419–1429. https://doi.org/10.1016/j.compositesa.2011.11.019
Du Maire P, Sert E, Deckert M, Johlitz M, Öchsner A (2022) Effects of reusing polyamide 12 powder on the mechanical properties of additively manufactured parts. In: Altenbach H, Johlitz M, Merkel M, Öchsner A (eds) Lectures notes on advanced structured materials. Springer International Publishing, Cham, pp 155–165
Faruk O, Bledzki AK, Fink H-P, Sain M (2012) Biocomposites reinforced with natural fibers: 2000–2010. Prog Polym Sci 37(11):1552–1596. https://doi.org/10.1016/j.progpolymsci.2012.04.003
Galantucci LM, Lavecchia F, Percoco G (2010) Quantitative analysis of a chemical treatment to reduce roughness of parts fabricated using fused deposition modeling. CIRP Ann 59(1):247–250. https://doi.org/10.1016/j.cirp.2010.03.074
Hague R, Mansour S, Saleh N (2003) Design opportunities with rapid manufacturing. Assembly Autom 23(4):346–356. https://doi.org/10.1108/01445150310698643
Ilyas RA, Sapuan SM, Harussani MM, Hakimi MYAY, Haziq MZM, Atikah MSN et al. (2021) Polylactic acid (PLA) biocomposite: processing, additive manufacturing and advanced applications. Polymer 13(8). https://doi.org/10.3390/polym13081326
Ilyas RA, Zuhri MYM, Aisyah HA, Asyraf MRM, Hassan SA, Zainudin ES et al (2022) Natural fiber-reinforced polylactic acid, polylactic acid blends and their composites for advanced applications. Polymer 14(1). https://doi.org/10.3390/polym14010202
Ilyas RA, Sapuan SM (2020) Biopolymers and biocomposites: chemistry and technology. Curr Anal Chem 16(5):500–503. https://doi.org/10.2174/157341101605200603095311
Joshi SV, Drzal LT, Mohanty AK, Arora S (2004) Are natural fiber composites environmentally superior to glass fiber reinforced composites? Compos Part A-Appl S 35(3):371–376. https://doi.org/10.1016/j.compositesa.2003.09.016
Li X, Lin Y, Liu M, Meng L, Li C (2023) A review of research and application of polylactic acid composites. J Appl Polym Sci 140(7). https://doi.org/10.1002/app.53477
Masood SH (2014) Advances in fused deposition modeling. In: Proceedings of the Society of Photo Optical Instrumentation. Elsevier, pp 69–91
Narayanan RG, Gunasekera JS (2019) Introduction to sustainable manufacturing processes. In: Narayanan RG, Gunasekera JS (eds) Sustainable material forming and joining: sustainable manufacturing: material forming and joining, pp 1–18
Ramesh M, Palanikumar K, Reddy KH (2017) Plant fibre based bio-composites: Sustainable and renewable green materials. Adv Mater Res-Switz 79:558–584. https://doi.org/10.1016/j.rser.2017.05.094
Reddy N, Yang Y (2005) Biofibers from agricultural byproducts for industrial applications. Trends Biotechnol 23(1):22–27. https://doi.org/10.1016/j.tibtech.2004.11.002

Sawpan MA, Pickering KL, Fernyhough A (2011) Improvement of mechanical performance of industrial hemp fibre reinforced polylactide biocomposites. Compos. Part A-Appl Sci 42(3):310–319. https://doi.org/10.1016/j.compositesa.2010.12.004

Trivedi AK, Gupta MK, Singh H (2023) PLA based biocomposites for sustainable products: a review. Adv Ind Eng Polym Res. https://doi.org/10.1016/j.aiepr.2023.02.002

Wang J, Bai J, Hua H, Tang B, Bai W, Wang X (2022) Characterization and scalable production of industrial hemp fiber filled PLA bio-composites. J Nat Fibers 19(16):13426–13437. https://doi.org/10.1080/15440478.2022.2095549

Xiao X, Chevali VS, Song P, He D, Wang H (2019) Polylactide/hemp hurd biocomposites as sustainable 3D printing feedstock. Compos Sci Techno l 184:107887. https://doi.org/10.1016/j.compscitech.2019.107887

Yan L, Kasal B, Huang L (2016) A review of recent research on the use of cellulosic fibres, their fibre fabric reinforced cementitious, geo-polymer and polymer composites in civil engineering. Compos Part B-Eng 92:94–132. https://doi.org/10.1016/j.compositesb.2016.02.002

Chapter 7
Material Modeling of Selective Laser Sintered Polyamid 12 for the Deviatoric Stress State

Dominik Hahne, Alexander Lion, and Michael Johlitz

Abstract This paper examines how polyamide 12 (PA12) behaves when exposed to shear stress. PA12 is a strong, versatile thermoplastic which is widely used in many industries. This research aims to better understand how PA12 reacts to shear-induced stress. Samples are produced on an industrial scale EOS P396 SLS-machine with a robust and well-proven parameter set. The study includes tests with two different shear specimens under several excitation speeds and relaxation tests, to characterize the viscoelastic properties. Digital image correlation is used for precise deformation measurements. A modeling approach combining viscoelasticity and endochronic plasticity is presented to map the inelastic effects. These models are implemented in MATLAB, providing precise simulations that confidently predict the mechanical material behavior.

7.1 Introduction

Selective laser sintering (SLS) is a sophisticated additive manufacturing method that employs a laser to selectively fuse powdered materials, like nylon or metal, layer by layer. This technique is extensively utilized in industrial production for its capacity to fabricate intricate and robust components with good precision and reproducibility (Breuninger et al. 2013; Schmid 2014). The material used in this production technology is polyamide 12 (PA12), which holds a market share of approximately 95% and possesses favorable calorimetric properties (Schmid 2015). With the increasing use of this material in prototypes and series parts, there is a growing need for material modeling using engineering methods to carry out load-bearing capacity verifications.

D. Hahne (✉) · A. Lion · M. Johlitz
Universität der Bundeswehr, Werner-Heisenberg-Weg 39, 85577 Neubiberg, Germany
e-mail: dominik.hahne@unibw.de

A. Lion
e-mail: alexander.lion@unibw.de

M. Johlitz
e-mail: michael.johlitz@unibw.de

H. Altenbach et al. (eds.), *Lectures Notes on Advanced Structured Materials 3*,
Advanced Structured Materials 226, https://doi.org/10.1007/978-3-031-77403-4_7

Consequently, a material model will be developed to represent the experimental findings on a phenomenological level. As the use of the material for prototypes and series parts increases, so does the need for comprehensive mechanical characterization of the material. For this reason, a number of studies have been carried out in recent years that deal with the experimental characterization and modeling of the material. Mechanical experiments on SLS parts that correspond well with the data obtained here can be found in Lammens et al. (2017). Studies on the failure behavior under cyclic loading are presented in Van Hooreweder et al. (2009) and Van Hooreweder et al. (2010). More recent investigations have also focused on reinforced PA12 grades produced using the SLS process. A comparison between glass bead reinforced PA12, which is laser sintered and injection molded, can be found in De Coninck et al. (2024), where the increased stiffness and the decrease in fatigue strength of the glass bead filled parts are explicitly worked out. With regard to the modeling of the experimental results, there are a number of studies that use the Chaboche model (Chaboche and Rousselier 1983) for mechanical modeling (Schob et al. 2020; Sagradov et al. 2020). Experimental results for specimens under shear loading, which are modeled using the Chaboche model, can be found in Schob et al. (2024), where a failure model based on Gurson–Tvergaard–Needleman (GTN) was also used. Apart from the implementation of a failure model, the work presented here pursues a similar goal, but with a different modeling approach. Two sample geometries suitable for shear loading are presented. Experimental characterization using local strain measurement of the specimens and modeling based on a modeling approach combining viscoelasticity and endochronic plasticity are performed.

7.2 Specimen Design and Experimental Results

7.2.1 Production Strategy

The production strategy follows a random approach as already presented in Hahne et al. (2024). All specimens are produced on an EOS P396 selective laser sintering machine and are placed randomly inside the volume. The parameters are summarized in Table 7.1. The mechanical tests are carried out on a universal testing machine Z020, manufactured by the company Zwick/Roell. The sensor for the torsional moment has a nominal value of 100 Nm (*Gassmann Theiss M100*). The sensor for axial forces has a nominal value of 2.5 kN (*ZwickRoell Xforce HP*). The advantage of torsional testing over simple shear testing is the possibility of measuring and controlling the axial forces. During the torsional experiments, the axial force was controlled to zero. During simple shear tests, the normal stresses in the x- and y-direction are not taken into account.

Table 7.1 Set of parameters for manufacturing the specimens

Parameter	Unit	EOS P396
Laserspeed	mm/s	4000
Laserpower	W	40
Hatch distance	mm	0.3
Layer height	μm	120
Energy density	J/mm^3	0.28
Powder base layer	mm	6
Height printjob	mm	≈60–200
Temperature building volume	°C	130
Temperature building chamber	°C	173
Volume capacity utilization	%	≈10

7.2.2 *Specimen Design*

Mechanical tests are carried out with two types of specimens to adapt and validate a purely deviatoric material model. Torsion tests utilized a tube sample that was previously described in a publication but was modified to allow for better access during local strain measurement (Hahne et al. 2024). Additionally, simple shear tests were performed using a Miyauchi-type plate-shaped sample geometry, which was originally developed for metallic sheets (Miyauchi 1984). Figure 7.1 shows both geometries. The sample for simple shear has been designed as a two-dimensional geometry with a thickness of 2 mm. Two plugs can be seen in Fig. 7.1 for the pipe specimen, which are manufactured separately and subsequently fitted as a transition fit to stiffen the flattened clamping area of the sample.

7.2.3 *Kinematics of Specimens*

To determine the strain behavior of the samples, local strain measurements were conducted using digital image correlation (DIC) (Sharpe 2008). The rotationally symmetrical structure of the pipe sample necessitated a three-dimensional strain measurement. Despite the two-dimensional measuring range, a stereo camera system from the company LiMess was used for all measurements. This comprehensive approach was followed by an in-depth analysis using Istra4D software (version 4.4.7). Figure 7.2 provides a clear depiction of the basic distortion mechanisms of both samples, while Fig. 7.3 explains the measurement principle. By applying the angle φ or displacement x, shear deformation is achieved in the measuring range, transforming the dotted rectangles, which represent the undeformed state of the sample, into a rhombus. With the displacements, the shear strain γ can be calculated

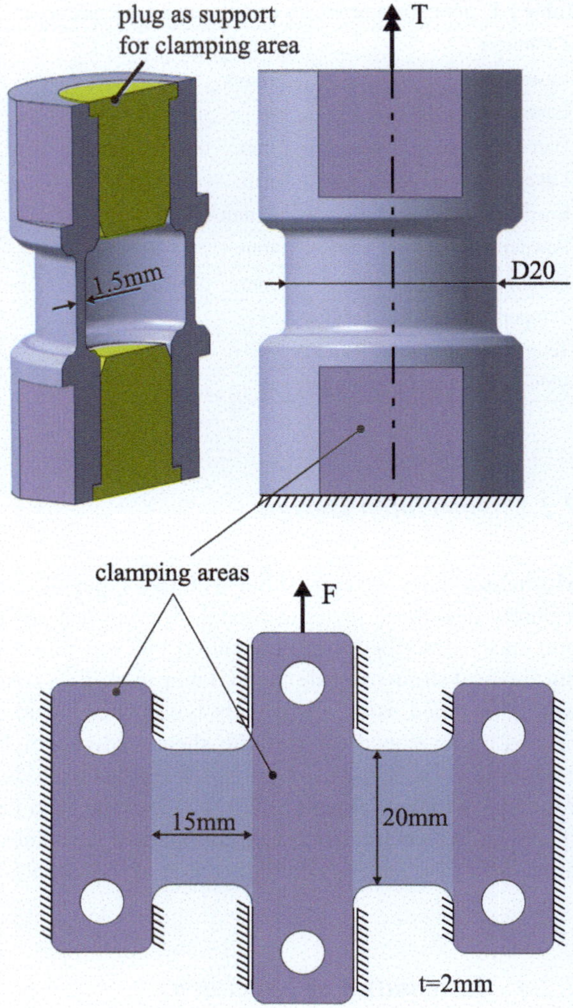

Fig. 7.1 Geometry of torsion and simple shear specimens

$$\gamma = \tan(\alpha) = \frac{u}{h}. \tag{7.1}$$

The local distortion measurement aims to establish a correlation between the path or angle of the machine crosshead and the local distortion in the area of the most considerable strain of the two sample geometries. All mechanical tests are carried out on the same machine mentioned at the beginning so that the relationship between the path and angle of the machine and the local strains is established based on a series of measurements and then assumed to be valid for all subsequent measurements. For this purpose, the local strain measurements are each repeated three times and averaged, provided that no unusual outliers are detected.

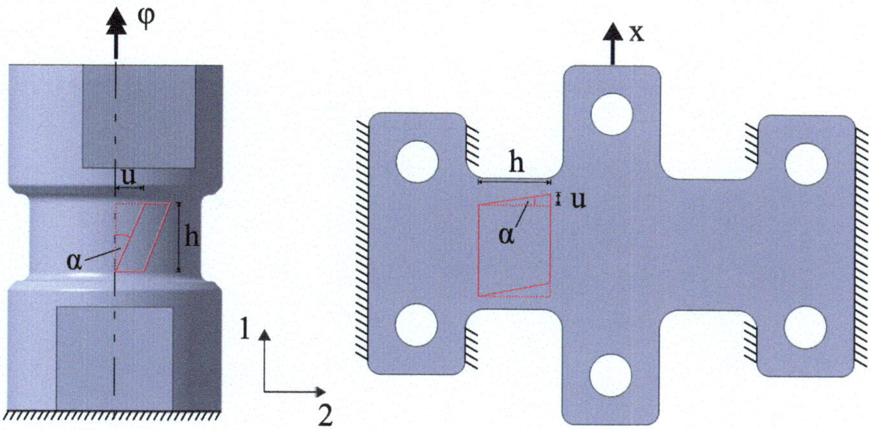

Fig. 7.2 Local shear deformations on both specimens

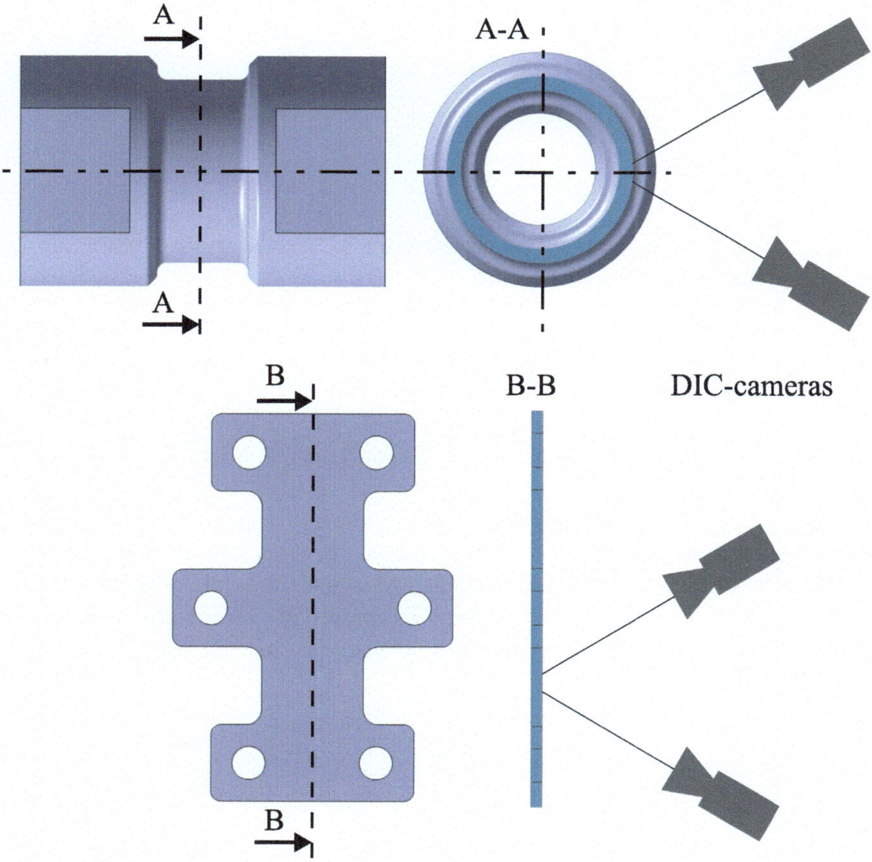

Fig. 7.3 Principle of DIC-measurement

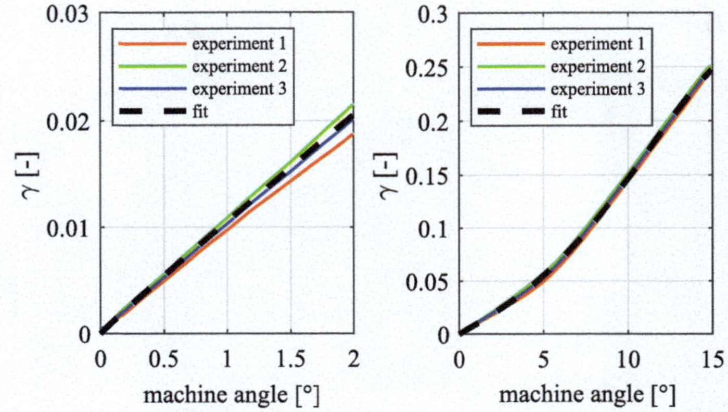

Fig. 7.4 Connection between machine angle φ and shear strain γ

A modified series expansion of the exponential function was used to establish the connection between the angle of rotation of the testing machine φ and local shear strain γ in the area of the tapered specimen region.

$$\gamma = \sum_{n=1}^{4} p_n \frac{\varphi^n}{n!} \tag{7.2}$$

The path of the traverse x can be connected to the local shear strain γ of the simple shear test via a third-order polynomial.

$$\gamma = \sum_{n=1}^{3} p_n \, u^n. \tag{7.3}$$

The result of the adjustment process for different strain levels is shown in Figs. 7.4 and 7.5.

Although the geometry of the shear specimen allows easier access for local strain measurement, the deviations in the measurements for large deformations are more significant than for the three-dimensional distortion field of the tube specimen. Based on the local strain measurement and the volume-preserving stress states, the stresses presented in the following experiments can be interpreted as true stresses.

7.2.4 *Results of Mechanical Tests*

Each mechanical test was repeated three times at several temperature levels between 0 °C ≤ T ≤ 90 °C. Mean value and standard deviation were analyzed.

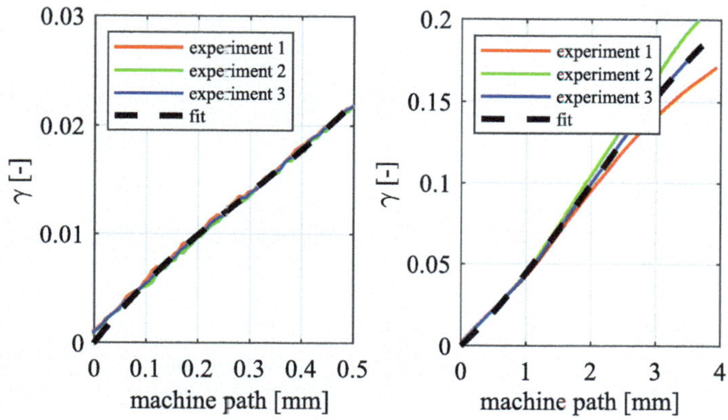

Fig. 7.5 Connection between machine path x and shear strain γ

7.2.4.1 Relaxation Tests

The reason for the time dependence of the mechanical properties of polymers is a spectrum of molecular relaxation processes with different relaxation times. These relaxation processes are described on a macroscopic level by the use of evolution equations for the internal state variables of the models. The relaxation experiment, in which a strain state is created at time t_0 and subsequently kept constant, is suitable for determining these internal state variables. The fading memory of the material can be observed through the stress response over time (Haupt 1977). The results of the relaxation tests for both specimens are shown in Figs. 7.6 and 7.7.

In addition to the time dependence of the material, a strong temperature dependence can also be recognized in the experimental results. In the semi-logarithmic plot in Figs. 7.6 and 7.7, two groups can be identified whose qualitative curves are similar (0–30 °C and 40–90 °C) and which are separated by the glass transition temperature,

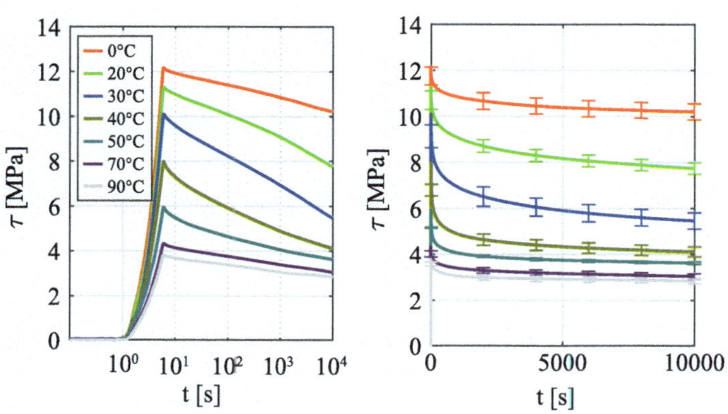

Fig. 7.6 Results of relaxation test for pipe specimen

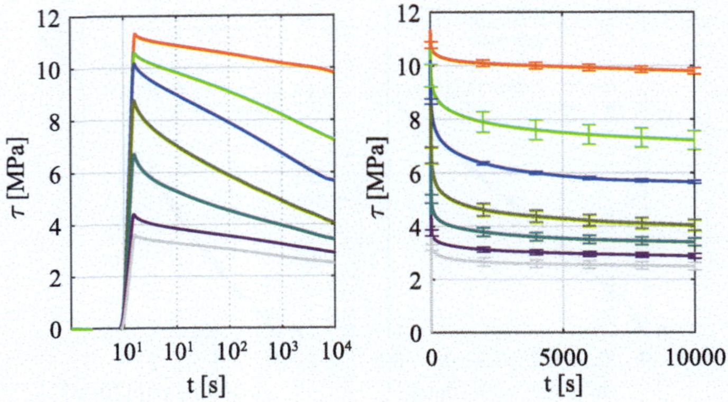

Fig. 7.7 Results of relaxation test for simple shear specimen

which is nominally 37 °C for PA12 (Bottenbruch and Binsack 1998). In terms of modeling the material, the glass transition temperature, which lies within the technical use range of the material, is a challenge. PA12 is used for pipes and hoses in the automotive and aircraft industries mainly because of its high wear, fatigue, and good chemical resistance (Bottenbruch and Binsack 1998). This means that the material is regularly in contact with fluids that have a temperature above the glass transition temperature. Therefore, the material should be modeled isothermally and above the glass transition temperature at 50 °C in the first step. The semi-logarithmic representation in Fig. 7.6 shows that a state of equilibrium has not yet been reached after the relaxation tests have been completed. Therefore, the modeling approach does not use a basic elasticity with the parameter G_{eq} for the deviatoric stress component.

7.2.4.2 Strain Rate Tests

In addition to fading memory, viscoelastic materials behave in a rate-dependent manner. Figure 7.8 shows torsion tests at a constant rate and varying temperatures. Figure 7.9 shows results where the rate was varied by three powers of ten at a constant temperature of 50 °C.

The validation experiments are carried out with the traverse speed of $\dot{x} = 10\ \frac{\text{mm}}{\text{min}}$ and are shown in Fig. 7.10.

After presenting these results, the theoretical framework for the modeling is explained below.

Fig. 7.8 Results of torsions tests at $\dot{\varphi} = 10 \frac{\circ}{\text{min}}$

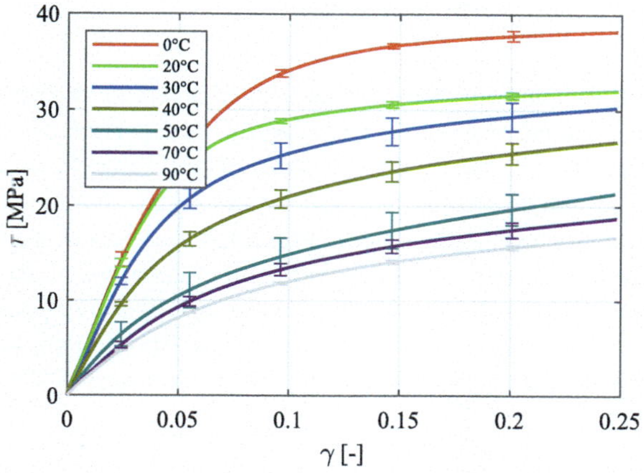

Fig. 7.9 Results of torsion test for varying traverse speeds at $T = 50\,°\text{C}$

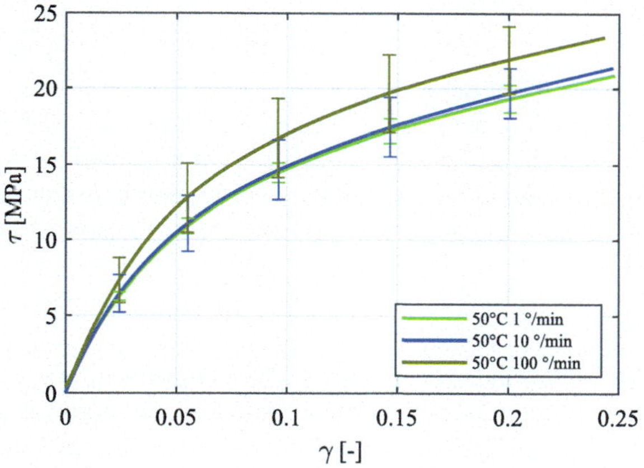

Fig. 7.10 Results of simple shear tests at $\dot{x} = 10 \frac{\text{mm}}{\text{min}}$

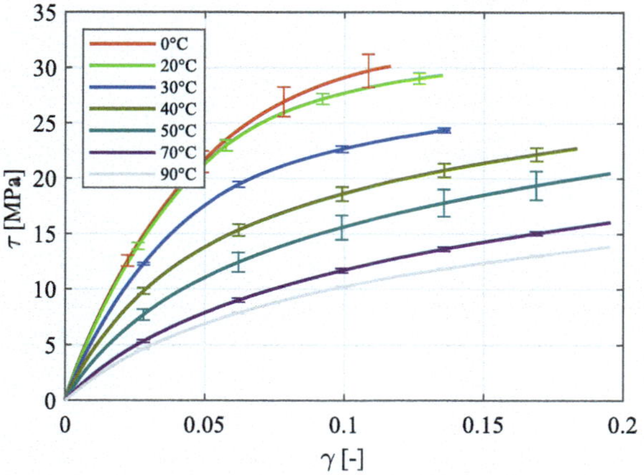

7.3 Material Theory

The relationship between stress and strain is fundamentally essential for solving mechanical problems. Depending on the observed material behavior, there are several possible approaches. In our case, the linear viscoelasticity theory is taken into account as polymers are fundamentally viscoelastic in behavior. Furthermore, the concept of endochronic plasticity is presented to describe the behavior of ductile materials without a clearly defined yield point. To symbolize the material, rheological models are used, which are designed by using serial and parallel connections of massless springs, dampers, and friction elements. This helps to formulate the constitutive laws and provides an illustrative approach (Tschoegl 1989; Krawietz 1986).

7.3.1 Viscoelasticity

A rheological model, according to Fig. 7.11, is used to explain viscoelastic material behavior for deviatoric stresses. It consists of n Maxwell elements connected in parallel (Maxwell 1867). Each Maxwell element consists of a Hookean spring and a Newtonian fluid. The global deformation γ is identical in all elements. For every Maxwell element, the stress of the spring G_k and of the damper η_k is identical.

$$\tau_{ov,k} = G_k \gamma_{el,k} = \eta_k \dot{\gamma}_{in,k}. \tag{7.4}$$

For very small shear rates $\dot{\gamma}$ the damper follows the shear deformation without generating stresses. So the total stress of the model is zero. For very high shear rates the dampers are locking and the total stress arises from the sum of all springs from the maxwell elements. For every other shear rate the dampers are relaxing within a certain time depending on their viscosity. With staggered viscosities for the dampers a continuous relaxation function can be depicted.

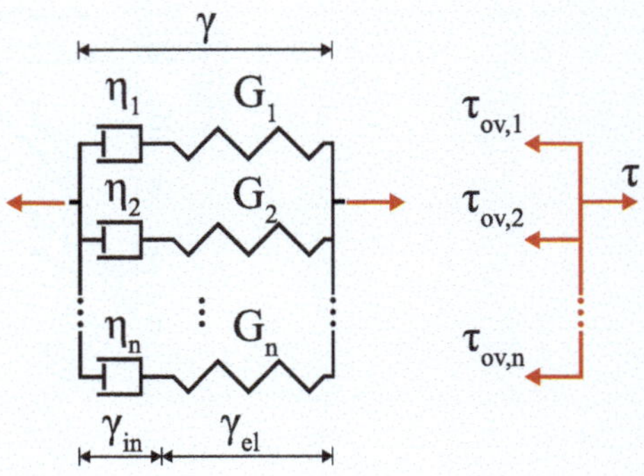

Fig. 7.11 Generalized Maxwell model

7.3.2 Endochronic plasticity

As seen from the experiments, the PA12 specimens are very ductile with unclear yield strength. For describing this kind of material behavior, the concept of endochronic plasticity will be used (Krawietz 1986; Valanis 1971). Starting from the already introduced Maxwell element, the equations for overstress τ_{ov} and the inelastic shear rate $\dot{\gamma}_{in}$ are

$$\tau_{ov} = G(\gamma - \gamma_{in}) \tag{7.5}$$

and

$$\dot{\gamma}_{in} = \frac{1}{\eta}\tau_{ov}. \tag{7.6}$$

The evolution equation for the inelastic shear rate of the model of endochronic plasticity is given by Eq. (7.7). In this model, the material parameter τ_F describes the maximum stress response of the model.

$$\dot{\gamma}_{in} = \frac{|\dot{\gamma}|}{\tau_F}\tau_{ov}\,. \tag{7.7}$$

Compared to the Maxwell element, this model is rate-independent and hence does not show any relaxation behavior. Increasing the shear rate leads to an increase in the stress rate by the same factor. The pass-through of the trajectories is faster, but the shape does not change. An alternative form of Eqs. (7.5) and (7.7) is

$$\dot{\tau}_{ov} = G\dot{\gamma} - |\dot{\gamma}|\frac{G}{\tau_F}\tau_{ov}. \tag{7.8}$$

Solving Eq. (7.8) analytically for a constant shear rate $\dot{\gamma} = |\dot{\gamma}| = \dot{\gamma}_0 > 0$ is straight forward resulting in the constitutive equations for the model of endochronic plasticity

$$\tau_{ov}(\gamma) = \tau_F\left(1 - e^{-\frac{G}{\tau_F}\gamma}\right). \tag{7.9}$$

The stress is a nonlinear function of the shear with the properties

$$\lim_{\gamma\to\infty} \tau_{ov}(\gamma) = \tau_F \tag{7.10}$$

$$\left.\frac{d\tau_{ov}}{d\gamma}\right|_{\gamma=0} = G. \tag{7.11}$$

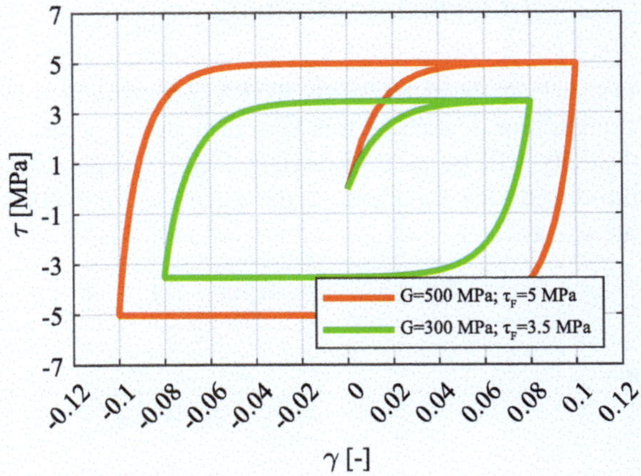

Fig. 7.12 Stress–shear curves of endochronic plasticity

In Fig. 7.12, the model of endochronic plasticity is analyzed with exemplary parameters G and τ_F, whose values are given in the figure. Sinusoidal functions with amplitudes of 0.1 and 0.08 are specified as shear strains. The frequency is not a relevant factor as the model is not rate-dependent. After passing through the first period of the sine wave, there are no changes in the stress response for further periods. The curves clearly show that the model is suitable for describing materials with undefined yield stress.

7.3.3 Combining Endochronic Plasticity and Viscoelasticity

The two effects of rate dependency and endochronic plasticity are combined into a new modeling approach where the evolution equations (7.6) and (7.7) are assembled additively resulting in

$$\dot{\gamma}_{in} = \left(\frac{1}{\eta} + \frac{|\dot{\gamma}|}{\tau_F}\right)\tau_{ov}. \tag{7.12}$$

Looking at two scenarios of extremely high and low strain rates the following stress rates using Eq. (7.8) can be calculated:

$$|\dot{\gamma}| << \frac{\tau_F}{\eta} \quad \Rightarrow \quad \dot{\tau}_{ov} = G\dot{\gamma} - \frac{G}{\eta}\tau_{ov} \tag{7.13}$$

$$|\dot{\gamma}| >> \frac{\tau_F}{\eta} \quad \Rightarrow \quad \dot{\tau}_{ov} = G\dot{\gamma} - |\dot{\gamma}|\frac{G}{\eta}\tau_{ov}. \tag{7.14}$$

The model interpolates between linear viscoelasticity and endochronic plasticity at high shear rates. It shows a nonlinear rate dependency, which vanishes at high shear rates—relaxation occurs at constant strain.

7.4 Parameter Identification with Pipe Specimens

By selecting the elastic strains of the springs in the overstressing elements as the process variable and Hooke's material law as the free energy function, the equation for the overstressing of a single element can be derived in volumetric-deviatoric division

$$\sigma_{neq} = 3K_{el}\varepsilon_{el}^{V} + 2G_{el}\varepsilon_{el}^{D}. \tag{7.15}$$

For the purely deviatoric shear strain, the symmetric strain tensor contains only two elements

$$\varepsilon = \varepsilon^{D} = \begin{bmatrix} 0 & \frac{\gamma}{2} & 0 \\ \frac{\gamma}{2} & 0 & 0 \\ 0 & 0 & 0 \end{bmatrix}. \tag{7.16}$$

With the relationship between the elastic and inelastic strains and since there is no volumetric strain component in shear tests, Eq. (7.15) yields

$$\sigma_{neq,i} = \tau_{neq,i} = 2G_{el,i}(\varepsilon^{D} - \varepsilon_{in,i}^{D}) \tag{7.17}$$

for every rheological element i. In order to identify the relaxation times, the relationship

$$t_{r,i} = \frac{\eta_i}{G_{el,i}} \tag{7.18}$$

will be used. In this way, the evolution equation (7.12) can be rewritten as

$$\dot{\gamma}_{in,i} = \left(\frac{2}{t_{r,i}} + \frac{2G_{el,i}|\dot{\gamma}|}{\tau_{F,i}}\right)(\varepsilon^{D} - \varepsilon_{in,i}^{D}). \tag{7.19}$$

Equation (7.19) can be solved straight forward using the backward Euler method (Hairer et al. 1993). The iteration rule in this case is

$$\gamma_{in}(t_{n+1}) - \gamma_{in}(t_n) = h\, f(\gamma_{in}(t_{n+1}), t_{n+1}). \tag{7.20}$$

The non-equilibrium stress using Eq. (7.17) can then be analyzed directly.

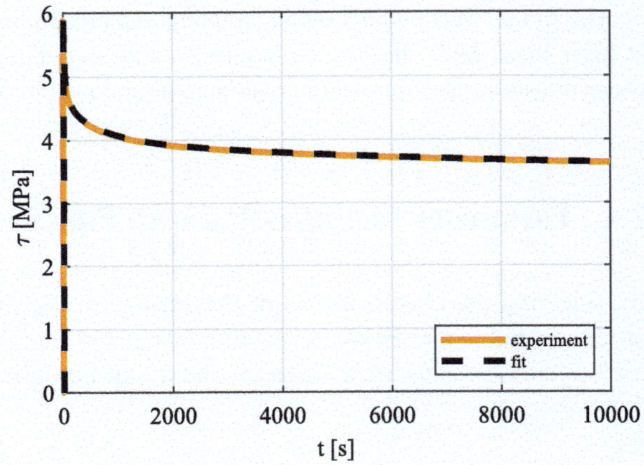

Fig. 7.13 Result of torsion relaxation test and model response at $T = 50\,°C$

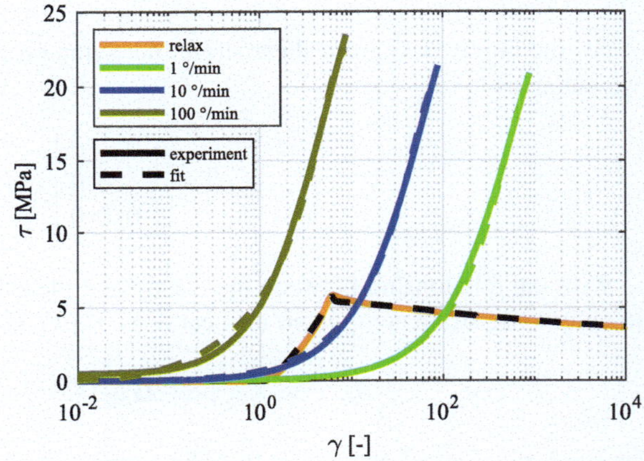

Fig. 7.14 Result of torsion test and model responses of relaxation and strainrate tests at $T = 50\,°C$

The relaxation experiments are available to identify the relaxation times $t_{r,i}$. In the long phase of constant strain, the strain velocity $|\dot{\gamma}|$ vanished; thus, the evolution equation (7.19) becomes the evolution equation of the Maxwell element. The number of required elements i is determined by staggering the relaxation times. This is an iterative process that starts with two elements per temporal decade. Once the relaxation times $t_{r,}$ have been identified, tests at several shear rates $\dot{\gamma}$ are used together with the relaxation tests to identify the shear moduli $G_{el,i}$ and the parameters $\tau_{F,i}$. The relaxation times are no longer varied. The results of the fitting process for the relaxation test can be seen in Fig. 7.13. The combined fitting process of relaxation and shear rate tests is depicted in Fig. 7.14. Finally, the strain rate tests plotted over the shear γ are shown in Fig. 7.15. Table 7.2 summarizes the parameters for the six rheological elements used.

Although the model has no equilibrium stress, the element with the longest relaxation time can be interpreted as such in the experiments' context.

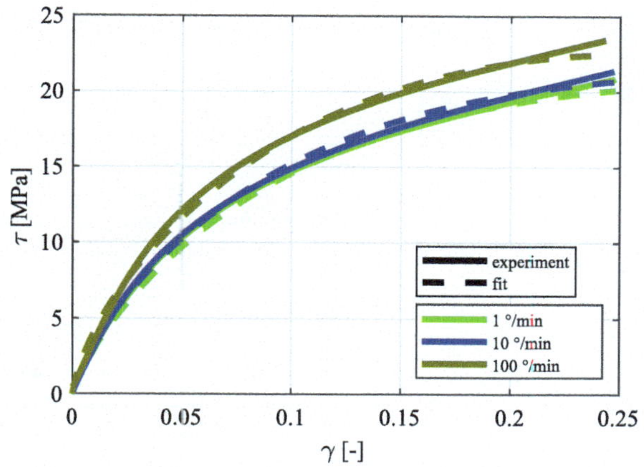

Fig. 7.15 Result of torsion strainrate tests and model responses at $T = 50\,°\mathrm{C}$

Table 7.2 Parameter set for the deviatoric model

Element No.	$t_{r,i}$	$G_{el,i}$	$\tau_{F,i}$
1	0.22	269.69	1000
2	9.43	17.97	0.5
3	80.68	89.67	0.78
4	627.71	49.96	0.71
5	5000.00	35.81	0.57
6	598019.21	204.03	19.83

7.5 Validation with Simple Shear Specimens

In order to validate the parameters identified (Table 7.2) on the basis of the torsion tests with the pipe sample, simple shear tests were carried out with Miyauchi-type samples (Figs. 7.7 and 7.10).

The calculation results of the relaxation test are less accurate than the torsion test results. However, due to the measurement inaccuracies of the force and distortion measurements, which add up in the results, the maximum deviations of 10% are considered acceptable. The calculation results of the rate test agree very well with the measurements. This is surprising at least to the extent that the standard deviation in the measurement series is more significant than in the relaxation tests (Fig. 7.16).

The calculation results of the rate test agree very well with the measurements (Fig. 7.17).

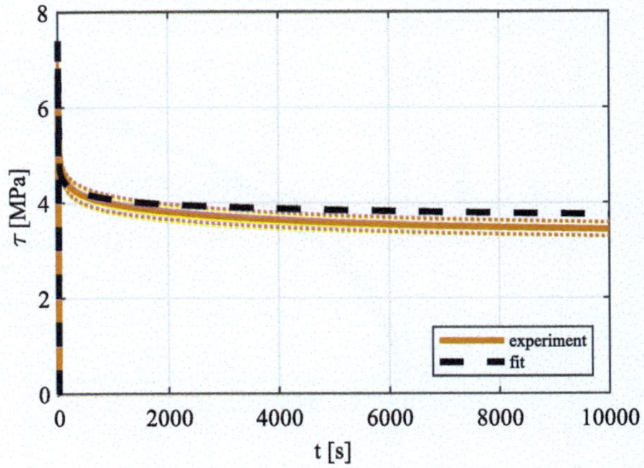

Fig. 7.16 Result of simple shear relaxation test with standard deviation and model response at $T = 50\,°\text{C}$

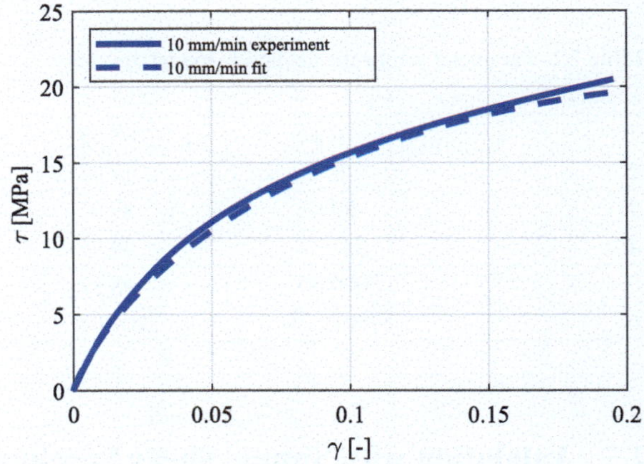

Fig. 7.17 Result of simple shear strainrate test and model responses at $T = 50\,°\text{C}$

7.6 Conclusion And Outlook

This research paper presented a modeling approach for deviatoric distortion states, combining the existing material theoretical approaches of viscoelasticity and endochronic plasticity into a new modeling approach. Relaxation tests and tests with different shear rates considering local strains and true stresses were performed with two different sample geometries. Torsion tests were used to identify the 18 model parameters. The validation of the parameters was carried out based on simple shear tests. The modeling approach for initially small strains reproduces the results of the mechanical tests with reasonable accuracy. The model must be extended to include volumetric stress components in the next step. In addition, implementing the model within software that performs three-dimensional component calculations based on the finite element method is planned.

Acknowledgements This research paper [project FLAB-3Dprint] is funded by dtec.bw—Digitalization and Technology Research Center of the Bundeswehr which we gratefully acknowledge. dtec.bw is funded by the European Union—NextGenerationEU.

References

Bottenbruch L, Binsack R (1998) Kunststoff Handbuch 3/4 Technische Thermoplaste Polyamide, 4 edn. Kunststoff-Handbuch / Hrsg. Gerhard W. Becker 3, Technische Thermoplaste / hrsg. von Ludwig Bottenbruch. Hanser, München

Breuninger J, Becker R, Wolf A, Rommel S, Verl A (eds) (2013) Generative Fertigung mit Kunststoffen: Konzeption und Konstruktion für Selektives Lasersintern. Springer Vieweg, Berlin and Heidelberg

Chaboche JL, Rousselier G (1983) On the plastic and viscoplastic constitutive equations–part i: rules developed with internal variable concept. J Press Vessel Technol 105(2):153–158. https://doi.org/10.1115/1.3264257

De Coninck H, Meyers S, Van Puyvelde P, Van Hooreweder B (2024) On the difference in mechanical behavior of glass bead-filled polyamide 12 specimens produced by laser sintering and injection molding. 3D Print Addit Manufact 11(2):419–433. https://doi.org/10.1089/3dp.2022.0160

Hahne D, Lion A, Johlitz M (2024) Selective laser sintered pipe specimen under torsional load: experimental investigations and material modeling. In: Altenbach H, Hitzler L, Johlitz M, Merkel M, Öchsner A (eds) Lectures notes on advanced structured materials 2, Advanced structured materials, vol 203, pp 123–140. Springer Nature Switzerland, Cham (2024). https://doi.org/10.1007/978-3-031-49043-9_7

Hairer E, Nørsett SP, Wanner G (1993) Nonstiff problems. Springer series in computational mathematics, vol 8, 2 edn. Springer, Berlin

Haupt P (1977) Viskoelastizität und Plastizität. Springer, Berlin, Heidelberg. https://doi.org/10.1007/978-3-662-13379-8

Krawietz A (1986) Materialtheorie. Springer, Berlin, Heidelberg

Lammens N, Kersemans M, De Baere I, Van Paepegem W (2017) On the visco-elasto-plastic response of additively manufactured polyamide-12 (pa-12) through selective laser sintering. Polym Test 57:149–155. https://doi.org/10.1016/j.polymertesting.2016.11.032

Miyauchi K (1984) Proposal of a planar simple shear test in sheet metals. Sci Pap Inst Phys Chem Res (Japan) 78(3):27–40

Sagradov I, Schob D, Roszak R, Maasch P, Sparr H, Ziegenhorn M (2020) Experimental investigation and numerical modelling of 3d printed polyamide 12 with viscoplasticity and a crack model at different strain rates. Mater Today Commun 25. https://doi.org/10.1016/j.mtcomm.2020.101542

Schmid M (2014) Am auf dem weg in die produktion. Kunststoff-Xtra 6–9

Schmid M (2015) Additive Fertigung mit Selektivem Lasersintern (SLS): Prozess- und Werkstofführblick. Essentials. Springer Vieweg, Wiesbaden

Schob D, Richter L, Kotecki K, Kurpisz D, Roszak R, Maasch P, Ziegenhorn M (2024) Characterization and simulation of shear-induced damage in selective-laser-sintered polyamide 12. Materials 17(1). https://doi.org/10.3390/ma17010038

Schob D, Sagradov I, Roszak R, Sparr H, Franke R, Ziegenhorn M, Kupsch A, Léonard F, Müller BR, Bruno G (2020) Experimental determination and numerical simulation of material and damage behaviour of 3d printed polyamide 12 under cyclic loading. Eng Fract Mech 229:106,841. https://doi.org/10.1016/j.engfracmech.2019.106841

Sharpe WN (2008) Springer handbook of experimental solid mechanics. Springer US, Boston, MA. https://doi.org/10.1007/978-0-387-30877-7

Tschoegl NW (1989) The phenomenological theory of linear viscoelastic behavior: an introduction; with 25 tables. Springer, Berlin and Heidelberg

Valanis K (1971) A theory of viscoplasticity without a yield surface. Arch Mech Stos 23(4):515–553
Van Hooreweder B, De Coninck F, Moens D, Boonen R, Sas P (2010) Microstructural characterization of sls-pa12 specimens under dynamic tension/compression excitation. Polym Test 29(3):319–326. https://doi.org/10.1016/j.polymertesting.2009.12.006
Van Hooreweder B, Sas P, Boonen R, Moens D, De Coninck F (2009) Experimental investigation of scaling laws for mechanical fatigue behaviour. In: 8th national congress on theoretical and applied mechanics 2009, vol 8, pp 659–663

Chapter 8
Temperature History in Laser Powder Bed Fusion: Analyzing the Effect of Heat Accumulation on the Microstructural State of Ti-6Al-4V

Johannes Rottler, Till K. Tetzlaff, Alexander Lion, Kristin Paetzold-Byhain, and Michael Johlitz

Abstract The nature of temperature history and its effect on the evolution of Ti-6Al-4V microstructure within the laser powder bed fusion process plays a driving role in process safety and component quality. Therefore, the temperature evolution is systematically evaluated and classified according to its influence on the formation and transformation of microstructure. The microstructural formation of the process-specific hierarchical α' martensite is discussed by reviewing the literature. For the transformation in case of insufficient heat removal, an approach to assess the influence of heat accumulation on microstructure dependent on reached holding temperatures and holding times is presented and experimentally investigated. For this purpose, the Vickers hardness was measured on heat-treated specimens up to a temperature range of 1000 °C. The observed hardness evolution and its underlying material mechanism were studied utilizing current research findings. The hardness increases significantly between 400 and 500 °C due to the refinement of α' martensite. Subsequently, a consistent decline in hardness is observed as the temperature rises from 500 to 850 °C, attributed to the degeneration of α' martensite and the precipitation of the β microstructure. Heating above 850 °C influences the precipitation of β, resulting in an increasing trend of hardness.

J. Rottler (✉) · A. Lion · M. Johlitz
Institute of Mechanics, University of the Bundeswehr Munich, Werner-Heisenberg-Weg 39, 85579 Neubiberg, Germany
e-mail: johannes.rottler@unibw.de

A. Lion
e-mail: michael.johlitz@unibw.de

M. Johlitz
e-mail: alexander.lion@unibw.de

T. K. Tetzlaff
Bundeswehr Research Institute for Materials, Fuels and Lubricants (WIWeB), Institutsweg 1, 85435 Erding, Germany
e-mail: TillTetzlaff@bundeswehr.org

K. Paetzold-Byhain
Institute of Machine Elements and Machine Design, TU Dresden, George-Bähr-Straße 3c, 01069 Dresden, Germany
e-mail: kristin.paetzold@tu-dresden.de

H. Altenbach et al. (eds.), *Lectures Notes on Advanced Structured Materials 3*,
Advanced Structured Materials 226, https://doi.org/10.1007/978-3-031-77403-4_8

8.1 Introduction

The advancement and improvement of additive manufacturing processes are becoming increasingly significant for research and industry. Laser powder bed fusion (PBF-LB/M) has been established as a method for rapid manufacturing, rapid tooling, and creating high-loaded prototypes made of high-performance materials such as Ti-6Al-4V. Especially with these materials, the extensive need for support structures limits the manufacturing advantages, such as the allowable freedom of geometric complexity of components (Weber et al. 2021; Coyne et al. 2021).

Support structures are solid elements that are not integral parts of the fabricated component and set a material bond between the component and the build platform. These additional elements are intended to ensure a safe process operation and prevent inadequate component quality, such as dimensional accuracy, surface quality, and crack formation, in the printed component. In literature, the purpose of the support structure is referred to as mechanical fixation in the powder bed against displacement and deformation driven by residual stresses (Kruth et al. 2004), ensuring the removal of the heat from the laser-scanned surface (Cao et al. 2020; Chen et al. 2017) and preventing the melt from sinking into the powder (Calignano 2014). Due to the small number of research studies, the influence of support structure concepts needs to be better understood in this context (Weber et al. 2021; Rottler et al. 2022). The effect of process heat with its temperature fields and gradients is identified as driving force, when comparing the purpose of support structures with physically related manufacturing processes, such as laser welding (Cao et al. 2020; Rottler et al. 2024).

In the case of Ti-6Al-4V, the temperature history provokes a lamellar and acicular α' martensite phase, which is aligned in preferred directions between the former β boundaries in a columnar shape in build-up direction (Thijs et al. 2010; Ren et al. 2021; Liu et al. 2022). Contrasting with conventionally utilized Ti-6Al-4V, in its hot-rolled state, featuring a granular $\alpha + \beta$ phase at room temperature. The observed distinction in microstructure state leads to anisotropy, markedly enhanced mechanical strength properties (yield strength and ultimate tensile strength), and reduced ductility (Beese and Carroll 2016; Ren et al. 2021; Yang et al. 2016). The current literature demonstrates that these properties rely on process parameters and scan strategies, which influence the alterations in the martensitic microstructure due to the effect of the thermal history (Liu et al. 2022; Thijs et al. 2010).

Owing the significant difference in thermal properties between the material in powder and solid state, the transfer of process heat from the laser-scanned surface to the environment is highly inhomogeneous in PBF-LB/M (Kruth et al. 2004; Cao et al. 2020; Chen et al. 2017). If the thermal energy cannot flow off during the breaks in the laser scanning sequence, the heat accumulates, which is susceptible to the process failures. Therefore, the component geometry and orientation within the powder bed affect the heat flow and, thus, the process safety and quality in the additive manufacturing process (Hagenlocher et al. 2022; Mohr et al. 2020; Ranjan et al. 2020). However, this work is intended to highlight the time-temperature progression occurring in metal additive manufacturing methods. Subsequently, the temperature history of the PBF-LB/M process and its effect on the process-specific microstructural state are the focus of this study. Therefore, a proposal to determine the characteristics of the PBF-LB/M temperature history according to its effect on the formation and transformation of microstructure is presented. From this aspect, the influence of insufficient heat removal and accumulation on the in-process component quality is analyzed on the Ti-6Al-4V microstructure.

8.2 Evaluating Temperature History of PBF-LB/M

8.2.1 Classification of PBF-LB/M Temperature History in Manufacturing Processes

In order to understand the highly dynamic nature of the transient PBF-LB/M temperature fields and their effect, the first step is to compare them within manufacturing processes. Like all additive manufacturing processes for metals, according to DIN 8580 (Deutsches Institut für Normung e.V. 2024), the PBF-LB/M process is classified in the division of primary shaping methods, whereby the nature of process physics has similarities with the established joining processes, especially fusion welding. In the context of metal fusion welding technologies, the driving effect of inhomogeneous, nonlinear, and transient temperature fields is widely acknowledged. Considering the primary mechanisms, the temperature field determines both the stress-distortion field through thermal stresses and the microstructural state through microstructural evolution. The stress–strain field is also influenced by the transformation stresses of the microstructure (Radaj 1988). Accordingly, process heat conduction is becoming more important in manufacturing processes with cyclical heat input, such as additive processes, especially for process optimization and thermal management. The first step in evaluating the dynamic behavior of the PBF-LB/M process heat is to classify comparable generative and multi-cyclic methods as well as methods with corresponding significance for component quality (shown in Fig. 8.1a). Direct energy deposition (DED) and electron beam laser melting (PBF-EB/M) are selected from the additive manufacturing processes for metals. Sintering is another temperature- and pressure-driven process from the division of primary forming processes with metal powers. Another laser-based process is laser beam welding, which is very similar to the DED process when adapted to the generation of components and is therefore not considered separately. Metal fusion welding in several short layers is considered as representative of the comparable joining processes. Due to the significance of the thermal effect on the component quality, post-processing heat treatments are schematically assessed as representative of the modifying properties division. Generally, the characteristics of a time–temperature cycle are the heating rate, the maximal reached temperature within the cycle, the holding times during which a material-specific temperature level is maintained, the cooling rate,

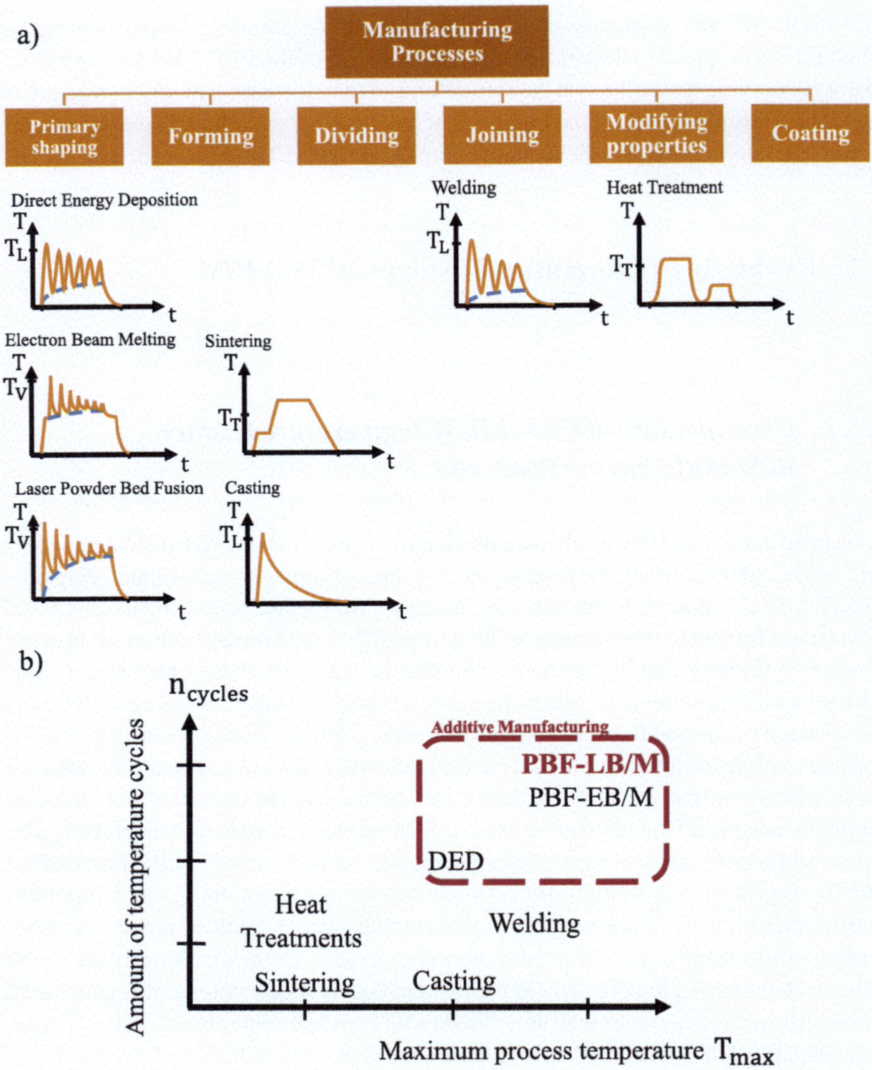

Fig. 8.1 Selection of manufacturing processes according to DIN 8580 (**a**), and evaluation considering the dynamic behavior of process heat (**b**) by the amount of reheating temperature cycles and maximum process temperature reached

and the minimum temperature after cooling. In terms of multi-cycle procedures, the number of cycles is relevant. The maximum temperature occurring in the process and the number of cycles are chosen to compare the processes' dynamic thermal behavior. Figure 8.1b illustrates the thermal characteristics of additive manufacturing methods. These methods are notable for their exceptionally high process temperatures, on par with laser welding technologies. However, what sets them apart is the significantly higher number of thermal cycles that occur within the component. The most significant number of temperature cycles occurs in PBF-LB/M printing, which is attributed to the ratio of the laser spot area to the sliced component area to be scanned by the laser and, therefore, the amount of laser scan tracks. Furthermore, the PBF-LB/M process exhibits the highest maximum temperature. This is a direct result of the high energy input in the laser focus on the powder bed, causing the temperatures to surpass the vaporization temperature of the alloying elements in the melt pool. As a result, the transient dynamics of the temperature fields of additive manufacturing processes outreaches those of traditional manufacturing processes. The temperature fields of PBF-LB/M are primarily distinguished by very high gradients and rates, which increases the need for the required knowledge to optimize the process control further and perform effective thermal management through laser scanning strategy adaptation and application of support structure concepts.

8.2.2 *Effect of PBF-LB/M Temperature History on Ti-6Al-4V Microstructure*

The intricate nature of temperature fields in PBF-LB/M presents challenges in measurement due to its dynamic behavior, the embedding of the component in an unprocessed powder bed, and the shielding within an inert gas atmosphere. In the case of experiments with IR cameras, investigations are limited to the currently processed layer or rather surficial temperature gradients. Another notable hurdle is the accurate recording of rapid cooling and heating with a suitable sample rate and sufficient spatial resolution (Rottler et al. 2024; Hooper 2018; Downing et al. 2021).

Figure 8.2 provides a detailed schematic of the unique temperature history of a point inside a component during PBF-LB/M printing (adapted from Liu et al. 2022; Downing et al. 2021). In the laser scanning process, the powder melts as the laser passes over it, reaching the maximum process temperature. Once the material is molten, it solidifies and rapidly cools. As the laser moves along nearby tracks, reheating occurs, whereby the maximum temperature is no longer exceeded, and primary cooling prevails (inter-layer cooling). After the layer processing time, involving laser scanning and recoating a new powder layer, the temperature decreases to a minimum (residual temperature). As the process continues layer by layer, the energy input area moves away from the observation point, preventing the maximum process temperature from exceeding in subsequent cycles. Once the printing sequence is finished, the component globally cools down. In interpretations that deal with PBF-LB/M temperature fields in component dimensions,

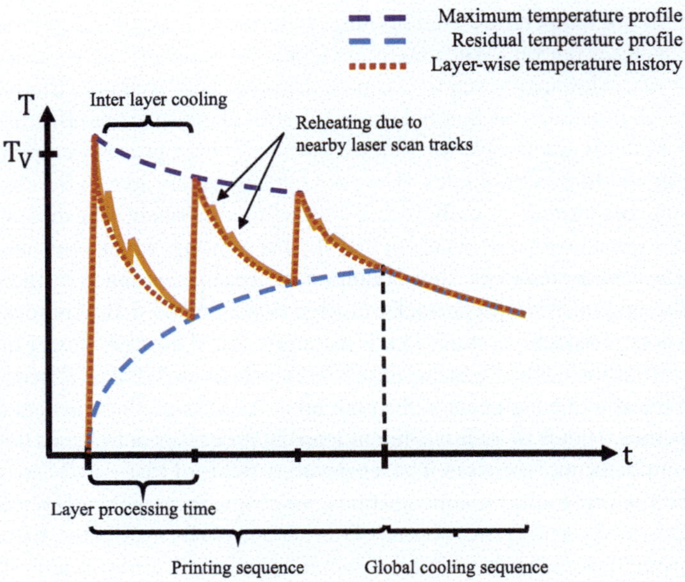

Fig. 8.2 Schematic progression and characteristics of the temperature history in the PBF-LB/M process

the layer-by-layer consideration of the energy input has proven to be efficient (Chaudry et al. 2022; Yang et al. 2016).

The difficulties of direct measurements have led to the investigation of the primary effect based on its influence on the microstructure of Ti-6Al-4V. Based on the procedure for multi-pass welding with short weld lines, the effect of the inhomogeneous temperature history can be classified according to the formation and transformation of the microstructure (Radaj 1988).

Ying et al. (2016) described the formation of the process-induced Ti-6Al-4V α' martensitic microstructure, which is arranged in columns of the former melt pool edges and former grain boundaries. A hierarchical order of the martensite needles according to their length and preferred orientation was observed and associated with process-specific temperature history. 95% of martensite needles have an angle of $-30°$ to $-60°$ or 30 to 60° between their morphological main axes and the direction of the building. The formation mechanism is determined by classifying the temperature cycles according to their maximum temperature and cooling rate. The cooling rate, a crucial factor, is stated to be $>10^5 \frac{\mathrm{K}}{\mathrm{s}}$, ensuring the consistent formation of martensite. Temperature cycles with a peak temperature above the melting point result in longer primary lamellar martensite needles measuring 10–70 μm. Secondary martensite needles measuring 1.0–2.0 μm are formed by cycles with peak temperatures between the solidus and the melting temperature. Cycles with temperature peaks between β transus and solid temperature provoke tertiary martensite and cycles with temperatures between martensite start and β transus temperature lead to quadric martensite. The main influencing parameter is the energy input through the laser, controlled by the laser power, scan velocity, hatching distance, and scan pattern (Yang et al. 2016; Liu et al. 2022).

The microstructure transformation describes the evolution of the formed martensite structure within the process. It is characterized by the residual temperature profile, determined by the minimal temperatures reached after each inter-layer cooling and the global cooling after the printing sequence (see Fig. 8.2). Its impact depends on material-specific effect temperatures and the time the temperature level is maintained (holding time). The amount of energy input and the ability to conduct the process heat through the component into the periphery governs the residual temperature profile. In additive manufacturing, only a few studies investigated the inhomogeneous heat removal or heat accumulation by the cooling behavior of currently laser-scanned layers by IR measurement (Downing et al. 2021; Chaudry et al. 2022; Mohr et al. 2020) or its local effect on the melt pool behavior (Hagenlocher et al. 2022; Ranjan et al. 2020). So, the influence of the component geometry was demonstrated, but not its effect on the process, the Ti-6Al-4V microstructure, and the part quality.

8.3 Effect of Heat Accumulation on Ti-6Al-4V Microstructure

8.3.1 Methodology

The research involved the use of pre-alloyed Ti-6Al-4V ELI-grade 23 powder with extra low interstitial quality obtained from EOS GmbH Electro Optical Systems. The powder was subjected to dynamic image analysis (ISO 13322-2 2024) using a CAMSIZER X2 manufactured by Microtrac RETSCH GmbH. The determined particle size distribution shows equivalent particle diameters of 25.4 μm (D_{10}), 37.6 μm (D_{50}-Median), and 46.4 μm (D_{90}), with a mean value of 36.8 μm. Additionally, the arithmetic-mean sphericity of the powder particle, following DIN ISO 9276-6 2008, was found to be 0.88. The pristine powder's chemical composition was analyzed using plasma emission spectrometry and IR spectrometry after melt extraction (refer to Table 8.1). The investigated specimens were generated using a TruPrint 1000 multilaser machine manufactured by Trumpf SE + Co. KG. This PBF-LB/M system is operating with two fiber lasers and f-theta-lenses optics, generating two continuous-wave lasers with a wavelength of 1070 nm and a laser spot diameter of 55 μm on the powder bed surface. The system is capable of delivering a maximum laser power of 200 W each. Preheating the build plate is not feasible. Furthermore, the processing of Ti-6Al-4V powder occurs under an inert argon atmosphere.

Cuboidal specimens 10 mm × 10 mm × 30 mm were manufactured to investigate the microstructural behavior dependent on different holding temperatures. Subsequently, cubic specimens with an edge length of 8 mm were manufactured to study the time-dependent behavior. All specimens were generated using the standard laser scanning parameter set and scanning strategy (Rottler et al. 2024) and machined by electrical dis-

Table 8.1 Analysis result of the chemical composition of pristine Ti-6Al-4V ELI powder

Element	Al	V	O	N	H	Ti
wt. %	6.23	3.76	0.119	0.035	0.004	bal.

charge machining. The specimens were then heat treated in a laboratory furnace under an air atmosphere at specific holding temperatures and holding times to investigate the effect of heat accumulations. To guarantee precise heat treatments, the heating behavior of the specimens on each furnace temperature was characterized in advance by one specimen equipped with a type-K thermoelement in the center. These pre-measurements were performed to ensure that the target holding temperature was reached within the specimen. The moment the specimen temperature deviates by less than 4 °C from the final holding temperature to be analyzed was defined as the start of the holding time. From this point onward, the specimens are regarded as fully heated. After the holding time has elapsed, the specimens are quenched in water. The specimens were embedded in two-component epoxy resin and ground to their center plane with SiC grinding paper until a grit size of 400 and polished with MD-Mol 6 μm and suspension. The Vickers hardness HV5 was measured using the Qness Q10A+ universal hardness device by ATM Qness GmbH with an according measurement force of 49.03 N maintained over 10 s. A minimum of 9 idents were tested in the center of each specimen.

8.3.2 *Results and Discussion*

The influence of various holding temperatures within heat accumulations was analyzed by heating specimens to 400, 500, 650, 850, and 1000 °C, maintained for a holding time of 1800 s, and tested together with as-built (no heat treatment) specimen. The results (represented in Fig. 8.3) show no significant difference between the as-built and the specimens hold at 400 °C. Beyond 400 °C, the values rise rapidly from 366 HV5 to a peak hardness of 399 HV5 at 500 °C. Afterward, the propagation of the values drops to a minimum of 355 HV5 at 850 °C. The value then rises slightly up to 1000 °C, which is the maximum temperature of the study.

In order to investigate the time-dependent behavior of the hardness changes and its mechanisms acting on the microstructure, the holding temperatures are selected based on the hardness peak at 500 °C shown in Fig. 8.3. Therefore, specimens were heated until 400 °C (before the peak), 475 and 500 °C (within the increase), and 650 °C (in the decrease). According to the experimentally determined heat-up time, specimens were water quenched after the 60, 120, 300, 600, 1800, 2700, and 3600 s in the furnace.

As Fig. 8.4 illustrates, at the initial state of 400 °C, the hardness shows a gradual, nearly linear rise over the isothermal holding time from 373 HV5 to 404 HV5. At the temperature levels of 475 and 500 °C, there is a rapid increase in hardness within 600 s, after which the hardness stabilizes at a value of 418 HV5. Upon reaching 650 °C, the val-

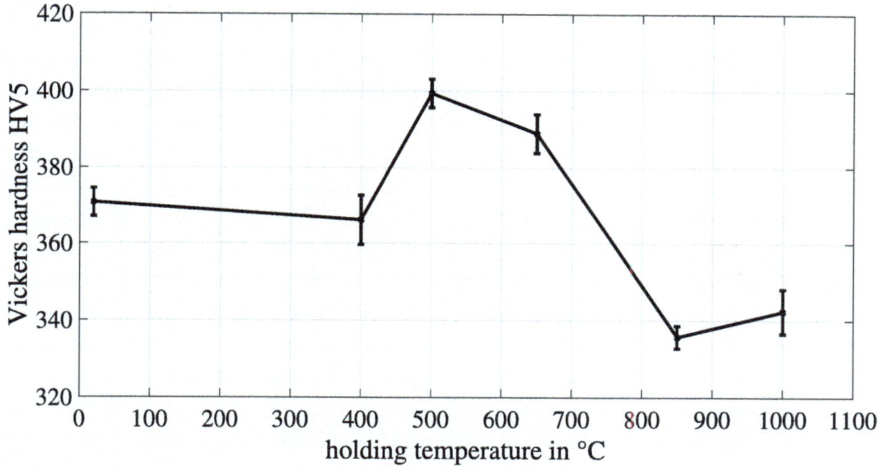

Fig. 8.3 Vickers hardness on different holding temperatures at 1800 s holding time, water quenched

ues exhibit a slight decrease from 394 HV5 to 374 HV5 after 3600 s. The represented hardness propagation provides insights into the Ti-6Al-4V microstructural transformation at temperatures and holding times in orders of magnitude that can occur in heat accumulation within a component during the PBF-LB/M process.

The impact can be divided into four temperature ranges up to 1000 °C, associated with respective evolution mechanisms of the Ti-6Al-4V PBF-LB/M printed microstructure. No effect on the microstructure can be observed in the first range up to about 400 °C. Therefore, the accumulated heat is not critical for undesired microstructural changes under these process conditions. By isothermal holding between 400 and 500 °C, a transformation mechanism becomes evident within the microstructure, which causes an increase in hardness. The microstructural transformation is significantly accelerated as

Fig. 8.4 Hardness propagation of heated up specimens at 400, 475, 500, 650 °C at holding times up to 3600 s

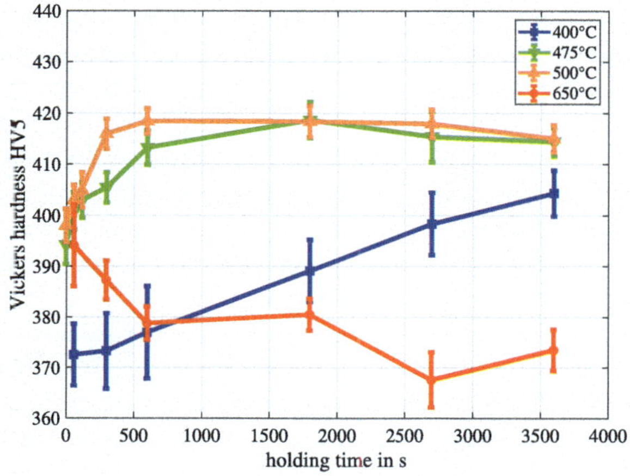

temperatures approach 500 °C. In literature, refinement effects within the α' martensite substructures are suspected to enhance the hardness (Wu et al. 2016). As the temperature surpasses 500 °C, a different mechanism comes into play, steadily reducing the hardness. This is attributed to the ongoing degradation of α' martensite to $\alpha + \beta$ and the precipitation of β microstructure (softer β in relation to α) (Haubrich et al. 2019). In addition, stress-relieving effects occur, which are typically expected above 600 °C and could explain changes in the reduction in hardness (Wu et al. 2016). A fourth section can be selected if temperatures surpass 850 °C. Upon heating and maintaining temperatures, the described transformation and precipitation of β occur, with a further formation of β content as temperature rises. The subsequent water quenching triggers the reversion of the formed β into a martensitic structure, thereby leading to the observed increase in hardness (Wu et al. 2016). The results show significant local differences in the Ti-6Al-4V microstructure can occur when printing with Ti-6Al-4V. Differences in hardness are already provoked by very short holding times at moderate temperatures far from the current laser-processed surface, resulting in local embrittlement and crack initiation in the component.

8.4 Conclusion

This contribution presents a unique perspective on the complex behavior of transient temperature fields in PBF-LB/M, underscoring its significance. By comparing temperature-driven manufacturing processes, the unique dynamic behavior of additive manufacturing techniques is highlighted, particularly the PBF-LB/M process. With its high process temperatures and frequent reheating cycles, this process underscores the critical need to understand and control heat flow. By detailed investigation of the PBF-LB/M temperature history and its effect, the evolution of Ti-6Al-4V microstructure within the PBF-LB/M process is divided into formation and transformation. Under regular process conditions with homogeneous heat flow away from the laser-processed layer, the formation of typical hierarchical α' martensitic microstructure was reviewed from the literature. The hierarchical stages of the lamellar and acicular martensite needle are characterized by their size and arrangement and are attributed to the maximum temperatures of individual temperature cycles. Whereby the cooling rates are assumed to be sufficient to form martensite. In the case of insufficient heat removal and accumulation, the residual temperature profile describes the Ti-6Al-4V microstructure transformation after its formation in the thermal cycles. The temperature profile is characterized by the temperatures reached after inter-layer cooling. This influence on the process depends on the material but can always be evaluated by the relevant holding temperatures and holding times.

The effect of heat accumulation on the microstructural state of Ti-6Al-4V was experimentally analyzed up to a temperature range of 1000 °C by hardness measurements (HV5). To determine the transient development of microstructural transformation, the Vickers hardness was investigated by process-relevant holding times of up to 3600 s. Under 400 °C, no effect on the microstructure was measured. In the transition from

400 to 500 °C, a microstructural refinement process results in a notable enhancement in hardness. The mechanism is notably accelerated as the temperature rises. Throughout the range of 500–850 °C, hardness consistently declines, attributed to the degeneration of α' martensite to $\alpha + \beta$ and the precipitation of β microstructure. Heating to 850 °C results in the precipitation and the further formation of β, provoking an increased proportion of martensite after quenching with water. This leads to an observable trend of increased hardness.

The results demonstrate the evolution of an inhomogeneous Ti-6Al-4V microstructural state within a component due to inhomogeneous heat removal and heat accumulations by PBF-LB/M manufacturing. Very short holding times at moderate temperatures, not in the laser's direct heat-affected zone, already prove differences in hardness. Particularly in geometric notches, where inhomogeneous heat removal occurs in the PBF-LB/M process, this leads to local embrittlement and crack initiation, which is also difficult to heal in post-processing. As a consequence, components would fail prematurely under cycling mechanical loading. The Vickers hardness could be used as an indicator to evaluate a homogeneous heat flow based on the microstructure. Finally, this investigation emphasizes the crucial role of global temperature fields in thermal process management to ensure sufficient component quality in PBF-LB/M. A pivotal step is the design of efficient heat removal among the selection and implementation of support structure concepts in critical geometric features. To address the challenge of managing process heat, our forthcoming research will expand the gained knowledge and investigate correlations between component geometry and global temperature propagation. This endeavor will contribute fundamental knowledge about tailored heat removal and its impact on PBF-LB/M component quality, facilitating the derivation of actionable recommendations.

Acknowledgements This research [project FLAB-3Dprint] is funded by dtec.bw—Digitalization and Technology Research Center of the Bundeswehr which we gratefully acknowledge. dtec.bw is funded by the European Union—NextGenerationEU.

References

Beese AM, Carroll BE (2016) Review of mechanical properties of Ti-6Al-4V made by laser-based additive manufacturing using powder feedstock. JOM 68(3):724–734. https://doi.org/10.1007/s11837-015-1759-z

Calignano F (2014) Design optimization of supports for overhanging structures in aluminum and titanium alloys by selective laser melting. Mater Des 64:203–213. https://doi.org/10.1016/j.matdes.2014.07.043

Cao Q, Zhang J, Chang S, Fuh JYH, Wang H (2020) The effect of support structures on maraging steel MS1 parts fabricated by selective laser melting at different building angles. Rapid Prototyping J 26(9):1465–1476. https://doi.org/10.1108/RPJ-11-2019-0287

Chaudry MA, Mohr G, Hilgenberg K (2022) Experimental and numerical comparison of heat accumulation during laser powder bed fusion of 316l stainless steel. Prog Addit Manuf 7(5):1071–1083. https://doi.org/10.1007/s40964-022-00282-x

Chen H, Gu D, Xiong J, Xia M (2017) Improving additive manufacturing processability of hard-to-process overhanging structure by selective laser melting. J Mater Process Technol 250:99–108. https://doi.org/10.1016/j.jmatprotec.2017.06.044
Coyne J, Barnes JE, King WE, Slattery K, Cummings C (2021) I want to break free: the journey towards reducing or eliminating support structures. METAL AM 7(3):155–163
Deutsches Institut für Normung e.V.: Fertigungsverfahren -Begriffe, Einteilung: Manufacturing processes -Terms and definitions, division
Deutsches Institut für Normung e.V.: Representation of results of particle size analysis - Part 6: Descriptive and quantitative representation of particle shape and morphology (ISO 9276-6:2008)
Downing David, Miller Jonathan, McMillan Matthew, Leary Martin, Wischeropp Tim, Emmelmann Claus, Brandt Milan (2021) The effect of geometry on local processing state in additively manufactured ti-6al-4v lattices. Integr Mater Manuf Innovation 10(3):508–523. https://doi.org/10.1007/s40192-021-00225-4
Hagenlocher C, O'Toole P, Xu W, Brandt M, Easton M, Molotnikov A (2022) Heat accumulation in laser additive Manufacturing processes and its effect on local solidification. Procedia CIRP 111:176–179. https://doi.org/10.1016/j.procir.2022.08.144
Haubrich J, Gussone J, Barriobero-Vila P, Kürnsteiner P, Jägle EA, Raabe D, Schell N, Requena G (2019) The role of lattice defects, element partitioning and intrinsic heat effects on the microstructure in selective laser melted Ti-6Al-4V. Acta Mater 167:136–148. https://doi.org/10.1016/j.actamat.2019.01.039
Hooper PA (2018) Melt pool temperature and cooling rates in laser powder bed fusion. Addit Manuf 22:548–559. https://doi.org/10.1016/j.addma.2018.05.032
International Organization for Standardization: Particle size analysis - Image analysis methods - Part 2: Dynamic image analysis methods
Kruth JP, Froyen L, van Vaerenbergh J, Mercelis P, Rombouts M, Lauwers B (2004) Selective laser melting of iron-based powder. J Mater Process Technol 149(1–3):616–622. https://doi.org/10.1016/j.jmatprotec.2003.11.051
Liu J, Li G, Sun Q, Li H, Sun J, Wang X (2022) Understanding the effect of scanning strategies on the microstructure and crystallographic texture of Ti-6Al-4V alloy manufactured by laser powder bed fusion. J Mater Process Technol 299:117,366 (2022). https://doi.org/10.1016/j.jmatprotec.2021.117366
Mohr G, Scheuschner N, Hilgenberg K (2020) In situ heat accumulation by geometrical features obstructing heat flux and by reduced inter layer times in laser powder bed fusion of AISI 316L stainless steel. Procedia CIRP 94:155–160. https://doi.org/10.1016/j.procir.2020.09.030
Radaj D (1988) Wärmewirkungen des Schweißens: Temperaturfeld, Eigenspannungen. Verzug. Springer, Berlin Heidelberg, Berlin, Heidelberg
Ranjan R, Ayas C, Langelaar M, van Keulen F (2020) Fast detection of heat accumulation in powder bed fusion using computationally efficient thermal models. Materials 13(20). https://doi.org/10.3390/ma13204576
Ren XP, Li HQ, Guo H, Shen FL, Qin CX, Zhao ET, Fang XY (2021) A comparative study on mechanical properties of Ti-6Al-4V alloy processed by additive manufacturing vs. traditional processing. Mater Sci Eng A 817:141,384. https://doi.org/10.1016/j.msea.2021.141384
Rottler J, Petroll C, Johlitz M, Paetzold K, Lion A (2022) A strategy to extend the limits of unsupported printing in laser powder bed fusion by optimized process parameters selection. In: Altenbach H, Johlitz M, Merkel M, Öchsner A (eds) Lectures notes on advanced structured materials, Advanced structured materials, vol 153, pp 119–130. Springer International Publishing, Cham. https://doi.org/10.1007/978-3-031-11589-9_7
Rottler J, Tetzlaff TK, Lion A, Paetzold-Byhain K, Johlitz M (2024) The importance of heat conduction and global temperature fields in the laser powder bed fusion process. In: Altenbach H, Hitzler L, Johlitz M, Merkel M, Öchsner A (eds) Lectures notes on advanced structured materials 2, Advanced structured materials, vol 203. Springer Nature Switzerland, Cham, pp 141–149

Thijs L, Verhaeghe F, Craeghs T, van Humbeeck J, Kruth JP (2010) A study of the microstructural evolution during selective laser melting of Ti-6Al-4V. Acta Mater 58(9):3303–3312. https://doi.org/10.1016/j.actamat.2010.02.004

Weber S, Montero J, Bleckmann M, Paetzold K (eds) (2021) Support-free metal additive manufacturing: a structured review on the state of the art in academia and industry. Cambridge University Press (2021). https://doi.org/10.1017/pds.2021.542

Wu SQ, Lu YJ, Gan YL, Huang TT, Zhao CQ, Lin JJ, Guo S, Lin JX (2016) Microstructural evolution and microhardness of a selective-laser-melted Ti-6Al-4V alloy after post heat treatments. J Alloys Compd 672:643–652. https://doi org/10.1016/j.jallcom.2016.02.183

Yang J, Yu H, Yin J, Gao M, Wang Z, Zeng X (2016) Formation and control of martensite in Ti-6Al-4V alloy produced by selective laser melting. Mater Des 108:308–318. https://doi.org/10.1016/j.matdes.2016.06.117

Chapter 9
Investigation of the Mechanical Properties of Polymer Interfaces in Fused Filament Fabrication Under Uniaxial Tensile Testing

Julian Klingenbeck, Alexander Lion, and Michael Johlitz

Abstract The fused filament fabrication (FFF) process is a well-established technique in 3D printing. This method results in a characteristic mesostructure, with visible lines and gaps, contributing to the anisotropic mechanical behaviour of FFF-produced structures. This study investigates the temperature-dependent influence of mesostructure interfaces on the mechanical properties of FFF printed Polyamide 6 using uniaxial tensile tests. Custom tensile samples were produced at three different chamber temperatures to alter the temperature history of the polymer interfaces and subjected to uniaxial tensile tests. The results showed that thin-walled samples could be reliably manufactured, exhibiting reproducible mechanical properties. Higher chamber temperatures during sample production were shown to have a significant influence on the mechanical behaviour.

9.1 Introduction

The fused filament fabrication Process (FFF) is an established manufacturing technique in the field of 3D printing. Fundamentally, the FFF process involves the melting of a prefabricated thermoplastic filament in the so-called extruder and the subsequent deposition of the molten material on a build-platform while rapidly solidifying. Through the relative movement between extruder and buildplate the geometry is fabricated in a line- and layerwise fashion (Gibson et al. 2015). This results in a characteristic mesostructure in the finished part, as depicted exemplarily in Fig. 9.1,

J. Klingenbeck (✉) · A. Lion · M. Johlitz
Faculty of Aerospace engineering, Institute of Mechanics, University of the Bundeswehr Munich, Werner-Heisenberg-Weg 39, 85577 Neubiberg, Germany
e-mail: julian.klingenbeck@unibw.de

A. Lion
e-mail: alexander.lion@unibw.de

M. Johlitz
e-mail: michael.johlitz@unibw.de

H. Altenbach et al. (eds.), *Lectures Notes on Advanced Structured Materials 3*,
Advanced Structured Materials 226, https://doi.org/10.1007/978-3-031-77403-4_9

Fig. 9.1 Visible mesostructure at the fracture surface of a FFF manufactured part

with the individual lines of the printing process and intermediate gaps clearly visible in the finished part.

It has been shown that structures produced with the FFF process can exhibit anisotropic mechanical behaviour, which is, at least in part, attributed to its mesostructure (Ahn et al. 2002; Dey and Yodo 2019; Gebisa and Lemu 2019; Gonabadi et al. 2020; Kiendl and Gao 2020). Based on previous research it may be concluded that this anisotropy has several sources related to the manufacturing process. The anisotropic behaviour in the mesostructure has itself two main sources: The intermediate gaps and interfaces forming between the individual polymer lines. The material behaviour of the interfaces is assumed to differ from that of the base polymer and exhibit a high dependency on the temperature history.

In the scope of this contribution the temperature-dependent influence of mesostrcuture-interfaces on the mechanical properties of FFF printed structures will be investigated in uniaxial tensile tests on the example of polyamide 6 (PA6). To this end, the relevant theoretical background as well as the potential influencing factors of the FFF process are briefly discussed in Sect. 9.1. In Sect. 9.2 the test specimen production and testing methodologies are discussed with the goal of isolating the effect of the interfaces and controlling for undesired influences. In Sect. 9.3, the results of the mechanical testing are discussed, followed by a summary in Sect. 9.4.

9.1.1 Characteristic Mesostructure

The anisotropic behaviour of the mesostructure has itself two main sources: intermediate gaps and interfaces (which will be the main focus of this contribution). The intermediate gaps form during the printing process as the polymer is not capable of filling the space between two printed lines fully. By the nature of the manufacturing process, these gaps are locally oriented inside the part, according to the path planning done during the slicing process. Interfaces, in the context of this paper, refer to the area of contact between two printed lines, as can be seen in Fig. 9.1. These interfaces are formed as the molten polymer of a newly printed line makes contact with previously deposited material. The polymer chains, driven by Brownian motion, begin to diffuse across the contact area, i.e. the interface, linking the previously unconnected polymer networks of the deposited lines. It is assumed that such diffusion processes can only take place above the glass transition temperature of the material. Given enough time and sufficient heat, the polymer networks across the interface may be indifferentiable from the base material itself, resulting in a perfect bond between the printed lines and mechanical properties equal to that of the base material (Butler et al. 1998; Doi and Edwards 1978).

Due to the rapid cooling rates in the FFF process, the diffusion process is typically prevented from completion. The resultant interfaces then exhibit significant differences in mechanical behaviour, such as a lower material strength (Butler et al. 1998; Doi and Edwards 1978). The influence of such interfaces on the material strength of FFF printed structures has been studied in recent years by Consul et al. (2021), Bartolai et al. (2018), and Coogan et al. (2020). They respectively based their work on research by Butler et al. (1998) regarding the fusion bonding process of polymers. Butler et al. (1998) developed an equation relating the material strength of a polymer-polymer interface to the process temperature and contact pressure during the fusion bonding process. They based their theory on research concerning diffusion processes at polymer interfaces by Kim and Wool (1983) and Wool and O'Connor (1981), who in turn relied on fundamental works regarding the dynamics of molecular movement in polymer melts by Doi and Edwards (1978) and DeGenns (1982). If, as in the simplification by Consul et al. (2021), the contact pressure is ignored, a relation for the tensile strength of the interface can be established according to Butler et al.:

$$\frac{\sigma}{\sigma_\infty} = \left(\frac{t}{t_R}\right)^{\frac{1}{4}} \tag{9.1}$$

With the material strength σ of the interface, the material strength of the base material σ_∞, the reptation time t_R, and the time t which the interface spent at a specified constant temperature. The reptation time represents a characteristic property of the polymer melt, quantifying the mobility of a polymer chain inside the polymer melt at a given temperature. The precise definition of the reptation time can be found in Wool et al. (1981).

In the FFF process the interfaces do not remain at a constant temperature, but rather exhibit a continuous thermal history including cooling and reheating over time. Additionally, the temperature history can vary locally due to complex geometries and the print paths required to create them. Furthermore, several process parameters defined during the slicing process can influence the temperature history as well as the interface dimensions.

In general, all process parameters may be divided into two categories: Firstly, geometric parameters (linewidth, layerheight), which determine the interface surface-area and volume of the individual polymer lines, which in turn determine the structures ability to retain heat. Secondly, process parameters influencing the supply of heat to the material (various temperature settings, movement speed).

9.1.2 Crystallinity

In the context of this contribution, it is necessary to discuss another process-dependent factor: The crystallinity of the polymer. Many of the staple FFF materials such as PLA, ABS, or PET, as well as the PA6 used in this contribution, are semi-crystalline thermoplastic polymers. With crystallinity referring to a locally ordered state of the molecular chains. The mechanical and physical properties of such polymers depend on the proportion of crystalline structure relative to the non-crystalline, i.e. amorphous, portion. The so-called degree of crystallinity can then be defined as the ratio of the mass of the crystalline portion and the total mass of the polymer. In general, it can be stated that a higher degree of crystallinity leads to a stiffer material and brittle fracture behaviour, while the same material with a lower degree of crystallinity will be comparatively weaker and exhibit ductile fracture behaviour. In general, most semi-crystalline polymers are neither purely amorphous nor crystalline, with the maximum degree of crystallinity dependent on the specific material (Ehrenstein et al. 2004).

The degree of crystallinity depends on the thermal history of the polymer. In the context of the FFF process this mostly relates to the cooling process, as all crystalline regions get dissolved during the melting in the printing process. The high cooling rates during the printing process lead to low degrees of crystallinity with the molecular chains swiftly losing their mobility, and therefore their ability to arrange themselves in crystalline structures, as the material loses temperature. Besides the thermodynamic properties of the material, the specific thermal history in any given region of a FFF printed part is dependent on the mass of already accumulated material. Bigger material accumulations cool slower, resulting in a higher degree of crystallinity. It is common practice in FFF printing to use infill structures instead of solid material, leading to the locally deposited amount of material varying widely. This may result in locally different material properties throughout the printed part due to variations in crystallinity.

9.1.3 Goal of this Contribution

The objective of the study is to quantify the effect of the thermal history on the tensile properties across the interfaces in the mesostructure, while simultaneously controlling the influence of crystallinity. To this end, samples are produced and tested under uniaxial tensile loading to investigate the influence of interfaces on the mechanical properties. Further, differential scanning calorimetry (DSC) is employed to control for the degree of crystallinity present in the material. This is achieved by the combination of two approaches.

Firstly, sample sets are produced at different chamber temperatures with the goal of slowing the cooling process in the interfaces allowing for greater interdiffusion between the polymer lines and thus stronger material properties.

Secondly, the use of specialized sample geometry described by Allum et al. (2020). In their 2020 paper, they presented a novel sample geometry for uniaxial tensile tests specific for the FFF process. They employed a custom piece of software called *Full-Control*, which allowed them to print boxes with a wall thickness of one extrusion line, while varying the linewidth during the printing process along the printed lines as well as across the printed layers, as depicted in Fig. 9.2. By subsequently cutting stripes from the printed boxes, they effectively produced tensile samples with a thickness of one extrusion line, while simultaneously retaining the tapered dog-bone/hourglass shape employed in standard testing samples as defined in the DIN ISO 527-2 standard (DIN EN ISO 527-2 2012). Allum et al. (2020) proved the applicability of their sample geometries by comparing them to injection moulded samples, with polylactic acid (PLA) as their chosen material.

In the context of this contribution, the main advantage of such specimen geometries lies in the avoidance of any internal mesostructure and the simplification of the temperature field during the printing process.

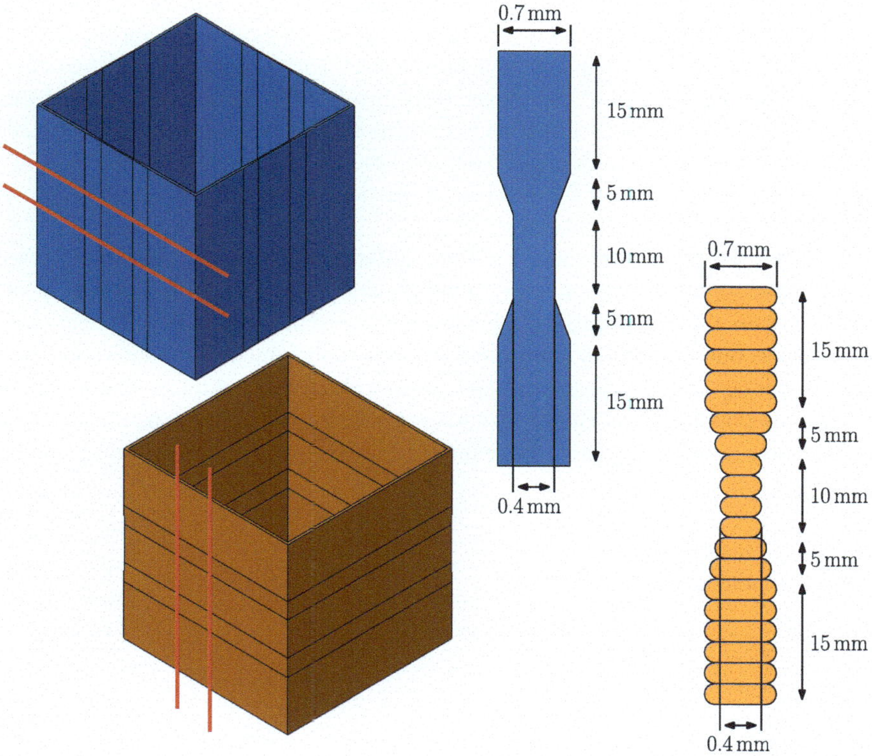

Fig. 9.2 Thin-walled samples based on Allum et al. (2020) with the 0° direction in *Blue* and the 90° direction in *Orange*

9.2 Sample Production and Testing Methodology

9.2.1 Sample Production

As discussed above, custom geometries for uniaxial tensile test, as described by Allum et al. (2020), are used in this study. To this end, boxes of 50 mm × 50 mm × 50 mm dimensions were printed on a *German RepRap X500* using a custom G-code generated in FullControl, realizing two versions of the boxes.

Firstly, a version in which the width of the individual polymer lines is varied across the layers. Resulting in the geometry depicted in orange in Fig. 9.2, suited for tensile testing across the polymer interfaces. Secondly, a version in which the width of the individual polymer lines is varied along the lines as they are printed. Resulting in the geometry depicted in blue in Fig. 9.2, suited for tensile testing parallel to the polymer interfaces. In both cases, this results in a tapered measurement region and a thicker clamping region to ensure failure in the central part of the sample, while reducing the influence of clamping pressure.

After printing, 10 mm wide stripes are cut from the boxes with the printed lines oriented either orthogonal (first box variety) or parallel (second box variety) to the loading direction present in the uniaxial tensile tests. These orientations are hereafter referred to as the 90° and 0° directions, respectively. It is further assumed that the material properties of the 0° direction may be representative of the base material as the polymer interfaces are oriented parallel to the loading direction, while the 90° direction is representative of the interface properties, as the load is distributed across the interfaces during tensile testing.

Each box variety was produced at three different chamber temperatures (room temperature, 50, 60 °C) with all other printing parameters (see Table 9.1) kept identical. Additionally to the built-in temperature sensor of the printer, the chamber temperature was monitored with an external thermocouple which was positioned close to the printing position and approximately 30 mm above the buildplate. For the room temperature condition, the chamber remained unheated. Due to the heat from both the buildplate and the extruder, the ambient temperature close to the print position

Table 9.1 Applied printing parameters

Constant printing parameters		
Layer height	0.2	mm
Movement speed	50	mm/s
Extrusion temp.	285	°C
Build plate temp.	60	°C
Varied printing parameters		
Line width	0.4–0.7	mm
Chamber temp.	40–60	°C

rose to approximately 40 °C, as measured by the external thermocouple, while the built-in sensor remained at approximately 30 °C. For 50 and 60 °C both sensors were in good accordance with each other.

The samples were produced with the PA6 filament *f3 Pa Pure Pro* produced by fiber three, which was dried according to manufacturer specifications prior to printing. After production the samples were stored under controlled climate conditions at 30°C and a relative humidity of less 2% to prevent moisture uptake in the material. Before mechanical testing a speckle pattern for contactless local strain measurement via digital image correlation (DIC) was applied to the samples.

9.2.2 Testing Methodology—DSC

Differential scanning calorimetry (DSC) is an established method for determining the degree of crystallinity of thermoplastic polymers. DSC samples were cut from the centre of the tapered measurement region and subjected to a heat/cool/heat cycle between 0 and 260 °C at a constant heating/cooling rate of 10 K/min (Frick and Stern 2013). Measurements were performed with a *TA Instruments DSC Q2000*. For the evaluation of crystallinity, only the first heating cycle is relevant, as it represents the material in the as is state. Measurements were taken for all utilized chamber temperatures and compared regarding the measured degree of crystallinity.

9.2.3 Testing Methodology—Tensile Tests

All samples were subjected to uniaxial tensile tests on a *Zwick Roell Z020* testing machine until either failure or significant necking occurred. Tests were performed at a constant deformation rate of 0.03 mm/min (which corresponds to a constant strain rate of 1 %/min for the clamping distance 30 mm) with a 2 N preload. Strains were measured locally using a *LIMESS Q400* DIC system with the *Intra4D* software. A sample size of 5 was chosen for each combination of direction and chamber temperature.

Since the custom sample geometry does not provide a rectangular cross section (see Fig. 9.3) like typical tensile specimen, the cross sections calculated from caliper measurements were adjusted with correction factors calculated for each of the 6 groups of samples by microscopic imaging. For the 0° direction the ratio between the actual cross section area and a rectangular cross section as measured by caliper was chosen as the correction factor. For the 90° direction the ratio between the width of the contact area and the maximum width of the printed lines was chosen as the correction factor.

Subsequently, the results were evaluated regarding their Young's modulus and tensile strength according to DIN ISO 527-1 (2012).

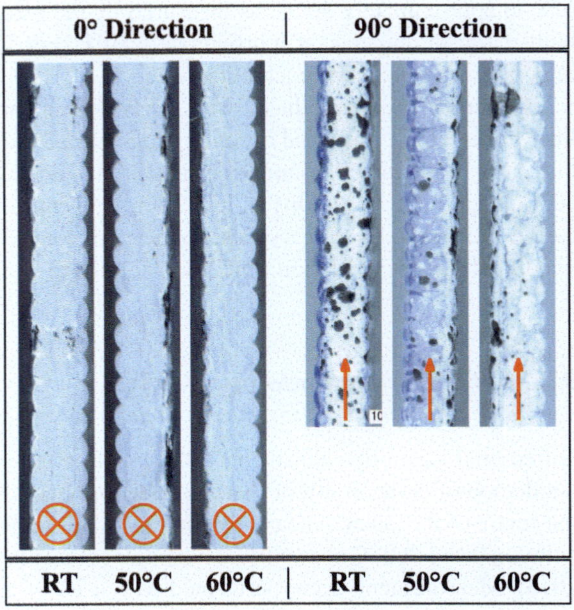

Fig. 9.3 Microscopy images of thin-walled specimen with the loading direction of uniaxial tensile test indicated by red arrows

9.3 Results

9.3.1 DSC

The first heating and cooling cycle measured in the DSC are displayed in Fig. 9.4 for the three chamber temperatures. The curves depict a behaviour typical for semi-crystalline thermoplastic polymers. The glass transition temperature can be identified at $72.86 \pm 1.03\,°C$ with a pronounced enthalpic recovery present for all measurements, correlating well with the datasheet value of 72 °C (Fiberthree GmbH 2021). Between 100 and 125 °C an exothermic recrystallization peak can be identified, indicating a low degree of crystallinity in the material prior to testing. The melting region, between 190 and 240 °C, exhibits a pronounced shoulder or plateau before reaching a peak at $235.54 \pm 0.62\,°C$, which in the case of PA6 is associated with the melting of the crystalline γ phase prior to the melting of the crystalline α phase around the peak temperature. The formation of such γ phases is associated with fast cooling rates from the melt (Kolesov et al. 2014), as they are present in FFF printing. This assumption can be substantiated by comparing the first heating cycle, in Fig. 9.4, to the second heating cycle in Fig. 9.5. After controlled cooling at 10 K/min, which is many times slower than the cooling during additive manufacturing, the γ plateau in the melting region is completely absent (Frick and Stern 2013; Kolesov et al. 2014).

The degree of crystallinity was evaluated according to Ehrenstein et al. (2004), by using separate baselines for the melting and recrystallization peak, following the formula:

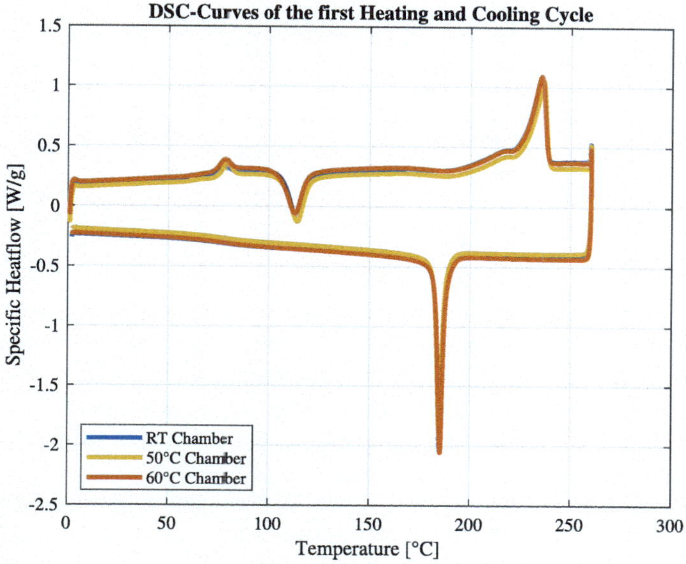

Fig. 9.4 DSC-curves of the first heating/cooling cycle for the three employed chamber temperatures

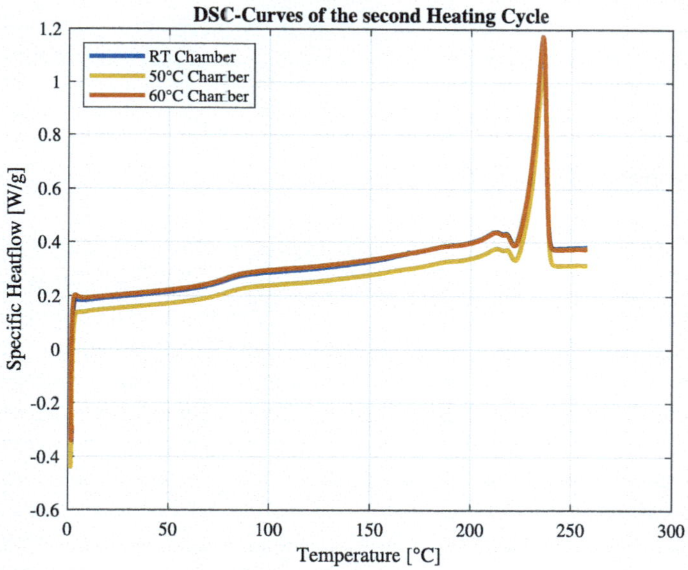

Fig. 9.5 DSC-curves of the second heating cycle for the three employed chamber temperatures

$$\chi_c = \frac{\Delta H_m - \Delta H_c}{\Delta H_{100\%}} \tag{9.2}$$

Table 9.2 Results of crystallinity evaluation

Chamber Temp.	ΔH_m	ΔH_c	$H_{100\%}$	χ_c
°C	J/g	J/g	J/g	%
RT	51.78	24.69	230	11.78
50	49.56	23.48	230	11.34
60	52.00	23.79	230	12.67

With the melting enthalpy H_m and the crystallization enthalpy H_c, which can be obtained from Fig. 9.4 by calculating the area enclosed between the measured curve and the constructed baseline from the melting and recrystallization peak, respectively. Using separate baselines for melting and recrystallization, ΔH_m andΔH_c were evaluated using the *TA Universal Analysis Software*. For the heat of fusion $\Delta H_{100\%}$ of PA6 a reference value of 230 J/g was chosen.

The results are summarized in Table 9.2. Considering that even under ideal conditions a measuring inaccuracy of 0.4 J/g or greater is to be expected (Frick and Stern 2013), it can be established that the variation in chamber temperature to the extent employed in this study had no effect on the degree of crystallinity, with a value of approximately 12%.

Therefore, it is assumed that any differences in the mechanical properties may be attributed solely to the polymer interfaces rather than the crystallinity of the material.

9.3.2 *Uniaxial Tensile Tests*

The results of the uniaxial tensile tests are summarized in Figs. 9.6 and 9.7. Figure 9.6 shows the averaged stress-strain curves for the six tested groups as well as the results of the individual samples, while Fig. 9.7 shows a comparison of the averaged curves for all tested groups. It needs to be noted that the averaged curves are only displayed up to the lowest strain at failure measured among the five repeat measurements. Further, it should be mentioned that the point of failure depicted in Figs. 9.6 and 9.7 represents the point at which the DIC pattern fails rather than the material itself.

For the samples of the 90° direction, these points of failure are identical, as all samples exhibited brittle failure behaviour and the DIC measurements could be evaluated up to the point of material failure. For the samples of the 0° direction, ductile failure with pronounced necking occurred. This led to a failure of DIC pattern prior to the failure of the samples. Therefore, these samples were only evaluated until the failure of the DIC pattern. An exemplary representation of the described behaviour is depicted in Fig. 9.8.

Regarding the 0° direction, no substantial change between room temperature and 50 °C samples can be observed, while samples manufactured at 60 °C samples show a slight increase in Young's modulus and tensile strength. Taking into account

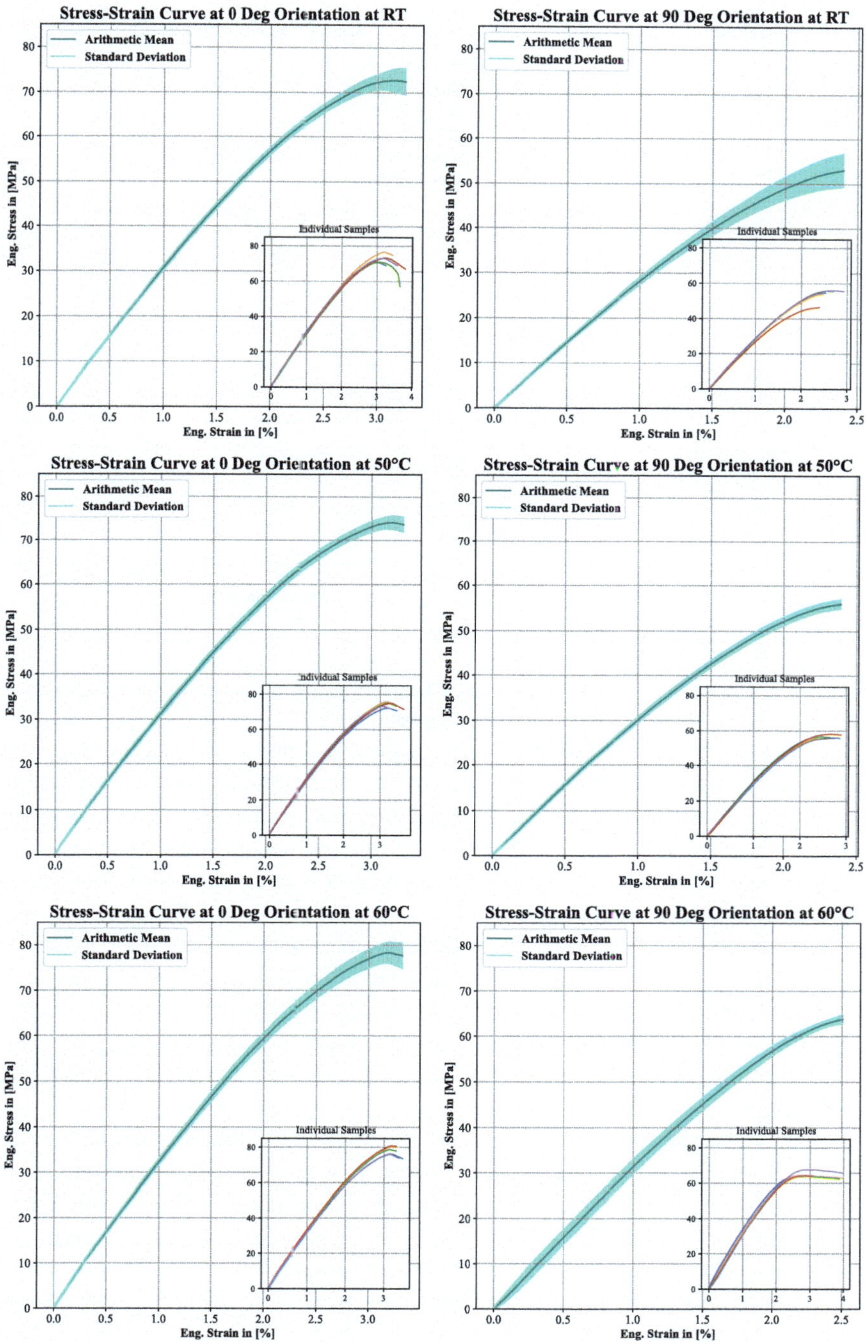

Fig. 9.6 Results of uniaxial tensile measurements for all tested groups

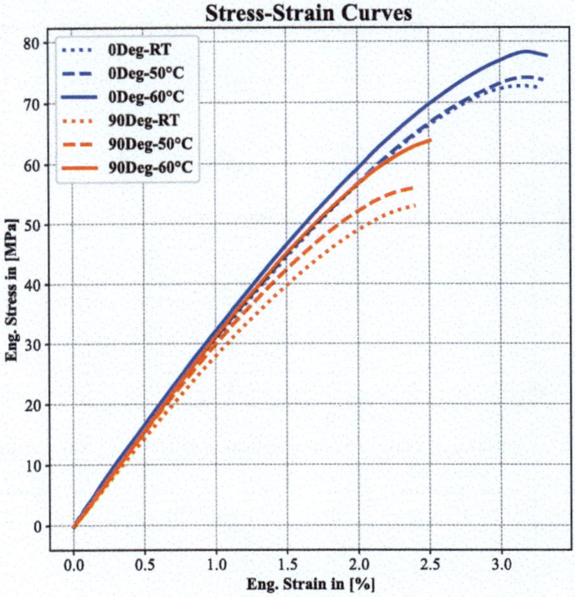

Fig. 9.7 Averaged stress-strain curves for all tested groups

the observed standard deviations, no clear trend regarding the chamber temperature can be established in 0° direction. This correlates well with the prior expectations, as for the 0° direction the polymer interfaces run parallel to the loading direction and are assumed to have little influence on the measurement. These samples are thus considered to be representative of the material properties. This assumption is substantiated by the measured Young's moduli and tensile strength, as listed in Table 9.3, which are comparable to the values given in the datasheet (Fiberthree GmbH 2021) of the manufacturer.

Regarding the 90°, direction a clear trend is visible. Higher chamber temperatures result in higher tensile strength, while the strain at break remains nearly identical. Although Young's moduli remain similar in the defined measurement region (0.25–0.5%), the samples fabricated at higher chamber temperatures retain a steeper incline at higher strains. These observations are in congruence with the theory discussed in Sect. 9.1.1.

Compared to the 0° direction all samples of the 90° direction display lower Young's moduli and far lower tensile strength as well as a lower strain at break. It can however be noted that the 90° samples produced at 60 °C show a curve progression similar to that of the 0° direction, while retaining a lower failure strain.

Regarding the clear difference in the failure strains between the two printing directions, it may be plausible to assume that this may not be solely due to the lesser strength of the polymer interfaces. The surface of the 90° direction samples displays a series of successive notches, which may lead to stress localizations and premature failure. The microscopic measurements of sample cross sections (see Fig. 9.3) did show, however, that the variation in chamber temperature did not have an effect on

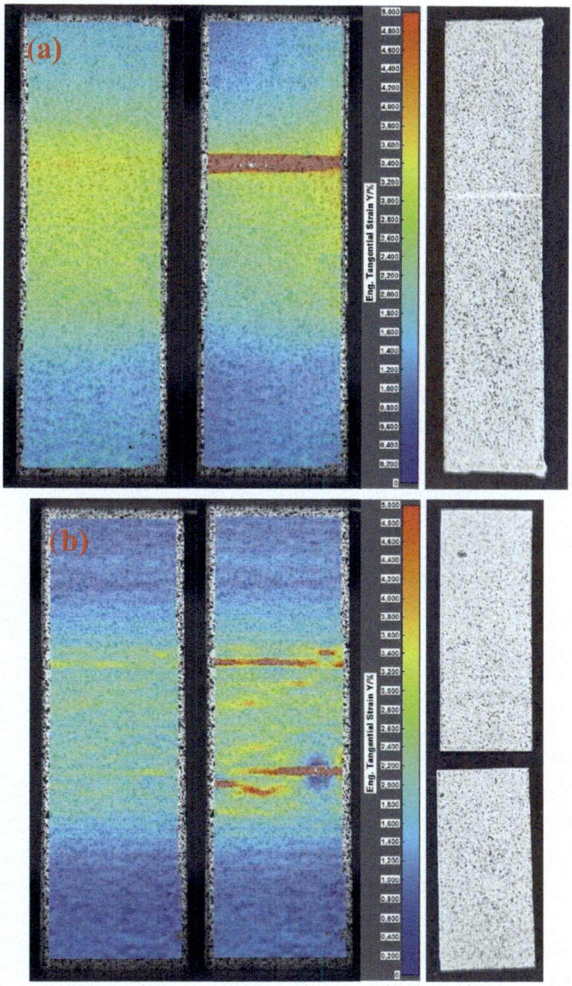

Fig. 9.8 Depiction of DIC images and samples post testing for exemplary specimen of **a** the 0° direction and **b** the 90° direction

Table 9.3 Results of tensile tests

Direction	**Chamber temp.**	**Young's modulus**	**Tensile strength**
0°	RT	3252 ± 213 MPa	73 ± 2 MPa
0°	50 °C	3266 ± 106 MPa	74 ± 1 MPa
0°	60 °C	3510 ± 88 MPa	78 ± 2 MPa
90°	RT	2947 ± 107 MPa	53 ± 4 MPa
90°	50 °C	3145 ± 111 MPa	56 ± 1 MPa
90°	60 °C	3133 ± 464 MPa	64 ± 1 MPa
Datasheet	–	3400 MPa	80 MPa

the local dimensions of the individual polymer lines. Thus, it can be assumed that the differences due to chamber temperature may be attributed to temperature-related changes in the polymer interfaces.

9.4 Summary and Outlook

In this contribution, the properties of polymer-polymer interfaces formed during the FFF printing of polyamide 6 were studied in uniaxial tensile tests. To this end, custom tensile samples inspired by Allum et al. (2020) were utilized. Furthermore, samples were produced at three different chamber temperatures with the intention of changing the temperature history in the polymer interfaces and thus altering their mechanical properties. DSC measurements were performed to control for any changes in the crystallinity of the material due to chamber temperature. Ultimately, the findings of this contribution can be summarized as follows:

- Thin-walled samples can be reliably manufactured and deliver reproducible results under uniaxial loading, comparable to literature values.
- Samples produced in the 90° direction exhibit lower Young's moduli, tensile strength, and strain at failure than samples of the 0° direction.
- Increased chamber temperatures during production lead to higher tensile strength in 90° direction samples, while no change in regard to the degree of crystallinity was measurable bis DSC.

As elaborated in Sect. 9.3.2, the influence of the notched sample surface on the mechanical measurements (especially in regard to the 90° direction) has to be further investigated in order to eliminate any potential influence of the geometry and isolate it from the temperature-dependent material properties of the polymer interfaces.

A further point of interest is the application of the thin-walled samples to testing methodologies for viscoelastic material behaviour, such as relaxation test or dynamic mechanical analysis, as according to theory the polymer interfaces should exhibit varying material properties in such tests as well.

Acknowledgements This research [project FLAB-3Dprint] is funded by dtec.bw—Digitalization and Technology Research Center of the Bundeswehr which we gratefully acknowledge. dtec.bw is funded by the European Union—NextGenerationEU.

References

Ahn SH, Montero M, Odell D, Roundy S, Wright PK (2002) Anisotropic material properties of fused deposition modeling abs. Rapid Prototyp J 8(4):248–257

Allum J, Moetazedian A, Gleadall A, Silberschmidt VV (2020) Interlayer bonding has bulk-material strength in extrusion additive manufacturing: new understanding of anisotropy. Addit Manuf 34:101297

Bartolai J, Simpson TW, Xie R (2018) Predicting strength of additively manufactured thermoplastic polymer parts produced using material extrusion. Rapid Prototyp J 24(2):321–332

Butler CA, Mccullough RL, Pitchumani R, Gillespie JW (1998) An analysis of mechanisms governing fusion bonding of thermoplastic composites. J Thermoplast Compos Mater 11(4):338–363

Consul P, Chaplin A, Tagscherer N, Zaremba S, Drechsler K (2021) Interlaminar strength in large-scale additive manufacturing of slow crystallizing polyaryletherketone carbon composites. Polym Int 70(8):1099–1108

Coogan TJ, Kazmer DO (2020) Prediction of interlayer strength in material extrusion additive manufacturing. Addit Manuf 35:101368

Dey A, Yodo N (2019) A systematic survey of fdm process parameter optimization and their influence on part characteristics. J Manuf Mater Process 3(3):64

DIN EN ISO 527-1 (2012) Kunststoffe - Bestimmung der Zugeigenschaften: Teil 1: Allgemeine Grundsätze

DIN EN ISO 527-2 (2012) Kunststoffe - Bestimmung der Zugeigenschaften: Teil 2: Prüfbedingungen für Form- und Extrusionsmassen

Doi M, Edwards SF (1978) Dynamics of concentrated polymer systems. part 1.—brownian motion in the equilibrium state. J Chem Soc Faraday Trans 2 74(0):1789–1801

Ehrenstein GW, Riedel G, Trawiel P (2004) Thermal analysis of plastics: theory and practice. Hanser, Munich

Fiberthree GmbH (2021) F3 PA Pure Pro / F3 PA Pure Pro Black, Technical datasheet (2021). https://www.fiberthree.com/downloads/tdb/filaments/TDS_F3_PA_Pure_Pro_English.pdf

Frick A, Stern C (2013) DSC-Prüfung in der Anwendung, 2nd edn. Hanser Verlag, München, Hanser eLibrary

Gebisa AW, Lemu HG (2019) Influence of 3d printing fdm process parameters on tensile property of ultem 9085. Procedia Manuf 30:331–338

de Gennes PG, Leger L (1982) Dynamics of entangled polymer chains. Annu Rev Phys Chem 33(1):49–61

Gibson I, Rosen D, Stucker B (2015) Additive manufacturing technologies: 3D printing, rapid prototyping and direct digital manufacturing, second, edition. Springer, New York and Heidelberg and Dodrecht and London

Gonabadi H, Yadav A, Bull SJ (2020) The effect of processing parameters on the mechanical characteristics of pla produced by a 3d fff printer. Int J Adv Manuf Tech 111(3–4):695–709

Kiendl J, Gao C (2020) Controlling toughness and strength of fdm 3d-printed pla components through the raster layup. Compos B Eng 180:107562

Kim YH, Wool RP (1983) A theory of healing at a polymer-polymer interface. Macromolecules 16(7):1115–1120

Kolesov I, Mileva D, Androsch R (2014) Mechanical behavior and optical transparency of polyamide 6 of different morphology formed by variation of the pathway of crystallization. Polym Bull 71(3):581–593

Wool RP, O'Connor KM (1981) A theory crack healing in polymers. J Appl Phys 52(10):5953–5963

Chapter 10
A Continuum Thermomechanical Approach for Modelling Rate-Dependent Material Behaviour of Laser Powder Bed Fusion AlSi10Mg Under Consideration of Damage Effects

Lukas Richter, Daniela Schob, Philipp Maasch, Robert Roszak, and Matthias Ziegenhorn

Abstract This work presents a continuum thermomechanical approach for modelling the behaviour of a laser powder bed fusion manufactured AlSi10Mg. Thermomechanical means that the analysis methods include the material's self-heating effect during inelastic deformations. This effect is directly coupled to the deformation process and can additionally be used for the parameter identification. The research introduces a method for the coupled thermomechanical experimental observation, the enhancement of the theoretical framework for a fully thermomechanical material description, and an approach for the parameter identification and validation. To describe the rate-dependent behaviour under consideration of damage effects, a coupled Chaboche–Gurson–Tvergaard–Needleman (GTN) model was utilised. For modelling high complex stress states with a good accuracy between the experiment and the simulation, the Chaboche–GTN model was enhanced with an approach that includes the effects of low stress triaxialities. In order to enable a fully coupled

L. Richter (✉) · D. Schob · P. Maasch · M. Ziegenhorn
Chair of Engineering Mechanics and Machine Dynamics, Brandenburg University of Technology, Cottbus-Senftenberg, Universitätsplatz 1, 01968 Senftenberg, Germany
e-mail: lukas.richter@b-tu.de

D. Schob
e-mail: daniela.schob@b-tu.de

P. Maasch
e-mail: philipp.maasch@b-tu.de

M. Ziegenhorn
e-mail: matthias.ziegenhorn@b-tu.de

R. Roszak
Institute of Applied Mechanics, Poznan University of Technology, Pl. M. Sklodowskiej-Curie 5, 60-965 Poznan, Poland
e-mail: robert.roszak@put.poznan.pl

H. Altenbach et al. (eds.), *Lectures Notes on Advanced Structured Materials 3*,
Advanced Structured Materials 226, https://doi.org/10.1007/978-3-031-77403-4_10

simulation of the deformation and temperature field for complex geometries, the model is implemented in the finite element programme Abaqus via a user material subroutine (UMAT).

10.1 Introduction

Additive manufacturing technologies enable the cost-efficient production of components and small series with complex structures and help to reduce the number of individual parts and assembly steps needed. This technology leads to a significant reduction in material requirements and the number of production steps as a result of the direct implementation of functional elements such as cooling channels, joints, and springs in the manufacturing process. Additive manufacturing is used in the fields of aerospace (Murr 2016), medical technology (Goyanes et al. 2015), toolmaking (Asnafi et al. 2020), and energy technology (Sun et al. 2021). Currently, there is a lack of specific, universally applicable standards for determining the material properties of AM materials. This means that existing standards that are unsuitable are often used (Forster 2015).

The laser powder bed fusion (LPBF) process and the direct energy deposition (DED) method dominate in the manufacturing of metal components (Ngo et al. 2018; Sames et al. 2016). With its ability to realise finer layers (10–50 m), LPBF is the leading additive manufacturing technology. In addition, this process impresses with its excellent flexibility in the production of complex structures and the achievement of outstanding mechanical properties (Cooke et al. 2020; D'Andrea 2023).

The main objective of this research is the numerical description of the material behaviour with respect to LPBF AlSi10Mg to make this material suitable for industrial applications. All information on the manufacture of the samples are available from Rümmler et al. (2024). To reach the main objective, the thermomechanical approach of Sparr (2023) and the coupled Chaboche–GTN model of Schob (2023) are combined and extended from a one-dimensional point problem to a three-dimensional volumetric problem. In order to make the model applicable to a large range of applications, the damage approach is modified. This extension of the GTN model is essential because it is unable to represent the damage in shear-dominated deformation processes (Nahshon and Hutchinson 2008; Reis et al. 2010; Gatea et al. 2017). In a superposition of tension and shear (low stress triaxialities), additional damage happens due to the rotation and deformation of the voids (Barsoum and Faleskog 2007). Literature specifies that shear localisation and shear crack formation lead to the failure of ductile materials (McClintock 1971) and that an increase in stress triaxiality reduces ductility (Johnson and Cook 1985). These aspects motivate the enhancement of the damage model according to the approaches of Nahshon and Hutchinson (2008).

For a methodological procedure, the directly determinable and indirectly determinable parameters for the material model are analysed first (Schob 2023). Directly determinable means in this case that the parameters can be identified in specific experiments. To determine the indirectly determinable parameters, thermomechanical tests are performed. The model parameters were identified based on the results of the thermomechanical tests and validated on a complex geometry in Abaqus applying a user material subroutine (UMAT).

The work is organised as follows. Sections 10.2.1–10.2.3 deal with the implementation and the results of the thermomechanical tests. Motivated by the experiments, Sects. 10.3.1–10.3.4 establish the theoretical framework. In Sects. 10.4.1–10.4.3, the results of parameter identification and validation are presented. This section additionally introduces the basic assumptions for the finite element (FE) simulations. At the end of this paper there is a brief outlook in Sect. 10.5.

10.2 Experimental Observation

10.2.1 Direct Determinable Parameters

For the determination of the direct parameters specific experiments are performed and documented in the following part. With these experiments the following parameters are identified:

1. the densities of the material ρ,
2. the specific heat capacity c^{σ},
3. the thermal expansion coefficient α,
4. the thermal conductivity k,
5. the initial void volume fraction $\hat{f}^{*}$, and
6. the critical value of void volume fraction before break $\hat{f}^{F}$.

The ABJ220-4NM analytical weighing instrument from KERN is used to determine the density of the material ρ. The weight of ten samples is recorded in order to statistically validate the measurements. The mass density ρ is calculated based on the ratio of the sample mass to the sample volume.

A calorimeter is applied to measure the specific heat capacity c^{σ} of the material (see Fig. 10.1). This calorimeter has a known heat capacity of $c^{\mathrm{k}} = 70\ \mathrm{J\ K^{-1}}$. A defined number of sample pieces with the mass m^{F} are heated first. In order to obtain a high temperature difference, the samples are heated overnight to a temperature $\Theta^{\mathrm{W}} = 250\,^{\circ}\mathrm{C}$. Cold, measured water with the mass m^{Fl} and the temperature $\Theta^{\mathrm{F}} = 7.6\,^{\circ}\mathrm{C}$ is then added to the calorimeter. Water is selected because its specific heat capacity c^{Fl} is known. The samples are then transported into the cold water and the temperature is detected with a PT100 sensor. The CASSY-Lab programme processes the data from the sensor. An experiment is completed when the measuring system no longer indicates a temperature increase and the mixing temperature Θ^{m} has been reached. The specific heat capacity $c\sigma$ of the material is calculated based on the temperature and mass differences according to Eq. (10.1).

Fig. 10.1 Schematic for determining the specific heat capacity c^{σ}

$$c^{\sigma} = \frac{\left(c^{\mathrm{Fl}} m^{\mathrm{Fl}} + c^{\mathrm{k}}\right)\left(\Theta^{\mathrm{m}} - \Theta^{\mathrm{Fl}}\right)}{m^{\mathrm{F}}\left(\Theta^{\mathrm{F}} - \Theta^{\mathrm{m}}\right)} \tag{10.1}$$

The coefficient of thermal expansion α can be determined on the basis of Fig. 10.2. The ARAMIS system and the IR camera (ImageIR® 8300 hp by INFRATEC) are used for the measurement. A heating plate enables the sample to be positioned at one end and heated by 30 K. The coupled measuring system is then analysed to determine the deformation change and the temperature change in a defined range. Three samples are tested for the investigation.

The thermal conductivity k is determined with the LaTima® measuring system from NANOTEST. Three samples are used to measure the parameter. The operating principle of the measuring system is explained in Fig. 10.3. A rectangular material sample is connected to two plates with heat-conducting paste. One plate is cooled in a controlled manner during the measurement and the other is heated in a controlled

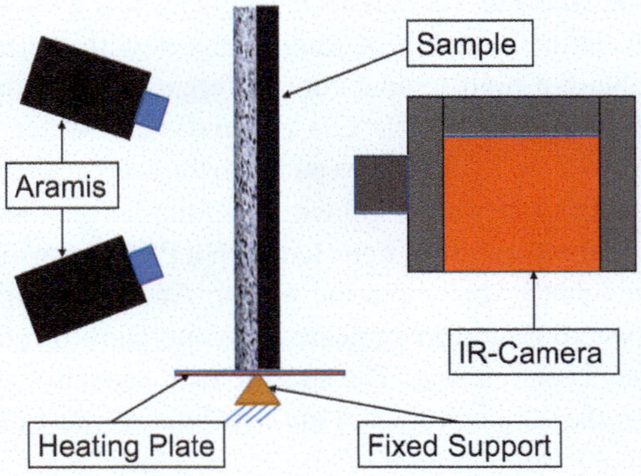

Fig. 10.2 Experimental setup for determining the thermal expansion coefficient α

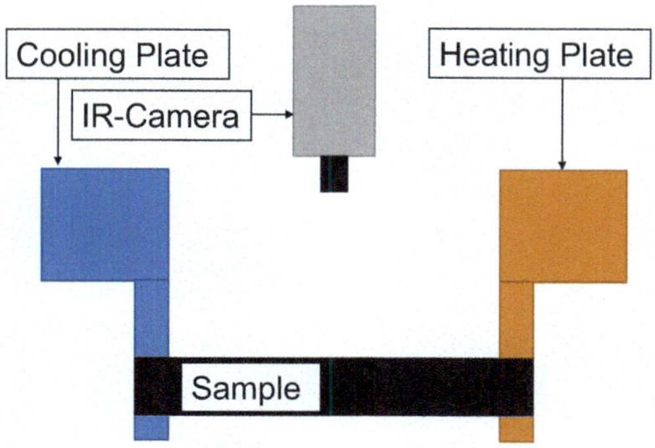

Fig. 10.3 Determination of thermal conductivity k

manner. By connecting the two plates to the sample, a heat flow $\dot{Q}$ is generated in the sample. The temperature differences can be determined with an IR camera. The software of the LaTima® measuring system determines the thermal conductivity from the thermogram.

In order to evaluate the existing material structure, the initial void volume fraction $\hat{f}^0$ and the critical damage value $\hat{f}^F$ are measured by experiment. Corresponding tests are performed with the GE vltomelx m 300/180 computer tomograph. The measuring system generates a 3D volume of the test specimen using a flat-panel detector. The detector works with a voltage of 180 kV and a current of 400 A. The initial void volume fraction $\hat{f}^0$ is determined for six samples. These are detected before the tests with a voxel size of 7.7 μm, an integration time of 200 ms, and 1500 projections. Ex-situ tests are used to determine the critical damage value $\hat{f}^F$. In these tests, three statically loaded samples are scanned again after damage.

10.2.2 Indirect Determinable Parameters

An identification of the indirectly determinable parameters is based on the analysis of material characteristics. In the case of thermomechanical material modelling, the experiment delivers the stress–strain and temperature response of the material under mechanical loading. The setup for the thermomechanical experiment is based on Golasinski et al. (2023) and Rose and Menzel (2020) and is shown in Fig. 10.4. The temperature field is measured with an IR camera and the deformation field is measured by the ARAMIS system. A servo-hydraulic testing machine from the ZWICK / ROELL HC series with a 25 kN load cell generates the mechanical loading.

After the measurement, the deformation field is calculated using the GOM Corrolate® software. The IRBIS® 3 analysis tool is used to generate the temperature field from the thermograms. The method for analysing the two fields is illustrated in Fig. 10.5. First, the ARAMIS system is converted to a half recording. This specification enables the realisation of recording frequencies $f = 200$ Hz. The damage point (Fig. 10.5, red rectangle at time $t = t_2$) corresponds to the evaluation range.

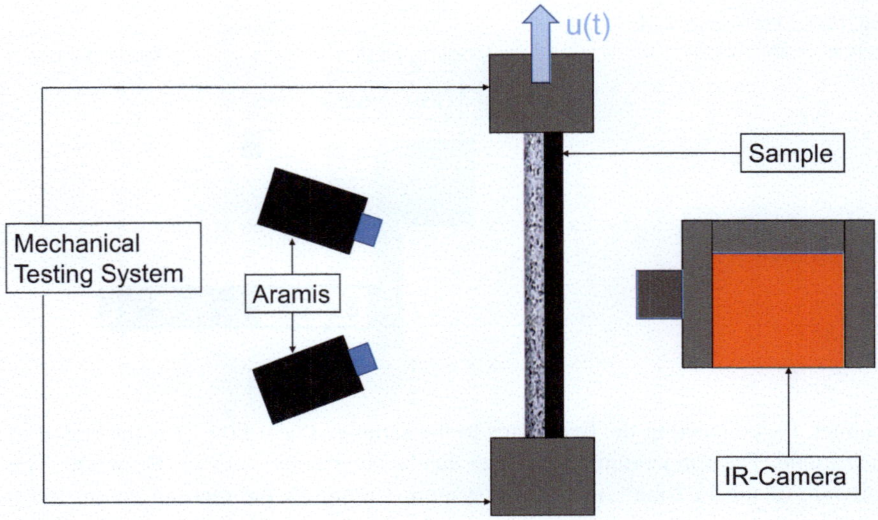

Fig. 10.4 Experimental setup with IR camera for a temperature field measurement and ARAMIS for a deformation field measurement

Both analysis tools calculate the mean value from the measurement results for this area. The shown process can be explained in more detail based on the thermograms or the images from the IR camera. The time $t = t_0$ represents the reference configuration in which a uniform surface temperature can be recognised on the sample. At time $t = t_1$, inelastic deformations are already observed. These are characterised as a result of the heating of the sample. The maximum value of the temperature rise can be identified at the point of failure. Finally, at time $t = t_2$, both images show the destruction of the material.

10.2.3 Overview of the Experimental Results

Experimental methods include simple and complex load cases in order to extensively test the material behaviour of AlSi10Mg. The simplest load case is the tensile test. In the context of this work, tensile tests are realised on circular samples. The geometry corresponds to DIN 50125-A 5 x 25 (Deutsches Institut für Normung e. V. 2022) and is shown in Fig. 10.6a. In the complex load case of stress triaxialities, the shear and tensile stress superposition is realised by the geometry of the specimen. A methodology for the design of shear-stressed components is described in ASTM B 831:2022 (ASTM 2022). If these standardised specimens are used, problems of buckling occur in the shear zone due to the presence of additional moments (Gatea et al. 2017). Since the model initially deals with the pure superpo-

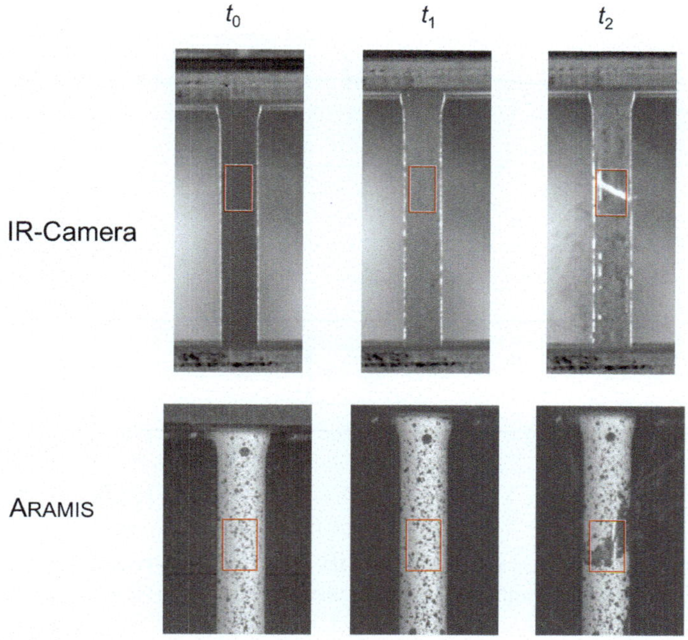

Fig. 10.5 Evaluation of the coupled measurement

sition of tensile and shear stresses, a modified ASTM model with the dimensions in Fig. 10.6b is selected.

The following section summarises and describes the results obtained from the experimental methods presented in Sects. 10.2.1 and 10.2.2. Table 10.1 shows the results of the individual investigations on the directly determinable parameters of Sect. 10.2.1.

Before the thermomechanical tests are performed, the material is classified according to Haupt (2002). Experiments are realised with the servo-hydraulic machine at a speed of $v = 0.3$ mm s^{-1} and additional holding times. During this holding time, the material is given the opportunity to achieve its equilibrium configuration. A holding time of $t = 1000$ s is selected in the tests. The experiments are realised on six circular samples with the geometry shown in Fig. 10.6a. For the experiments, a pure intermittent load with a stress ratio of $R = -0.2$ was selected. Figure 10.7 presents the results of the material classification.

Detail a shows the definition of the displacement-controlled load path and detail b shows the stress–strain response of the material. In the stress–strain response of the material, a clear decrease in stress is recognised during the holding times. This phenomenon is interpreted in the literature as relaxation. As a result, the material behaviour is attributed to viscoplasticity by showing an equilibrium hysteresis.

After material classification, the thermomechanical tests will follow. The series of experiments begins with a pure displacement-controlled tensile test at a loading

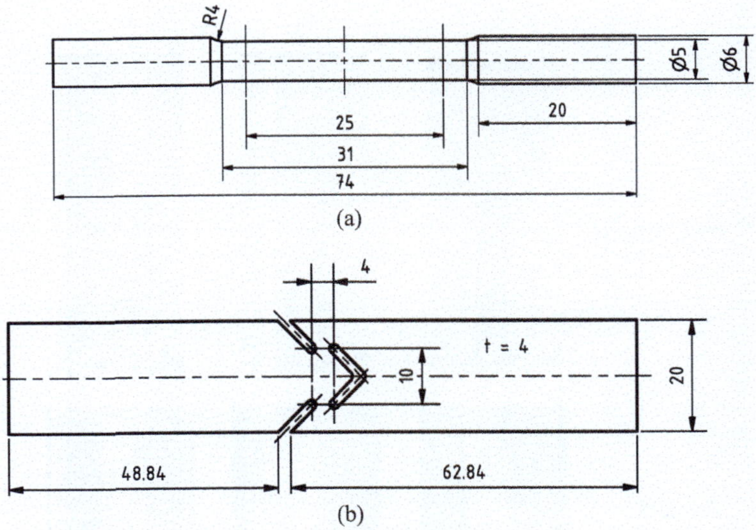

Fig. 10.6 Samples geometries **a** for tensile test DIN 50125-A 5 x 25 (Deutsches Institut für Normung e. V. 2022) **b** for tests with low stress triaxialites (Gatea et al. 2017)

Table 10.1 Direct determinable parameters

Parameter	Unit	Magnitude
Density ρ	$g \cdot cm^{-3}$	2.6018
Thermal expansion coefficient α	K^{-1}	$21.5 \cdot 10^{-6}$
Thermal conductivity k	$W \cdot (m \cdot K)^{-1}$	135
Specific heat conduction c^{σ}	$J \cdot (kg \cdot K)^{-1}$	853
Initial void volume fraction $\hat{f}^0$	%	0.0563
Critical value before break $\hat{f}^F$	%	0.0652

speed of $v = 3$ mm s^{-1} on the circular sample in Fig. 10.6a. This high loading speed is based on modelling the complete thermomechanical material response. In order to identify the parameters at a later stage with reference to the temperature curve of the material, the temperature curve results mainly from the effects of the material. The choice of a high loading rate leads to an approximately adiabatic experiment in which heat conduction effects are neglected (Golasinski et al. 2023; Sela et al. 2023).

Figure 10.8 presents the full thermomechanical material response for the uniaxial tensile test. The experiment was performed on six samples and a mean value curve is calculated for the numerical part of this work. In the diagram a the six experimental curve $e_1, \ldots, e_6$ and the mean value curve of the material's stress characteristic are shown. These experimental curves indicate a maximum standard deviation of 6.19% for the stress signal and 2.65% for the time signal. Diagram b represents the temperature evolution of the sample as a result of the deformation process. The maximal

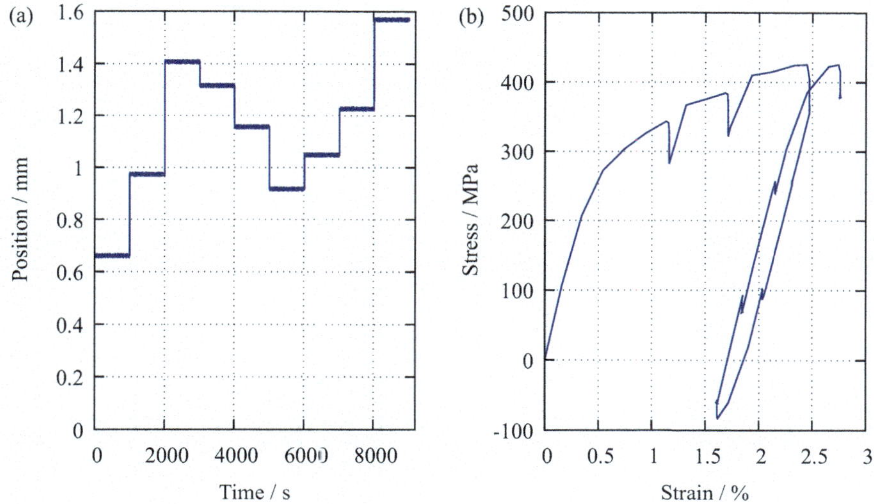

Fig. 10.7 Pre-Tests for material classification according to Haupt (2002) **a** definition of the displacement-controlled loading path **b** material's stress–strain relation

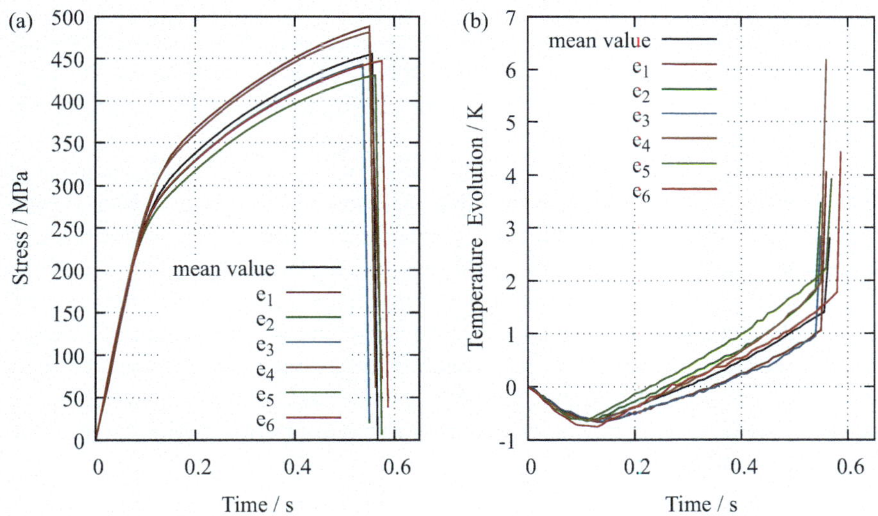

Fig. 10.8 Tensile test with $v = 3\ \text{mm s}^{-1}$ **a** stress relation **b** temperature relation

standard deviation of the temperature signal is 2.07%. Based on the coupled material response, the phenomena observed will now be explained. At the start of the deformation process, the material has an elastic behaviour. In the stress characterisation a, this area is recognised as approximately linear. During the elastic range, the temperature of the sample decreases b. This effect is known as the thermoelastic effect and is equivalent to the Kelvin effect (Bröcker and Matzenmiller 2013). As the material's

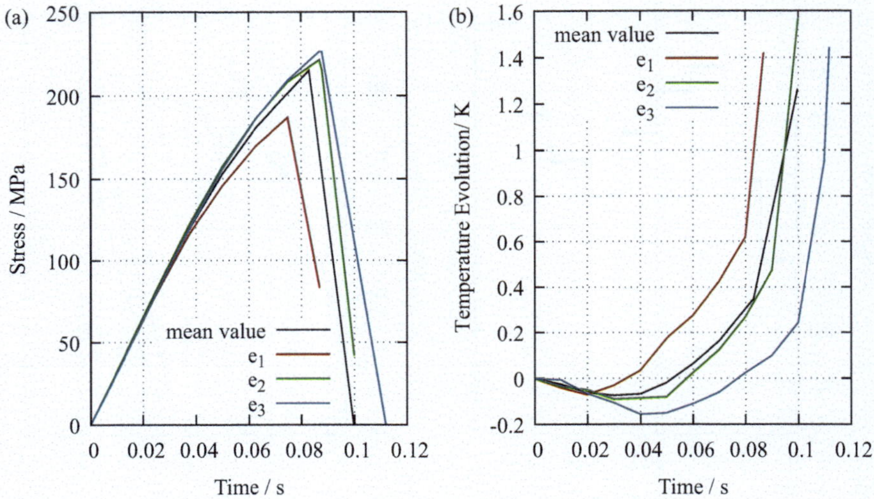

Fig. 10.9 Experiment with low stress triaxialities and with $v = 3$ mm s^{-1} **a** stress relation **b** temperature relation

yield point is passed, inelastic deformation processes occur. These lead to a change in the microstructure of the material (Haupt 2002), which is recognised on the level of the stress response as hardening and on the level of the temperature characteristic as heating of the material.

In order to generate additional information about the material behaviour based on the experiments, a test is performed that generates a superposition of tensile and shear stresses resulting from the sample's geometry. The experiments were performed on the sample in Fig. 10.6b with a speed of $v = 3$ mm s^{-1}. Figure 10.9 shows the material's stress relation in diagram a, and the material's temperature characteristic in b. For this complex loading path, the mean value curves were calculated by three different samples. In case of the stress relation, the maximal standard deviation of the stress signal is 6.25%. At the end of the test, the damage of the material is very dispersed, resulting in a maximum standard deviation in time of 12.5% and in the temperature evolution of 1.18%.

10.3 Theoretical Framework

Before starting the definition of a thermomechanical material model, the most important information of the material behaviour in Sect. 10.2.3 is summarised. The experimental observation shows the following:

1. Rate dependent material behaviour with an equilibrium hysteresis in Fig. 10.7
2. Evolution of the effective void volume fraction caused by the difference between the initial void volume fraction $\hat{f}^0$ and the critical value before break $\hat{f}^F$ (see Table 10.1)
3. Coupled displacement–temperature behaviour in Figs. 10.8 and 10.9
4. Influence on the material behaviour caused by introducing a complex stress state in Fig. 10.9.

The material behaviour is dependent on the microstructure of the material. Because the experiments have verified a change in the porosity, a numerical approach is selected that considers damage to the material. To include the damage phenomena, a Gurson–Tvergaard–Needleman (GTN) (Gurson 1977; Chu and Needleman 1980; Tvergaard and Needleman 1984) model is chosen and enhanced with a shear damage approach (Nahshon and Hutchinson 2008). As the experiments show a rate-dependent material behaviour, a linear elastic viscoplastic material model by Chaboche (Lemaitre and Chaboche 1990) is coupled with the GTN model according to Schob (2023).

10.3.1 Damage Model

The GTN damage model includes the numeric description of the initial porosity, void growth, void nucleation, and void coalescence. In order to represent the phenomena correctly, the model works with a case differentiation according to Eq. (10.2). Consequently, the effective void volume fraction $\hat{f}^*$ is defined as the initial void volume fraction $\hat{f}^0$ before testing, the total void volume fraction $\hat{f}$, and the coalescence criteria after reaching the coalescence limit $\hat{f}^c$.

$$\hat{f}^* = \begin{cases} \hat{f}_0 & \text{if } t_0 = 0 \\ \hat{f} & \text{if } \hat{f} \leq \hat{f}^c \\ \hat{f}^c + \frac{\frac{1}{q_1} - \hat{f}^c}{\hat{f}^F - \hat{f}^c}(\hat{f} - \hat{f}^c) & \text{if } \hat{f}^c < \hat{f} \leq \hat{f}^F \end{cases} \tag{10.2}$$

An evolution of the total void volume $\hat{f}$ starts after passing the yield point of the material and is defined in terms of the void growth $\hat{f}_{\text{gr}}$ and the nucleation $\hat{f}_{\text{nuc}}$:

$$\dot{\hat{f}} = \dot{\hat{f}}_{\text{gr}} + \dot{\hat{f}}_{\text{nuc}}, \tag{10.3}$$

with

$$\dot{\hat{f}}_{\text{gr}} = (1 - \hat{f}^*)\text{tr}(\dot{\boldsymbol{\varepsilon}}^{vp}) \tag{10.4}$$

and

$$\dot{\hat{f}}_{\mathrm{nuc}} = \frac{\hat{f}^N \dot{\overline{\varepsilon}}^{vp}}{S^N \sqrt{2\pi}} \exp\left(-\frac{1}{2}\left(\frac{\overline{\varepsilon}^{vp} - \varepsilon^N}{S^N}\right)^2\right). \tag{10.5}$$

Equation (10.4) describes a dependence on the evolution of the void growing $\hat{f}_{\mathrm{gr}}$ to the evolution of the viscoplastic strains $\boldsymbol{\varepsilon}^{\mathrm{vp}}$. The evolution of the nucleation in Eq. (10.5) introduces the mean value of inelastic strain ε^N, the standard deviation of the inelastic strain S^N, and the void nucleation volume fraction $\hat{f}^N$ as material parameter.

An enhancement of the damage approach to take account of shear stresses according to Nahshon and Hutchinson (2008) is achieved by the additive supplementation of the total void volume $\hat{f}$ by a shear part $\hat{f}_{\mathrm{Shear}}$:

$$\dot{\hat{f}}_{\mathrm{Shear}} = k_{\omega} \frac{\hat{f}^* \, \omega_{\mathrm{Shear}}}{T^{\mathrm{v}}} \boldsymbol{T}^{\mathrm{D}} : \dot{\boldsymbol{\varepsilon}}^{\mathrm{vp}}, \tag{10.6}$$

with the shear parameter k_{ω}, the equivalent stress T^{v}, the deviatoric stress tensor $\boldsymbol{T}^{\mathrm{D}}$, and the measuring value ω_{Shear}. The ω_{Shear} is defined by Eq. (10.7) and determines the current stress state. This means that ω_{Shear} is equal to zero in case of a pure shear stress state and decreases towards one in case of pure shear. For more information about this parameter, see (Nahshon and Hutchinson 2008). The parameter ω_{Shear} is defined by

$$\omega_{\mathrm{Shear}} = 1 - \left(\frac{27 J_3}{2\,(T^{\mathrm{v}})^3}\right)^2, \tag{10.7}$$

with the third invariant of the deviatoric stress tensor $J_3 = \det\left(\boldsymbol{T}^{\mathrm{D}}\right)$.

10.3.2 *Elasto Viscoplastic Chaboche Model with Damage Approach*

Before introducing the Chaboche model, the mechanical system is constrained to the technically relevant range of small deformations. As a consequence, the displacement gradient tensor norm is much smaller than 1 and the deformation is defined by the linearised deformation–displacement relationship. This description leads to the additive segmentation of the total strain tensor $\boldsymbol{\varepsilon}$ to an elastic strain tensor $\boldsymbol{\varepsilon}^{\mathrm{el}}$ and a viscoplastic strain tensor $\boldsymbol{\varepsilon}^{\mathrm{vp}}$. It is a common practice to transform this relationship according to the elastic strain tensor $\boldsymbol{\varepsilon}^{\mathrm{el}}$ (Kamlah and Haupt 1997):

$$\boldsymbol{\varepsilon}^{\mathrm{el}} = \boldsymbol{\varepsilon} - \boldsymbol{\varepsilon}^{\mathrm{vp}}. \tag{10.8}$$

The mechanical stress evolution based on the restriction of small deformation in Eq. (10.8) reads to

$$\dot{\boldsymbol{T}} = \mathbb{E} : (\dot{\boldsymbol{\varepsilon}} - \dot{\boldsymbol{\varepsilon}}^{\mathrm{vp}}), \tag{10.9}$$

with the fourth-order elasticity tensor $\mathbb{E}$.

The viscoplastic strain tensor $\boldsymbol{\varepsilon}^{\mathrm{vp}}$ is defined under consideration of the flow direction $\partial f/\partial \sigma$ and the viscoplastic multiplier (El Khatib et al. 2023):

$$\dot{\boldsymbol{\varepsilon}}^{\mathrm{vp}} = \lambda \frac{\partial f}{\partial \boldsymbol{\sigma}} = \left(\frac{f}{K}\right)^n \left(\frac{\partial f}{\partial \sigma^v}\boldsymbol{N} + \frac{1}{3}\frac{\partial f}{\partial \sigma^H}\boldsymbol{I}\right), \tag{10.10}$$

where f is the flow function, K is the viscosity factor, n is the viscosity exponent, $\boldsymbol{I}$ is the second-order identity tensor, and $\boldsymbol{N}$ is the second-order normal tensor. The normal tensor $\boldsymbol{N}$ is calculated by

$$\boldsymbol{N} = \frac{3}{2T^{\mathrm{v}}}\boldsymbol{T}^{\mathrm{D}}. \tag{10.11}$$

For a separation between the elastic and the viscoplastic domain, the Chaboche model applied a flow function f. This flow function f combines the von Mises approach with the Gurson damage concept:

$$f = \left(\frac{T^{\mathrm{v}} - \left(1 - \hat{f}^*\right) X^{\mathrm{v}}}{\sigma^{\mathrm{M}}}\right)^2 - 1 + 2q_1\hat{f}^*\cosh\left(\frac{3q_2}{2}\frac{T^{\mathrm{H}}}{\sigma^{\mathrm{M}}}\right) - q_1^2(\hat{f}^*)^2 \geq 0, \tag{10.12}$$

with the equivalent back stress X^{v}, the non-linear isotropic hardening σ^{M}, the hydrostatic stress T^{H}, and the GTN damage model coefficients q_1 and q_2. In case of viscoplastic material behaviour, the evolution of the hardening variables starts. In case of the non-linear isotropic hardening σ^{M} the evolution equation is depended on the yield stress σ^{y}, the accumulated plastic strain $\bar{\varepsilon}^{\mathrm{vp}}$, and the hardening parameter Q^{∞} and b:

$$\sigma^{\mathrm{M}} = \sigma^{\mathrm{y}} + Q^{\infty}\left(1 - \exp\left(-b\bar{\varepsilon}^{\mathrm{vp}}\right)\right) + K\left(\dot{\bar{\varepsilon}}^{\mathrm{vp}}\right)^{1/n}, \tag{10.13}$$

with

$$\dot{\bar{\varepsilon}}^{\mathrm{vp}} = \frac{\boldsymbol{T} : \dot{\boldsymbol{\varepsilon}}^{\mathrm{vp}}}{\left(1 - \hat{f}^*\right)\sigma^{\mathrm{M}}}. \tag{10.14}$$

In addition to the non-linear isotropic hardening, the flow function f in Eq. (10.12) is depended on the non-linear kinematic hardening $\boldsymbol{X}$. The kinematic hardening tensor $\boldsymbol{X}$ is defined by the sum of two internal kinematic hardening variables $\boldsymbol{X}^{(1)}$ and $\boldsymbol{X}^{(2)}$:

$$\dot{\boldsymbol{X}} = \sum_{i=1}^{2} \left(\frac{2}{3} C^{(i)} \dot{\boldsymbol{\varepsilon}}^{\mathrm{vp}} - \gamma^{(i)} \varphi^{(i)} \boldsymbol{X}^{(i)} \dot{\bar{\varepsilon}}^{\mathrm{vp}} - R^{\mathrm{kin}(i)} \boldsymbol{X}^{(i)} \right) \tag{10.15}$$

and

$$\varphi = \sum_{i=1}^{2} \left(\varphi_{\infty}^{(i)} + (1 - \varphi_{\infty}^{(i)}) \exp(-\omega^{(i)} \bar{\varepsilon}^{\mathrm{vp}}) \right), \tag{10.16}$$

where C is the strain hardening parameter, γ is the dynamic recovery factor, R^{kin} is the static recovery factor, ω is the decay constant, and φ_∞ is the saturation parameter.

10.3.3 Dissipation

The self-heating phenomena base on the fact that inelastic deformation processes are caused by high dissipative changes in the material's microstructure (Bröcker and Matzenmiller 2013). With the theory of the internal variables in the presented Chaboche–GTN model the observed macroscopic effects of the material are described numerically. For a theoretical expression of the coupled thermomechanical effect, thermodynamic quantities are identified for the calculation. An expression for the dissipation Δ results in a theoretical coupling of the internal variables and the self-heating effect. To enable an introduction to this field, a general conceptualisation of dissipation is presented. The basis of the thermomechanical description is the Clausius–Duhem inequality that results due to the combination of the first and second laws of thermodynamics:

$$\Delta = \Theta \dot{s} - \dot{e} + \frac{1}{\rho} \boldsymbol{T} : \dot{\boldsymbol{\varepsilon}} - \frac{1}{\rho \Theta} \boldsymbol{q} : \boldsymbol{g} \geq 0, \tag{10.17}$$

where Θ is the surface temperature, $\dot{s}$ is the entropy rate, $\dot{e}$ is the free energy rate, ρ is the mass density, $\boldsymbol{T}$ is the Cauchy stress tensor, $\dot{\boldsymbol{\varepsilon}}$ is the linearised Green–Lagrange strain tensor, $\boldsymbol{q}$ is the specific heat flux vector, and $\boldsymbol{g}$ is the temperature gradient.

Equation (10.17) presents a dependence of the internal energy e evolution on the entropy s and the strain $\boldsymbol{\varepsilon}$ evolution. Measuring entropy requires a very complex experimental setup. To describe a thermodynamic potential in terms of measurable quantities, the Legendre transformation allows to switch between the potentials. For this purpose, the free energy ψ, depending on the temperature and strain evolution, and the free enthalpy G, depending on the temperature and stress evolution, are considered (Haupt 2002). As the presented Chaboche–GTN model defines its internal variables in terms of stress, the thermomechanical system is formulated in a physically consistent way by the thermodynamic potential of the free enthalpy G (Kamlah 1994). After realising the Legendre transformation, Eq. (10.17) reads

$$\Delta = -\frac{1}{\rho}\boldsymbol{\varepsilon}^{\mathrm{el}} : \dot{\boldsymbol{T}} + \dot{G} - s\dot{\Theta} + \frac{1}{\rho}\boldsymbol{\sigma} : \dot{\boldsymbol{\varepsilon}}^{\mathrm{vp}} - \frac{1}{\rho\Theta}\boldsymbol{q} : \boldsymbol{g} \geq 0. \tag{10.18}$$

With the assumption that the internal variables are expressed by the thermodynamic forces $\boldsymbol{A}^1, \ldots, \boldsymbol{A}^n$ (Kamlah and Haupt 1997), the free enthalpy G is defined as a function by

$$G = G\left(\boldsymbol{T}, \Theta, \boldsymbol{g}, \boldsymbol{A}^1, \ldots, \boldsymbol{A}^n\right). \tag{10.19}$$

Considering Eq. (10.19) in Eq. (10.18) leads to

$$\begin{aligned}\Delta = \left(\frac{\partial G}{\partial \boldsymbol{T}} - \frac{1}{\rho}\boldsymbol{\varepsilon}^{\mathrm{el}}\right)\dot{\boldsymbol{T}} + \left(\frac{\partial G}{\partial \Theta} - s\right)\dot{\Theta} + \frac{1}{\rho}\boldsymbol{T} : \dot{\boldsymbol{\varepsilon}}^{\mathrm{vp}} \\ + \sum_{j=1}^{n}\frac{\partial G}{\partial \boldsymbol{A}^j}\dot{\boldsymbol{A}}^j - \frac{\partial G}{\partial \boldsymbol{g}}\dot{\boldsymbol{g}} - \frac{1}{\rho\Theta}\boldsymbol{q} : \boldsymbol{g} \geq 0.\end{aligned} \tag{10.20}$$

The meaning of the dissipation inequality (10.20) is explained first before analysing this. A thermodynamic system is reversible if the equation is equal to zero. The term reversible means that the system is able to return to its initial state (Haupt 2002). One theoretical example of this is the adiabatic elastic process. In case the inequality is greater than zero, the process is irreversible. As a consequence, the system is no longer capable of returning to its initial state. This situation is characterised by inelastic deformation processes and heat conduction. For the evaluation of Eq. (10.20), it is a common procedure to focus on the reversible processes first (Coleman and Noll 2024). This leads to the dissipation inequality:

$$\Delta = \left(\frac{\partial G}{\partial \boldsymbol{T}} - \frac{1}{\rho}\boldsymbol{\varepsilon}^{\mathrm{el}}\right)\dot{\boldsymbol{T}} + \left(\frac{\partial G}{\partial \Theta} - s\right)\dot{\Theta} - \frac{\partial G}{\partial \boldsymbol{g}}\dot{\boldsymbol{g}} - \frac{1}{\rho\Theta}\boldsymbol{q} : \boldsymbol{g} \geq 0. \tag{10.21}$$

Equation (10.21) is solved by analysing each single summand (Coleman and Noll 2024). Following this assumption, the last summand is analysed with Fourier's model of heat conduction in terms of thermal conductivity k:

$$\boldsymbol{q} = -k\boldsymbol{g}. \tag{10.22}$$

In case of the adiabatic boundary conditions, the temperature gradient $\boldsymbol{g}$ becomes zero and Eq. (10.21) is satisfied. The other summands solve Eq. (10.21) when

$$\boldsymbol{\varepsilon}^{\mathrm{el}} = \rho\frac{\partial G}{\partial \boldsymbol{T}}, \quad s = \frac{\partial G}{\partial \Theta}, \quad \frac{\partial G}{\partial \boldsymbol{g}} = 0. \tag{10.23}$$

It is a common practice that the consequences of Coleman and Noll in Eq. (10.23) are established for inelastic irreversible processes (Kamlah 1994). The dissipation inequality for irreversible processes is derived by implementing Eq. (10.23) in Eq. (10.20):

$$\Delta = \underbrace{\frac{1}{\rho}\boldsymbol{T} : \dot{\boldsymbol{\varepsilon}}^{\mathrm{vp}}}_{e^{\mathrm{vp}}} - \underbrace{\sum_{j=1}^{n} \frac{\partial \psi}{\partial \boldsymbol{A}^{j}} : \dot{\boldsymbol{A}}^{j}}_{e^{\mathrm{s}}} - \frac{1}{\rho\theta}\boldsymbol{q} : \boldsymbol{g} \geq 0. \tag{10.24}$$

The dissipation structure for an inelastic system is defined by Eq. (10.24). In the inelastic deformation process, a viscoplastic stress power e^{vp} is induced in the system. As the material hardening in the inelastic range results in a change of the material's microstructure (Haupt 2002), the entire stress power is not dissipated. This stored energy of cold work e^{s} defines the amount of energy responsible for the microstructural change. Both terms are often referred to as the intrinsic dissipation δ^{intr}. And the third term of Eq. (10.24) is called thermal dissipation δ^{th} (Kamlah 1994). Applying these principles, the intrinsic dissipation δ^{intr} of the coupled Chaboche–GTN model is obtained:

$$\delta^{\mathrm{intr}}\rho = \boldsymbol{T} : \dot{\boldsymbol{\varepsilon}}^{\mathrm{vp}} + \frac{1}{b}\dot{R} - \frac{3}{2}\boldsymbol{X} : \dot{\boldsymbol{X}} \geq 0, \tag{10.25}$$

with

$$R = -\left(\sigma^{\mathrm{y}} - Q^{\infty}\right)\exp\left(-b\bar{\varepsilon}^{\mathrm{vp}}\right). \tag{10.26}$$

10.3.4 Heat Equation

The section on dissipation presents a numerical description of the self-heating effect for the thermomechanical system. In a discussion between experiment and numerics, the temperature evolution is important for the entire deformation process. To derive the heat conduction equation, the first law of thermodynamics is applied:

$$\dot{e} = -\frac{1}{\rho}(\boldsymbol{q}) + r + \frac{1}{\rho}\boldsymbol{T} : \dot{\boldsymbol{\varepsilon}}. \tag{10.27}$$

The entropy balance of the system is generally analysed to represent the temperature increase due to dissipation. For the definition of the entropy balance of the free enthalpy, the Gibbs relation is analysed (Haupt 2002):

$$\dot{G} = \frac{1}{\rho}\boldsymbol{\varepsilon} : \dot{\boldsymbol{T}} + s\dot{\Theta}. \tag{10.28}$$

Implementing Eq. (10.28) with the help of the Legendre transformation in Eq. (10.27) results in the entropy balance:

$$\Theta\dot{s} = -\frac{1}{\rho}(\boldsymbol{q}) + r. \tag{10.29}$$

Based on the Coleman–Noll approach in Eq. (10.23) the evaluation of the entropy potentials leads to the heat conduction equation:

$$\underbrace{\theta\frac{\partial^2 G^{\mathrm{el}}}{\partial\theta^2}}_{c^\sigma}\dot{\theta} = \underbrace{-\theta\frac{\partial^2 G^{\mathrm{el}}}{\partial\theta\partial\boldsymbol{T}}}_{e^{\mathrm{el}}}\dot{\boldsymbol{T}} + \underbrace{\frac{1}{\rho}\boldsymbol{T} : \dot{\boldsymbol{\varepsilon}}^{\mathrm{vp}}}_{e^{\mathrm{vp}}} \underbrace{-\theta\sum_{j=1}^{n}\left[\frac{\partial^2 G}{\partial\theta\partial\boldsymbol{A}^j}\left(\frac{\partial G}{\partial\boldsymbol{A}^j}\right)^{\cdot}\right]}_{e^{\mathrm{s}}} \underbrace{-\frac{1}{\rho}(\boldsymbol{q}) + r}_{e^{Q}}, \tag{10.30}$$

with the elastic part of the free enthalpy G^{el}, the specific heat for the constant stress case c^σ (Haupt 2002), the thermoelastic effect e^{el}, and the heat conduction process e^{Q} including the heating by an external heat source. To establish the elasto-plastic material behaviour, the thermoelastic effect e^{el} is determined in more detail. In tension, the sample is cooling and in compression is heating (Bröcker and Matzenmiller 2013). In order to describe this effect in numerical models, a linear Hooke's thermoelasticity is applied (Kamlah 1994):

$$\rho G^{\mathrm{el}} = \frac{1}{2}\frac{1}{2\mu}\boldsymbol{T}^{\mathrm{D}} : \boldsymbol{T}^{\mathrm{D}} + \frac{1}{2}\frac{1}{9\kappa}(\mathrm{tr}(\boldsymbol{T}))^2 + \alpha^0(\mathrm{tr}(\boldsymbol{T}))\left(\theta - \theta^0\right) + \frac{\rho}{2}\frac{c^\sigma}{\theta^0}\left(\theta - \theta^0\right)^2, \tag{10.31}$$

where μ defines the shear modulus, κ defines the bulk modulus, α^0 defines the thermal expansion coefficient, and Θ^0 defines the reference temperature.

10.4 Results

10.4.1 Parameter Determination

On the basis of the thermomechanical Chaboche–GTN model presented in Sect. 10.3, the model is integrated into the mathematical program MATLAB for the three-dimensional adiabatic calculation in a node. This process is motivated by the access to the MATLAB optimisation toolbox. To identify the indirectly determinable parameters, an objective function is generated with the complete thermomechanical material response from Figs. 10.8 and 10.9. On the basis of Rothermel et al. (2022) observations, the MultiStart function is applied for parameter identification. This function generates random start points in the defined parameter space and searches for local minima applying the lsqnonlin algorithm (T. M. Inc 2022). In the case of this work, the optimisation function converges to the global minimum after sampling 13 different starting points. Table 10.2 presents the results of the parameter identification, including direct and indirect determinable parameters.

Table 10.2 Model parameters

Symbol	Unit	Value	Symbol	Unit	Value	Symbol	Unit	Value
Chaboche parameters								
ν		0.33	E	MPa	43000	σ^{y}	MPa	141
K	MPa	3.894	n		9.385	$C^{(1)}$	MPa	14810
$C^{(2)}$	MPa	12997	$\gamma^{(1)}$		296.25	$\gamma^{(1)}$		421
Q^{∞}		94.96	b		22.994	$R^{\text{kin}(1)}$	s^{-1}	0.001
$R^{\text{kin}(2)}$	s^{-1}	0.0023	$\varphi_{\infty}^{(1)}$		0.431	$\varphi_{\infty}^{(2)}$		0.242
$\omega^{(1)}$		69.315	$\omega^{(2)}$		57.344			
GTN parameters								
$\hat{f}_0$	%	0.0563	q_1		1.532	q_2		1.021
$\hat{f}^F$	%	0.0652	$\hat{f}^N$	%	0.118	S^N		0.191
ε^N		0.275	$\hat{f}^C$	%	0.0625			
Thermal parameters								
ρ	g cm^{-3}	2.602	α^0	K^{-1}	$21.5 \cdot 10^{-6}$	c^{σ}	$\text{J kg}^{-1}\,\text{K}^{-1}$	853
k	$\text{W m}^{-1}\,\text{K}^{-1}$	135						

10.4.2 Concept for Thermomechanical FE-Model

For the implementation of a fully coupled thermomechanical simulation with heat conduction effects on complex geometries, the model is transferred to the FE software Abaqus via a user material subroutine (UMAT). Under consideration of an 8 node thermally coupled, trilinear displacement and temperature elements (C3D8T) (Smith 2009), the thermomechanical calculation is realised. For a better understanding of the generated results, Fig. 10.10 presents the defined boundary conditions in the FE model. In this case the geometry of the tests with low stress triaxiality is shown. The simulation on the tension tests are performed with the same assumptions. Detail a presents two displacement boundary conditions. The condition on the lower clamped area blocks all displacement directions and the upper clamped area defines a displacement in the y-direction. Because it is a coupled simulation, thermal boundary conditions are defined too. The clamped area is numerically described with a surface film boundary condition as shown in detail b, where the film coefficient is $h = 2.9\,\text{J s}^{-1}\,\text{m}^{-2}\,^{\circ}\text{C}^{-1}$ (Cengel 2016) and the sink temperature for the tensile test is $\Theta_t = 20.6\,^{\circ}\text{C}$ and for the tests with low stress triaxiality is $\Theta_s = 25.1\,^{\circ}\text{C}$. In addition to the thermal boundary conditions of the machine clamps, the surrounding air needs to be included in the calculation as presented in detail c. For this specification a second surface film condition is defined with a film coefficient $h = 1.5\,\text{J s}^{-1}\,\text{m}^{-2}\,^{\circ}\text{C}^{-1}$ (Cengel 2016) and the same sink temperature as in detail b.

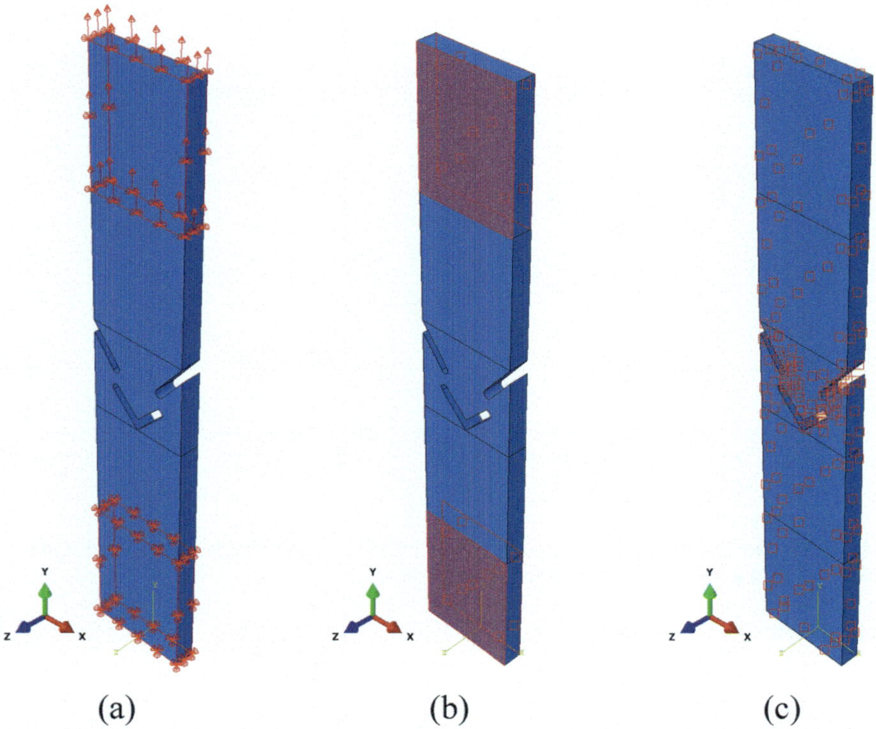

Fig. 10.10 Boundary conditions for the tests with the low stress triaxiality **a** displacement boundary condition **b** thermal boundary condition as surface film condition for modelling the clamping of the machine **c** thermal boundary condition as surface film condition for modelling the surrounding air

10.4.3 Numerical Results and Discussion

With the defined material parameters in Table 10.2 and the coupled displacement temperature boundary conditions in Fig. 10.10, the numerical simulations are performed. Figure 10.11 presents the results of the simulation for the tensile test and compares them with the experimental observed mean value curve of Fig. 10.8. In general, the numerical simulation describes the thermomechanical material behaviour with a high degree of accuracy. Detail a presents a maximum difference at the end of the loading path with an absolute value of 10 MPa. The maximal relative value is in the region of the yield point at 141 MPa with an amount of 8%. In case of the material's temperature characteristic, the highest difference is located after a loading time of 0.1 s with an absolute error of 0.15 K that corresponds to a relative error of 0.8%.

To validate the determined parameters, a more complex stress state is obtained by testing with low stress triaxiality (see Fig. 10.9). The objective of this procedure is demonstrating the prediction quality of the material model. Figure 10.12 presents the

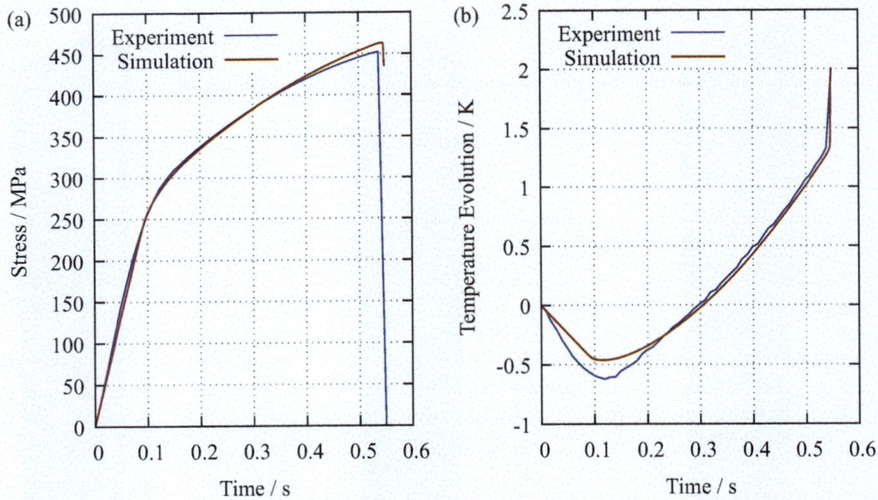

Fig. 10.11 Thermomechanical material behaviour for tensile test with $v = 3$ mm s^{-1}, experiment and simulation **a** stress response **b** temperature response

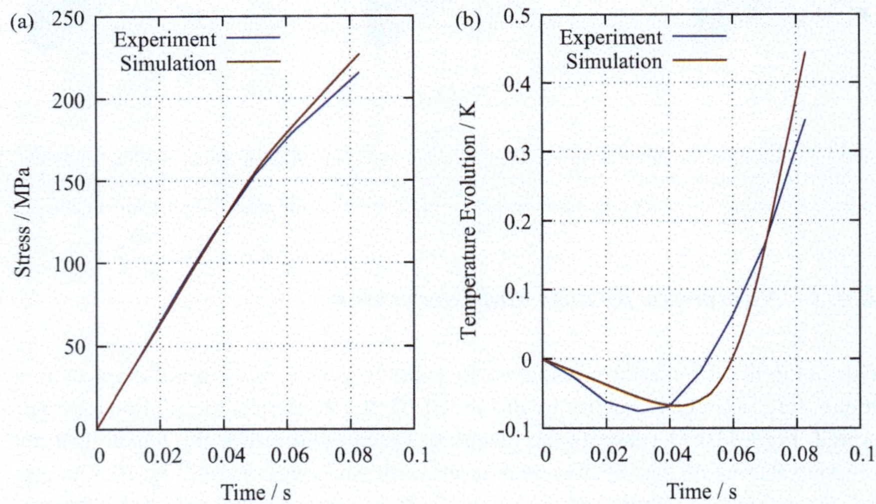

Fig. 10.12 Thermomechanical material behaviour for tests with low stress triaxiality with $v = 3$ mm s^{-1}, experiment and simulation **a** stress response **b** temperature response

numerical results. The simulation is aborted before the stress decay. This means that the FE model is able to forecast the failure of the material. Both details in Fig.10.12 show a good accuracy between the simulation and the experiment. The highest difference between the numerical result and the experiment in detail a is located at the end of the loading time with an absolute error of 8 MPa, which represents a relative error of 4%. In case of the temperature characteristic in detail b, the highest difference is shown at the end of the loading path. Here the absolute error amounts 0.09 K, that equals to a relative error of 0.4%. The main difference in the first 0.07 s in the temperature is caused by the energy assumptions for the derivation of the intrinsic dissipation δ^{intr} for the Chaboche–GTN model. In Bröcker and Matzenmiller (2013), a clear study of the sensitivity to changes in energy assumptions of the temperature development is shown. This research motivates a closer analysis of the energy evolution for stress states with shear components. A subsequent step for the estimation of the energy approaches is the simplification of the stress state into a pure shear state. The result of this procedure is a segmented analysis of shear and tensile stress before they are coupled again by introducing the low stress triaxialities.

10.5 Conclusion and Outlook

The presented research shows a technique for the identification of directly and indirectly determinable parameters in complex numerical models. In the case of the indirect determinable parameters, an enhancement of the classical continuum mechanics by the description of self-heating effects is shown. This extension enables a more specific analysis of the material behaviour and results in the generation of precise objective functions in optimisation tasks. The enhancements introduced to the one-dimensional point problem to a fully three-dimensional volumetric problem with the additional consideration of shear stresses on the damage behaviour allow the access of the presented model to calculations of complex geometries in the context of FE analysis.

In the case of the low stress triaxiality, the high differences in the temperature characteristic shown in Sect. 10.4.3 lead to additional consideration of the energy approaches. For a validation of this energy analysis, pure shear tests can be performed.

Additional load scenarios can be integrated in the analyses to examine the identified parameters of the model. At the current state, information is missing on the ability to represent rate dependency. This can be realised by means of slower loading speeds for the two presented loading scenarios. The model includes effects that are modelled by cyclical tests.

The implementation of in-situ CT investigations provides an additional option for extending the directly determinable parameters. These measurements assist in the evaluation of additional GTN parameters and allow an understanding of the complex phenomena related to changes in the microstructure of the material.

Acknowledgements The realisation of this work has been made possible by the Project-related exchange programme on the topic of advanced visualisation and numerical methods for additive manufacturing (LPBF) application with the Technical University of Wroclaw. This project enables the support and collaboration of Grzegorz Ziólkowski and Emilia Grochowska at the Department of Laser Technology, Automation and Production Organisation of the Technical University of Wroclaw. The Dedication and expertise in performing the CT scans were essential in the successful completion of this research. The authors wish to express their thanks to the German Aerospace Centre (DLR) within the programme "CertiFlight" (No. 20D1925F) for provision of the material.

References

Asnafi N, Rajalampi J, Aspenberg D, Alveflo A (2020) Production tools made by additive manufacturing through laser-based powder bed fusion. BHM Berg- Huettenmaenn. Monatshefte 165:125–136. https://doi.org/10.1007/s00501-020-00948-3

ASTM B 831:2022 (2022) Test method for shear testing of thin aluminum alloy products

Barsoum I, Faleskog J (2007) Rupture mechanisms in combined tension und shear-Micromechanics. Int J Solids Struct 44:5481–5498. https://doi.org/10.1016/j.ijsolstr.2007.01.024

Bröcker C, Matzenmiller A (2013) An enhanced concept of rheological models to represent nonlinear thermoviscoplasticity and its energy storage behavior. Continuum Mech Thermodyn 25:749–778. https://doi.org/10.1007/s00161-013-0295-5

Cengel YA (2016) Heat transfer: a practical approach

Chu CC, Needleman A (1980) Void nucleation effects in biaxially stretched sheets. J Eng Mater Technol 102:249–256. https://doi.org/10.1115/1.3224807

Coleman BD, Noll W, The thermodynamics of elastic materials with heat conduction and viscosity. In: Heat conduction

Cooke S, Ahmadi K, Willerth S, Herring R (2020) Metal additive manufacturing: Technology, metallurgy and modelling. J Manuf Process 57:978–1003. https://doi.org/10.1016/j.jmapro.2020.07.021

D'Andrea D (2023) Additive manufacturing of AISI 316L stainless steel: a review. Metals 13:1370. https://doi.org/10.3390/met13071370

Deutsches Institut für Normung e. V. (2022) DIN 50125 : Prüfung metallischer Werkstoffe - Zugproben

El Khatib O, Hütter G, Pham R-D, Seupel J, Kuna M, Kiefer B (2023) A non-iterative parameter identification procedure for the non-local Gurson-Tvergaard-Needleman model based on standardized experiments. Int J Fract 241:73–94. https://doi.org/10.1007/s10704-023-00660-5

Forster AM (2015) Materials testing standards for additive manufacturing of polymer materials: state of the art and standards applicability. Technical Report NIST IR 8059, National Institute of Standards and Technology. https://doi.org/10.6028/NIST.IR.8059

Gatea S, Ou H, Lu B, McCartney G (2017) Modelling of ductile fracture in single point incremental forming using a modified GTN model. Eng Fract Mech 186:59–79. https://doi.org/10.1016/j.engfracmech.2017.11.005

Golasinski KM, Staszczak M, Pieczyska EA (2023) Energy storage and dissipation in consecutive tensile load-unload cycles of Gum Metal. Materials 16:3288. https://doi.org/10.3390/ma16093288

Goyanes A, Wang J, Buanz A, Martínez-Pacheco R, Telford R, Gaisford S, Basit AW (2015) 3D printing of medicines: engineering novel oral devices with unique design and drug release characteristics. Mol Pharm 12:4077–4084. https://doi.org/10.1021/acs.molpharmaceut.5b00510

Gurson AL (1977) Continuum theory of ductile rupture by void nucleation and growth: part I-yield criteria and flow rules for porous ductile media. J Eng Mater Technol 99:2–15. https://doi.org/10.1115/1.3443401

Haupt P (2002) Continuum mechanics and theory of materials. Advanced texts in physics. Springer, Berlin, Heidelberg

T. M. Inc (2022) MATLAB version: 9.11.0. (R2021b)

Johnson GR, Cook WH (1985) Fracture characteristics of three metals subjected to various strains, strain rates, temperatures and pressures. Eng Fract Mech 21:31–48. https://doi.org/10.1016/0013-7944(85)90052-9

Kamlah M (1994) Zur Modellierung des Verfestigungsverhaltens von Materialien mit statischer Hysterese im Rahmen der phänomenologischen Thermomechanik. PhD Thesis, University of Kassel, FB 15, Maschinenbau

Kamlah M, Haupt P (1997) On the macroscopic description of stored energy and self heating during plastic deformation. Int J Plast 13:893–911. https://doi.org/10.1016/S0749-6419(97)00054-5

Lemaitre J, Chaboche J (1990) Mechanics of solid materials. Cambridge University Press

McClintock FA (1971) Plasticity aspects of fracture. In: Engineering fundamentals and environmental effects, pp 47–225, Elsevier. https://doi.org/10.1016/B978-0-12-395645-0.50008-5
Murr LE (2016) Frontiers of 3D printing/additive manufacturing: from human organs to aircraft fabrication. J Mater Sci Technol 32:987–995. https://doi.org/10.1016/j.jmst.2016.08.005
Nahshon K, Hutchinson J (2008) Modification of the Gurson Model for shear failure. Eur J Mech A Solids 27:1–17. https://doi.org/10.1016/j.euromechsol.2007.08.002
Ngo TD, Kashani A, Imbalzano G, Nguyen KT, Hui D (2018) Additive manufacturing (3D printing): a review of materials, methods, applications and challenges. Compos B Eng 143:172–196. https://doi.org/10.1016/j.compositesb.2018.02.012
Reis FJP, Malcher L, Andrade Pires FM, César De Sá JMA (2010) A comparison of shear mechanisms for the prediction of ductile failure under low stress triaxiality. Int J Struct Integr 1:314–331. https://doi.org/10.1108/17579861011083511
Rose L, Menzel A (2020) Optimisation based material parameter identification using full field displacement and temperature measurements. Mech Mater 145:103292. https://doi.org/10.1016/j.mechmat.2020.103292
Rothermel R, Panfilenko W, Sharma P, Wald A, Schuster T, Jung A, Diebels S (2022) A method for determining the parameters in a rheological model for viscoelastic materials by minimizing Tikhonov functionals. Appl Math Sci Eng 30:141–165. https://doi.org/10.1016/j.jmapro.2020.07.021
Rümmler C, Neumann GRH, Groh W, Hähnel F, Grumbt G, Tropschuh R, Schiemann T, Markmiller JFC (2024) Study of the influence of additive manufacturing applicable surface treatment methods on mechanical part properties for use in aerospace applications. CEAS Aeronaut J. https://doi.org/10.1007/s13272-023-00601-5
Sames WJ, List FA, Pannala S, Dehoff RR, Babu SS (2016) The metallurgy and processing science of metal additive manufacturing. Int Mater Rev 61:315–360. https://doi.org/10.1080/09506608.2015.1116649
Schob D (2023) Experimentelle Untersuchung und numerische Simulation des Material- und Schädigungsverhaltens von 3D gedrucktem Polyamid 12 unter quasistatischer und zyklischer Beanspruchung. PhD Thesis, BTU Cottbus - Senftenberg
Sela A, Ortiz-de Zarate G, Soler D, Germain G, Gallegos L, Arrazola P (2023) Adiabatic self-heating determination for Ti6Al4V at different temperatures. Int J Heat Mass Transf 204:123747. https://doi.org/10.1016/j.ijheatmasstransfer.2022.123747
Smith M (2009) ABAQUS/standard user's manual, Version 6.9. Dassault Systèmes Simulia Corp, United States
Sparr H (2023) Thermomechanische Analyse von inelastischen Deformationsvorgängen bei komplexer Beanspruchung in Modellierung und Experiment. PhD Thesis, BTU Cottbus - Senftenberg
Sun C, Wang Y, McMurtrey MD, Jerred ND, Liou F, Li J (2021) Additive manufacturing for energy: a review. Appl Energy 282:116041. https://doi.org/10.1016/j.apenergy.2020.116041
Tvergaard V, Needleman A (1984) Analysis of the cup-cone fracture in a round tensile bar. Acta Metall 32:157–169. https://doi.org/10.1016/0001-6160(84)90213-X

Chapter 11
Experimental and Numerical Study on the Temperature-Dependent Tensile Properties of AlSi10Mg Fabricated by Laser-Based Powder Bed Fusion

Stefan Brenner and Vesna Nedeljkovic-Groha

Abstract In recent years, additive manufacturing (AM) processes, especially the laser-based powder bed fusion of metals (PBF-LB/M), have evolved from rapid prototyping to the production of end-use parts. During fabrication, PBF-LB/M parts experience repeated heating and cooling cycles, leading to residual stresses in the finished parts. In addition, the parts are often exposed to superposed mechanical and thermal loads during operation. Designers and AM engineers use sophisticated software tools like finite element analysis (FEA) to generate topology-optimized parts that are only feasible with AM processes. The success of the FEA depends on factors like the modeling assumptions and the quality of the material model. However, in commercially available FE software, heavily simplified material models are implemented to enable reasonable calculation times. The availability of temperature-dependent material properties is limited to low and medium temperatures and the constitutive relationships are simplified to bilinear elasto-plastic models, which cannot capture nonlinear material responses. Here, we show that the highly nonlinear behavior of AlSi10Mg observed in uniaxial tensile tests at temperatures up to 440 °C can be excellently described by a Chaboche nonlinear kinematic hardening material model. An FEA verification of the tensile tests matches the linear-elastic region of the stress–strain curves and the strain hardening in the range of the experimental scatter. The ultimate tensile strength, which drops from 470 MPa at room temperature to 15 MPa at 440 °C, can be reproduced in the simulation with errors of 0–4%.

S. Brenner (✉) · V. Nedeljkovic-Groha
Institute for Design and Production Engineering, University of the Bundeswehr Munich, Werner-Heisenberg-Weg 39, 85577 Neubiberg, Germany
e-mail: stefan.brenner@unibw.de

V. Nedeljkovic-Groha
e-mail: vesna.nedeljkovic-groha@unibw.de

H. Altenbach et al. (eds.), *Lectures Notes on Advanced Structured Materials 3*, Advanced Structured Materials 226, https://doi.org/10.1007/978-3-031-77403-4_11

11.1 Introduction

Additive manufacturing (AM), and laser-based powder bed fusion of metals (PBF-LB/M) in particular, has transitioned from rapid prototyping to a valuable production process for manufacturing end-use parts. Its advantages, like the capability to produce highly complex, lightweight components from CAD data without the need for tools other than a laser beam, make PBF-LB/M an attractive choice for the manufacturing industry.

Sectors such as aerospace, automotive, defense, and biomedical leverage PBF-LB/M to fabricate critical components subjected to various thermal and mechanical loads during operation. Heat exchangers in aerospace applications are a current example. The complex geometries achievable with PBF-LB/M allow for optimized internal structures that enhance heat transfer efficiency and can withstand significant thermal and mechanical loads (Careri et al. 2023). These demanding operating conditions require materials that offer an optimal balance of thermal conductivity, mechanical strength, and durability. This is where Al alloys can play an essential role because of their high thermal conductivity, high strength-to-weight ratio, and proven processability in PBF-LB/M, especially the Al-Si alloys (Ponnusamy et al. 2020).

To be able to design such complex components, designers take advantage of tools like finite element (FE) software to find optimized topologies. Furthermore, the repeated heating and cooling cycles during the PBF-LB/M process induce undesirable residual stresses that can be evaluated prior to manufacturing by means of process simulation. Necessary prerequisites are profound knowledge of the material's behavior and reliable material models for the simulation tasks.

Ponnusamy et al. provide a comprehensive literature review of the current understanding of the mechanical properties of PBF-LB/M fabricated aluminum alloys. They reveal a need for knowledge about the mechanical performance of AM aluminum alloys at elevated temperatures, even for AlSi10Mg, which is one of the most widely used materials in PBF-LB/M (Ponnusamy et al. 2020). The number of recent studies on this topic is still limited and not sufficient for a comprehensive understanding. Uzan et al. carried out creep and tensile tests with vertically built AlSi10Mg specimens between room temperature and 400 °C. The specimens were stress-relief heat-treated prior to the tests (Uzan et al. 2018). Cao et al. used horizontally oriented AlSi10Mg specimens without heat treatment for tensile tests between room temperature and 400 °C (Cao et al. 2021). Bao et al. tested horizontally oriented, stress-relief heat-treated AlSi10Mg specimens at room temperature and at 250 °C (Bao et al. 2022). Lehmhus et al. used horizontally, 45°, and vertically oriented AlSi10Mg specimens in stress-relief heat-treated condition for tensile tests at temperatures from room temperature to 450 °C (Lehmhus et al. 2022).

Only a few studies focus on temperature-dependent material models for AlSi10Mg. Schuch et al. carried out compression tests with vertically oriented, heat-treated specimens at room temperature and at 400 °C with high strain rates but did not have enough data to include the temperature dependence in their constitutive model based on the Orowan equation (Schuch et al. 2021). Stanczak et al. used six different material conditions (horizontally, 45° and vertically oriented, each with and without heat treatment) for compression test at a wide range of strain rates, but a limited temperature range between room temperature and only 200 °C. They fitted the parameters of a modified Johnson–Cook material model to describe the strain rate and temperature dependency of the thermo-visco-plastic behavior (Stanczak et al. 2022). Aktürk et al. conducted tensile tests on both horizontally and vertically oriented specimens without heat treatment. These tests were performed at various strain rates at room temperature and at a constant strain rate at temperatures up to 300 °C. They fitted the Johnson–Cook material model parameters to the tests and achieved deviations of 7–8% compared to an FE tensile test simulation (Aktürk et al. 2021). In the current release 2024R1 of the commercial FE program Ansys, a bilinear isotropic hardening material model for AlSi10Mg is available that covers the temperature range up to 250 °C.

The characteristic roundness of the stress–strain response known from aluminum alloys cannot be adequately captured by bilinear material models. Therefore, more advanced models are required for sophisticated numerical simulations (Yun et al. 2021). Besides the already mentioned Johnson–Cook model, the Voce (Nammalvar Raja Rajan et al. 2022; Roth et al. 2021) and the Hockett-Sherby (Irmak and Tröster 2019) hardening laws as well as a Chaboche cyclic plasticity model (Sausto et al. 2022) were used for AlSi10Mg from the PBF-LB/M process. These models are well suited, but none of the latter is reported to include temperature dependence.

In this study, the temperature-dependent tensile properties of AlSi10Mg from the PBF-LB/M process are investigated in a wide temperature range. Stress–strain curves are evaluated from tensile tests at elevated temperatures of up to 440 °C, and the mechanical properties such as elastic modulus, yield strength, and ultimate tensile strength are determined. Further investigations into the deformation behavior and fracture surfaces are carried out in order to deepen the understanding of the temperature dependence. The stress–strain curves are used to fit the parameters of a Chaboche nonlinear kinematic hardening material model. The applicability of the fitted model is verified by simulating the tensile tests with the commercial FE software Ansys.

11.2 Material and Methods

11.2.1 PBF-LB/M Fabrication of AlSi10Mg Specimens

Cylindrical rods with a diameter of 12 mm and a height of 80 mm were produced via PBF-LB/M in vertical orientation under an argon atmosphere using an SLM 125HL PBF-LB/M system (Nikon SLM Solutions AG, Germany). The rods were cut from the baseplate and machined on a CNC lathe to produce standardized tensile test specimens in accordance with DIN 50,152-B 6 × 30 with a diameter *d* of 6 mm and

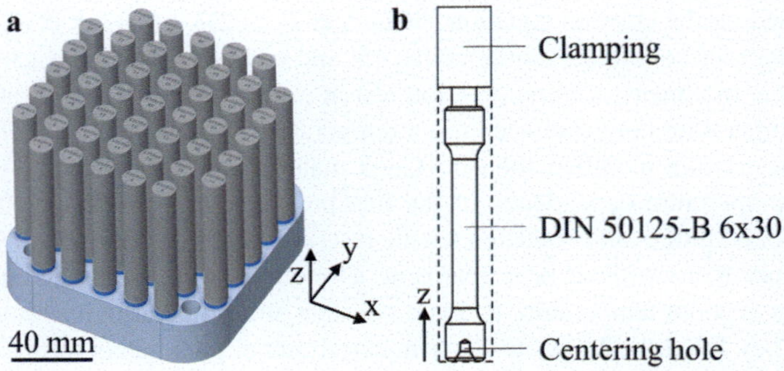

Fig. 11.1 **a** Build job layout with 45 rods on the SLM 125HL PBF-LB/M system. The shielding gas flows in negative *x*-direction, and the recoater moves bidirectional along the *y*-axis. **b** Simplified sketch of the rods that are clamped on the top side during CNC machining

Table 11.1 Main PBF-LB/M processing parameters for AlSi10Mg

Processing parameters	
Metal powder	AlSi10Mg (20–63 μm)
Laser power	350 W
Scan speed	1650 mm/s
Hatch distance	0.13 mm
Layer thickness	0.03 mm
Stripe rotation angle after each layer	67°
Baseplate temperature	150 °C

a gauge length L_0 of 30 mm. For the tensile tests, 45 specimens from one build job were available, which were not heat-treated before testing. In Fig. 1a, the build job layout is shown as a rendering of the rods (gray) and the 2.7 mm solid supports (blue) on which they were built as connection to the baseplate (azure). The shielding gas flows in negative *x*-direction, and the recoater moves bidirectional along the *y*-axis. Figure 1b shows a simplified sketch of the rods from which the tensile specimens are machined.

The AlSi10Mg powder (20–63 μm) was in recycled condition and was processed with the main process parameters listed in Table 11.1 using a stripe scanning strategy. Only the volume parameters are shown, as the contour of the rods was subsequently removed by machining and is therefore not relevant in this study.

11.2.2 Uniaxial Tensile Tests at Elevated Temperatures

The uniaxial tensile tests at elevated temperatures of 50, 100, 150, 200, 320, 370, and 440 °C were carried out in accordance with DIN EN ISO 6892-2:2018-02. Additional

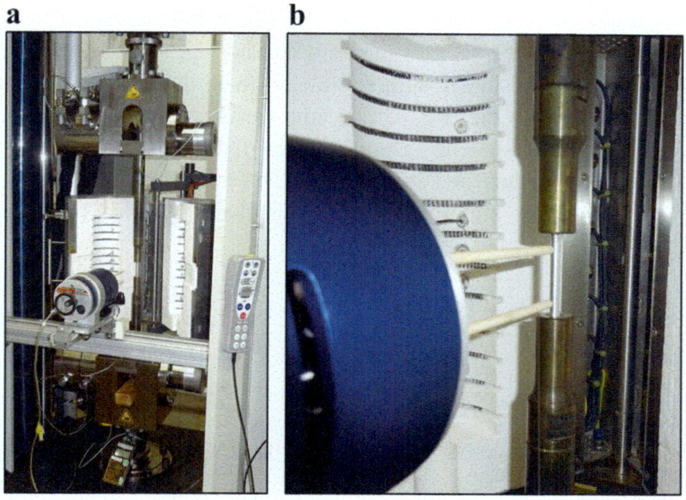

Fig. 11.2 **a** Universal testing machine with a mounted tensile specimen and open furnace, **b** detail view of the extensometer arms attached to the specimen

reference tests at room temperature (20 °C) were carried out in accordance with DIN EN ISO 6892-1:2019.

For better readability, the temperature levels are indicated hereinafter as a three-digit index number, e.g. T_{020} for 20 °C, T_{100} for 100 °C, etc.

The tests were performed at the Bundeswehr Research Institute for Materials, Fuels and Lubricants (WIWeB) in Erding, Germany using a Z250 universal testing machine (ZwickRoell GmbH & Co. KG, Germany) equipped with an HTO-08 three-zone round furnace (Maytec GmbH, Germany). The experimental setup is depicted in Fig. 11.2.

The specimens were pre-loaded between 100 and 500 N depending on the temperature level and heated to the specified temperature at which they were held for 10 min before being tested at a constant strain rate of 0.00025 s^{-1}. As primary test results, engineering stress–strain curves ($\sigma_{eng}-\varepsilon_{eng}$-curves) were recorded for a sample size of $n = 3$–6 for each temperature level. The temperature-dependent tensile properties, like the elastic modulus (E), the yield strength (YS), and the ultimate tensile strength (UTS), were determined using the ZwickRoell software. The percentage reduction of area (Z) was determined by measuring the broken specimens after the tests in accordance with DIN EN ISO 6892–1:2019.

It must be mentioned that the extensometer arms had to be removed shortly before the specimens broke to prevent the arms from being damaged. Therefore, the elongation at break could not be recorded and the stress–strain curves do not show the end of the elongation range. In addition, the ZwickRoell software could not output E and YS for T_{370} and T_{440}. Since E is a required parameter for the material model and for plastic strain calculation, it was determined from true stress–strain curves as described below.

The input required to fit the parameters of the material model, as described in Sect. 2.3, are true stress-true plastic strain curves ($\sigma_{true}-\varepsilon_{true}^{pl}$-curves) (ANSYS 2024a). Therefore, the pre-loads were subtracted from the engineering stresses to shift the start of curves to the origin. The engineering strain and the engineering stress obtained from the tensile tests can be converted to true strain and true stress (Tu et al. 2020). For this, the true strains ε_{true} also referred to as logarithmic strain or Hencky strain, and the true stresses σ_{true} were calculated by Eqs. (11.1) and (11.2)

$$\varepsilon_{true} = \ln\left(1 + \varepsilon_{eng}\right) \tag{11.1}$$

$$\sigma_{true} = \sigma_{eng}\left(1 + \varepsilon_{eng}\right) \tag{11.2}$$

for each recorded tensile test. The σ_{true}-ε_{true} curves were averaged, and standard deviations (*SD*) for the stresses were calculated using the *Average/Merge Multiple Curves* analysis function in Origin 2024 software (OriginLab Corporation, USA). The averaged σ_{true}-ε_{true} curves were used for the evaluation of E_{true} and the proportional limit (*PL*). For each temperature level, a straight line was fitted to the linear-elastic region of the curve according to Hooke's law and its slope E_{true} was adjusted so that the sum of squared errors between the straight line and the curve was minimized. Figure 11.3 shows this fitting of E_{true} and the determination of *PL*.

Since stress–strain curves of aluminum alloys typically do not show a clearly defined end of the linear-elastic region (Yun et al. 2021), the squared errors increase when the straight line and the curve deviate from each other with increasing strain. *PL* was chosen to be the stress where the squared error exceeded 10 MPa2, corresponding to an absolute error of $\sqrt{10}$ MPa. Note that *PL* is only used as an initial value for fitting the material model (see Sect. 2.3) and is therefore replaced in the material

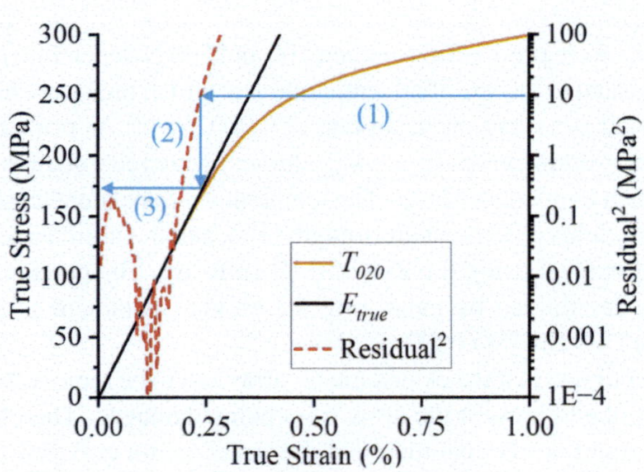

Fig. 11.3 Procedure to find E_{true} and *PL* from the linear-elastic section of the true stress–strain curves. E_{true} (black line) is fitted to the experimental curve (orange line) by minimizing the sum of squared errors. The point at which the squared residuals (red dotted line) exceed the control level of 10 MPa2 (1) is read from the experimental curve (2) to determine the value of *PL* (3)

model by the optimization result for the initial yield stress Y_0. The plastic strain ε_{true}^{pl} was calculated from the σ_{true}-ε_{true}-curves. Using the additive decomposition of the total true strain, ε_{true}^{pl} was calculated by Eq. (11.3)

$$\varepsilon_{true}^{pl} = \varepsilon_{true} - \frac{\sigma_{true}}{E_{true}} \tag{11.3}$$

in the strain hardening range of the curves after *PL*.

To evaluate the fracture surfaces, microscopy images were captured using the focus variation method of a Keyence VK-X3000 3D Surface Profiler (Keyence Corporation, Japan) with a motorized stage for automatic stitching. Overview images were captured using a 5 × lens with a height measurement accuracy of $\pm 1 + L_m/100$ μm or less, where L_m is the measuring length.

11.2.3 Fitting of Chaboche Material Model Parameters

It is well-known that the temperature-dependent plastic deformation of AlSi10Mg from the PBF-LB/M process shows nonlinear behavior (Cao et al. 2021; Lehmhus et al. 2022; Aktürk et al. 2021). This study utilizes a rate-independent nonlinear kinematic hardening model based on the Chaboche formulation to account for this nonlinear plasticity. The Chaboche model integrates linear isotropic elasticity, a von Mises yield function, and an associated flow rule, making it capable of accurately simulating complex cyclic plastic behaviors such as the Bauschinger effect, cyclic hardening, softening, ratcheting, and shakedown (ANSYS 2024b). While the Chaboche model can address a wide range of behaviors, its application here is focused solely on monotonic strain hardening.

The Chaboche model is implemented in the Ansys 2024R1 FE software in such a way that up to five superposed kinematic back-stress models α_i can be used per temperature level (ANSYS 2024a). This study uses three models ($i = 1, 2, 3$). For each back-stress model, the slope parameter C_i and the decay parameter γ_i must be identified. Together with the initial yield stress Y_0, there are seven material model parameters per temperature level that need to be fitted. In commercial FE software packages, like Ansys or Abaqus (Wójcik and Skrzat 2021; Kalnins et al. 2015), there are nonlinear numerical optimization procedures implemented to fit the parameters of the back-stress evolution law in Eq. (11.4), which is formulated as an exponential equation of the form

$$\alpha_i = \frac{C_i}{\gamma_i}\left(1 - e^{\left(-\gamma_i \varepsilon^{pl}\right)}\right) \tag{11.4}$$

Here, ε^{pl} is the plastic strain and the ratio $\frac{C_i}{\gamma_i}$ is the asymptote that is approached. To fit the model parameters to the σ_{true}-ε_{true}^{pl} curves, the Ansys Workbench *Curve Fitting*

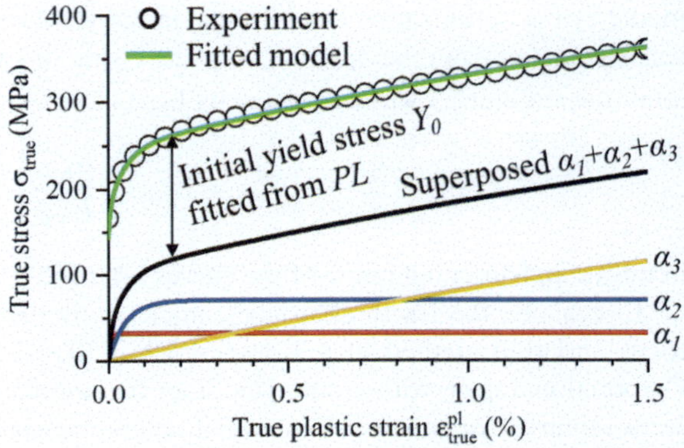

Fig. 11.4 Chaboche material model with three superposed back-stresses α_1, α_2, and α_3 to describe the nonlinear plasticity of an exemplary experimental stress–strain curve

tool available in the *Engineering Data* was used in two iterations. The first set of parameters was calculated from the initial values $C_i = \gamma_i = 1$ and $Y_0 = PL$, where Y_0 was fixed. For the second iteration, the calculated parameters for C_i and γ_i were used as initial values and Y_0 was made variable as well. This procedure results in a total of 56 fitted Chaboche model parameters for eight temperature levels.

It should be noted that only the pre-necking domain of the σ_{true}-ε^{pl}_{true} curve is suitable as input data for numerical modeling (Tu et al. 2020). There are methods to address the challenges of the post-necking regime at large strains, but that is not the focus of this study. Therefore, only the sections of the σ_{true}-ε^{pl}_{true} curves up to *UTS* are used for the fitting.

Figure 11.4 shows exemplary stress–strain curves, where the back-stress models α_1, α_2, and α_3 are superposed (black line) to describe the shape of the experimental curve (o markers), and the initial yield stress Y_0 shifts the Chaboche model curve (green line) to fit the experiment.

As Chaboche and Rousselier suggest, the respective back-stress models should serve different purposes to find the best fit (Chaboche and Rousselier 1983). Back-stress models with large slopes are meant to represent the initial nonlinearity of the curve. Here, α_1 and α_2 capture this region. Other back-stress models should cover the larger strain region where the first terms would flatten. This is achieved here by α_3, whose slope is approximately one order of magnitude lower than that of α_1 and α_2.

As a measure of the fitting quality, the resulting residuals are given as an output value of the *Curve Fitting* tool. Since the formula for calculating the residuals is not known to the authors, an average curve fit error (*CFE*), as used for the same purpose by Kalnins et al. (Kalnins et al. 2015), is defined by Eq. (11.5) and calculated in a spreadsheet:

$$\boldsymbol{CFE} = \sqrt{\frac{1}{\boldsymbol{L}-1}\sum_{l=1}^{L}\left(\boldsymbol{\alpha}_{\boldsymbol{Exp},l} - \boldsymbol{\alpha}_{\boldsymbol{Fitted},l}\right)^2} \tag{11.5}$$

The terms α_{Exp} and α_{Fitted} denote the experimental σ_{true}-ε_{true}^{pl} curves and the fitted Chaboche model curves and L is the number of recorded data points. With the unit megapascal, this is a physically meaningful measure of the fitting accuracy.

11.2.4 Simulation of Tensile Tests with the Ansys FE Software

The tensile tests at elevated temperatures were modeled and simulated to verify the usability of the temperature-dependent Chaboche material model in Ansys. The simulation was conducted using a static structural finite element analysis (FEA) approach to evaluate the material's response under tensile loading. Figure 11.5 shows the FEA setup.

The model employed 2D axisymmetry to accurately represent the specimen while ensuring computational efficiency. For the meshing, higher-order axisymmetric *PLANE183* elements with a mesh size of 0.5 mm at the top and bottom ends and 0.25 mm in the gauge length were used (Fig. 5a). A displacement-controlled load u_y in the y-direction was applied at the top edge while the bottom edge was fully constrained. The displacement was adjusted depending on the temperature to simulate sufficient strain. Figure 5b and c shows contour plots of normal y-stresses (σ_y) at half of u_y and at full u_y, respectively. The number of substeps was chosen to define an incremental displacement of 0.005 mm per substep. Large-deflection effects were

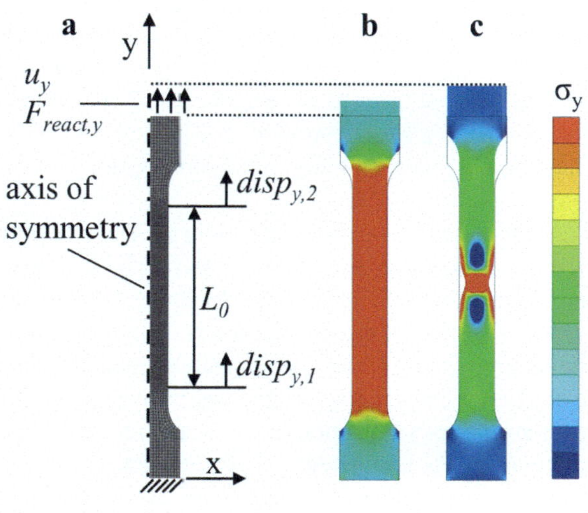

Fig. 11.5 **a** 2D axisymmetry FE model of the tensile test to verify the temperature-dependent Chaboche material model for AlSi10Mg. The bottom edge is fully constrained and the top edge is loaded with the y-displacement u_y. The resulting nodal y-displacements are tracked at the locations of the extensometer arms ($disp_{y,1}$ and $disp_{y,2}$). The reaction force in y-direction $F_{react,y}$ is tracked at the top edge. **b** Contour plot of normal y-stresses (σ_y) at half of u_y and **c** at full u_y

included to incorporate geometric nonlinearity due to significant deformations, especially in the necked region. The temperature-dependent Chaboche material model was assigned and the reference temperature was set according to the temperature levels of the tensile tests. To obtain engineering stress–strain curves from the simulations ($\sigma_{eng,sim}$-$\varepsilon_{eng,sim}$-curves) that are comparable to the experiments, simulation results that best match the measurements were evaluated. At the bottom and top end of the gauge length L_0, the nodal y-displacements $disp_{y,1}$ and $disp_{y,2}$ were tracked for each substep. The *y*-reaction force $F_{react,y}$ was tracked at the top edge. $\varepsilon_{eng,sim}$ and $\sigma_{eng,sim}$ were calculated by Eqs. (11.6) and (11.7) as follows

$$\varepsilon_{eng,sim} = \frac{disp_{y,2} - disp_{y,1}}{L_0} \tag{11.6}$$

$$\sigma_{eng,sim} = \frac{F_{react,y}}{A_0} \tag{11.7}$$

where A_0 is the nominal cross-section.

11.3 Results and Discussion

11.3.1 Tensile Properties and Fracture Behavior

The resulting σ_{eng}-ε_{eng} curves are given in Fig. 11.6, where Fig. 6a shows the measured strain ranges (which end before the break) and Fig. 6b shows an enlarged detail view of the beginning of the curves. There, the transition from the initial linear-elastic region to the nonlinear "knee" region, which is a typical, continuously rounded curve for aluminum alloys (Yun et al. 2021) is visible. The curves scatter only slightly, indicating a high repeatability, and the temperature levels can be clearly distinguished from each other.

Trends of decreasing strength and increasing elongation can be observed from the curves as expected. T_{020} and T_{050} show similar behavior in the linear-elastic region and the beginning of strain hardening, followed by lower *UTS* and a slightly larger strain for T_{050}. Up to T_{100}, the elongation increases, but does not increase further up to T_{200}. At higher temperature levels, the elongation increases significantly. Also increasing strain hardening and necking are present from T_{020} up to T_{200}. The high temperature levels show an early transition to the "knee" region and plastic behavior.

Figure 11.7 summarizes the temperature-dependent tensile properties and shows the mean values and standard deviations of the elastic modulus, the ultimate tensile strength, the yield strength, and the percentage reduction of area (Z). *E* and *YS* could only be determined up to 320 °C.

The strength properties show a clear downward trend with increasing temperature. *UTS* decreases almost linearly from 469.8 ± 1.8 MPa at T_{020} to 15.4 ± 0.4 MPa at

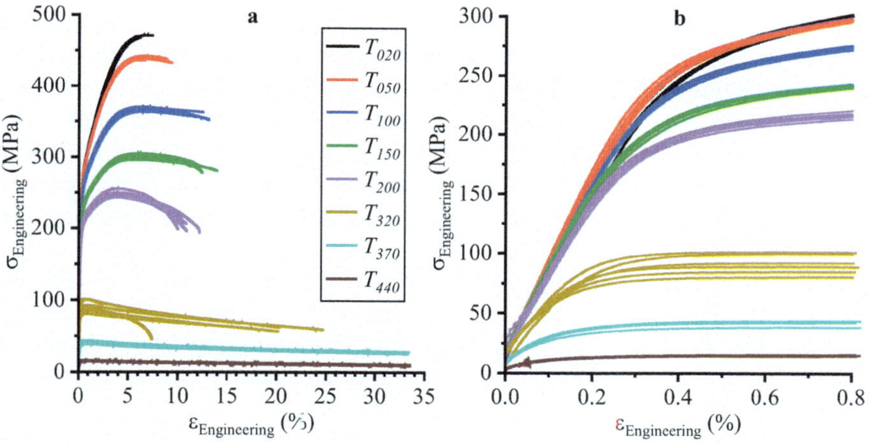

Fig. 11.6 **a** Engineering stress–strain curves of the eight temperature levels between room temperature and 440 °C. Note that the curves end before the break and therefore do not represent the entire strain range of the tensile tests. **b** Detail view of the beginning of the curves

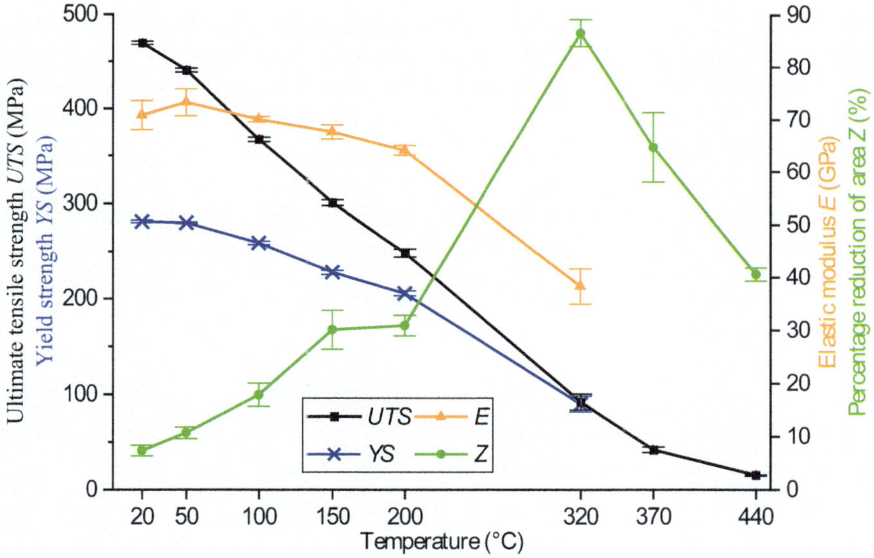

Fig. 11.7 Temperature-dependent tensile properties of AlSi10Mg from the PBF-LB/M process. Ultimate tensile strength *UTS* (left-hand side, lhs), yield strength *YS* (lhs), elastic modulus *E* (right-hand side, rhs), and percentage reduction of area *Z* (rhs) are plotted over the temperature levels. The error bars indicate ±1 *SD*, but are partly too small to be visible. *E* and *YS* could only be determined up to 320 °C

T_{440}. *YS* decreases from 281.6 ± 1.5 MPa at T_{020} to 90.0 ± 8.2 MPa at T_{320}, where it is almost equal to *UTS* and therefore leading to an increase in the yield ratio *YS*/*UTS* from 0.60 to 0.98. *E* is relatively stable from 70.8 ± 2.7 GPa at T_{020} to 64.1 ± 1.0 GPa at T_{200}. At T_{320}, *E* drops to 38.4 ± 3.3 GPa. *Z* increases rapidly with increasing temperature from 7.4 ± 2.6% at T_{020} up to 86.6 ± 2.6% at T_{320}. With further increasing temperature, *Z* decreases to 40.6 ± 1.3% at T_{440}. One specimen did not break at T_{440} because the testing machine reached the pre-defined elongation limit.

Figure 8a–h displays one exemplary specimen per temperature level after the tensile tests, representing the temperature-dependent differences in deformation behavior. The T_{020}, T_{050}, and T_{100} specimens show uniform plastic deformation with no or only slight local necking (Fig. 8a–c). From T_{150} to T_{320}, necking significantly intensifies (Fig. 8d–f), which was also observed in the decline of stress after *UTS* in Fig. 6a. The T_{370} and T_{440} specimens (Fig. 8gi, h_i) show less pronounced local necking at the fracture location, but a high degree of uniform plastic deformation and several cracks and ruptured surfaces near the fracture location (Fig. 8gii, h_{ii}).

It is remarkable that almost all specimens, independent of the temperature level, broke in the upper half of the parallel length near the upper end of the gauge length. Hitzler et al. report a similar location of failure for vertically printed AlSi10Mg specimens. They attribute this to deteriorating material properties with increasing build height (Hitzler et al. 2017). The high packing density of 45 rods in the build job in this study is also a possible cause for porosity with increasing build height and overall reduced mechanical properties (Haeckel et al. 2017). A further, superimposed effect in this study is that the machined specimens were slightly tapered towards the top due to a CNC machining inaccuracy, but still within the shape tolerances of DIN 50,125. It cannot be ruled out that the necking near the extensometer arm impaired

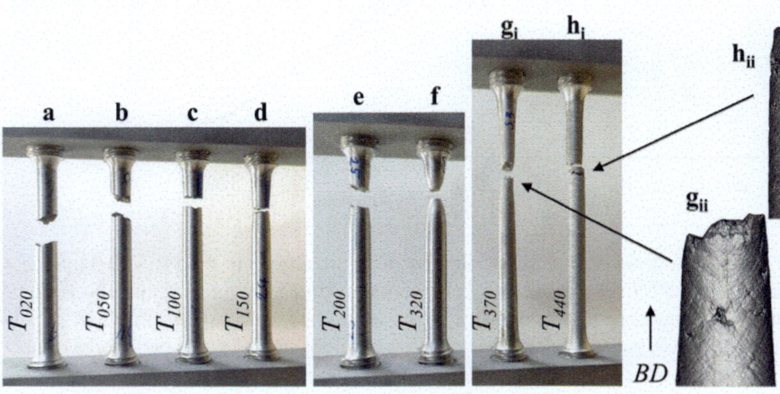

Fig. 11.8 Post-test tensile specimens showing varying deformation behaviors depending on the temperature level of the tensile test. Almost all specimens broke in the upper half in the build direction (*BD*). The detailed views of the T_{370} and T_{440} specimens (g_{ii} and h_{ii}) show cracks and ruptured surfaces near the fracture location

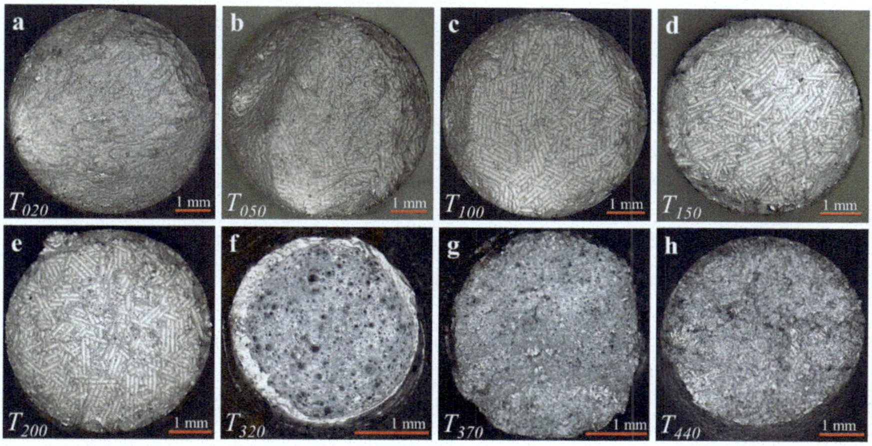

Fig. 11.9 Focus variation microscopy overview images (*5* × magnification) of typical fracture surfaces from specimens tested at temperature levels between 20 and 440 °C. The fracture surfaces show a different appearance depending on the temperature level of the tensile test

the strain measurement. Therefore, the strain measurements in the post-*UTS* region should be viewed with care.

To further evaluate the fracture behavior, focus variation microscopy images of the fracture surfaces of the lower halves of exemplary specimens were analyzed and are shown in Fig. 11.9.

The fracture surfaces of T_{020} and T_{050} are smooth and show no noticeable pores or other significant defects (Fig. 9a, b). For T_{100}, T_{150}, and T_{200}, the melt pool boundaries (MPB) of the scan tracks are clearly visible in the fracture surfaces (Fig. 9c–e). The T_{320} specimens show high porosity in the fracture surfaces (Fig. 9f). At the highest temperature levels T_{370} and T_{440}, which are shown in Fig. 11.9g and h, the surfaces are rough and rugged.

Together with the increasing strain shown in Fig. 6a, the observed deformation behavior (Fig. 11.8) and the fracture surfaces suggest the gradual transition from brittle to ductile fracture behavior. The temperature-dependent tensile properties and the brittle–ductile transition are comparable with the results reported by Cao et al. (2021), who also used non-heat-treated AlSi10Mg specimens in their study. In addition, they analyzed the microstructural evolution with increasing temperature. They found that the microstructure remains almost unchanged when the samples are kept at 100 or 200 °C for 30 min. For higher temperatures of 300 and 400 °C, the Si particles become coarser, precipitation of the Si phase from the Al matrix happens, and the density of dislocations decreases. In high magnification fracture surface microscopy images from specimens tested up to 100 °C, they find cleavage planes and elongated dimples. At higher temperatures, the dimples get larger and equiaxed. This suggests a transition from a mixed brittle-ductile fracture mechanism up to 200 °C to ductile fracture for higher temperatures. These microstructural phenomena support the results of this study, in which the transition from brittleness to ductility is inferred macroscopically on the basis of elongation and necking behavior.

The visible MPBs in the fracture surfaces (Fig. 9c–e) suggest that the fracture path follows the MPBs, as it is also reported in the literature for vertically oriented tensile specimens (Lehmhus et al. 2022; Patakham et al. 2021; Xiong et al. 2019). In this orientation, the MPBs perpendicular to the tensile force direction have a more efficient load bearing capacity (Patakham et al. 2021).

The high degree of porosity in the fracture surfaces at T_{320} (Fig. 9f) could result from an effect known as thermally induced porosity (TIP). Strumza et al. describe TIP as a process in which entrapped gas pores expand due to increasing internal pressure. When this pressure overcomes the material's yield strength, plastic deformations occur. They observed this phenomenon with vertically oriented AlSi10Mg at 240 °C (Strumza et al. 2019). The drop in *YS* to only 90 MPa at T_{320} makes the occurrence of TIP appear likely.

The plastic behavior at T_{370} and T_{440}, accompanied by the rugged fracture surfaces, is attributed to microstructural changes in the material. Considering the low strain rate of 0.00025 s^{-1} and the high strains of 33%, the specimens had a total dwell time of approximately 1 h at the highest temperatures, not including the heating time to reach the steady state temperature before the test. This represents a heat treatment of the specimens. Lehmhus et al. discuss the effects of elevated temperatures during the tensile tests as a heat treatment that leads to significant changes in the microstructure. Especially at temperatures above the usual heat treatment temperatures of up to 350 °C, the eutectic phase's original cellular structure dissolves, forming larger, more globular Si particles within the Al matrix. The dissolution of the original network will facilitate plastic deformation, as the ductile Al matrix dominates the deformation behavior (Lehmhus et al. 2022).

11.3.2 Fitted Chaboche Material Model Parameters

Tables 11.2 and 11.3 summarize the temperature-dependent values for E, E_{true}, PL, and Y_0 and the fitted, temperature-dependent Chaboche material model parameters C_i and γ_i as well as the average curve fit error (CFE), respectively.

The elastic moduli E_{true} determined from the σ_{true}-ε_{true}-curves are higher than the E values determined by the ZwickRoell software from the σ_{eng}-ε_{eng}-curves. The residuals were minimized by the *Curve Fitting* tool by reducing the initial values of PL to the fitted values of Y_0. The fit between the three superposed back-stress models and the experimental σ_{true}-ε_{true}^{pl}-curves is accurate for all temperature levels. The resulting *CFE*s are between 0.1 and 2.7 MPa, which is in the same range as Kalnins et al. reported for the fitting of a Chaboche model for a SA-312 TP304 steel. They used one to five back-stress models with resulting *CFE*s between 11.47 and 0.15, respectively (Kalnins et al. 2015). This indicates a slight potential for improvement by using more back-stress models, especially at lower temperature levels.

Table 11.2 Overview of the temperature-dependent values for E, E_{true}, PL, and Y_0. E_{true} and Y_0 are used in the material model, E and PL are shown for comparison

Temp (°C)	E (MPa)	E_{true} (MPa)	PL (MPa)	Y_0 (MPa)
20	70,772.0	71,301.3	167.0	143.9
50	73,209.4	73,226.3	164.5	157.1
100	70,028.8	72,455.6	143.5	126.1
150	67,664.1	69,232.7	119.1	108.6
200	64,080.1	65,429.2	113.2	99.9
320	38,363.2	43,844.0	46.8	41.4
370	–	28,855.4	17.9	15.4
440	–	11,903.4	8.1	6.8

Table 11.3 Overview of the fitted, temperature-dependent Chaboche material model parameters C_i and γ_i and the average curve fit error (CFE)

Temp (°C)	C_1	γ_1	C_2	γ_2	C_3	γ_3	CFE (MPa)
20	527,169.0	16,433.6	150,506.5	2134.4	9915.1	34.6	2.7
50	138,051.3	3243.4	138,047.9	3553.2	10,546.4	46.6	2.7
100	434,307.4	11,596.0	131,448.4	1959.3	7340.9	47.6	1.7
150	193,335.3	8620.6	130,294.7	1826.3	5421.5	51.5	1.4
200	299,530.9	10,546.0	88,890.2	1619.5	4223.5	69.2	0.9
320	47,878.4	5805.9	41,018.1	2337.5	9395.1	736.5	0.8
370	14,796.1	3642.5	14,758.9	3628.1	8344.7	715.4	0.3
440	2756.5	1878.8	2756.4	1883.4	1403.5	428.8	0.1

The fitted parameters were implemented in Ansys Workbench using a *Commands object* written in the scripting language *Ansys Parametric Design Language* (*APDL*) for communication with the Ansys Mechanical solver. The temperature dependency of the elastic modulus can be achieved with the *APDL* commands `MPTEMP` and `MPDATA`, while sequences of `TB,CHABOCHE`, `TBTEMP`, and `TBDATA` commands are used for the temperature-dependent Chaboche model.

11.3.3 Tensile Test Simulation Results

The results of the simulated tensile tests at all temperature levels are evaluated by direct comparison with the experiments. The engineering stress–strain curves ($\sigma_{eng,sim}$-$\varepsilon_{eng,sim}$-curves) are plotted together with the experimental curves in Fig. 11.10. The pre-load was added to $\sigma_{eng,sim}$ to shift the simulated curves to

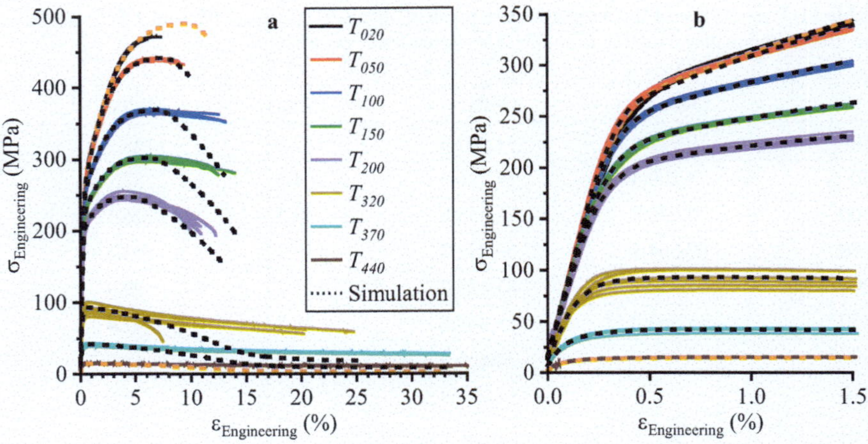

Fig. 11.10 Experimental σ_{eng}-ε_{eng}-curves (solid lines) of the eight temperature levels between room temperature and 440°C are compared to the simulated $\sigma_{eng,sim}$-$\varepsilon_{eng,sim}$-curves (dotted lines) obtained using the fitted Chaboche material model. **a** Full strain range and **b** detailed view of the beginning of the curves. Note that the experimental σ_{eng}-ε_{eng}-curves have already been shown in Fig. 11.6 and are reproduced here for direct comparison with the simulations

initial values in the unloaded state that correspond to the experiments. Note that the curves from the simulations are only shown up to a strain comparable to the experiments and would otherwise extend further, as there is no damage criterion in the material model.

The linear-elastic region and the strain hardening are represented very accurately in the simulation. Up to *UTS*, the simulated curves lie within the scatter of the experimental curves (Fig. 10a). Only at T_{020} does the simulation slightly exceed the results of the experiments at *UTS*. With this exception, *UTS* are predicted with errors of less than 1.5 MPa compared to the mean values of the experiments. For T_{020} the error is 20.2 MPa, which corresponds to 4.1%. Interestingly, the simulation of T_{050} matches the experiments almost perfectly with an *UTS* error of 0.3%, although the largest *CFE* of 2.7 MPa (see Table 11.3) is also present here, as with T_{020}. The nonlinear "knee" transition is represented very accurately in the simulation for all temperature levels (Fig. 10b).

The minor *UTS* errors express the reliability of the material model developed in this study, especially with respect to models reported in the literature. For the maximum stresses in tensile tests with vertically oriented specimens at room temperature, 150, and 300 °C, Aktürk et al. report deviations of 7–10% from the corresponding simulations using their fitted Johnson–Cook material model (Aktürk et al. 2021).

In the post-necking region, the experimental and simulated curves deviate. In the simulations, necking is more pronounced than observed in the experiments. A mesh convergence study with mesh sizes between 0.5 and 0.1 mm in the gauge length ruled out the dependence on the mesh size. Tu et al. emphasize the complexity of accurately modeling post-necking behavior. After reaching *UTS*, uniform deformation transitions into localized deformation in the necking zone and the stress state is no longer uniaxial (Tu et al. 2020). Since the material model was fitted to the σ_{true}-ε_{true}^{pl}-curves only up to *UTS*, which represents the valid range, as described in Sect. 2.3, the simulation response is likely to be less accurate than in the fitted region. Additionally, the measured stress–strain curves might be impaired after *UTS* due to the necking location close to the upper extensometer arm, as described in Sect. 3.1. Therefore, the necking region should be considered with caution in both the experiments and the simulations.

11.4 Conclusion

The present study contributes to increasing the limited knowledge about the high temperature mechanical properties of AlSi10Mg produced via PBF-LB/M. Tensile tests were performed at elevated temperatures to determine the temperature-dependent tensile properties of vertically oriented as-built specimens without prior heat treatment. The temperature levels tested were room temperature, 50, 100, 150, 200, 320, 370, and 440 °C. The stress–strain curves, the deformation behavior, and the fracture surfaces show clear tendencies of transition from brittle to ductile with increasing temperatures in response to uniaxial, monotonic loading. The stress–strain curves were analyzed, interpreted, and used as a target for fitting a Chaboche nonlinear kinematic hardening material model with three back-stress models for each temperature level. The main results and conclusions are summarized as follows:

1. The tensile tests showed high repeatability, which indicates a reliable experimental basis for the investigations.
2. As expected, clear tendencies of decreasing strength and increasing elongation with increasing temperatures were observed. The ultimate tensile strength decreases almost linearly from 470 MPa at room temperature to 15 MPa at 440 °C. The yield strength decreases from 282 MPa at room temperature to 90 MPa at 320 °C. Pronounced necking was observed at 320 °C and significant plastic deformation at 370 and 440 °C.
3. Seven temperature-dependent Chaboche material model parameters were fitted to the true stresses-true plastic strain curve for each temperature level, and minimal fitting errors were obtained
4. To verify the applicability of the Chaboche material model, a finite element analysis of the tensile tests was performed, showing excellent agreement between the experimental and numerical results in the linear-elastic and strain hardening regions. The ultimate tensile strength could be predicted with errors between 0 and 4%. Higher deviations were found in the simulation of the post-necking behavior, where the simulation tends to underestimate the experiments in the region of high strains.

5. The results in the post-necking region should be considered with caution in both the experiments and the simulations, as the strain measurements may be erroneous. This region must be better taken into account when fitting the material model.

The availability of reliable, experimental results for AlSi10Mg produced via PBF-LB/M will facilitate the use of additively manufactured parts in industrial applications, especially when both mechanical and thermal loads are present. Furthermore, users will benefit from the advanced material model when simulating the PBF-LB/M process or related processes such as welding.

Future research will focus on the optimization potential of advanced material models for use in the PBF-LB/M process simulation. The ability to simulatively predict residual stresses and distortions as well as numerical efficiency play important roles in enabling high-quality components without the need for trial-and-error iterations in industrial manufacturing.

Acknowledgements We would like to thank Mr. Till Tetzlaff, Mr. Mathias Zimmermann, and Mr. Stefan Irl from the Bundeswehr Research Institute for Materials, Fuels and Lubricants (WIWeB) in Erding, Germany for their support and efforts in carrying out the tensile tests and compiling the data.

This paper is partially funded by dtec.bw—Digitalization and Technology Research Center of the Bundeswehr, which we gratefully acknowledge [project FLAB-3Dprint]. dtec.bw is funded by the European Union—NextGenerationEU.

References

Aktürk M, Boy M, Gupta MK, Waqar S, Krolczyk GM, Korkmaz ME (2021) Numerical and experimental investigations of built orientation dependent Johnson-Cook model for selective laser melting manufactured AlSi10Mg. J Mater Res Technol. https://doi.org/10.1016/j.jmrt.2021.11.062

Bao J, Wu Z, Wu S, Hu D, Sun W, Wang R (2022) The role of defects on tensile deformation and fracture mechanisms of AM AlSi10Mg alloy at room temperature and 250 °C. Eng Fract Mech. https://doi.org/10.1016/j.engfracmech.2021.108215

Cao Y, Lin X, Wang QZ, Shi SQ, Ma L, Kang N, Huang WD (2021) Microstructure evolution and mechanical properties at high temperature of selective laser melted AlSi10Mg. J Mater Sci Technol. https://doi.org/10.1016/j.jmst.2020.04.066

Careri F, Khan RH, Todd C, Attallah MM (2023) Additive manufacturing of heat exchangers in aerospace applications: a review. Appl Therm Eng. https://doi.org/10.1016/j.applthermaleng.2023.121387

Chaboche JL, Rousselier G (1983) On the plastic and viscoplastic constitutive equations—part i: rules developed with internal variable concept. J Pressure Vessel Technol. https://doi.org/10.1115/1.3264257

Haeckel F, Meixlsperger M, Burkert T (2017) Technological challenges for automotive series production in laser beam melting. In: Proceedings of the 28th annual international solid freeform fabrication symposium. University of Texas at Austin, Austin, Texas, pp 2682–2698. https://doi.org/10.26153/tsw/16985

Hitzler L, Janousch C, Schanz J, Merkel M, Heine B, Mack F, Hall W, Öchsner A (2017) Direction and location dependency of selective laser melted AlSi10Mg specimens. J Mater Process Technol. https://doi.org/10.1016/j.jmatprotec.2016.11.029

ANSYS, Inc. (2024) Theory reference. 2024R1

ANSYS, Inc. (2024) Material reference. 2024R1

Irmak EA, Tröster T (2019) Fracture prediction of additively manufactured AlSi10Mg materials. Procedia Struct Integr. https://doi.org/10.1016/j.prostr.2019.12.101

Kalnins A, Rudolph J, Willuweit A (2015) Using the nonlinear kinematic hardening material model of chaboche for elastic-plastic ratcheting analysis. J Pressure Vessel Technol. https://doi.org/10.1115/1.4028659

Lehmhus D, Rahn T, Struss A, Gromzig P, Wischeropp T, Becker H (2022) High-temperature mechanical properties of stress-relieved AlSi10Mg produced via laser powder bed fusion additive manufacturing. Materials. https://doi.org/10.3390/ma15207386

Patakham U, Palasay A, Wila P, Tongsri R (2021) MPB characteristics and Si morphologies on mechanical properties and fracture behavior of SLM AlSi10Mg. Mater Sci Eng A. https://doi.org/10.1016/j.msea.2021.141602

Ponnusamy P, Rahman Rashid RA, Masood SH, Ruan D, Palanisamy S (2020) Mechanical properties of SLM-printed aluminium alloys: a review. Materials. https://doi.org/10.3390/ma13194301

Nammalvar Raja Rajan A, Krochmal M, Wegener T, Biswas A, Hartmaier A, Niendorf T, Moeini G (2022) Micromechanical modeling of alsi10mg processed by laser-based additive manufacturing: from as-built to heat-treated microstructures. Materials. https://doi.org/10.3390/ma15165562

Roth CC, Tancogne-Dejean T, Mohr D (2021) Plasticity and fracture of cast and SLM AlSi10Mg: high-throughput testing and modeling. Addit Manuf. https://doi.org/10.1016/j.addma.2021.101998

Sausto F, Tezzele C, Beretta S (2022) Analysis of fatigue strength of L-PBF AlSi10Mg with different surface post-processes: effect of residual stresses. Metals. https://doi.org/10.3390/met12060898

Schuch M, Hahn T, Bleckmann M (2021) The mechanical behavior and microstructure of additively manufactured AlSi10Mg for different material states and loading conditions. Mater Sci Eng A. https://doi.org/10.1016/j.msea.2021.141134

Stanczak M, Rusinek A, Broniszewska P, Fras T, Pawłowski P (2022) Influence of strain rate and temperature on the mechanical behaviour of additively manufactured AlSi10Mg alloy–experiment and the phenomenological constitutive modelling. Bull Pol Acad Sci Tech Sci. https://doi.org/10.24425/bpasts.2022.141983

Strumza E, Yeheskel O, Hayun S (2019) The effect of texture on the anisotropy of thermophysical properties of additively manufactured AlSi10Mg. Addit Manuf. https://doi.org/10.1016/j.addma.2019.06.013

Tu S, Ren X, He J, Zhang Z (2020) Stress–strain curves of metallic materials and post-necking strain hardening characterization: a review. Fatigue Fract Eng Mat Struct. https://doi.org/10.1111/ffe.13134

Uzan NE, Shneck R, Yeheskel O, Frage N (2018) High-temperature mechanical properties of AlSi10Mg specimens fabricated by additive manufacturing using selective laser melting technologies (AM-SLM). Addit Manuf. https://doi.org/10.1016/j.addma.2018.09.033

Wójcik M, Skrzat A (2021) Identification of Chaboche-Lemaitre combined isotropic–kinematic hardening model parameters assisted by the fuzzy logic analysis. Acta Mech. https://doi.org/10.1007/s00707-020-02851-z

Xiong ZH, Liu SL, Li SF, Shi Y, Yang YF, Misra R (2019) Role of melt pool boundary condition in determining the mechanical properties of selective laser melting AlSi10Mg alloy. Mater Sci Eng A. https://doi.org/10.1016/j.msea.2018.10.083

Yun X, Wang Z, Gardner L (2021) Full-range stress-strain curves for aluminum alloys. J Struct Eng. https://doi.org/10.1061/(ASCE)ST.1943-541X.0002999

Chapter 12
Recycling Potential of Niche Products: Case Studies on Climbing Ropes and Agricultural Binding Twines

Philippe du Maire, Felix Gärtner, Matthias H. Deckert, Michael Johlitz, and Andreas Öchsner

Abstract Halting climate change represents the most significant challenge facing humanity in the current century. With the escalating challenges caused by global warming, reducing greenhouse gas emissions through effective recycling strategies for plastic components has become imperative. Besides the large plastic waste producers, niche products also offer significant opportunities for reducing greenhouse gas emissions and conserving resources. This research provides an overview of the recycling techniques and explores the potential of recycling niche products exemplarily focussing on old climbing ropes and agricultural binding twines. Climbing ropes made from polyamide 6 (PA6) were successfully recycled into granulate, exhibiting superior tensile strength and elastic modulus compared to the reference virgin PA6. Similarly, agricultural binding twines made from polypropylene (PP) were recycled, with detailed analysis revealing mechanical and rheological properties degradation over multiple cycles. The findings highlight the hidden potential of niche product recycling.

P. Maire (✉) · M. H. Deckert · A. Öchsner
Faculty of Mechanical and Systems Engineering, Esslingen University of Applied Sciences, Kanalstraße 33, 73728 Esslingen, Germany
e-mail: philippe.du-maire@hs-esslingen.de

M. H. Deckert
e-mail: matthias.deckert@hs-esslingen.de

A. Öchsner
e-mail: andreas.oechsner@hs-esslingen.de

F. Gärtner
Faculty 7: Engineering Design, Production Engineering and Automotive Engineering, Stuttgart University, Pfaffenwaldring 9, 70569 Stuttgart, Germany
e-mail: st183004@stud.uni-stuttgart.de

M. Johlitz
Faculty of Aerospace Engineering, Institute of Mechanics, Universität der Bundeswehr München, Werner-Heisenberg-Weg 39, 85577 Neubiberg, Germany
e-mail: michael.johlitz@unibw.de

H. Altenbach et al. (eds.), *Lectures Notes on Advanced Structured Materials 3*,
Advanced Structured Materials 226, https://doi.org/10.1007/978-3-031-77403-4_12

12.1 Introduction

The effects of global warming can be seen and felt more apparent than ever before. Everywhere humans are facing serious challenges, including extreme weather events, rising sea levels, glacier retreat, and other impacts caused by global warming Allen et al. (2018). In order to achieve the 1.5 °C target set out in the Paris Climate Agreement and attain climate neutrality by 2050 there is a high need to take immediate action Allen et al. (2018). To stop global warming greenhouse gas emissions must be significantly reduced, for instance through sustainable production and a closed-loop process for the products fabricated. Consequently, research into effective recycling processes is therefore of high importance.

Plastic production, which heavily relies on crude oil and energy, significantly contributes to greenhouse gas emissions. Annually, Europe produces 49 million tonnes of plastic, primarily used in packaging, construction, automotive parts, and electronics, generating 25.8 million tonnes of waste. Despite various recycling methods, only less than 30% of plastic waste is collected for recycling, with a significant amount exported to countries with lower environmental standards Volk et al. (2021); Singh et al. (2017).

In addition to the significant greenhouse gas emissions generated by larger waste producers, smaller niche products also offer considerable potential to reduce these emissions, contribute to fulfil the climate goals, and stop global warming. Despite their relatively minor role in the waste production landscape, these niche products are of great value from a circular economy perspective, as they are largely composed of pure materials. The purity of the material facilitates recycling, as no complex separation of the individual materials is required.

Several studies have already shown the hidden recycling potential of these niche products. Van Straten et al. showed in their work the viability of reprocessing surgical plastic waste into a high-quality raw material through mechanical recycling techniques van Straten et al. (2021). The study was focussed on the polypropylene wrapping paper used in the medical sector to wrap surgical instruments. In the Netherlands alone, an estimated amount of 1.3 million kg of plastic waste from surgical wrapping paper is generated annually van Straten et al. (2021). The fact that this is just one of the many plastic products used in the medical sector illustrates the enormous potential for saving energy and emissions through consistent recycling.

Furthermore, the agriculture sector also exhibits a high potential for plastic waste recycling. As stated by Mayanti and Helo, plastics are predominantly utilised in agriculture as films for applications such as greenhouses, mulching, low tunnels, silage, and bale wrap Mayanti and Helo (2022). Their research examined the economic and environmental implications of recycling agricultural plastic waste, specifically focussing on bale wrap waste in Finland. Notably, Finland produces approximately 12'000 tonnes of agricultural plastic waste annually, with bale wraps constituting around 7'000 tonnes Mayanti and Helo (2022). This also serves to illustrate the considerable quantity of waste generated by a supposedly niche product alone. Mayanti and Helo found that the majority of this waste is disposed of in landfills (70%), with only a small portion being recycled (10%) Mayanti and Helo (2022); Briassoulis et al. (2013). This further illustrates that improper management of this waste leads to significant environmental and economic issues. This includes the loss of potential financial gains from recycling and the environmental impact of increased raw material demand and landfill practices.

Another illustrative example can be found in the mountaineering sector. The research conducted by Bradford et al. indicates that Switzerland alone has approximately 226 t of old ropes with the potential to save greenhouse gas emissions in the amount of 1,170 t CO_2-equivalent Bradford et al. (2021). The present study should provide a brief overview of the recycling techniques. Furthermore, the potential for recycling niche products is demonstrated in detail by two case studies on recycling in the agricultural and mountaineering sectors. The study demonstrates the wide range of applications for recycling techniques and their hidden potential in everyday life.

12.2 Recycling Techniques

For the recycling of plastic waste, a wide range of recycling techniques are available Damayanti et al. (2022). In general, these techniques can be divided into mechanical and chemical recycling processes. Both methods have their advantages. Mechanical recycling is cost-effective, whereas chemical recycling offers potential economic benefits Damayanti et al. (2022). More details about the advantages and disadvantages of both techniques can be found in Sects. 12.2.1 and 12.2.2. The choice of the respective recycling process depends on various parameters, including the type of polymer, the intended later application of the recyclate, and contamination of the plastic waste. Thermoplastics, for example, can be remelted and reshaped, which makes them ideally suited for mechanical recycling. Cross-linked polymers such as thermosets, polyurethanes, or rubbers can be pyrolysed, gasified, or otherwise processed using chemical or feedstock processes. The resulting raw materials can then be used as a naphtha substitute in crackers for the production of basic chemicals Orth et al. (2022); Rahimi and Garcia (2017). Nevertheless, at the outset of each recycling process, it is necessary to ensure the purity of the material and therefore to separate plastic waste into its constituent base materials. This preprocessing can be achieved through various technologies, including gravity separation, electrostatic separation, or sensor-based sorting Serranti and Bonifazi (2019).

12.2.1 Mechanical Recycling

Thermo-mechanical recycling using an extrusion process is a commonly used mechanical recycling technique Ragaert et al. (2017); Schyns and Shaver (2020). The process itself requires the utilisation of sorted and regrind polymeric waste as a feed material either provided as ground flakes or as agglomerates or regranulates

which have to be obtained through previous processing steps. During the agglomeration process, the particles of the regrind are bonded together by surface melting in order to standardise the particle shape and size and achieve higher bulk densities, which ensures a better processability in the extruder. The actual recycling process is then conducted with the help of an extruder, resulting in the formation of homogeneous regranulates. If necessary, the material properties of the recyclate can be adjusted by the compoundation with additives such as pigments, reinforcing fibres, and stabilisers Orth et al. (2022). Finally, the homogenised melt passes through melt filters to remove any impurities before being solidified by cooling and pelletised to the desired granulate size.
The main advantages that make mechanical recycling processes the preferred method of choice in industries are their cost effectiveness, solvent-free operation, and wide applicability to many polymers Schyns and Shaver (2020). In addition, extrusion processes can be easily adapted from small laboratory scale to large industrial scale applications. In contrast, thermo-mechanical recycling techniques have the disadvantage that heat conduction and viscous shear through the extrusion process have a negative influence on the material properties of the polymer. This is due to the degradation of the molecules, which can occur as a result of thermo-oxidative and shear-induced chain scission or cross-linking Oblak etv al. (2015).

12.2.2 Chemical Recycling

Chemical recycling is defined as the depolymerisation of the polymer, whereby the macromolecules are broken down into monomers or other basic chemical building blocks (oligomers) Dorigato (2021). The depolymerisation can be achieved through the following three techniques Orth et al. (2022):

- Solvent-based processes (dissolution, solvolysis);
- Thermochemical (gasification, pyrolysis);
- Enzymolysis.

Solvent-based processes are employed to recover plastics in a physical-chemical process, whereby the polymer structure is retained. With the aid of various solvents like dichloromethane, methylethylketone, tetrahydrofuran, or xylene, the polymers are selectively dissolved from the waste (dissolution). After dissolving, liquid and solid components are separated by a multistage process, during which any foreign polymers, additives, or aggregates are extracted. Subsequently, this allows the objective polymer to be precipitated and freed from the solvent. Once the recovered polymers have been dried, they can be finally processed into granulate with an extrusion process, i.e., compounding as described in Sect. 12.2.1. Similar to thermo-mechanical recycling, the quality of the processed starting material is mainly decisive for the later quality of the recyclate Orth et al. (2022).
Solvolysis describes the chemical reaction of a dissolved substance with a solvent such as alcoholysis, aminolysis, or ammonolysis. However, the process is limited to

polycondensates and polyadducts that contain heteroatoms such as nitrogen or oxygen in the backbone such as polyester (PL), polyamides (PA), polycarbonates (PC), and polyurethanes (PU). As the C-C bond is retained, an application on polyethene (PE), polypropylene (PP), polystyrene (PS), etc. is not possible. The final resulting reaction products are monomers, dimers, or oligomers, which can be further processed into polymers Neumüller (1988).

The advantage of solvent-based processes is that polymers can also be recovered from composites or plastic waste in very pure quality. But on the other hand, the availability of suitable solvents for the respective separation problem is limited. Another critical point to consider is the safety when using solvents and chemicals, in particular regarding ecology, flammability, toxicity, and teratogenicity. Furthermore, the solvents must be removed from the polymer and residual materials, and the final solvent preparation or disposal must be carried out Orth et al. (2022).

A thermochemical process is pyrolysis. Pyrolysis refers to the chemical conversion of substances at high temperatures, accompanied by the breaking of chemical bonds. The resulting hydrocarbons are structurally modified and do not constitute the actual building blocks of the decomposed starting material. Further modifications may be caused by the reaction with components of the pyrolysis atmosphere, particularly oxygen Neumüller (1988).

Also, the gasification can be categorised under the thermochemical processes. Gasification represents the most complete possible conversion of carbon carriers into combustible gas mixtures, in particular, synthesis gas under very high temperatures. This conversion is primarily achieved through the reaction with gasification agents, such as air, oxygen, water vapour, or mixtures thereof, and is carried out at very high temperatures Orth et al. (2022); Neumüller (1988).

Enzymolysis describes the chemical breakdown of an organic substance into simple building blocks under the influence of enzymes, usually of bacterial origin Orth et al. (2022). At present, enzymolysis is still not a fully researched process and is not yet ready to be applied on a larger industrial scale. Nevertheless, it has the potential to offer a sustainable and environmentally friendly alternative to chemical recycling techniques Zhu et al. (2022).

12.3 Recycling of Old Climbing Ropes

A case study for the recycling of old climbing ropes can be found in du Maire (2022, 2024). The research investigated the feasibility of recycling old climbing ropes made from polyamide 6 (PA6) through a thermo-mechanical process. For the recycling process old climbing ropes were collected from a local climbing club of the German Alpine Association. The designation "old" is not employed in the context of a time scale, but rather in reference to the extensive utilisation within the climbing gym environment, accompanied by the accumulation of wear and tear. These are expressed by a misalignment of the ropes core and sheath as well as mechanical damages such as abrasion, burns, or hardening. For climbing ropes, these signs of wear indicate that the rope needs to be sorted out for safety reasons although they can still carry the loads.

At the outset of this research, a thermal analysis of the rope material was conducted with the objective of determining whether core and sheath of the ropes are composed of the same base material. To this end, three distinct ropes were subjected to examination, with samples obtained from both the core and the sheath. This was realised with the aid of dynamic differential scanning calorimetry (DSC) and thermogravimetric analysis (TGA). The DSC and TGA analyses revealed that, for the first and third ropes, both the core and the sheath are made of the same material, specifically PA6. However, the second rope showed a difference in the measured DSC curves between the core and sheath, mainly in the peaks corresponding to the melting temperature. The evaluation of the different melting temperatures led to the conclusion that the core was composed of PA66, while the sheath was made of PA6. These findings are significant as they indicate that the majority of the collected old climbing ropes consist of the same material and can be recycled without the need for complex separation processes for the core and the sheath. The thermal properties of the materials, including the melting temperatures (221 °C for PA6 and 263 °C for PA66), were also determined, which are crucial for the recycling process with the twin screw extruder.

Once the purity of the material had been confirmed, the ropes were cut into smaller pieces and shredded using a shredding granulator. This method was found to be the most effective for reducing the rope material to a size suitable for extrusion. Initial shredding experiments were conducted using liquid nitrogen and water in combination with liquid nitrogen. However, these attempts were found to be insufficient. Subsequently, the shredded material was dried before processing in a twin screw extruder, wherein the granulate is produced. Within the scope of this study, two batches of recyclates were fabricated, one comprising cleaned rope material and the other without any pretreatment. The cleaning process involved washing the rope at 40 °C. In the subsequent stage of the study, the granulate was once again dried before further processing in an injection moulding machine, with the objective of producing tensile test specimens. A variety of specimens were produced with different material compositions, including 100% recyclate and a composition of 50% recyclate and 50% virgin PA6. The virgin polyamide employed as a reference material is a Zytel PA6 from Dupont, a standard polyamide developed for injection moulding and extrusion. An analysis of the process parameters associated with the fabrication of the specimen revealed a deterioration in the quality of the recycled material as the injection pressure and the dosing time increased in proportion to the rising content of recycled material in the compound.

Finally, the mechanical properties of the test specimens were evaluated through tensile tests conducted in accordance with the standard DIN EN ISO 527-1. The results indicated that the recycled material exhibited higher tensile strength and elastic modulus compared to the virgin PA6 from Dupont. Especially, the ultimate tensile strength of the recycled material surpassed that of the virgin material, although its elongation at break decreased, suggesting a loss in plasticity. This reduction in plasticity is attributed to thermal degradation during the extrusion process, which nevertheless does not significantly diminish the material's overall strength, which makes it a viable substitute for a virgin standard PA6 material. The findings indicate that the crystallinity of the recycled material is higher, contributing to the increased elastic modulus and tensile strength observed. The reasons can be found in the crystalline zones within the polymer, which act as cross-linking points, thereby stiffening the material and enhancing its mechanical properties. Furthermore, the study emphasises the significance of the cleaning process prior to recycling. Cleaned recycled material exhibits slightly better mechanical properties in comparison to non-cleaned material, indicating that the removal of impurities and contaminants can enhance the quality of the recycled material.

In conclusion, the research demonstrated that thermo-mechanical recycling of old climbing ropes into PA6 granulate is feasible. The recycled material, exhibiting mechanical properties that in some cases exceed those of the virgin reference material, can potentially be used in standard injection moulding applications requiring lower technical standards. This closed-loop recycling process not only offers a sustainable solution for climbing rope waste but also contributes to reducing environmental impact by providing an effective method to recycle plastic waste. The study serves as a foundational feasibility analysis, suggesting that further optimization of the recycling process could lead to even higher quality recycled materials, thereby supporting broader adoption of such recycling practices in the industry.

12.4 Recycling of Old Binding Twines from Agriculture

Another research, which addressed the critical issue of enhancing plastic waste, can be found in Maire et al. (2024). This study was specifically focussing on the recycling potential of polypropylene (PP) binding twines used in agriculture. These twines, predominantly made from PP, are typically disposed of after a single use, presenting a suitable opportunity for recycling due to their purity. Moreover, the degradation behaviour of the material properties due to multiple recycling of the same material was examined in detail within this work.

For this study, old binding twines were collected from local farmers in their dirty and coiled state. The term "old" means in this case that the twines were no longer in use and sorted out. In order to confirm the purity of the material and ensure that the twines are made exclusively of PP, infrared spectroscopy was carried out. The measurements confirmed PP as the base material of the collected twines, allowing to proceed the study. To compare the mechanical and rheological properties with a virgin material, Moplen HP500W from LyondellBasell is used as reference material. This standard PP is developed for the application in injection moulding and in technical compounds as flow modifier. The initial step involved washing one portion of the twines, while maintaining the other portion unwashed. This was done to investigate the influences of dirt on the recycling process and the material properties.
Afterwards, the binding twine was shredded to a wool-like mass with a suitable size for further processing in an extruder. The actual recycling process was then carried

out with the aid of a twin screw extruder, where the shredded twine material was recycled to granulate. In order to investigate the ageing behaviour, the granulate was repeatedly processed through the extruder, which simulated the multiple usage of the material. The recycling was done up to ten times for the soiled twines and only once for the cleaned material. In the next step, tensile test specimens were fabricated from the recycled material batches of the ageing generation (Gen.) 1, Gen. 5, and Gen. 10. For the purpose of comparison, specimens were also produced from the virgin PP and Gen. 1 of the cleaned binding twines. Subsequently, mechanical properties were evaluated in accordance with DIN EN ISO 527-1 through quasi-static tensile tests. The results indicated a general degradation in mechanical properties, such as ultimate tensile strength (UTS), as well as a significant decrease in elongation at break, reflecting a loss in material plasticity. However, the most significant deterioration occurred in the initial recycling cycles, which shows that early intervention in recycling processes is crucial to mitigate the loss in strength.

Rheological properties were analysed through the measurement of the melt volume rate (MVR). In contrast to the tensile tests, in which only Gen. 1, 5, and 10 were analysed, the MVR measurements were carried out on samples of all ageing states. The investigations revealed a clear trend of reduced viscosity with each ageing stage. This can be attributed to molecular chain scissioning and a decrease in molecular weight. Nevertheless, the overall increase in MVR remained below the comparative results for virgin material. Despite the observed degradation, the material remained processable, indicating the potential for continued recycling loops with appropriate adjustments in processing parameters.

In conclusion, the study highlights the feasibility of recycling agricultural binding twines and emphasises the necessity of developing effective recycling practices for smaller niche products. Moreover, the study provided comprehensive information about the ageing behaviour and the associated degradation process for repeated use of the material. The findings demonstrated that while the mechanical properties of the recycled material deteriorated over time, the material's processability remained intact, allowing for multiple recycling loops.

12.5 Conclusion

In summary, the present work gives an overview of the substantial potential of recycling plastic waste from niche products, specifically focussing on old climbing ropes and agricultural binding twines. The findings underscore the importance of advancing recycling techniques to mitigate the environmental impact of plastic waste and contribute to achieving climate neutrality goals.

The feasibility of recycling old climbing ropes made from PA6 through a thermo-mechanical process was demonstrated. Recycled PA6 exhibited superior tensile strength and elastic modulus compared to virgin PA6, making it a viable substitute for standard injection moulding applications. The study highlighted the significance of cleaning the material before recycling to enhance the quality of the recyclate. Additionally, PP binding twines used in agriculture were effectively recycled, with a detailed analysis of the degradation behaviour due to multiple recycling cycles. Mechanical properties such as ultimate tensile strength and elongation at break degraded over time, but the material remained processable, indicating potential for continued recycling with appropriate adjustments in processing parameters. The study emphasised the critical need for early intervention in recycling processes to mitigate the loss in material strength.

These case studies illustrate that niche products, despite their smaller contribution to overall plastic waste, hold significant potential for reducing greenhouse gas emissions and conserving resources through effective recycling practices. By focussing on materials with high purity, such as climbing ropes and agricultural twines, complex separation processes can be avoided, making recycling more efficient and economically viable. Nevertheless, the environmental impact of the recycling processes themselves, including energy consumption and emissions, needs to be thoroughly assessed to ensure overall sustainability.

In conclusion, the recycling of niche plastic products, such as climbing ropes and agricultural twines, presents a promising avenue for reducing plastic waste and contributing to climate goals. By advancing recycling techniques and focussing on material purity, significant environmental benefits can be achieved. This research provides a foundation for further studies and innovations in recycling practices, ultimately supporting a more sustainable and circular economy.

Acknowledgements The authors acknowledge the Studienstiftung des deutschen Volkes for its comprehensive support of this research work through the doctoral scholarship. Also many thanks to the staff of the laboratory for polymer technology for the kind support to realise the experiments. Finally, our thanks go to the students who actively supported the experiments and helped to carry out this research.

References

Allen MR, Dube OP, Solecki W, Aragón-Durand F, Cramer W, Humphreys S, Kainuma M, Kala J, Mahowald N, Mulugetta Y, Perez R, Wairiu M, Zickfeld K (2018) Summary for policymakers. In: Masson-Delmotte V et al (eds) Global warming of 1.5Â°C. An IPCC special report on the impacts of global warming of 1.5Â°C above pre-industrial levels and related global greenhouse gas emission pathways, in the context of strengthening the global response to the threat of climate change. Sustainable development, and efforts to eradicate poverty, vol 49. Cambridge University Press, Cambridge, pp 1–24

Volk R, Stallkamp C, Steins JJ, Yogish SP, Müller RC, Stapf D, Schultmann F (2021) Techno-economic assessment and comparison of different plastic recycling pathways. A German case study. J Ind Ecol 25:1318

Singh N, Hui D, Singh R, Ahuja IPS, Feo L, Fraternali F (2017) Recycling of plastic solid waste: a state of art review and future applications. Compos B Eng 115:409

van Straten B, van der Heiden DR, Robertson D, Riekwel C, Jansen FW, van der Elst M, Horeman T (2021) Surgical waste reprocessing: Injection molding using recycled blue wrapping paper from the operating room. J Clean Prod 322:129121

Mayanti B, Helo P (2022) Closed-loop supply chain potential of agricultural plastic waste: economic and environmental assessment of bale wrap waste recycling in Finland. Int J Prod Econ 244:108347

Briassoulis D, Hiskakis M, Babou E (2013) Technical specifications for mechanical recycling of agricultural plastic waste. Waste Manag 33:1516–1530

Bradford S, Rupf R, Stucki M (2021) Climbing ropes—environmental hotspots in their life cycle and potentials for optimization. Sustainability 13:707

Damayanti D, Saputri DR, Marpaung DSS, Yusupandi F, Sanjaya A, Simbolon YM, Asmarani W, Ulfa M, Wu H-S (2022) Current prospects for plastic waste treatment. Polymers 14:3133

Orth P, Burder J, Rink M (2022) Kunststoffe im Kreislauf. Vom Recycling zur Rohstoffwende. Springer, Wiesbaden

Rahimi A, Garcia J (2017) Chemical recycling of waste plastics for new materials production. Nat Rev Chem 1:0046

Serranti S, Bonifazi G (2019) Techniques for separation of plastic wastes. In: Pacheco-Torgal F et al (eds) Use of recycled plastics in eco-efficient concrete. Woodhead Publishing, pp 9–37

Ragaert K, Delva L, van Geem K (2017) Mechanical and chemical recycling of solid plastic waste. Waste Manag 69:24–58

Schyns ZOG, Shaver MP (2020) Mechanical recycling of packaging plastics: a review. Macromol Rapid Commun 42(3):e2000415

Oblak P, Gonzalez-Gutierrez J, Zupančič B, Aulova A, Emri I (2015) Processability and mechanical properties of extensively recycled high density polyethylene. Polym Degrad Stab 114:133–145

Dorigato A (2021) Recycling of polymer blends. Adv Ind Eng Polym Res 4:53–59

Neumüller OA (1988) Römpps Chemie-Lexikon, 8th edn. Franckh, Stuttgart

Zhu B, Wang D, Wei N (2022) Enzyme discovery and engineering for sustainable plastic recycling. Trends Biotechnol 40:22–37

du Maire P, Deckert M, Johlitz M, Öchsner A (2022) Thermo-mechanical recycling of climbing ropes: a case study on a closed loop process for PA6. Appl Res e202200091

du Maire P, Friebertshäuser M, Deckert MH, Johlitz M, Öchsner A (2024) Thermal analysis of old climbing ropes. In: Altenbach H et al (eds) Lectures notes on advanced structured materials 2. Advanced structured materials, vol 203. Springer, Cham, pp 367–376

du Maire P, Gärtner F, Deckert M, Johlitz M, Öchsner A (2024) Investigation of the ageing behaviour of multiple reused polypropylene binding twines. Appl Res e202400090. https://doi.org/10.1002/appl.202400090

Chapter 13
Sorption Mechanisms of Oils and Fuels in Thermoplastic Polyurethanes

Yvonne Breitmoser, Tobias Förster, Alexander Lion, Michael Johlitz, and Sebastian Eibl

Abstract Thermoplastic polyurethanes (TPUs) are characterized by their unique elastic and thermoplastic properties, which make them indispensable for a wide range of applications in industry. The present study focuses on investigating the interactions of TPUs with oils and fuels, particularly on the analysis of sorption mechanisms and the role of chemical structure for material compatibility. Two TPUs with different soft segments, ether-based and ester-based, were tested for their compatibility with these media. The influence of the structural composition on material compatibility was demonstrated by combining sorption tests with mechanical and chemical analyses. It was found that ether-based TPUs have a higher absorption and diffusion rate of non-polar media, indicating worse compatibility with these substances. In addition, aromatics were found to be preferentially absorbed by the TPUs, with those based on ester having a higher affinity for these compounds due to stronger polar interactions. These results provide important insights for developing chemically resistant TPUs that should retain their mechanical and elastic properties. By precisely controlling these interactions, TPUs can be used more effectively in demanding environments, leading to improved performance and extended applicability of TPU-based products.

Y. Breitmoser (✉) · A. Lion · M. Johlitz
Institute of Mechanics, University of the Bundeswehr Munich,
Werner-Heisenberg-Weg 39, 85579 Neubiberg, Germany
e-mail: yvonne.breitmoser@unibw.de

A. Lion
e-mail: alexander.lion@unibw.de

M. Johlitz
e-mail: michael.johlitz@unibw.de

Y. Breitmoser · T. Förster · S. Eibl
Bundeswehr Research Institute for Materials, Fuels and Lubricants (WIWeB),
Institutsweg 1, 85435 Erding, Germany
e-mail: tobias1foerster@bundeswehr.org

S. Eibl
e-mail: sebastianeibl@bundeswehr.org

H. Altenbach et al. (eds.), *Lectures Notes on Advanced Structured Materials 3*,
Advanced Structured Materials 226, https://doi.org/10.1007/978-3-031-77403-4_13

13.1 Introduction

Thermoplastic elastomers (TPEs), especially thermoplastic polyurethanes (TPUs), have become increasingly important because of their unique elastic and thermoplastic properties. These materials have proven indispensable in various applications, including the automotive industry (Scholz and Gehringer 2021), medical technology (Asefnejad et al. 2011; Haryńska et al. 2018), packaging industry (Zahid et al. 2020), and mechanical engineering (Endo et al. 2018). Their versatility is based on the exceptional balance between flexibility, durability, and processability that distinguishes TPUs from other polymers.

TPUs are characterized by their molecular structure, consisting of a sequence of hard and soft segments. These segments allow TPUs to be elastic at room temperature while processable at elevated temperatures. The ability of TPUs to retain their physical properties over a wide temperature range makes them particularly attractive for applications that require high flexibility and dimensional stability simultaneously. In addition, the chemical diversity of the soft segments, which mainly consist of ether or ester polyols, enables the production of TPUs with customized properties to meet specific requirements (Frick and Rochman 2004; Prisacariu 2011; Scholz and Gehringer 2021).

These two types of soft segments have different properties that affect the overall performance of the TPUs. Ether-based soft segments provide excellent hydrolysis resistance and retain flexibility at lower temperatures (Frisch and Dieter 1975; Prisacariu 2011; Scholz and Gehringer 2021). Therefore, TPUs with ether-based soft segments are suitable for applications exposed to moisture or weathering, such as outdoor sportswear, marine products, and medical devices (Scholz and Gehringer 2021). On the other hand, TPUs with ester-based soft segments exhibit superior mechanical properties and exceptional elasticity. It typically shows higher tensile strength and better abrasion resistance, making it ideal for engineering components, automotive parts, and industrial belting applications (Prisacariu 2011).

A crucial aspect that determines the range of applications of TPUs is their compatibility with oils and fuels. This property is essential for the automotive and aviation industries and oil and gas production applications. The interaction of TPUs with these substances can lead to significant changes in the physical and chemical properties, which can affect the performance and service life of the materials.

Although TPUs offer numerous advantages in compatibility with oils and fuels, there are potential challenges to consider. Some harsh chemicals, such as solvents, additives, or highly aromatic compounds, can negatively affect TPUs, leading to swelling, loss of mechanical properties, or premature failure (Oertel 1994). Aromatic compounds have unique properties due to the delocalized electron system in the ring structure, are present in certain oils and fuels, and strongly affect the swelling of elastomers (Blivernitz et al. 2019) and, thus, also of TPUs. Especially small aromatic molecules exhibit a high swelling potential. Due to their low molecular weight, they can interact easily with the polymer chains of TPUs through various intermolecular forces, such as van der Waals forces or π-π-interactions, compared to longer-chain oil and fuel components. These interactions can disrupt the internal structure of the polymer and allow aromatic molecules to diffuse into the TPU matrix, expanding it and facilitating access for larger molecules, resulting in increased swelling (Demmel et al. 2022; Graham et al. 2006).

The complexity of the interactions between TPUs and chemical substances requires an in-depth understanding of the underlying mechanisms that control these relationships. This includes studying the sorption processes by which oils and fuels penetrate the TPU matrix and the associated physical and chemical changes to the material. The ability to precisely control and predict these processes is essential for developing TPUs designed explicitly for high resistance to chemical substances without compromising their mechanical and elastic properties.

The present work aims to comprehensively understand the sorption mechanisms of oils and fuels in TPUs. By analyzing the interactions at the molecular level, insights will be gained into the factors that influence the compatibility of TPUs with different chemical substances.

Therefore, two chemically different TPUs were studied for their compatibility with oils and fuels. Using sorption tests in combination with mechanical and chemical analyses, the influence of structural composition could be correlated with compatibility, depending on the ester or ether structure used.

This knowledge will help to improve the performance and longevity of TPU-based products and identify new ways to apply these versatile materials in critical and demanding environments.

13.2 Experimental Part

13.2.1 Materials and Fluids

The experiments presented investigated BASF Elastollan grades 1170A, ether-based, and C78A, ester-based. S2 tension rods, following DIN 53504:2017-03 (Deutsches Institut für Normung 2017), were punched out of injection-molded sheets of 2 mm thickness. Deviating from the standard, these were sliced to 1 mm thickness to achieve sorption equilibrium more rapidly. Table 13.1 shows the mechanical properties of the

Table 13.1 Properties of investigated TPUs provided by the manufacturer (BASF 2022)

	Elastollan C78A (ester-based)	Elastollan 1170A (ether-based)
Density ($g\,cm^{-3}$)	1.18	1.08
Tensile strength (MPa)	50	30
Elongation at break (%)	650	850
Hardness (Shore A)	80	71

two types according to the manufacturer's specifications (BASF 2022). The reference oil IRM 903 and commercial diesel fuel were used as testing fluids.

The calibration standards for measuring the aromatic content in gas chromatography/mass spectrometry (GC/MS) assisted sorption experiments were prepared by mixing 10 to 90 vol% AroMix in an aromatic-free Fischer-Tropsch fuel from Sasol. AroMix, consisting of the aromatic solvents Solvesso 100, 150, and 200 supplied by Exxon in concentrations of 25, 53, and 22 vol%, was used to simulate the aromatic composition (DeWitt et al. 2008).

13.2.2 Chemical Characterization

Fourier transform infrared spectroscopy (FTIR) was used to characterize the polymers using the Bruker Tensor 27 FTIR spectrometer and a Bruker Platinum ATR diamond single reflection module. For one spectrum, 32 individual scans were averaged between 4000 and 400 cm^{-1}.

13.2.3 Experiments

Each sample was tested in a triplicate measurement. First, the samples were dried at 80 °C for one hour to obtain a uniform moisture content. Then, the mass and density of each sample were determined. Samples were then placed in screw-top jars containing approximately 150 mL of oil or diesel fuel and removed after sorption times of 0.5 ($\approx$ 42 $s^{0.5}$), 2, 6, 8, 24, 48, 72, 168, and 336 h ($\approx$ 1100 $s^{0.5}$, two weeks), with an additional 672 h for oil sorption. Following Fick's law of diffusion, sorption times are expressed on the abscissa as the square root of time $t^{0.5}$. After the samples were taken out, they were briefly immersed in low boiling 40/60 benzine to remove the adhering medium, dried with a lint-free paper, and weighed for mass uptake and density measurement. Small pieces were punched out of the tensile test specimen head and subsequently extracted in vials containing 1 mL heptane for GC/MS measurements. After that, tensile strength and elongation at break were determined.

In the gravimetric sorption experiments, the relative mass change Δm_{g} of the elastomer samples due to the sorption of oil or diesel fuel with absolute mass m_t at time t was determined by $\Delta m_{\mathrm{g}} = \left(\frac{m_t - m_0}{m_0}\right) \cdot 100$ wt%, with m_0 being the mass of pure elastomer after drying, i.e., $t = 0$ s.

For the GC/MS experiments, the punched samples were extracted for five days at room temperature. They were then removed from the solvent and dried at room temperature for seven days. The extract obtained was used for the GC/MS measurements, the re-dried samples were weighed, and the measured values were included in further calculations.

The masses of oil or fuel absorbed were determined using an external calibration and then normalized to the weight m_0 of the pure re-dried elastomer pieces to obtain the relative mass change $\Delta m_{GC/MS}$, equivalent to Δm_g (gravimetric).

Using Eq. (13.1), the explicit solution of Fick's second law for one-dimensional diffusion into a plane sheet, the theoretical sorption curves, and the diffusion coefficients D can be fitted iteratively from measured data (Crank 1975). Here, Δm_t and Δm_{eq} are the relative uptakes of the media at time t and equilibrium and $h = 1[mm]$ is the sample thickness.

$$\frac{\Delta m_t}{\Delta m_{eq}} = 1 - \frac{8}{\pi^2}\sum_{n=0}^{\infty}\left(\frac{1}{(2n+1)^2}\exp\left(\frac{-(2n+1)^2\pi^2 Dt}{h^2}\right)\right) \qquad (13.1)$$

An external calibration was also performed using standards with different aromatics content (AroMix) to determine the absorbed aromatics content in the sample. The mass spectrum signals for aliphatic ($m/z_{\text{aliphatic}} = 41, 55, 57, 69, 71, 83, 85, 97, 99$) and aromatic ($m/z_{\text{aromatic}} = 91, 105, 115, 119, 120, 128, 142$) hydrocarbons were analyzed and summed in the retention time range of the storage medium (Demmel et al. 2022). The ratio of the aliphatic signals to the sum of the aromatic and aliphatic signals was plotted against the aromatic content of the standard solutions to create a calibration curve. This allows the aromatic content in the oil and fuel to be quantified. To measure mass and density, a SARTORIUS Secura® 225D 1S precision balance with the VF 4601 density kit was used. The tensile tests were determined on a Zwick Roell 1445 universal testing machine with a force sensor of 500 N and an optical strain sensor at ambient temperature, a preload of 0.1 MPa, and a strain rate of 200 mm min^{-1} following DIN 53504:2017-03 (Deutsches Institut für Normung 2017). An Agilent 7890A gas chromatograph with a 30 m DB-5MS column (0.25 mm inner diameter, 0.25 μm film thickness) coupled with an Agilent 5975 MSD was used for the GC/MS studies. The GC oven was heated from 50 to 320 °C at 50 K min^{-1} and held at 320 °C for 5 min.

13.3 Results and Discussion

13.3.1 Chemical Characterization

The chemical characterization was performed by FTIR, and the spectra are shown in Fig. 13.1. The urethane groups in the hard segments of the TPUs have four typical regions in the spectrum. The N-H stretching vibration at 3330 cm^{-1}, the free and hydrogen bonded stretching vibration of the C=O at 1730 cm^{-1} and 1700 cm^{-1}, the band at 1529 cm^{-1} is attributed to the C-N stretching and N-H bending vibration, and the C-N stretching and N-H bending vibration at 1220 cm^{-1}. The bands at 1600, 1415, 815, 770, and 510 cm^{-1} indicate methylene diphenyl diisocyanate (MDI) as a

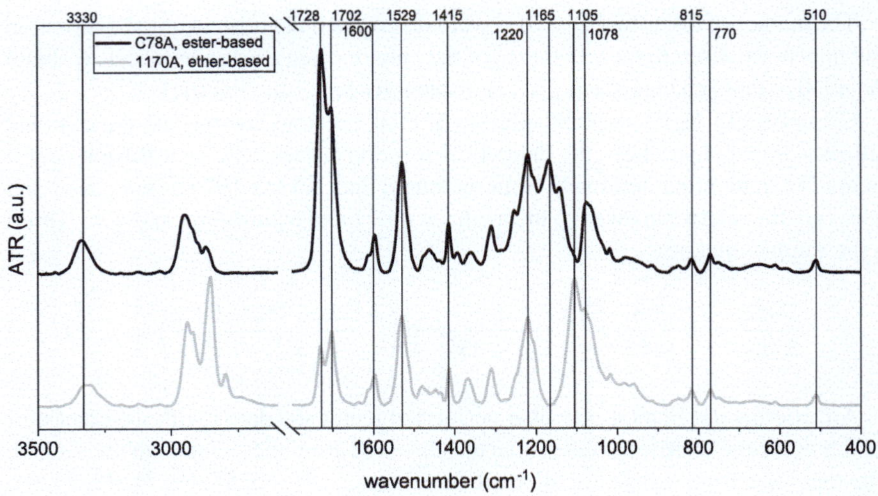

Fig. 13.1 FTIR spectra of the two tested Thermoplastic Polyurethanes. Black: ester-based TPU; Gray: ether-based TPU

reactant in the polymerization (Hesse et al. 2012; Hummel and Scholl 1988; Jiang et al. 2018; Kasprzyk et al. 2019; Zhang et al. 2008).

Typical bands for identifying the soft segments are the ether band at 1105 cm^{-1} and the ester bands at 1730 cm^{-1} and 1165 cm^{-1}. In the spectra of the TPU C78A, it is noticeable that the ether band at 1105 cm^{-1} is not present. Instead, a significant superposition is found, which combines the stretching vibration of the carbonyl group (C=O) of the ester with the corresponding band of the urethane group. This results in a confluent band with a peak at about 1730 cm^{-1}. The presence of the band at 1165 cm^{-1} also indicates ester groups. The absence of the ether band in combination with the visible ester bands emphasizes that the TPU C78A is ester-based. In contrast, TPU 1170A exhibits an ether band without showing ester bands, suggesting an ether-based polyol in the soft segments (Hesse et al. 2012; Hummel and Scholl 1988; Jiang et al. 2018; Kasprzyk et al. 2019).

13.3.2 Behavior in Contact with Oils and Fuels

The polyols' influence in the TPUs is also evident in the sorption tests. After immersion in reference oil and diesel fuel, the masses of both TPUs increased due to media absorption until equilibrium was reached, as shown in Fig. 13.2a and b.

The ether-based TPU exhibited significantly higher mass uptake in both media than the ester-based TPU, reaching an equilibrium mass uptake of approximately 16% in reference oil and diesel fuel. In contrast, the ester-based TPU achieved an equilibrium mass uptake of 3.5% in reference oil and 5% in diesel fuel, indicating

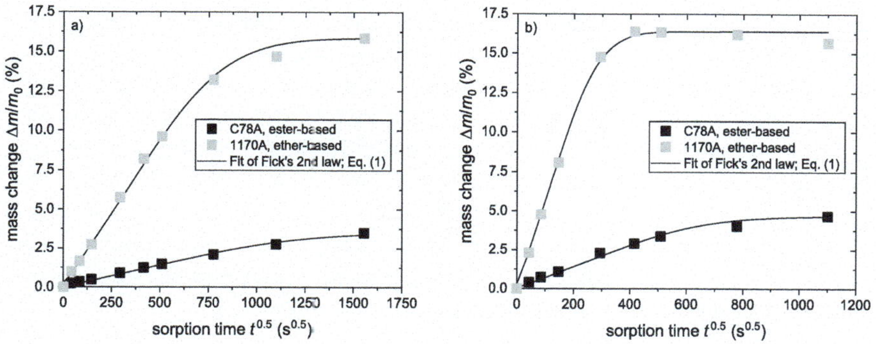

Fig. 13.2 Mass change of the tested TPUs in reference oil IRM 903 (**a**) and diesel fuel (**b**) dependent on sorption time $t^{0.5}$. Samples: black squares (■): ester-based TPU C78A, gray squares (■): ether-based TPU 1170A

Table 13.2 Calculated diffusion coefficients *D* of the investigated TPUs in their surrounding media

Medium	TPU	D $(10^{-6}\text{mm}^2\ \text{s}^{-1})$
Reference oil IRM 903	Ester-based	1.3 ± 0.1
	Ether-based	2.8 ± 0.1
Diesel fuel	Ester-based	4.4 ± 0.3
	Ether-based	23.2 ± 1.2

a less pronounced absorption capacity. This is in the range of the swelling behavior observed for commonly used NBRs, which typically show a mass increase between 3% and 13% depending on the acrylonitrile content (Coronado et al. 2014; Förster and Blivernitz 2021; Haseeb et al. 2010; Nihmath and Ramesan 2021; Zhu et al. 2015).

The analyses revealed that the ether-based TPU not only absorbs more media but also absorbs it faster, as the calculated diffusion coefficients *D* show. These coefficients *D*, derived by iteratively fitting the experimental data using Eq. (13.1) and presented in Table 13.2, show a direct correlation between the chemical structure of the TPU and its interaction with non-polar substances such as oil and diesel fuel. Fick's diffusion coefficient *D* of the ether-based TPU in oil was almost twice as high as *D* of the ester-based TPU. This discrepancy was even more pronounced in diesel fuel, where the coefficient *D* of the ether-based TPU was nearly six times higher than *D* of the ester-based TPU.

The reason for the different mass uptakes at equilibrium and the different diffusion rates is, on the one hand, due to the polarity of the materials. Both the reference oil and the diesel fuel are non-polar. In contrast, TPUs are generally more polar, albeit to varying degrees, due to the functional groups of the polyols they contain. Ester groups are more polar than ether groups, so the ether-based TPUs are less polar and, therefore, more compatible with non-polar materials, based on the "like dissolves like" principle. This difference in polarity facilitates more efficient interaction and sorption of the non-polar media into the ether-based TPUs, which explains their higher equilibrium mass uptake and faster diffusion rate.

Additionally, analysis by GC/MS confirmed that the average molecular size of the diesel fuel is smaller than that of the reference oil IRM 903. This difference in molecular size also contributes to the observed differences in absorption behavior, such that the smaller molecules of the diesel fuel penetrate the TPU faster than the molecules of the reference oil IRM 903, thus explaining the higher diffusion coefficients. The smaller molecular size of diesel fuel allows for more rapid diffusion and quicker interaction with the TPU, further enhancing the absorption rate compared to the larger molecules of the reference oil.

13.3.3 GC/MS-Assisted Sorption Tests

The gravimetric analysis is followed by determining the amount of absorbed medium using GC/MS. Figure 13.3 shows the gravimetric mass uptake and the mass uptake calculated based on GC/MS analyses. It can be seen that the gravimetric and GC/MS-based methods are very good comparable and provide similar results considering experimental errors. However, the advantage of the GC/MS measurement is the higher information content. In addition to quantifying the mass of the absorbed oil or fuel, it is also possible to identify which components have been absorbed by the material.

A particular focus here is on the in-depth analysis of certain substances, especially aromatics. Figure 13.4 shows the aromatic content in extracts of absorbed media by the TPUs over the sorption time. The aromatic content of the surrounding medium is included in the diagrams via a linear line. For the liquids, the measurements showed

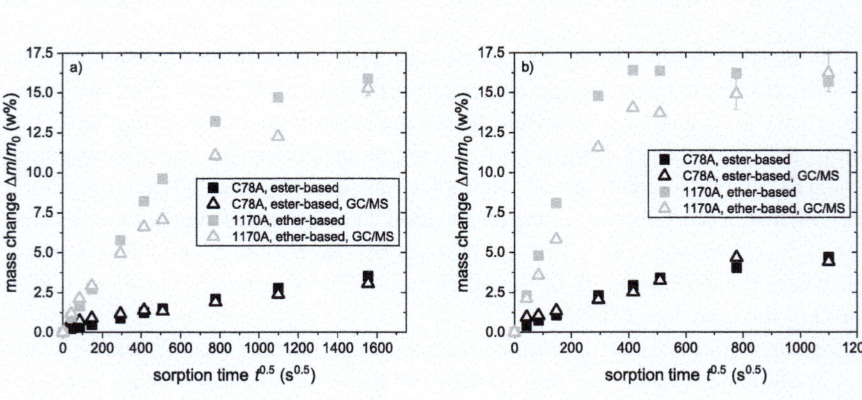

Fig. 13.3 Comparison of the gravimetric mass uptake and the mass uptake calculated with GC/MS of the tested TPUs in reference oil IRM 903 (**a**) and diesel fuel (**b**) dependent on sorption time $t^{0.5}$. Samples: black squares (■): ester-based TPU, gravimetric; black triangles (△): ester-based TPU, GC/MS; gray squares (■): ether-based TPU, gravimetric; gray triangles (△): ether-based TPU, GC/MS

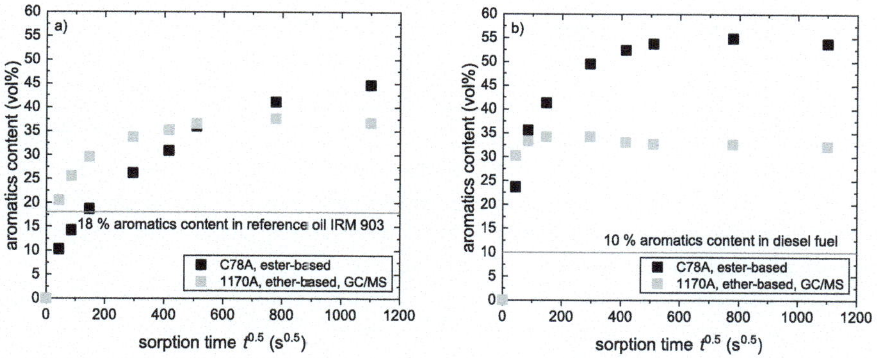

Fig. 13.4 Extracted aromatic content absorbed by the TPUs in reference oil IRM 903 (**a**) and diesel fuel (**b**) dependent on sorption time $t^{0.5}$. Samples: black squares (■): ester-based TPU, gray squares (■): ether-based TPU

an aromatics content of 18 vol% for the reference oil IRM 903 and 10 vol% for the diesel fuel. For both TPU materials, it can be determined that more aromatics accumulate in the material over time than the surrounding medium provides.

The ester-based TPU extract stored in oil contains 47 vol% aromatics, and the ether-based TPU extract contains 37 vol%. This effect is emphasized by the storage of the TPUs in diesel fuel. The ether-based TPU absorbs 32 vol% aromatics, and the ester-based TPU absorbs 54 vol%. This shows that the aromatics preferentially penetrate the materials and accumulate there (Blivernitz 2019).

The absorption of aromatics into polyester-based TPU and polyether-based TPU is significantly influenced by the interactions between the π-electrons of the aromatics and the chemical structures of the two polymers. Polyester-based TPU contains ester groups, which, due to their polarity, can form strong interactions with the π-electrons of aromatic compounds. These interactions facilitate the absorption of aromatics into the polymer because the π-π-interactions and polar attractions between the ester groups of the polyester and the delocalized π-electrons of the aromatics increase the solubility and penetration of the aromatics into the polymer. In contrast, polyether-based TPU exhibits lower polarity and offers fewer opportunities for strong interactions with the π-electrons of aromatics. Polyether structures tend to enable weaker van der Waals forces and less pronounced π-π-interactions, leading to a lower affinity and absorption of aromatic compounds. The polyester-based TPU is more susceptible to the absorption of aromatics due to its stronger polar interactions and the ability to engage in effective π-π-interactions with the π-electrons of aromatic compounds, compared to polyether-based TPU.

13.3.4 Mechanical Testing

The physical and mechanical measurement results of the elastomers as a function of the sorption time $t^{0.5}$ are summarized in Fig. 13.5. The values at $t^{0.5} = 0s^{0.5}$ reflect the properties before immersion in the media. The TPUs differ in their density of 1.08 g cm^{-3} and 1.17 g cm^{-3} for ether-based and ester-based TPUs, respectively, as well as in their elongation at break of 785% and 631%. The stress at break is similar for both materials at approximately 44.5 MPa.

As the sorption time progresses ($t^{0.5} > 42s^{0.5}$), the density decreases due to the absorption of the surrounding medium. The reference oil IRM 903 has a density of 0.92 g cm^{-3}, and the diesel fuel of 0.83 g cm^{-3}. At sorption equilibrium, a 0.2% decrease in density was observed with the reference oil IRM 903 in the ester-based TPU and 1.3% in the ether-based TPU. The diesel fuel showed a more significant reduction in density of 1.0% for the ester-based TPU and 3.1% for the ether-based TPU. This is also due to the different polarities of the materials: more medium is absorbed by the ether-based TPU than by the ester-based TPU. The tensile tests show a strong scatter in identifying the elongation and stress at break. No trend is discernible for the samples stored in reference oil. The results of the tensile tests do not allow a clear interpretation. Defects that occur naturally during manufacture or further processing are a possibility.

13.4 Conclusions and Remarks

This study investigated the interactions of TPUs with oils and fuels to gain an in-depth understanding of the sorption mechanisms. It was shown that the chemical composition of TPUs, especially the type of soft segments, ester- or ether-based, significantly influences their compatibility with oils and fuels. Ether-based TPUs show a higher absorption and a faster diffusion rate in these media than ester-based TPUs due to the different polarities of the materials. In addition, aromatics were found to be preferentially absorbed by the TPUs, with ester-based TPUs showing a stronger tendency to absorb aromatics due to the stronger polar interactions and the ability to form effective π-π-interactions with the π-electrons of the aromatics. The findings from this work emphasize the necessity of understanding and controlling the chemical compatibility of TPUs with oils and fuels to improve their performance and longevity in demanding environments. The ability to precisely control and predict these interactions is crucial for developing TPUs designed explicitly for high chemical resistance without compromising their mechanical and elastic properties. This knowledge opens up new possibilities for applying these versatile materials in critical and demanding environments, leading to the development of products that meet the increasing demands in innovative industries.

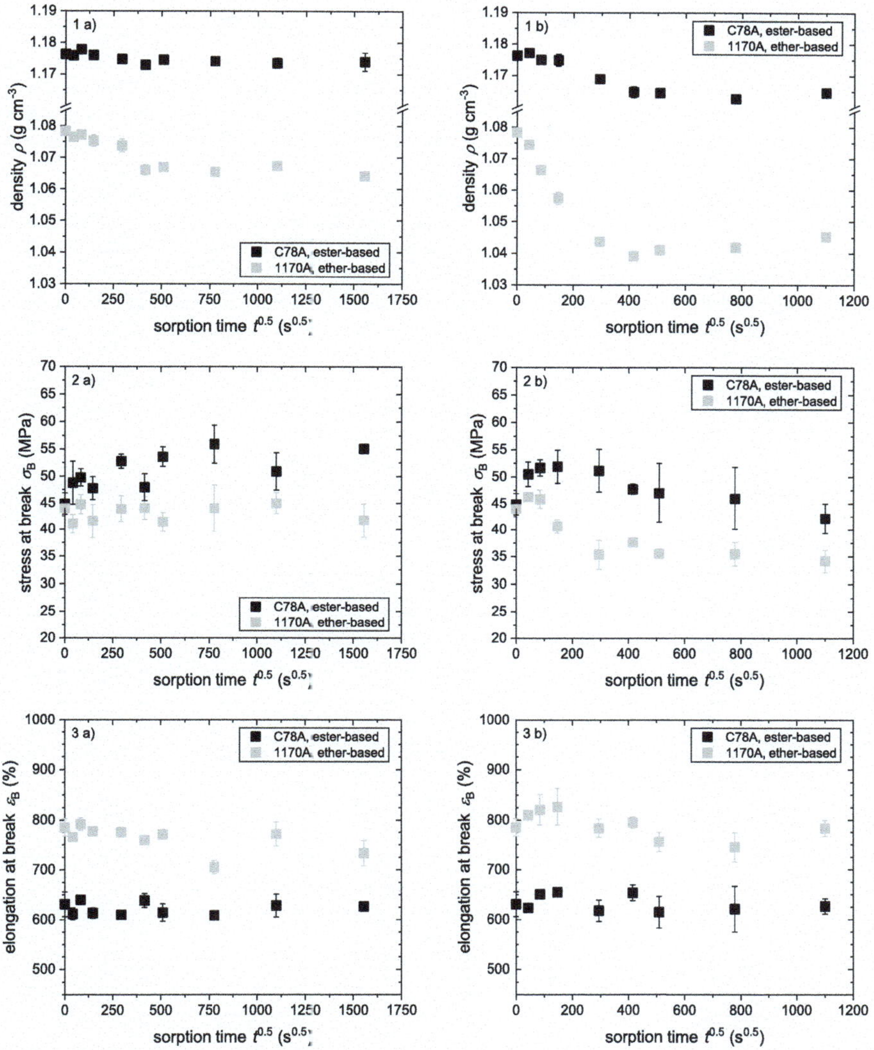

Fig. 13.5 Changes in density (1), stress at break (2), and elongation at break (3) of the tested TPUs in reference oil IRM 903 (**a**) and diesel fuel (**b**) dependent on sorption time $t^{0.5}$. Samples: black squares (■): ester-based TPU, gray squares (■): ether-based TPU

Acknowledgements The scientific cooperation between the University of the Bundeswehr, Faculty of Aerospace Engineering, Department of Mechanics and the Bundeswehr Research Institute for Materials, Fuels, and Lubricants (WIWeB) is gratefully acknowledged.

References

Asefnejad A, Khorasani MT, Behnamghader A, Farsadzadeh B, Bonakdar S (2011) Manufacturing of biodegradable polyurethane scaffolds based on polycaprolactone using a phase separation method: Physical properties and in vitro assay. Int J Nanomedicine 6:2375–2384. https://doi.org/10.2147/IJN.S15586

BASF SE: Thermoplastic Polyurethane Elastomers (TPU), Elastollan® - Product Range (2022)

Blivernitz A (2019) Untersuchung der Verträglichkeit von Elastomeren mit synthetischen Flugturbinenkraftstoffen anhand ablaufender Diffusionsprozesse. PhD thesis, Universität der Bundeswehr München

Blivernitz A, Förster T, Eibl S (2019) Chemometric prediction of swelling potentials of aviation fuels towards elastomers using various spectroscopic data. Polym Test 76(March):261–272. https://doi.org/10.1016/j.polymertesting.2019.03.009

Coronado M, Montero G, Valdez B, Stoytcheva M, Eliezer A, García C, Campbell H, Pérez A (2014) Degradation of nitrile rubber fuel hose by biodiesel use. Energy 68:364–369. https://doi.org/10.1016/j.energy.2014.02.087

Crank J (1975) The mathematics of diffusion. Clarendon Press, Oxford

Demmel B, Förster T, Eibl S (2022) Sorption dynamics of synthetic fuels and corresponding model mixtures in aged acrylonitrile-butadiene rubber. Polym Test 115(May):107719. https://doi.org/10.1016/j.polymertesting.2022.107719

Deutsches Institut für Normung e.V.: DIN 53504:2017-03, Testing of rubber - Determination of tensile strength at break, tensile stress at yield, elongation at break and stress values in a tensile test (2017)

DeWitt MJ, Corporan E, Graham J, Minus D (2008) Effects of aromatic type and concentration in Fischer-Tropsch fuel on emissions production and material compatibility. Energy Fuels 22(4):2411–2418. https://doi.org/10.1021/ef8001179

Endo H, Kato S, Watanebe M, Kikuchi T, Kawai M, Mitsumata T (2018) Magnetically tunable vibration transmissibility for polyurethane magnetic elastomers. Polymers 10(1):104. https://doi.org/10.3390/polym10010104

Förster T, Blivernitz A (2021) Migration of mineral oil in elastomers. J Rubber Res 24(2):257–269. https://doi.org/10.1007/s42464-021-00087-7

Frick A, Rochman A (2004) Characterization of TPU-elastomers by thermal analysis (DSC). Polym Test 23(4):413–417. https://doi.org/10.1016/j.polymertesting.2003.09.013

Frisch KC, Dieter JA (1975) An overview of urethane elastomers. Polym Plast Technol Eng 4(1):1–21. https://doi.org/10.1080/03602557509409244

Graham JL, Striebich RC, Myers KJ, Minus DK, Harrison WE (2006) Swelling of nitrile rubber by selected aromatics blended in a synthetic jet fuel. Energy Fuels 20(2):759–765. https://doi.org/10.1021/ef050191x

Haryńska A, Gubanska I, Kucinska-Lipka J, Janik H (2018) Fabrication and characterization of flexible medical-grade TPU filament for fused deposition modeling 3DP technology. Polymers 10(12):1304. https://doi.org/10.3390/polym10121304

Haseeb A, Masjuki H, Siang C, Fazal M (2010) Compatibility of elastomers in palm biodiesel. Renew Energy 35(10):2356–2361. https://doi.org/10.1016/j.renene.2010.03.011

Hesse M, Meier H, Zeeh B, Bienz S, Bigler L, Fox T (2012) Spektroskopische Methoden in Der Organischen Chemie. Georg Thieme Verlag, Stuttgart, New York

Hummel DO, Scholl FK (1988) Atlas Der Polymer- Und Kunststoffanalyse. Hanser, München

Jiang L, Ren Z, Zhao W, Liu W, Liu H, Zhu C (2018) Synthesis and structure/properties characterizations of four polyurethane model hard segments. R Soc Open Sci 5(7):180536. https://doi.org/10.1098/rsos.180536

Kasprzyk P, Sadowska E, Datta J (2019) Investigation of thermoplastic polyurethanes synthesized via two different prepolymers. J Polym Environ 27(11):2588–2599. https://doi.org/10.1007/s10924-019-01543-7

Nihmath A, Ramesan MT (2021) Comparative evaluation of oil resistance, dielectric properties, AC conductivity, and transport properties of nitrile rubber and chlorinated nitrile rubber. Prog Rubber Plast Recycl Technol 37(2):131–147. https://doi.org/10.1177/1477760620925490

Oertel G (1994) Polyurethane handbook: chemistry, raw materials, processing application, properties, 2nd edn. Carl Hanser Verlag GmbH & Co. KG, Munich

Prisacariu C (2011) Polyurethane elastomers. Springer Vienna, Vienna. https://doi.org/10.1007/978-3-7091-0514-6

Scholz G, Gehringer M (2021) Thermoplastische Elastomere. De Gruyter, Berlin, Boston. https://doi.org/10.1515/9783110740066

Zahid M, Del Río Castillo AE, Thorat SB, Panda JK, Bonaccorso F, Athanassiou A (2020) Graphene morphology effect on the gas barrier, mechanical and thermal properties of thermoplastic polyurethane. Compos Sci Technol 200:108461. https://doi.org/10.1016/j.compscitech.2020.108461

Zhang C, Ren Z, Yin Z, Qian H, Ma D (2008) Amide II and amide III bands in polyurethane model soft and hard segments. Polym Bull 60(1):97–101. https://doi.org/10.1007/s00289-007-0837-y

Zhu L, Cheung C, Zhang W, Huang Z (2015) Compatibility of different biodiesel composition with acrylonitrile butadiene rubber (NBR). Fuel 158:288–292. https://doi.org/10.1016/j.fuel.2015.05.054

Chapter 14
Elastomeric Engine Mounts – Numerical Studies for E-Mobility Applications

Tomas Gejgus, Ondrej Farkas, Alexander Lion, and Michael Johlitz

Abstract This work focuses on the modeling of elastomeric vibration isolators considering the mass of the bushing. The elastomeric component is represented with the help of a hyperelastic model combined with linear viscoelasticity. The complete experimental and numerical characterization of the industrial elastomeric blend is described, including experimental suggestions and numerical models. Next, the parameters of used models are identified and listed to enable the community to access the model material, characterized with a focus on dynamic loading, easily and quickly. Once the material is fully described two finite element simulations in the time and frequency domain are presented. In the efficient frequency domain simulation, the isolator is harmonically excited, and the simulated dynamic stiffnesses are validated by experiment on the test bench for dynamic stiffness. The discussion of the simulation outcomes deals with the critical aspects of numerical investigation of such dynamic occasions. Lastly, the proposal for future research to solve the issues the work is certainly facing is expressed.

14.1 Introduction

As electric vehicles are overall smoother and more pleasant to drive, the driver's and passengers' attention is focused on the comfort aspects of the driving experience. A decisive quality feature here is the noise emitted by wind, tires, or the less intense but still present distortions caused by the vehicle's traction machine.

The spectrum of electric motors includes structural resonances, rotational mechanical orders, or bearing orders known from internal combustion engines (ICE). Furthermore, the spectrum is filled with several effects stemming from the motor controller in particular the pulse width modulation done by the inverter or from so-called slot locking as explained in following works Neapolitan and Nam (2019), Hanselman

T. Gejgus (✉) · O. Farkas · A. Lion · M. Johlitz
Institute of Mechanics, University of the Bundeswehr Munich, 85577 Neubiberg, Germany
e-mail: tomas.gejgus@unibw.de

O. Farkas
e-mail: ondrej.farkas@unibw.de

H. Altenbach et al. (eds.), *Lectures Notes on Advanced Structured Materials 3*,
Advanced Structured Materials 226, https://doi.org/10.1007/978-3-031-77403-4_14

(2003), Le Besnerais et al. (2018) or Gejgus et al. (2024). These effects are present and carry sufficient energy also at higher frequencies which are very well hearable according to the threshold of audibility for the humans' ear (Stevens 1951). Typical operational measurements of synchronous machines can be found in Le Besnerais et al. (2018) or Gejgus et al. (2024).

Due to these effects present well beyond 500 Hz there is a demand to simulate the behavior of elastomeric vibration isolators under harmonic loading. Such simulations should reduce development costs by reducing the amount of prototyping loops, augment the quality of transportation, and simultaneously increase safety by reducing the fatigue of the driver caused by unpleasant disturbing background noise during the drive.

Within this contribution, the reader is guided through the experimental and 1D numerical characterization of industrial elastomeric blends. In the Sect. 14.2.1, first for hyperelastic material behavior and in the Sect. 14.2.2 for linear viscoelasticity.

In Sect. 14.3, the finite element simulation (FEM) is discussed. Section 14.3.1 deals with the time domain simulation validated on the relaxation experiment. Further, Sect. 14.3.2 discusses the frequency domain simulation of the harmonically excited viscoelastic body concerning the simulation details, its difficulties, challenges, and the validation experiment on the test bench for dynamic stiffness.

Finally, the last section summarizes the outcomes and formulates the vision of future focus.

14.2 Experimental and 1D Numerical Investigation

All of the investigated specimens are produced of the same rubber blend, which contains 90% of natural rubber, 10% of butadiene rubber, antioxidants and is filled with 22 phr of carbon black. In this section, the compound will be characterized via static and dynamic tests performed with specimens S2 and S3 according to DIN 53504:2009-10 and specimen for uniaxial compression similar to type B according to DIN ISO 815-1:2010-09. Additionally, parameters of established material models for hyperelastic and linear viscoelastic behavior are identified within a 1D numerical simulation of the normalized tests.

14.2.1 Quasistatic Experiments and Numerical Characterization

All samples regardless of loading or test type were subjected to cyclic preconditioning, to eliminate strain-dependent stress-softening, i.e., Mullins effect (Mullins 1948). Figure 14.1a depicts the aforementioned preconditioning process. To enhance the sense of the impact of this effect, the first Piola-Kirchhoff stress (P_{11}) has been

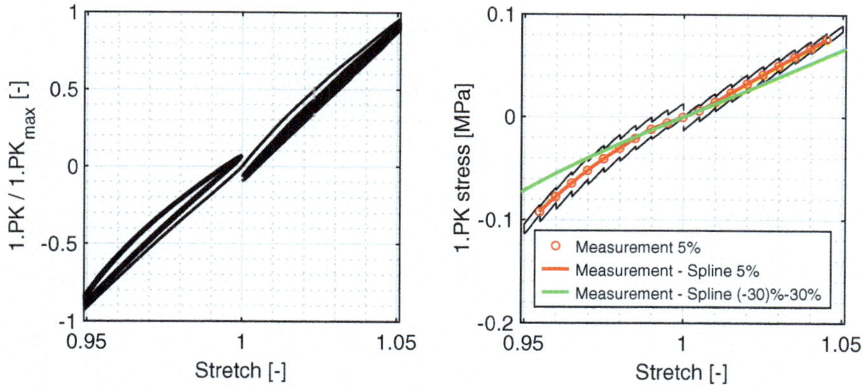

(a) Preconditioning under uniaxial tension and uniaxial compression - first Piola-Kirchhoff stress normed to the maximal stress in the particular test.

(b) Equilibrium first Piola-Kirchhoff stress for different preconditioning levels.

Fig. 14.1 Strain-dependent stress-softening—Mullins effect

normalized to the $P_{11_{max}}$ for both uniaxial tension and compression. It is legit to state, that for quasi-static load cases, there is a slight strain-dependent stress-softening caused possibly by the damage of the filler clusters. For this reason, it is necessary to precondition the samples used for the characterization of the elastomeric compound. The strain level of preconditioning should be chosen according to applications which are to be simulated with the parametrized material model.

Figure 14.1b shows the difference in the equilibrium stress caused by different strain amplitudes in the preconditioning process. For purposes of this contribution the multistep relaxations up to 5% with the step of 0.5% were utilized for both load cases. The strain rate in all experiments was set to 100%/min. and the relaxation time between the steps was 30 min. and for the turning point 90 min. After the identification of equilibrium stress the parameters of the strain energy function were optimized with the help of least square nonlinear (lsqnonlin) and particle swarm optimization (PSO) algorithms available in Matlab. For the error function, the relative deviation between the measured and simulated stress was defined for the least squares algorithm to avoid underestimation of small deformations. Similarly, for the PSO algorithm, the error function was defined to the particular measured value, and finally, the Euclidean norm was used on the vector of error function values and this scalar value was minimized (Fig. 14.2).

For the representation of hyperelastic behavior for this particular application, the 2-parametric Mooney-Rivlin strain energy function, as in Eq. (14.1), was chosen. This model contains the first and the second invariants of the right Cauchy-Green strain tensor **C**, since in bushing shear stress states are present. The constitutive law for the first Piola-Kirchhoff stress **P** in incompressible material is then formulated in Eq. (14.2) and reduces for uniaxial tension and compression to the scalar Eq. (14.3).

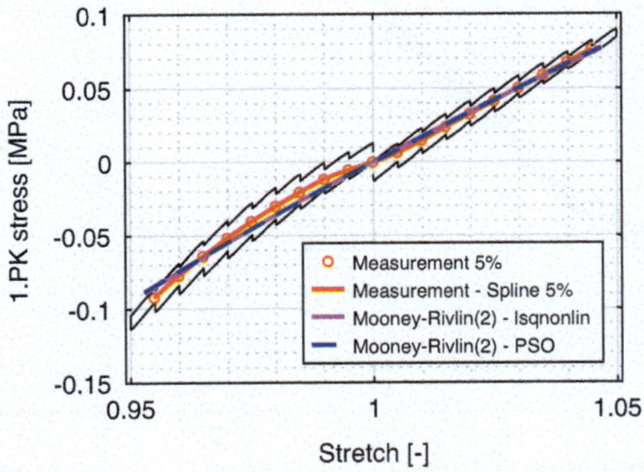

Fig. 14.2 Multi-step relaxation of preconditioned (5%) samples under uniaxial tension and uniaxial compression. Identified equilibrium stress. Fitted 2-parametric Mooney-Rivlin

Table 14.1 Hyperelastic parameters for 2-parametric Mooney-Rivlin

Parameters	C_{10}	C_{01}
Least square nonlinear	0.25393	0.33016
Particle swarm optimisation	0.19664	0.09599

Derivation of these equations as well as further information regarding the Mooney-Rivlin model can be found, e.g., in the textbook by Holzapfel (2000).

$$\rho_0\psi = C_{10}(\mathrm{I_C} - 3) + C_{01}(\mathrm{II_C} - 3), \tag{14.1}$$

$$\mathbf{P} = 2\,C_{10}\,\mathbf{F} + 2\,C_{01}(\mathrm{I_C}\mathbf{F} - \mathbf{FC}) - p\mathbf{FC}^{-1}, \tag{14.2}$$

$$P_{\mathrm{UT,UC}} = 2\,C_{10}\left(\lambda - \frac{1}{\lambda^2}\right) + 2\,C_{01}\left(1 - \frac{1}{\lambda^3}\right). \tag{14.3}$$

From the two utilized optimisation algorithms, one obtains two different sets of hyperelastic parameters which are listed in Table 14.1.

Table 14.1 shows that no global minimum was found and still the residuum (described above) satisfies the stopping criterion defined in each optimization algorithm. Since the strain within the experiment is only $\pm 5\%$ lots of parameter combinations satisfy the stopping criterion. These sets of parameters can be directly used as input for the hyperelastic Mooney-Rivlin model in Abaqus.

14.2.2 Dynamic Mechanical Analysis and Viscoelastic Properties

In the later simulation of dynamic excitation applied on the rubber component, the viscoelastic properties of the elastomeric blend are essential. For the determination of these properties, the classical stress relaxation is not suitable since the response is to be simulated at the excitation of several hundreds of Hertz.

For these purposes, dynamic mechanical analysis (DMA) is usually used since it provides information about elastic behavior in terms of storage modulus and energy dissipation in the material in terms of loss modulus. A very good insight into the topic of dynamic mechanical analysis is offered for example by Ehrenstein's book Ehrenstein et al. (2004) or in ISO-6721-1:2019. In the case of Abaqus can these moduli then be used as direct input for the case of frequency domain viscoelasticity, or also be used within this study, the parameters of the Prony series can be identified to approximate the frequency sweep obtained from DMA. The benefit of this approximation is the improvement of the curve smoothness, which is obvious by the look at the master curve for preconditioned (black) versus pristine (blue) material depicted in Fig. 14.3.

In this study, the master curves smoothed by the Prony series were used to compensate for the shifting artifacts which are visible on black master curves. These parameters can then be used also for time-domain viscoelasticity. At this place, the master curve and its development should be mentioned. As in detail explained by Farkas (2024), single frequency sweeps can be shifted horizontally and vertically, with respect to the reference temperature according to the time-temperature superposition principle (TTS) to obtain a master curve which provides information about material behavior in the very wide frequency range.

Figure 14.3 shows the master curve of Fig. 14.3a storage modulus, Fig. 14.3b loss modulus and their ratio Fig. 14.3c loss factor. One can observe two master curves in each subfigure, one (blue) for the pristine material in the post-production condition

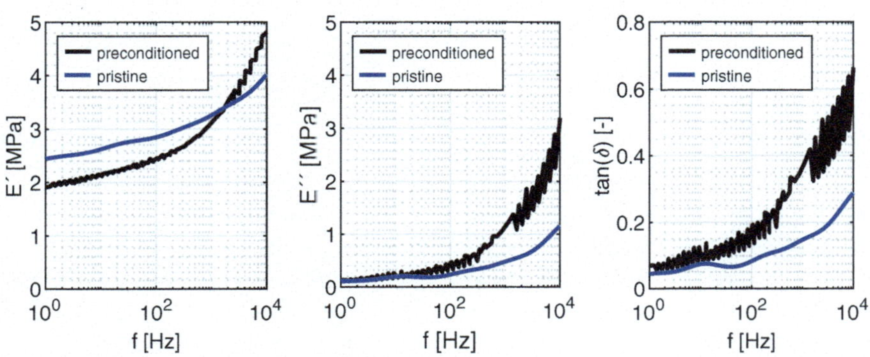

(a) Storage modulus (b) Loss modulus (c) Loss factor

Fig. 14.3 Master curves for preconditioned and pristine material

Table 14.2 Parameters of Prony series

Maxwell element	g_i [–]	τ_i [s]
1	0.486203665	2.1E-07
2	0.422428636	1E-06
3	0.045017137	1E-05
4	0.018818838	0.0001
5	0.006156207	0.0008
6	0.004248484	0.005
7	0.002236819	0.08
8	0.001369286	1
9	0.000260828	5

and a second (black) for material which was prior to the DMA test preconditioned cyclically with an amplitude of only 5% to eliminate the above-mentioned Mullins effect. Important to notice is the significant difference in the ratio of loss and storage modulus for frequencies over 100 Hz which are of interest in the main simulation within this study.

In Table 14.2 the parameters of the Prony series are listed with ascending relaxation time τ_i. The times were fixed while the stiffnesses μ_i were identified to fit the mastercurve in this case of preconditioned material. This will be discussed in Sect. 14.3.1 where these parameters are used for the definition of linear viscoelastic properties within time-domain simulation. Before use in Abaqus, the stiffnesses μ_i have to be normed to the instantaneous response G_0 to obtain g_i. The equations for storage and loss moduli represented by the Prony series denote:

$$G' = G_\infty + \sum_{i=1}^{n} \mu_i \frac{(\omega\tau_i)^2}{1 + (\omega\tau_i)^2}, \tag{14.4}$$

$$G'' = \sum_{i=1}^{n} \mu_i \frac{\omega\tau_i}{1 + (\omega\tau_i)^2}. \tag{14.5}$$

$$g_i = \frac{\mu_i}{G_0}. \tag{14.6}$$

Here the G, g and μ refer to the shear. Since the DMA was conducted under uniaxial tension a conversion has to be done as shown in Eqs. (14.7)–(14.9). The G_∞ refers to the equilibrium modulus linearized for the pre-strain used within the DMA measurement in the case of nonlinear elastic material definition. Also, the $'$ and $''$ as well as * refers to real (storage), imaginary (loss) and complex modulus.

$$G^* = \frac{3K^* E^*}{9K^* - E^*}. \tag{14.7}$$

Here the E, K and G are uniaxial, bulk and shear moduli. In case of incompressible material, Eq. (14.7) simplifies and solved for G' and G'' denotes:

$$G' = \frac{E'}{3}, \tag{14.8}$$

$$G'' = \frac{E''}{3}. \tag{14.9}$$

14.3 Numerical Investigation of the Simple Component

The simple elastomeric component investigated in this section is a cylinder to which are at the top and bottom vulcanized bolt with a round steel plate as depicted in Fig. 14.4. The elastomeric section with a height of 20.25 mm and width of 25.3 mm is modeled with an axisymmetric approach within FEM software and the metal elements are assumed to behave rigidly. Unfortunately as can be seen in Fig. 14.4 the positioning of the metal components is due to production tolerances not exactly centric.

The used mesh will be identical for both time and frequency domain simulations and its density is determined with respect to the speed of sound in elastomers mostly containing natural rubber.

14.3.1 *Time Domain Simulation*

This simple simulation of quasi-static relaxation should serve as validation for the calibration of the chosen material model. Additionally, it is shown, that even a very simple experiment, is due for machine control necessary PID controllers, in reality indeed nonlinear occasion. The bottom surface of the axisymmetry-modeled cylindrical specimen is in the whole simulation hold still. On the top surface is applied displacement of 0.5 mm in the vertical direction as a linear function of time.

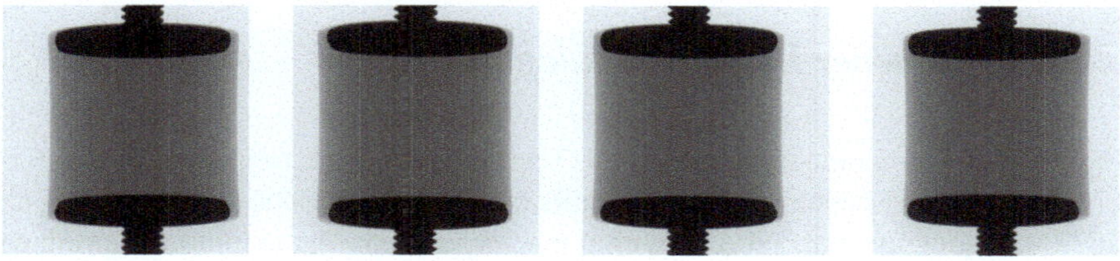

(a) 0° (b) 90° (c) 180° (d) 270°

Fig. 14.4 Cylindrical specimen—high energy μCT imaging

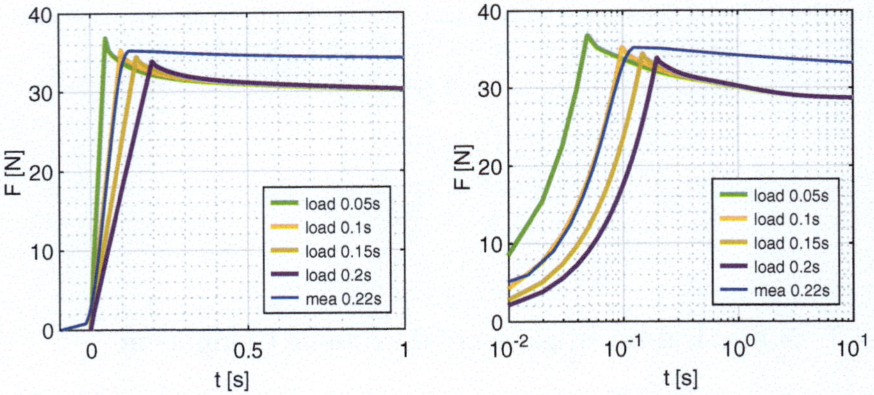

(a) 1st second of the relaxation

(b) 1st 10 seconds of the relaxation - semi-logarithmic

Fig. 14.5 Short-time relaxation—validation measurement versus FEM simulation (Prony series fitted to a master curve)

Figure 14.5 depicts in different time scales four simulation results for different loading rates and the measurement in blue. The experiment was conducted on the specimen preconditioned up to 5% quasi-statically. By analogy, the Mooney-Rivlin parameters as well as parameters of the Prony series were identified based on the experiments conducted on preconditioned samples with matching strain amplitude. For the calibration of the Prony series, the master curve stemming from DMA measurement was utilized, but the relaxation response was simulated. As can be seen in Fig. 14.5, the simulated spontaneous response fits the experiment quite well, nevertheless the relaxation process in the experiment is more gradual. This is plausible since if one looks at the Prony parameters in Table 14.2, the longest relaxation time which could be identified for the black master curve shown in the Fig. 14.3 with reasonable stiffness is 5 s. The contribution of this Maxwell element fades after ca. 10 s away. For this material blend, the DMA experiment simply can not provide information about long-time relaxation properties. Unless one would decide to extrapolate the master curve, which already is an extrapolation (Maiti 2016).

14.3.2 Frequency Domain Simulation

The frequency domain simulation in this section mimics the frequency sweep performed on the test bench for dynamic stiffness (TBDS) introduced in Gejgus et al. (2022) and Gejgus et al. (2022). Within this experiment, the cylindrical specimen as in Fig. 14.4 was tested in a post-production (pristine) state, and the carbon black clusters were only damaged by the highest displacement amplitude in the experiment which belongs to the lowest frequency. Since the target value for the closed-loop con-

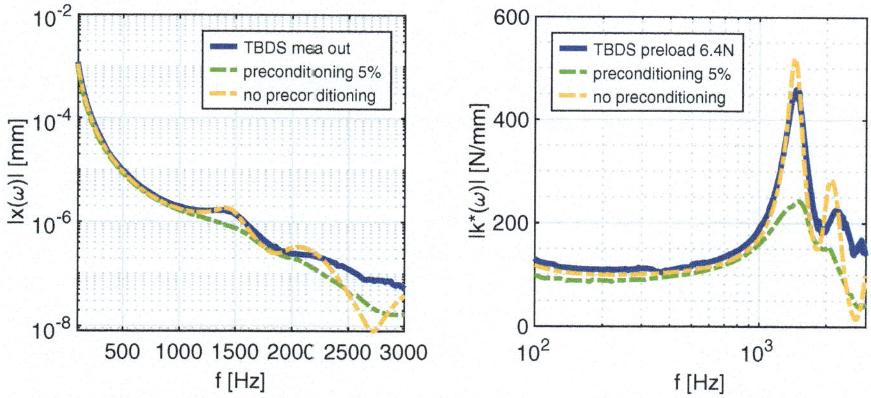

(a) Displacement of the seismic mass

(b) Dynamic stiffness of the elastomeric component

Fig. 14.6 Frequency domain simulation versus TBDS experiment—impact of viscoelastic properties

trol of the excitation was the constant amplitude of excitation velocity for the whole frequency sweep. This simulation is divided into two separate steps, the first—static step in the time domain, which neglects viscoelastic properties and uses only hyperelastic material definition for the equilibrium—static preload of the specimen. In this step the gravitation is applied to the specimen and the top surface is loaded by the desired static load while the bottom surface is fixed. The second step—steady-state dynamics is a linearization at the operation point. Here the material definition is linear viscoelastic and the frequency sweep is simulated in the frequency domain which has major benefits in simulation efficiency. For this step, the bottom boundary condition changes, and the kinematic excitation is applied in the form of the same frequency sweep as in the TBDS experiment.

The results of this simulation are presented in Fig. 14.6, where Fig. 14.6a depicts the top surface displacement of the specimen which can be interpreted as the chassis or passive part, and Fig. 14.6b depicts the amplitude of the dynamic stiffnesses $|k^*(\omega)|$. The blue curve represents again the measurement and the dashed curves are simulation results. The two representative simulations differ only in viscoelastic definition in this case via G' and G″ which were determined for the post-production state of the sample according to DIN 815 type B and for the preconditioned sample where the strain amplitude of only 5% was used. The frequency domain simulation of the viscoelastic specimen is capable of capturing at least the first structural resonance in the elastomeric material and the dynamic stiffness in the stable region prior to the first resonance can be simulated with decent precision. Unfortunately, such simulations are rather unreliable since preconditioning utilizing such small strain amplitude results in tremendous deviation of the simulation results especially in the range of structural resonances.

14.4 Conclusion and Outlook

The introduction to this contribution motivates due to the characteristics operational spectrum of PMSM, the demand for structural dynamics simulations of elastomeric components used for vibration isolation. The simulation performed in Sect. 14.3.2 has shown that even small changes in the preconditioning of samples used for the calibration of viscoelastic parameters result in significant changes in simulation results.

In the applications of these elastomeric isolators, the geometry of the used components becomes much more complex than the specimen shown in Fig. 14.4. Thus, the damage state caused by preconditioning during the assembly process and the history of the operation, combined with the complex geometry becomes highly inhomogeneous.

For components like the bushing in Fig. 14.7a the simulation with the material model used in this work would not be reliable since the viscoelastic parameters would have to be calibrated only for one preconditioning state. The authors are convinced that for the improvement of simulation results especially for more complex components the implementation of the material model which includes also the influence of the Mullins effect is essential. Several approaches of modeling stress-softening in elastomeric materials are presented exemplary in Diercks (2015).

To that point, there are two possibilities to estimate the dynamic stiffness of industrial vibration isolators. Each of these two methods has its benefits and drawbacks.

The inverse substructuring method was applied while the bushing was connected to the transmission simulators similar to those presented in the work of Haeussler et al. (2020). Two virtual points on each side of the bushing were defined and by the inverse of the frequency response function (FRF) matrix, the dynamic stiffness was determined. The matrix inversion in this method indicates that the method is sensitive to the signal-noise ratio of the measured signals. This drawback causes that the calculated dynamic stiffnesses to be only usable up to 2000–2500 Hz. The measurement

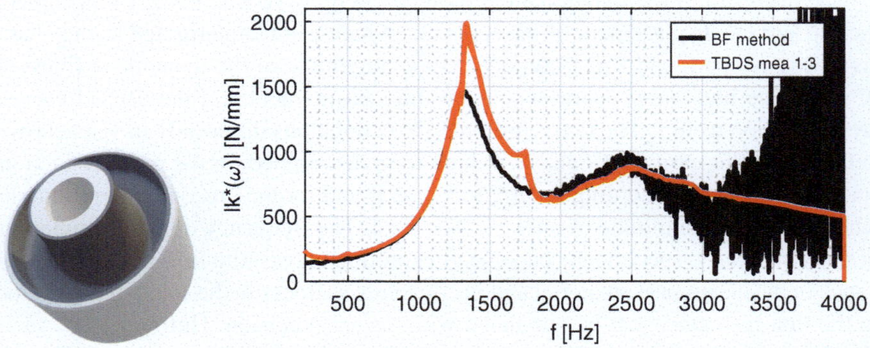

(a) Elastomeric bushing 3D model

(b) Magnitude of the dynamic stiffness of the elastomeric bushing - TBDS vs. Inverse substructuring method.

Fig. 14.7 Industrial elastomeric vibration isolator (bushing) requires rather a complex specimen attachment, which has to be carefully designated concerning its structural dynamics. Also, this experiment is more time-consuming than TBDS on the contrary provides the dynamic stiffness in 6 degrees of freedom (DOF). The TBDS measurement requires three separate experiments, one for each translational DOF and can provide dynamic stiffness only for translational DOFs. The red curve in Fig. 14.7b are in fact three separate measurements, this demonstrates the repeatability of the TBDS measurement. Unfortunately, the of phase parts of the measurement show significant deviation, which is caused by phase error in TBDS measurement which is currently not compensated and gains on impact when the ratio between the specimen mass and the mass of the seismic mass is poor. This issue is for the measurement depicted in Fig. 14.6b negligible.

Future work should focus on the modeling of the Mullins effect and its implementation in FEM so that the viscoelastic characteristics can be adapted after static preload of the component in every material point for reliable dynamic simulation. Optionally the implementation of experimentally determined FRF model for elastomeric isolators into simulation environments like Adams or Nastran should be explored.

References

Diercks N (2015) The dynamic behaviour of rubber under consideration of the Mullins and the Payne effect. Verlag Dr, Hut GmbH, Muenchen

Ehrenstein GW, Riedel G, Trawiel P (2004) Thermal analysis of plastics: theory and practice. Carl Hanser Verlag, Munich

Farkas O, Gejgus T, Lion A, Johlitz M (2024) From raw data to master curve—improving the time-temperature superposition process using geometrical properties of temperature-frequency sweeps. In: Collaborative research advancing engineering solutions for real-world challenges: the 2023 postgraduate seminar in Esslingen. Springer Nature Switzerland

Gejgus T, Farkas O, Lion A, Johlitz M (2022) Experimental and numerical investigation of NVH behaviour of NR—BR blend used for engine mounts of electric vehicles. In: Constitutive models for rubber XII. CRC Press, London. https://doi.org/10.1201/9781003310266-33

Gejgus T, Farkas O, Loos K, Lion A, Johlitz M (2024) Frequency-domain analysis of elastomeric vibration isolators: experimental and numerical investigations under dynamic loading. In: Constitutive models for rubber XIII. CRC Press, London

Gejgus T, Schröder J, Loos K, Lion A, and Johlitz M (2022) Advanced characterisation of soft polymers under cyclic loading in context of engine mounts. Polymers 14(3):429. Basel, Switzerland. https://doi.org/10.3390/polym14030429

Haeussler M, Klaassen S, Rixen D (2020) Experimental twelve degree of freedom rubber isolator models for use in substructuring assemblies. J Sound Vib. https://doi.org/10.1016/j.jsv.2020.115253

Hanselman DC (2003) Brushless permanent magnet motor design. The Writers' Collective, Orono, ME, USA

Holzapfel GA (2000) Nonlinear solid mechanics: a continuum approach for engineering. John Wiley & Sons Ltd, New York

Le Besnerais J, Dendiviel C, Régniez M, Boussard P (2018) Sound quality optimization of electric powertrains for e-mobility applications. In: CFA2018 proceedings, Le Havre, France

Maiti A (2016) A geometry-based approach to determining time-temperature superposition shifts in aging experiments. Rheol Acta 55. https://doi.org/10.1007/s00397-015-0898-z
Mullins L (1948) Effect of stretching on the properties of rubber. Rubber Chem Technol. 10(5254/1):3546914
Neapolitan RE, Nam KH (2019) AC motor controland electrical vehicle applications, 2nd edn. Taylor & Francis, Boca Raton
Stevens SS (1951) Handbook of experimental psychology. John Wiley & Sons Inc, New York

Chapter 15
Influence of Antiozonants on the Evolution of Crack Depth in Natural Rubber

Caroline Treib, Michael Johlitz, and Alexander Lion

Abstract Natural rubber (NR) is employed in a multitude of industries beyond the manufacture of tyres, including the production of earthquake-resistant isolators, bridge bearings, gaskets and hoses. However, NR is susceptible to environmental aggressors such as temperature, oxygen, ozone and UV light, which can lead to degradation and cracking. In particular, ozone induces rapid visible cracking in NR due to reactions with its double bonds. Antiozonants, such as diamines and paraffin wax, are commonly employed to mitigate cracking, yet challenges persist, particularly under cyclic deformation. This study examines the impact of antiozonants on crack evolution in NR subjected to ozone exposure. Artificial ageing processes are developed to simulate accelerated ageing at ground-level conditions, varying ozone concentrations and ageing times. Crack depth evolution is monitored using optical microscopy. The results demonstrate that while antiozonants delay cracking, prolonged exposure leads to increased crack depths, particularly under cyclic strain. Simulation models predict crack growth as a function of ageing time and ozone concentration, revealing the effectiveness of antiozonants at different environmental conditions. The study highlights the complexity of NR ageing dynamics and offers insights for optimising antiozonant usage to enhance material durability. Further research recommendations are outlined to address extrapolation challenges to lower ozone concentrations.

C. Treib (✉) · M. Johlitz · A. Lion
Institut für Mechanik, Fakultät für Luft- und Raumfahrttechnik, Universität der Bundeswehr München, Werner-Heisenberg-Weg 39, 85577 Neubiberg, Germany
e-mail: caroline.treib@unibw.de
URL: https://www.unibw.de/lrt4

M. Johlitz
e-mail: michael.johlitz@unibw.de

A. Lion
e-mail: alexander.lion@unibw.de

H. Altenbach et al. (eds.), *Lecture Notes on Advanced Structured Materials 3*,
Advanced Structured Materials 225, https://doi.org/10.1007/978-3-031-77403-4_15

15.1 Introduction

A multitude of industries utilise natural rubber (NR) in addition to its primary use in tyres. Applications encompass isolators below constructions in earthquake zones beneath expansion joints, and bearings of bridges to regular gaskets, sealings and hoses (Gu and Itoh 2010; Iwase et al. 2017). The material is susceptible to a variety of ageing sources present in the surrounding environment of the component. Some of these sources that have been discussed in the literature include temperature, oxygen, ozone and UV light. These may lead to a gradual change in material properties or macroscopic damage, including cracking (Gu and Itoh 2010; Lemaire et al. 1993; Mars and Fatemi 2005; Verdu 2012).

Ozone can lead to fast cracking, which may be observed by the unaided eye. Some of the earliest publications on this topic were published in the mid-20th century (Newton 1945; Buckley and Robison 1956; Andrews and Braden 1961). Plain NR is highly susceptible to ozone in the absence of antiozonants, evidenced by the findings of Radhakrishnan et al. (2006). Ozone molecules react with the double bond in the monomer *cis-1,4-polyisoprene*, which is the primary component of natural rubber (Bhala 1997). This reaction results in the cleavage of the double bond and consequently the cleavage of the main chain, forming products such as carbonyl compounds (Criegee 1975; Lemaire et al. 1993). A minimal tensile strain is necessary to initiate and propagate cracks, a concept established in the 1960s by Andrews and Braden (1961). Cataldo (2019) attributes the initiation of cracks to the combination of the cleavage of double bonds within the rubber main chain and strain, which departs the molecular chain ends. This is consistent with the observation that ozone ageing in the absence of applied strain does not result in the formation of cracks. Previous research utilising FTIR microscopy has demonstrated that carbonyl groups evolve within a depth of maximum 20 μm from the surface of the samples as referenced in Treib et al. (2022). The significant impact of strain level on crack formation is not only apparent as a threshold for initiation but also influences the orientation of cracks, leading from fissures to macroscopic cracks as evidenced by Andrews et al. Andrews and Braden (1961).

Although antiozonants are commonly employed with beneficial effects, cracking nevertheless persists and reduces lifespan of components, as reviewed by Mars and Fatemi (2005). Two types of antiozonants are regularly applied in industry: 6PPD[1] and paraffin wax, as described for example by Lake and Mente (1992); Cataldo (2019). Other protection mechanisms are possible as coating systems (Lewis 1986; Früh et al. 2018; Treib and Johlitz 2022). However, these are limited by delamination effects and abrasion requirements. The paraffin wax moves to the surface to create a protective barrier against ozone exposure. In contrast, the diamine 6PPD interacts with ozone molecules, functioning as a scavenger. It may also leave behind reaction products that provide some barrier effect, as cited by Cataldo (2019). However, under

[1] *N*-(1,3-dimethylbutyl)-*N'*-phenyl-p-phenylenediamine.

cyclic deformation, which occurs in most applications, both methods are ineffective in preventing rubber from cracking when exposed to ozone (Cataldo 2019; Treib and Johlitz 2022).

The appearance of cracks and their depth can be quantified in samples after artificial ageing. This is a common approach to measuring damage, as evidenced by Kamaruddin and Muhr (2018). Zuev and Pradvednikova (1962) propose an indirect proportionality between the ageing time and the ozone concentration until crack appearance, which is dependent on a material factor (Dolezel 1978; Nonnenmacher 2008). Besides a threshold of strain necessary for cracks to evolve, the strain value is not considered to be a factor in the evolution of cracks. Following the nucleation of cracks, their growth is linearly proportional to the ozone concentration according to Braden et al. (1962) and Spreckels (2013). In contrast, Buckley and Robison (1956) found that crack growth in butyl rubber is proportional to the square root of the ozone concentration. The evolution of damage is assessed by the crack depth in the cross section in a former publication (Treib et al. 2024). Influence of the type of strain, ageing time and ozone concentration were measured and will be published in Treib and Johlitz (2024). Under dynamic strain, the crack growth has been approximated by Gent and Lindley (2012), Lindley (1973) in dependence on an ozone concentration factor, the time under load in each cycle and the frequency applied. In the case of static strain and dymanic strain, the actual strain value is not part of the simulation methodology. However, different material compounds and different strain types lead to different evolution equations.

The following analysis documents the influence of antiozonants on the cracking that occurs due to ozone exposure. Furthermore, the evolution of crack depth is simulated. A method of accelerated ageing is developed in order to reduce the ageing times of real components. The influence of ozone concentration and ageing times on the evolution of cracks is investigated by tracking their development at the surface and in the cross-section using optical microscopy. The analysis is focused on the differences in the evolution of cracks in an unprotected compound and in compounds protected by antiozonants. The limitations of extrapolating damage prediction to ageing parameters outside the measured range are discussed, with particular focus on the different evolution equations that can be employed. Finally, this leads to a number of open questions and recommendations for further research.

15.2 Experimental

15.2.1 Material

Three variants of natural rubber compounds were subjected to investigation, each differing only in the types of antiozonants that were added. It is conventionally produced using mixing ratios outlined in Table 15.1. A vulcanisation system

Table 15.1 Natural rubber compounds

Component [phr[a]]	C1	C2	C3
NR CV 60		100	
Carbon Black N330		40	
Zinc Oxide		3	
Stearic Acid		1	
Accelerator TBBS		1	
Sulphur		2	
6PPD	3	3	0
Paraffin Wax	3	0	0

[a] Parts per hundred rubber

incorporating zinc oxide, stearic acid, sulphur, and the accelerator TBBS[2] is employed. During this process, sulphur bridges are formed between the molecular chains of natural rubber, facilitated by the inorganic activator zinc oxide and the organic compound stearic acid. The compound characterised by a consistent Mooney viscosity of 60, designated as *NR CV 60*, is reinforced with *high abrasion furnace* carbon black N330, which conducts heat more effectively than the rubber phase. In the absence of ageing prevention additives, the natural rubber compound primarily comprises *NR CV 60* and carbon black *N330*, in addition to the vulcanisation agents. Among the antiozonants under investigation are two compounds that are commonly used in industry, as noted by Huntink et al. (2004) and Cataldo (2019). The two compounds under investigation were 6PPD and paraffin wax. Paraffin wax migrates to the surface, forming a physical barrier against ozone attack. In contrast, 6PPD reacts with ozone molecules, acting as a scavenger and leaving behind reaction products that offer an additional physical barrier (Cataldo 2019). In addition to 6PPD, the compound C1 includes paraffin wax, which provides a physical barrier. Compound C2 contains solely the chemical antiozonant, whereas compound C3 is devoid of any antiozonants.

15.2.2 Ground-Level Ozone

Industrial components are typically exposed to elevated concentrations of ground-level ozone, and much higher values inside or within the vicinity of production facilities. The formation of ground-level ozone is a result of an equilibrium process whereby UV radiation or electric discharge splits molecular oxygen into oxygen atoms, which then combine to form ozone (Murphy and Orr 1975). In situ photochemistry, air pollution (particularly from volatile organic compounds (VOCs)), carbon dioxide and nitric oxide significantly influence ozone production (Crutzen

[2] *N*-tert-Butyl-2-benzothiazolesulfenamide.

et al. 1998). Pollution from hydrocarbon emissions has the potential to raise ozone levels by up to 50 pphm from 2 pphm to 5 pphm in unpolluted areas, according to De et al. (2001). In contrast, the temperature at ground level has minimal impact on ozone formation. Ground-level ozone concentrations are monitored extensively by environmental stations, with the majority of reports focusing on 1- and 8-hour average levels and annual exceedances of limits. However, stationary governmental monitoring systems will never fully capture local variations (Bundesumweltamt 2023; United States Environmental Protection Agency 2021). Further, differences are observed between marine and continental regions, as shown by the findings of Crutzen et al. (1998); Salisbury et al. (2001). Ozone concentrations reach peaks of about 15 pphm on summer afternoons and nights, fluctuating around 3 pphm to 5 pphm daily (Bundesumweltamt 2023).

15.2.3 Artificial Ageing

The regulation of an ozone climate simulator necessitates a prompt response to adjust the current ozone concentration to the desired level. Ozone generation is achieved using the *SIM7500-TH* device manufactured by *Anseros® Klaus Nonnenmacher GmbH* by means of the corona discharge method. The ozone concentration is quantified through UV absorption spectroscopy utilising the property of ozone to block certain UV wavelengths. In contrast, meteorological stations employ slower indirect measurement methods for reporting 1- to 8-hour-average values. The differing measurement approaches result in varying levels of precision, with faster response times in the ozone simulator sacrificing some accuracy. In general, direct measurement methods have detection limits ranging from 2.9 pphm to 3.6 pphm according to Silviera Petruci et al. (2022), with a stated deviation of ±5 pphm in the ozone chamber used.

The concentrations employed in the study commence at 25 pphm, which is approximately 1.65 times higher than the ambient values observed at environmental stations on summer afternoons. In order to prevent any alteration to the chemical reaction of ozone with natural rubber, the concentrations employed in the study do not exceed 75 pphm which corresponds to 5 times the ambient maximal value measured. A logarithmic distribution is selected in order to align with the logarithmic timescale of many ageing processes (Gillen et al. 2016).

Further details regarding the selected ageing times and concentration values can be found in Table 15.2.

15.2.4 Crack Analysis

Following the artificial ageing process, samples are prepared for crack depth evaluation using digital microscopy outlined in former studies (Treib et al. 2024). To

Table 15.2 Ageing parameters

Parameter	Unit	Value
Ageing times	/h	4, 12, 37, 111
Ozone concentrations	/pphm	25, 36, 52, 75
Strain levels during ageing	/%	20, 30 ± 10

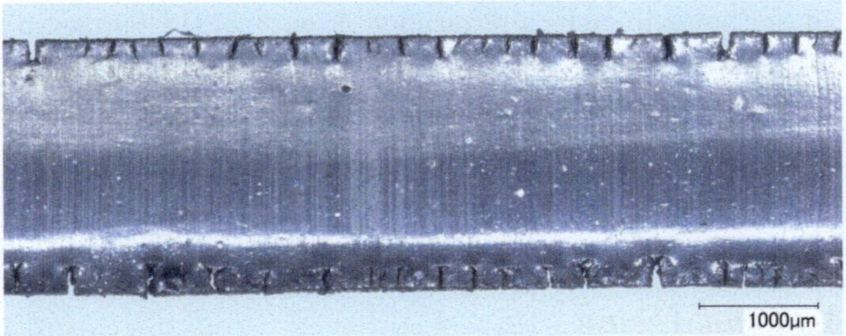

Fig. 15.1 Opened cracks in the cross-section of a sample of C2 aged at 30 (±10)% strain at 52 pphm for 111 h

determine the crack depth within the cross-section, samples are sectioned by punching narrow strips, extracted specifically from the central region of the aged material plates to ensure consistent one-dimensional ageing. Given the slightly tacky nature of NR, the sample strips are gently strained to reveal open cracks under an optical microscope, specifically the *KEYENCE VHX-2000* model. The microscope's software incorporates calibrated length measurement tools to quantify the crack depth of the samples, as illustrated in Fig. 15.1. To assess the accuracy of the measurements, 100 readings of a single crack were recorded, yielding a standard deviation of 12.4 μm.

15.2.5 Experimental Results

The resulting crack depths of ozone ageing under 20 % static strain are presented in Figs. 15.2 and 15.3. In the case of static strain during ageing, compound C1 did not show any cracks in the cross-section. The measured crack depths resulting from cyclic loading at a strain of 30 ± 10% during ageing are presented in Figs. 15.4, 15.5 and 15.6.

Due to a corrupted ozone measurement at 25 pphm for sample C2, repeated tests were performed at three different ageing times. Comparing the two cracking compounds, C2 and C3, except at 75 pphm ozone concentration the crack depth is slightly shallower in C2. This may be attributed to the mode of action of 6PPD, which requires

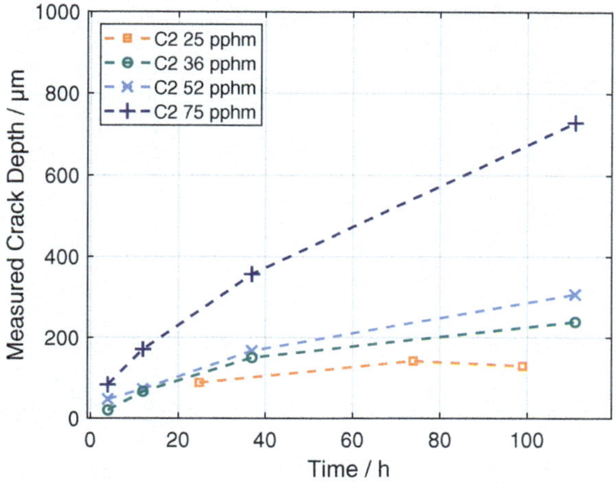

Fig. 15.2 Measured cracks in the cross-section of samples of compound C2 aged at 20% strain

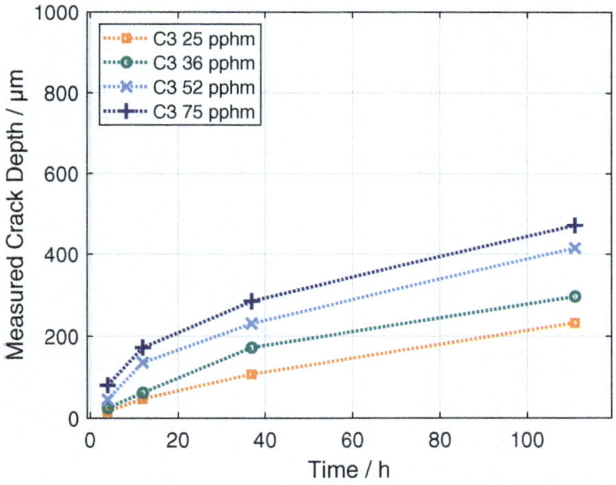

Fig. 15.3 Measured cracks in the cross-section of samples of compound C3 aged at 20% strain

migration to the surface to react with ozone rather than the double bond in the rubber backbone. Two factors limit the effect of protection: the migration efficiency and the consumption rate at the surface. The greater the concentration of ozone, the greater the quantity of 6PPD is required on the surface to prevent local depletion, which appears to partially be the case at 75 pphm. At lower concentrations, which are mostly present in the ambient conditions of compounds in use, the use of 6PPD appears to reduce the crack depth. However, only the protective layer of paraffin wax is able to prevent cracks completely.

In the context of cyclic loading during the ageing process, it can be observed that both compounds, including antiozonants, exhibit a reduction in crack depth compared to C3. However with longer ageing times the crack depth approaches and eventually surpasses that of the compound without antiozonants. Migration of the protectors is expected to be enhanced due to the cyclic loading, which might lead to

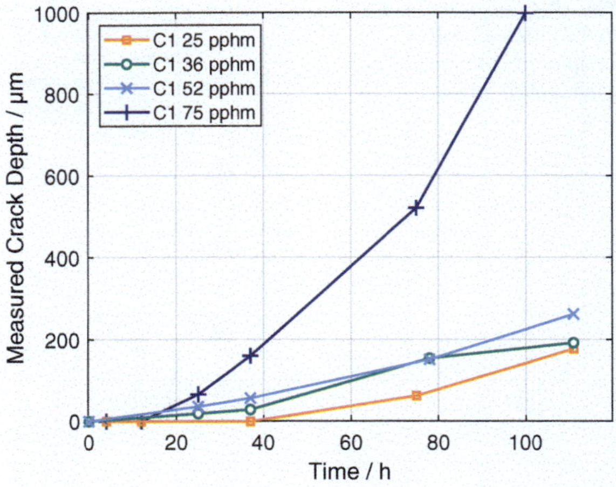

Fig. 15.4 Measured cracks in the cross-section of samples of compound C1 aged at 30(±10)% strain

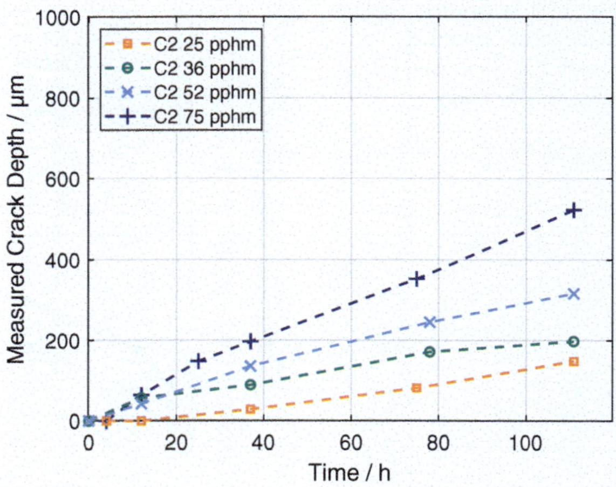

Fig. 15.5 Measured cracks in the cross-section of samples of compound C2 aged at 30(±10)% strain

primary reduction in the cracking rate. However, this reduction in cracking rate is not sustained, and the cracking rate accelerates for C1 to a greater extent than for C2. To facilitate the visualisation of the differences in crack evolution the measured depths are normalised to the longest ageing time. The average of all ozone concentrations for each compound is shown in Fig. 15.7. While the level of crack depth in C3 appears to remain constant over time, C2 exhibits linear growth, whereas in C1 the crack growth rate increases with ageing time.

Even at the ozone concentration of 25 pphm after 111 h, the crack depth in compound C1 reaches the one in C3. Yet, the use of paraffin wax is able to prevent cracking at least within a time limit wether under static or cyclic strain. Given the low ozone concentrations at ground-level of 3 pphm to 5 pphm, this may be sufficient to prevent cracking. Nevertheless, if cracking commences, the growth rate will exceed the depth of the cracks that have been measured in the absence of any antiozonants.

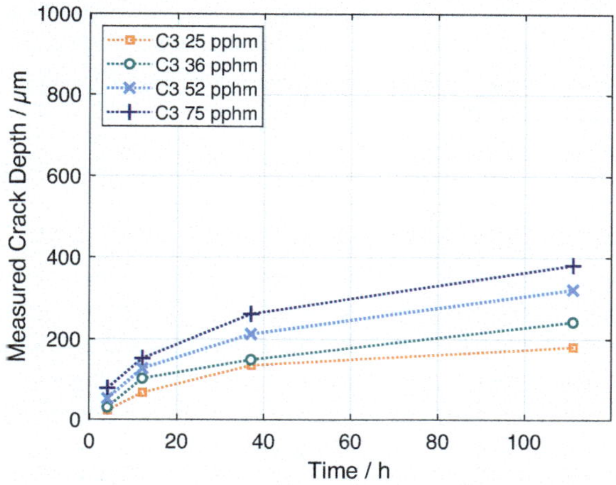

Fig. 15.6 Measured cracks in the cross-section of samples of compound C3 aged at 30(±10)% strain

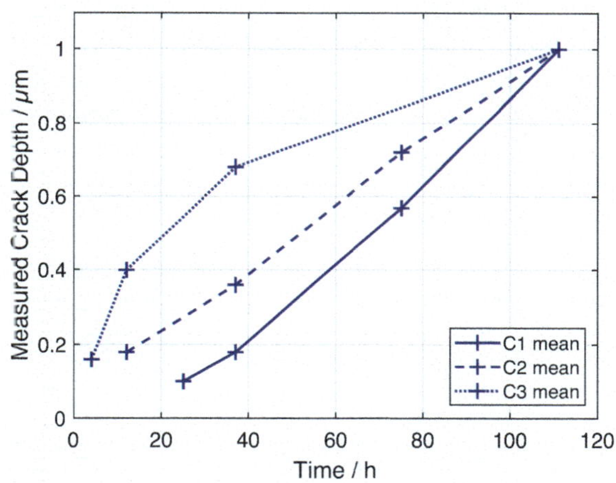

Fig. 15.7 Normalised evolution of crack depth for the three compounds C1, C2 and C3

15.3 Simulation of Crack Growth

Feasible ozone concentrations in the laboratory require higher than ground-level measurements, thus the evolution of the crack depth must be simulated in order to make predictions for lower concentrations.

15.3.1 Evolution over Ageing Time

The time until cracks appear seems to be negligible in the case of static strain, as cracks are present in both compounds C2 and C3 after only 4 h of ageing time. In contrast, the compounds including antiozonants do show a delayed onset of cracking for 25 pphm ozone concentration. Thus, for the cyclic strain an initial time t_{init} in h is considered. The onset is indirectly proportional to the ozone concentration as recommended by Zuev and Pradvednikova (1962), depending on a material parameter β in h times pphm:

$$t_{\text{init}} = \frac{\beta}{c_{\text{ozone}}}. \tag{15.1}$$

Measurements with longer ageing times revealed an increase in crack depth over that measured at 111 h. Thus, the evolution equation based on a kinetic approach as published in Treib et al. (2024) is replaced by new equations based on the presented measurements. Over ageing time the evolution is similar for static strain but a distinct difference in the evolution occures if cyclic loading is applied during ageing (see Fig. 15.7). Considering the normalised evolution of cracks over time, three different equations for the crack depth (h) in µm are used to fit the measurement data over ageing time in h with parameter α_1 in µm and dimensionless γ:

- Equation (15.2) squareroot function for static strain in compound `C2` and `C3` and for cyclic strain in `C3`.
- Equation (15.3) linear approach for cyclic strain in `C2`.
- Equation (15.4) power function for cyclic strain in `C1`.

$$h(t) = \alpha_1 \cdot \sqrt{t/\text{h}}, \tag{15.2}$$

$$h(t) = \alpha_1 \cdot t/\text{h}, \tag{15.3}$$

$$h(t) = \alpha_1 \cdot (t/\text{h})^{\gamma}. \tag{15.4}$$

15.3.2 Evolution over Ozone Concentration

As stated in the literature, two relationships are considered between the ozone concentration and the crack growth: in Eq. (15.5) a linear one and in Eq. (15.6) a proportionality to the square root of the ozone concentration both multiplied with a factor α_2 in µm:

$$h(c_{\text{ozone}}) = \alpha_2 \cdot c_{\text{ozone}}/\text{pphm}, \tag{15.5}$$

$$h(c_{\text{ozone}}) = \alpha_2 \cdot \sqrt{c_{\text{ozone}}/\text{pphm}}. \tag{15.6}$$

15.3.3 Combined Crack Depth Evolution

To combine the influencing parameters ageing time and ozone concentration, their evolution equations are combined by multiplication. The factors α_1 and α_2 are combined into one factor α in µm. Essentially, Eqs. (15.2), (15.3), (15.4) are expanded by multiplying either directly the ozone concentration or the square root of the ozone concentration. In addition, an initial time to cracking starts is considered for cyclic loading in C1 and C2. These relations result in six equations:

$$h(t) = \alpha \cdot \sqrt{t/\mathrm{h}} \cdot \frac{c_{\mathrm{ozone}}}{\mathrm{pphm}}, \tag{15.7}$$

$$h(t) = \alpha \cdot \sqrt{t/\mathrm{h}} \cdot \sqrt{\frac{c_{\mathrm{ozone}}}{\mathrm{pphm}}}, \tag{15.8}$$

$$h(t) = \alpha \cdot (t - t_{\mathrm{init}})/\mathrm{h} \cdot \frac{c_{\mathrm{ozone}}}{\mathrm{pphm}}, \tag{15.9}$$

$$h(t) = \alpha \cdot (t - t_{\mathrm{init}})/\mathrm{h} \cdot \sqrt{\frac{c_{\mathrm{ozone}}}{\mathrm{pphm}}}, \tag{15.10}$$

$$h(t) = \alpha \cdot ((t - t_{\mathrm{init}})/\mathrm{h})^{\gamma} \cdot \frac{c_{\mathrm{ozone}}}{\mathrm{pphm}}, \tag{15.11}$$

$$h(t) = \alpha \cdot ((t - t_{\mathrm{init}})/\mathrm{h})^{\gamma} \cdot \sqrt{\frac{c_{\mathrm{ozone}}}{\mathrm{pphm}}}. \tag{15.12}$$

The following equations are used to simulate the measured cracks for the sets of test data sets:

- Equations (15.7) and (15.8) for static strain in compound C2 and C3 and for cyclic strain in C3.
- Equations (15.9) and (15.10) for cyclic strain in C2.
- Equations (15.11) and (15.12) for cyclic strain in C1.

A genetic algorithm is applied to fit the parameters α, β and γ using MATLAB optimtool. The residual is based on a least-squares algorithm and weighted by the deviation of the ozone measurement of ±5 pphm with respect to the ozone concentration at the sampling point.

In particular, for the cyclic strain data in compounds C2 and C1 depicted in Figs. 15.5 and 15.6, a strong change in cracking is visible at the ozone concentration of 75 pphm. Therefore, all fits are performed in- and excluding this ozone concentration of 75 pphm with the results listed in Table 15.3.

To compare the goodness of the fits, the root mean square error (RMSE) in μm^2 and the absolute mean error (MAE) in µm are listed with the correlating parameters in Table 15.3. Since the MAE is case sensitive, indications are given whether the simulation over- or underestimates the measured data on average. With MAEn in µm, Eq. (15.13), the sum of the absolute errors divided by the number of measurements n is added to the goodness of fit comparison values.

Table 15.3 Parameters α, β and γ with the correlating RMSE, MAE and MAEn for different measurement data

Strain	Compound	Equations	Maximum ozone concentration	α/μm	β/h × pphm	γ/–	RMSE/μm^2	MAE/μm	MAEn/μm
Static	C3	Eq. (15.7)	75	6.54E-01			31.20	–5.20	26.07
			52	7.51E-01			**21.17**	**7.15**	**17.43**
		Eq. (15.8)	75	5.12E+00			28.76	14.98	25.14
			52	4.95E+00			30.33	16.82	27.83
Static	C2	Eq. (15.7)	75	7.66E-01			57.81	29.02	46.73
			52	5.65E-01			**18.57**	**0.09**	**16.40**
		Eq. (15.8)	75	5.75E+00			96.09	45.73	80.24
			52	3.72E+00			22.29	8.31	17.27
Cyclic	C3	Eq. (15.7)	75	5.52E-01			29.07	–14.80	24.16
			52	6.34E-01			**18.47**	**–5.90**	**16.01**
		Eq. (15.8)	75	4.33E+00			21.09	2.53	17.68
			52	4.20E+00			19.39	2.88	16.12
Cyclic	C2	Eq. (15.9)	75	6.14E-02	1.00E-05		23.29	3.28	19.65
			52	5.71E-02	1.00E-05		**21.33**	**–0.48**	**19.84**
		Eq. (15.10)	75	5.11E-01	3.60E+02		53.30	6.79	43.95
			52	3.84E-01	9.09E+01		35.69	4.09	31.15
Cyclic	C1	Eq. (15.11)	75	1.04E-01	2.15E+03	1.08E+00	89.32	–15.59	67.55
			52	6.04E-03	1.00E-05	1.44E+00	21.62	–0.20	15.40
		Eq. (15.12)	75	2.69E+00	3.56E+03	8.89E-01	108.80	–45.24	80.26
			52	2.91E-01	6.74E+02	1.05E+00	**15.51**	**0.18**	**12.04**

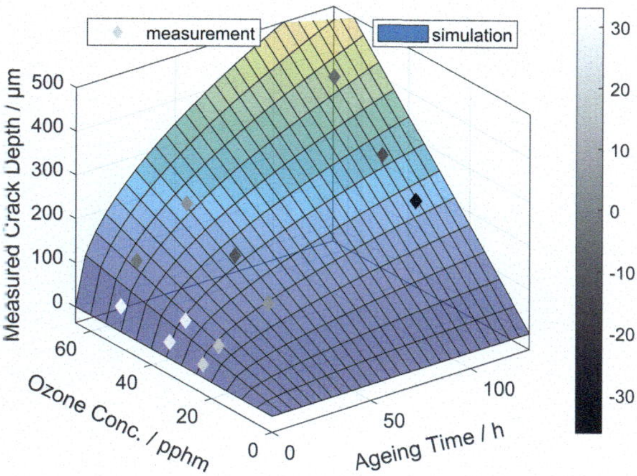

Fig. 15.8 Simulation for static strain during ageing in C3 up to 52 pphm ozone concentration using Eq. (15.7)

$$MAEn = \frac{\sum |h_{sim} - h_{meas}|}{n}. \tag{15.13}$$

By comparing RMSE, MAE and MAEn in Table 15.3, the equations are selected to simulate the different parameter sets and compounds. With Eq. (15.7) the simulation of static strain during ageing and cyclic strain in C3 is closer to the measured values than with a proportionality to the square root of the ozone concentration. For all three cases the results improve when the highest ozone concentration is not considered. However, the difference in the goodness of the fit is less pronounced for C3. The fits giving the smallest RMSE, MAE and MAEn are plotted in Figs. 15.8, 15.9 and 15.10. The following plots show the absolute error (AE) calculated as $AE = h_{sim} - h_{meas}$, delineating the distance of the simulation to the measured crack depths. Regarding cyclic loading during ageing in compound C2, the linear correlation with the ozone concentration also results in better fits (see Fig. 15.11). The inclusion of measured data for 75 pphm does not significantly change the results with Eq. (15.7). Compound C1 is the only one that is better simulated with the square root of the ozone concentration being proportional to the crack depth and ozone concentrations below 75 pphm, as shown in Fig. 15.12.

15.3.4 Prediction for Ground-Level Ozone Concentration

Using the adjusted parameters for the different compounds and deformations during ageing, it is possible to calculate the break-even points at which the use of antiozonants adversely affects crack depth at lower ozone concentrations. Assuming an ambient ground-level ozone concentration of 4 pphm and static loading, cracking could be completely prevented with paraffin wax. However, if the antiozonant is not applicable (e.g. due to wear), the addition of 6PPD in C2 will reduce the size of the

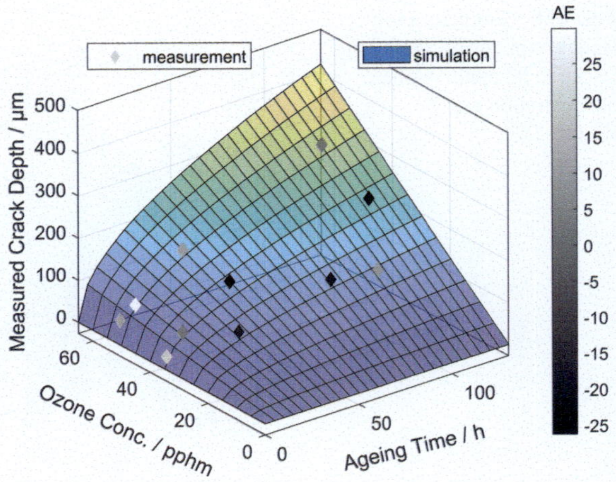

Fig. 15.9 Simulation for static strain during ageing in C2 up to 52 pphm ozone concentration using Eq. (15.7)

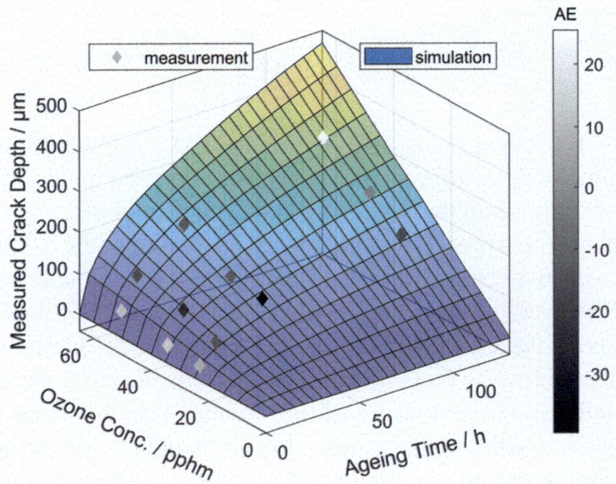

Fig. 15.10 Simulation for cyclic strain during ageing in C3 up to 52 pphm ozone concentration using Eq. (15.7)

cracks. At such low concentrations the use of antiozonants is beneficial under static load conditions. In contrast, under cyclic loading during ageing at 4 pphm, the simulation shows deeper cracks in C2 after 124 h and in C1 after 222 h, see Fig. 15.13. The influence of cyclic loading markedly changes the resulting crack evolution with the presence of antiozonants, whereas without protection the crack evolution in NR is similar with some reduction in crack depth for cyclic loading.

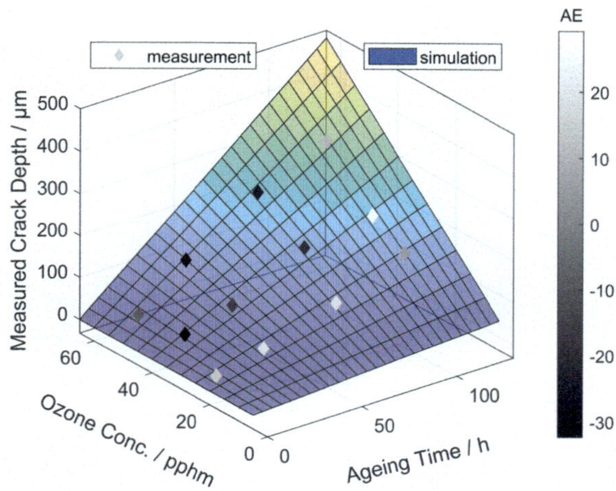

Fig. 15.11 Simulation for cyclic strain during ageing in C2 up to 52 pphm ozone concentration using Eq. (15.9)

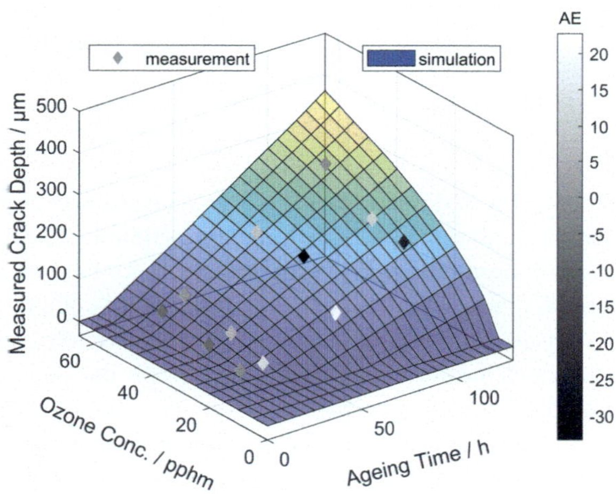

Fig. 15.12 Simulation for cyclic strain during ageing in C1 up to 52 pphm ozone concentration using Eq. (15.12)

15.4 Conclusion and Outlook

In this study, artificially aged natural rubber samples show an accelerated development of cracks, allowing a microscopic analysis of the cross-section to identify characteristic crack properties in a reasonable amount of time. The resulting crack depth evolutions are simulated for compounds containing antiozonants 6PPD and paraffin wax.

The difference in crack evolution under static loading is negligible with the addition of 6PPD, although it may reduce the crack depth. No cracks were measured when paraffin wax was also added. In contrast, different characteristics in the evolution with and without antiozonants are dominant for cyclic strain during ageing. In compound C3 the crack evolution remains similar for static and cyclic loading.

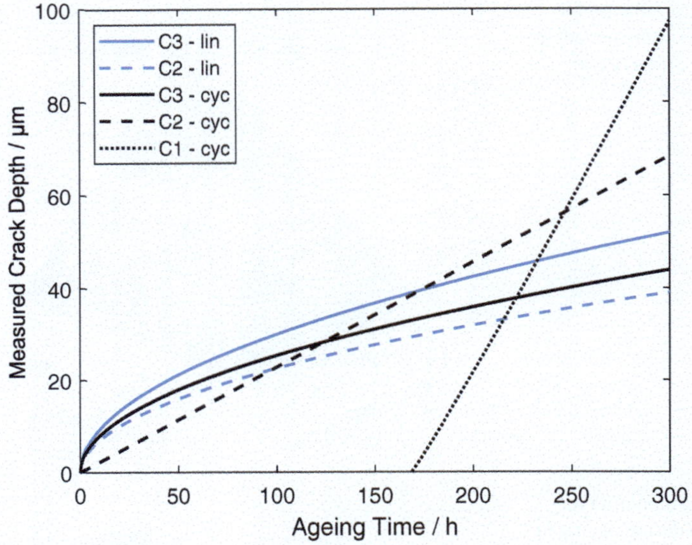

Fig. 15.13 Simulation for 4 pphm ozone concentration

Whereas the addition of 6PPD in C2 changes the evolution to a quasi-constant crack rate. If paraffin wax is added, there is a time for crack initiation depending on the ozone concentration. However, the crack growth rate accelerates with ageing time and the crack depth exceeds that of the other compounds at ageing times of a few days at realistic ground-level ozone concentrations. The cyclic loading appears to counteract the formation of a wax film on the surface, although the deformation should promote migration.

Both antiozonants are dependent on the concentration of ozone attacking the NR, as they must migrate to the surface to either react with or block the ozone. A notable change in the goodness of the fit, excluding data for 75 pphm ozone, is observed when antiozonants are present. In contrast, the difference is less pronounced for the unprotected compound. Therefore, it is recommended to use a maximum concentration of 50 pphm. The validity of extrapolation to ozone concentrations below the minimum measured concentration of 25 pphm should be addressed in further research, including the challenge of finding a method to artificially control and measure very low ozone concentrations in a laboratory setup over a long period of time without fluctuations. As a next step, sequential exposure to different ozone concentrations (as encountered in the natural environment) would improve understanding and studies on ozone ageing.

Acknowledgements Parts of this research paper are funded by dtec.bw—Digitalization and Technology Research Center of the Bundeswehr which we gratefully acknowledge [project FLAB3Dprint].

References

Andrews EH, Braden M (1961) The reaction of ozone with surfaces of natural rubber, and its dependence upon strain. J Polym Sci 55(162):787–798

Bhala MJ (1997) Non-migratory antiozonant system for natural rubber. PhD thesis, Loughborough University

Braden M, Gent A (1962) The attack of ozone on stretched rubber vulcanizates. III. Actions of antiozonants. J Appl Polym Sci 6(22):449–455

Buckley DJ, Robison SB (1956) Ozone attack on rubber vulcanizates. J Polym Sci

Bundesumweltamt: Aktuelle Luftdaten. Accessed 02 Apr 2024 (2023). www.umweltbundesamt.de/daten/luft/luftdaten/karten

Cataldo F (2019) Protection mechanism of rubbers from ozone attack. Ozone Sci Eng 41(4):358–368

Criegee R (1975) Mechanism of ozonolysis. Angew Chem Int Ed 14(11):745–752

Crutzen PJ, Lawrence MG, Pöschl U (1998) On the background photochemistry of tropospheric ozone. Tellus B 51(1):123–146

De SK, White JR, Naskar K (2001) Rubber technologist's handbook, vol 2. Rapra Technology Limited

Dolezel B (1978) Die Beständigkeit von Kunststoffen und Gummi. Carl Hanser Verlag

Früh A, Egelhaaf HJ, Hintz H (2018) PMMA as an effective protection layer againt the oxidation of P3HT and MDMO-PPV by ozone. J Mater Res 33(13):1891–1901

Gent AN (2012) Engineering with rubber. Carl Hanser Verlag, München

Gillen KT, Bernstein R, Celina M (2016) Die Herausforderungen beschleunigter Alterungsverfahren für die Vorhersage der Elastomerlebensdauer - Teil 1. Gummi Fasern Kunstst

Gu HS, Itoh Y (2010) Ageing behaviour of natural rubber and high damping rubber materials used in bridge rubber bearings. Adv Struct Eng 13(6):1105–1113. https://doi.org/10.1260/1369-4332.13.6.1105

Huntink NM, Datta RN, Noordermeer JWM (2004) Addressing durability of rubber compounds. Rubber Chem Technol 77(3):476–511

Iwase Y, Shindo T, Kondo H, Ohtake Y, Kawahara S (2017) Ozone degradation of vulcanized isoprene rubber as a function of humidity. Polym Degrad Stabil 142:209–216

Kamaruddin S, Muhr AH (2018) Investigation of ozone cracking on natural rubber. J Rubber Res 21(2):73–93

Lake GJ, Mente PG (1992) Ozone cracking and protection of elastomers at high and low temperatures. J Nat Rubb Res 7(1):1–13

Lemaire J, Gardette Jl, Lacoste J (1993) Use of micro-ftir spectrophotometry in the determination of polymer photoageing mechanisms. Makromol Chem Macromol Symp 431:419–431

Lewis PM (1986) Effect of ozone on rubbers: countermeasures and unsolved problems. Polym Degrad Stab 15(1):33–66

Lindley PB (1973) Relation between hysteresis and the dynamic crack growth resistance of natural rubber. Int J Fract 9(4):449–462. https://doi.org/10.1007/bf00036325

Mars WV, Fatemi A (2005) Multiaxial fatigue of rubber: part II: experimental observations and life predictions. Fatigue Fract Eng Mater Struct 28(6):523–538

Murphy JS, Orr JR (eds) (1975) Ozone chemistry and technology: a review of literature 1961–1974. The Franklin Insitute Press, Philadelphia

Newton RG (1945) Mechanism of exposure-cracking of rubbers. with a review of the influence of ozone. Rubber Chem Technol 18(3):504–556

Nonnenmacher K (2008) Dynamic rubber test by ozone. KGK. Kautschuk, Gummi, Kunststoffe 61(5):230–232

Radhakrishnan CK, Alex R, Unnikrishnan G (2006) Thermal, ozone and gamma ageing of styrene butadiene rubber and poly (ethylene-co-vinyl acetate) blends. Polym Degrad Stabil 91(4):902–910. https://doi.org/10.1016/j.polymdegradstab.2005.06.013

Salisbury G, Rickard AR, Monks PS, Allan BJ, Bauguitte S, Penkett SA, Carslaw N, Lewis AC, Creasey DJ, Heard DE, Jacobs PJ, Lee JD (2001) Production of peroxy radicals at night via reactions of ozone and the nitrate radical in the marine boundary layer. J Geophys Res 106(D12):12669–12687. https://doi.org/10.1029/2000JD900754

da Silveira Petruci JF, Barreto DN, Dias MA, Felix EP, Cardoso AA (2022) Analytical methods applied for ozone gas detection: a review. TrAC Trends Anal Chem 149:1–9. https://doi.org/10.1016/j.trac.2022.116552

Spreckels J (2013) Ein Beitrag zur Berücksichtigung von Alterungs- und Umwelteinflüssen bei der Lebensdauerprüfung von Elastomerbauteilen. PhD thesis, TU Hamburg-Harburg

Treib C, Johlitz M (2022) Ozone ageing of polydimethylsiloxane coated natural rubber. In: Consitutive models for rubber XII. CNC Press Taylor and Francis Group, pp 408–414

Treib C, Johlitz M (2024) Evolution of crack depth in natural rubber due to ozone loading. Manuscript submitted for publication

Treib C, Johlitz M, Lion A (2024) Crack evaluation of ozone-aged elastomers. In: Lectures notes on advanced structured materials 2, chap. Springer, Cham

Treib C, Lion A, Johlitz M (2022) FT-IR microscopic analysis of the chemical change in the molecular structure after ozone ageing of natural rubber. In: Lectures notes on advanced structured materials. Springer, pp 241–252

United States Environmental Protection Agency: Ground-level Ozone Basics [Webpage] (2021). https://www.epa.gov/ground-level-ozone-pollution/ground-level-ozone-basics

Verdu J (2012) Oxidative ageing of polymers. John Wiley & Sons

Zuev YS, Pradvednikova SI (1962) Influence of concentration of ozone upon cracking of vulcanized rubber. Rubber Chemistry and Technology

Chapter 16
A Model for Thermal Ageing of Elastomers: Parameter Identification Using Continuous Relaxation and Intermittent Uniaxial Tests

Ondrej Farkas, Michael Johlitz, and Alexander Lion

Abstract In order to design and manufacture components that can reliably withstand loading in different environments, the material properties of the components and their changes under the respective environmental influences must be analysed in detail. This study uses the acquired experimental data to determine the chemical changes in the material structure with the focus on thermal ageing of elastomers. The mechanisms of thermal ageing are briefly described, and based on them, a very simple approach to model the ageing behaviour is introduced and the model parameters are identified using commercially available optimisation algorithms. Afterwards, the numerical results are compared with the measured data and the model performance is discussed along with the possible model weak points and their improvements.

16.1 Motivation

The products such as tyres, sealings, or fuel supply lines or even pencil erasers are just a few examples of elastomers that can be stumbled upon in everyday life. If these are exposed to different environments, for instance high temperature fields, fuels, ozone, or salt for extended time periods, the structure of the material can change, and the change is, unfortunately, negative. The material properties are deteriorating—the material ages. Ageing can compromise the integrity of the material, which can lead to premature component failure, and therefore unnecessary costs for repairs and replacements. In essence, understanding how elastomers age helps to predict and

O. Farkas (✉) · M. Johlitz · A. Lion
Institute of Mechanics, University of the Bundeswehr Munich, Werner-Heisenberg-Weg 39, 85579 Neubiberg, Germany
e-mail: ondrej.farkas@unibw.de

M. Johlitz
e-mail: michael.johlitz@unibw.de

A. Lion
e-mail: alexander.lion@unibw.de

H. Altenbach et al. (eds.), *Lectures Notes on Advanced Structured Materials 3*,
Advanced Structured Materials 226, https://doi.org/10.1007/978-3-031-77403-4_16

improve the product lifespan and reduces the total environmental impact while also reducing the costs by scheduling the maintenance intervals with respect to estimated component service life.

Due to the reasons stated above, this contribution focuses on the investigation of one of the above-mentioned effects of ageing, namely thermal ageing caused by the exposure of components to higher temperatures. The thermal ageing is first described and divided into two processes, which are represented with a suitable mathematical model. The model parameters are then identified using particle swarm optimisation algorithm so that the numerical stresses match the experimental data as precisely as possible. Lastly, the results of the identification process and the model performance are discussed along with the limitations and drawbacks of the proposed ageing model.

16.2 Thermal Ageing Mechanisms and Modelling Approach

The thermal ageing of elastomers causes permanent change in the chemical structure of the material and can be characterised with the assumption of two main processes. The first one is the network degeneration, where the original material structure degrades due to chain scission or breakage of chemical bonds—the current material network breaks down due to increased ambient temperature or solvent presence. The second process, namely the network regeneration, focuses on the newly originated connections that make the material stiffer. These processes are visualised in Fig. 16.1. Network degeneration and regeneration can overlap, and either one of them can be dominant throughout any moment of the observation depending on the ambient conditions. The internal variables representing these ageing processes are directly linked

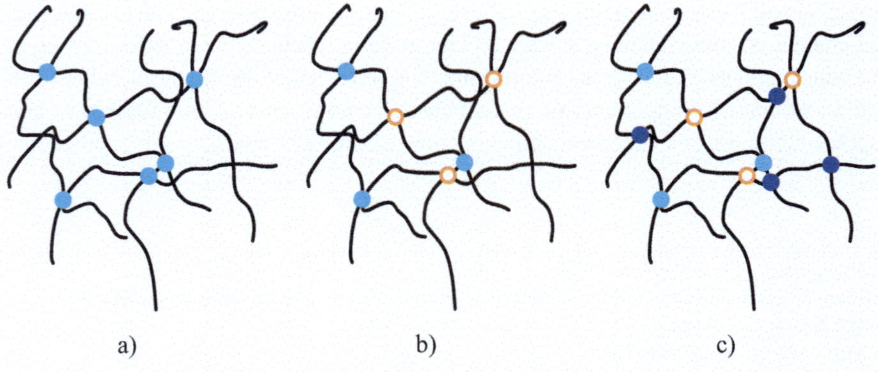

Fig. 16.1 Changes in material network (adapted from Ehrenstein and Pongratz (2007)): **a** Network of an unaged elastomer crosslinked with sulphur vulcanization (light blue dots). **b** Degeneration of the connections between the monomers (orange circles). **c** Origin of the new network connections (dark blue dots)

to changes in the chemical structure of the material—if these are properly characterised, the proposed model, while very simple, can reliably represent the material behaviour at any desired ageing state. The chemical ageing is further described in the publication issued by Johlitz (2012).

To determine the properties of both networks, the total stress P present in the material has to be additively decomposed into degenerative P^- and regenerative P^+ contribution:

$$P = P^- + P^+. \tag{16.1}$$

Following the argumentation of Lion and Johlitz (2012), only the changes in base elasticity of the material induced by ageing are to be modelled as the viscoelastic behaviour typically acts on much shorter time scales and is less pronounced. In case of interest in the influence of ageing on the viscoelastic material properties, one is referred to the work of Rabanizada et al. (2015) or Petríková et al. (2012). Assuming a linear dependence of stress P on strain ε, the degenerative stress contribution P^- is defined as follows:

$$P^- = E_1(q_d)\,\varepsilon, \quad \text{where} \quad E_1(q_d) = E_1(1 - q_d) \tag{16.2}$$

and the contribution of the newly formed network is defined using the hypoelastic (strain rate-dependent) approach

$$\dot{P}^+ = E_2(q_r)\,\dot{\varepsilon}, \quad \text{where} \quad E_2(q_r) = E_2 q_r. \tag{16.3}$$

The E_i represents Young's moduli of the degeneration and regeneration of the material network, which are then scaled with the internal variables q_d and q_r. A hypoelastic modelling approach is chosen because of the fact that the new network is created in a stress-free state; therefore, any dependence on strain would increase the measured and the simulated stress response—this is, however, false. Moreover, since the ageing of the material can be accelerated by increasing its temperature, the equations above suggest that the internal ageing variables $q_{d,r}$ are dependent not only on the time t but on the temperature θ as well. The evolution equation is for simple cases of ageing defined as

$$\dot{q}_{d,r} = \nu_{d,r}\, e^{-\frac{E_{d,r}}{R\theta}}\,(1 - q_{d,r}), \tag{16.4}$$

whereas its solution leads to the following:

$$q_{d,r}(t,\theta) = 1 - e^{-\frac{t}{\tau_{d,r}(\theta)}}, \quad \text{with} \quad \tau_{d,r}(\theta) = \left(\nu_{d,r}\, e^{-\frac{E_{d,r}}{R\theta}}\right)^{-1}. \tag{16.5}$$

Here, the $\tau(\theta)$ is a temperature-dependent chemical relaxation time correlating with the ageing-dependent decrease of stress. It can be represented with an Arrhenius equation (Arrhenius 1889) containing the parameters R, $E_{d,r}$, and $\nu_{d,r}$. The

$R = 8.314\,\mathrm{J/molK}$ represents the universal gas constant, whereas the $E_{d,r}$ and $\nu_{d,r}$ are the activation energies and the rates of the chemical reactions responsible for the corresponding degradation and reformation processes in the material structure, respectively. In case of the ageing behaviour being a bit more complicated, more sets of the exponential functions could be added together. The internal variables $q_{d,r}$ have to be restricted to maintain the thermodynamical consistency of this modelling approach; hence, the restrictions for the degenerative function are defined as

$$\dot{q}_d \geq 0, \quad 0 \leq q_d \leq 1, \quad \frac{\partial E_1(q_d)}{\partial q_d} \leq 0, \tag{16.6}$$

along with the limitations for the regenerative function represented similarly as

$$\dot{q}_r \geq 0, \quad 0 \leq q_r \leq 1, \quad \frac{\partial E_2(q_r)}{\partial q_r} \geq 0. \tag{16.7}$$

Considering these assumptions, a virgin material follows the relation $P = E_1\,\varepsilon$ ($q_d = 0, q_r = 0$), and a material that has completely aged can be represented using the relation $P = E_2\,\varepsilon$ ($q_d = 1, q_r = 1$). Additionally, the internal ageing variables can be strain (stretch) dependent although (Tobolsky et al. 1944) has proved that the deformation dependence can be omitted up until strains of 200%.

Since the material to be described is an elastomer, a filled natural rubber to be precise, an elastic approach is no longer sufficient to represent the material under large deformations. For this very reason, a hyperelastic approach is considered instead. To showcase this problem, one of the simplest models representing the nonlinear stress-strain behaviour is selected—a temperature-independent two-parameter Mooney-Rivlin model (MR2). The only change in comparison with the linear model is that instead of two different Young's moduli $E_{1,2}$, the used hyperelastic model will have two different sets of model parameters C_{10}, C_{01} and D_{10}, D_{01}, which will be, of course, dependent on the ageing state of the material. The stresses for the case of uniaxial tension are defined for the degeneration and the regeneration networks as follows:

$$P_{11}^{-}|_{q_d=0} = C_{10}\left(\lambda - \frac{1}{\lambda^2}\right) + C_{01}\left(1 - \frac{1}{\lambda^3}\right), \tag{16.8}$$

$$P_{11}^{+}|_{q_r=1} = D_{10}\left(\lambda - \frac{1}{\lambda^2}\right) + D_{01}\left(1 - \frac{1}{\lambda^3}\right). \tag{16.9}$$

The details on the derivation of the formulas above can be found in the publication issued by Holzapfel (2000). To formulate the total stress response P as a function of internal ageing variables, the partial stresses are scaled with the ageing functions and then simply added together

$$P_{11}(\lambda, t, \theta) = (1 - q_d(t,\theta))\; P_{11}^{-}(\lambda) + q_r(t,\theta)\; P_{11}^{+}(\lambda). \tag{16.10}$$

The equation above now contains the complete information about the ageing state of the material and is therefore suitable to represent it for any deformation, ageing temperature, and ageing time.

16.3 Experimental Data

The material examined in this study is natural rubber filled with 60 phr of carbon black. It was used to produce samples of prismatic shape with a cross-section of 2.5 mm x 5 mm at a length of 50 mm, which were then tested to provide experimental data for the ageing model calibration. The two sets of data that are absolutely necessary to characterise the ageing behaviour of elastomers are:

(1) Continuous relaxation tests
The continuous relaxation tests essentially describe the time-dependent decrease of the measured stress P_{11} due to the degeneration of the original polymer network in the samples with respect to different deformation states λ and, of course, different ageing temperatures θ. The samples described above were observed at ageing temperatures of 60 °C, 80 °C, and 100 °C under uniaxial deformation of 20, 35, and 50% of their initial length. The continuous relaxation measurements were stopped if the ageing time exceeded 550 h, which is roughly 23 days. The results are shown in Fig. 16.2. As one can see, the ageing temperature heavily reduces the time necessary for material degeneration to take place. On the other hand, the sample deformation λ does not seem to influence the ageing of the material in any way—this fact is in an agreement with the previously mentioned results of Tobolsky et al. (1944).

(2) Intermittent monotonic tests
The intermittent monotonic tests are used to determine the properties of the newly formed material network and the rate of its formation. Prior to monotonic tests, the rubber samples shall be aged at elevated temperatures in a stress-free state. In this case, the samples were aged at the same ageing temperatures θ as in continuous relaxation measurements for different ageing times t. It is important to note that the ageing rate is different at every temperature as the ageing processes are heavily temperature-dependent—the general tendency is that lower ageing temperatures θ require longer ageing times t to achieve a somewhat significant difference in the measured stress, which is why the ageing times are not consistent across the chosen ageing temperatures. The results depicted in Fig. 16.3 show a clear stiffening of the material with respect to the increasing ageing time regardless of the ageing temperature.

Ideally, one could perform the multi-step relaxation test of the unaged sample to properly characterise the base elasticity of the material, respectively, of the degeneration network. If this test is not available, the base elasticity can be included as an optimisation parameter in the fitting procedure. The data presented in this work were taken from the work of Johlitz et al. (2014).

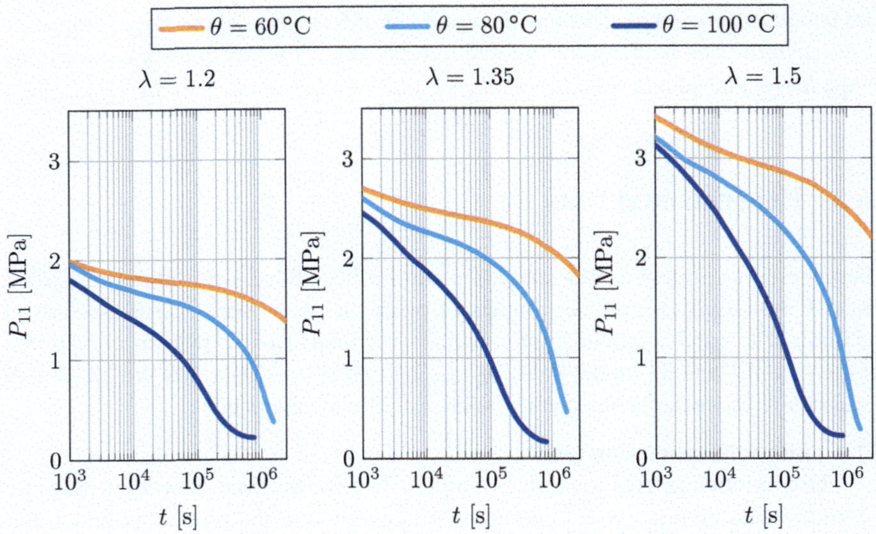

Fig. 16.2 Continuous chemical relaxation tests at different sample deformation states λ and ageing temperatures θ

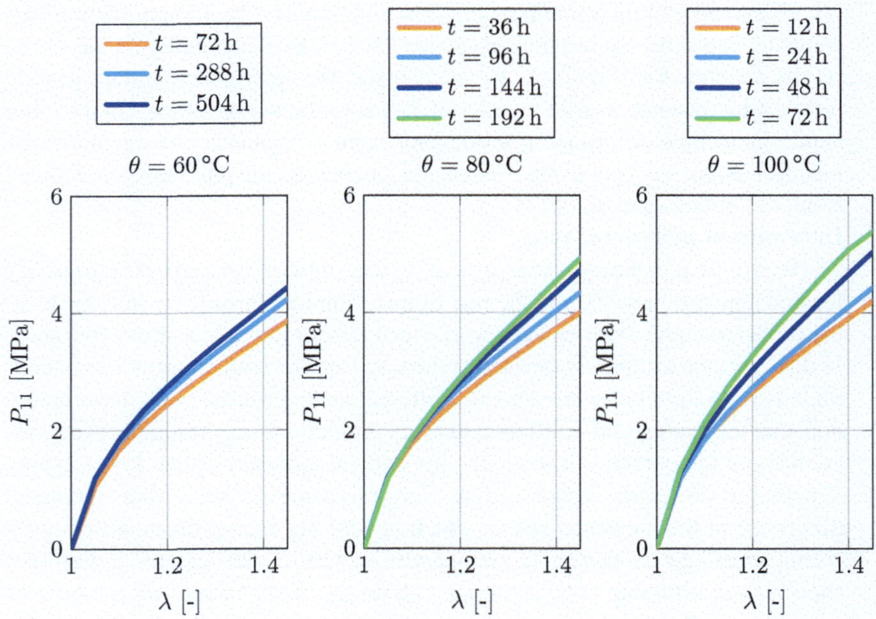

Fig. 16.3 Intermittent monotonic tests of samples aged at different ageing temperatures θ for various ageing times t

16.4 Model Fitting and Performance

The experimental data introduced in the previous section will now be used to identify the parameters of the ageing model. Since the usual approach is the variation of the model parameters followed by the evaluation of the difference between the numerical and measured stresses, the fitting procedure is categorised as an optimisation problem. In general, the goal is to find a parameter set for which the value of the objective function is minimum.

Particle Swarm Optimisation (PSO) Algorithm

In this case, both the base elasticity parameters and the parameters related to the degeneration and the regeneration functions were identified using particle swarm optimisation algorithm. The decision-making responsible for the algorithm progress is motivated by the behaviour of the bird flocks—once a single individual finds the source of food, the entire flock will likely follow him shortly after. A simple procedure capitalising on this observation was programmed by Kennedy and Eberhart (1995) and Eberhart and Kennedy (1995)—a flock of birds, respectively agents, randomly distributed in a planar space with the goal of reaching a fixed predefined position, with the velocity of every individual being the key parameter for adjusting the behaviour of the flock. The minimal evaluated distance was then stored for every bird in every movement step, and its movement and velocity in subsequent iterations were evaluated and modified. An example of such a velocity-based decision-making process is shown in the figure below. The objective functions in all of the optimisation steps are rather simple—the measured data is compared with the respective model part and the difference between them is minimised (Fig. 16.4).

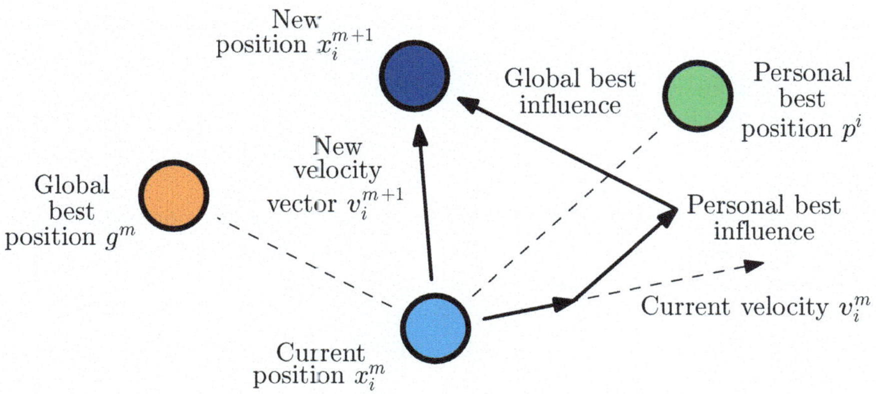

Fig. 16.4 Example of velocity adjustment in the flock behaviour (Farkas 2022)

Optimisation Process and Results

The hypoelastic definition of stress in regeneration network is of great benefit when it comes to the definition of the optimisation process. Since the regeneration network is created in a stress-free state and the strain at the chemical relaxation tests is constant (strain rate is equal to 0), the regeneration network does not contribute to the measured stress. This assumption leads to the fact that the stress measured in chemical relaxation tests is equal to the stress present in the degeneration network—the two ageing processes are thus separated. The first parameter to be identified is the chemical relaxation time of the degeneration network. As there is no data available for the base elasticity of the material, the equilibrium stress is defined as the second optimisation parameter. The objective function uses a simplified version of Eq. (16.10) defined above

$$P_{11}(\lambda, t, \theta) = P_{11}^{-}(\lambda)\, e^{-\frac{t}{\tau_d(\theta)}} + res. \tag{16.11}$$

The left side of this equation represents the results of the chemical relaxation tests, while the right side consists of the model stress and the residual value res. The smaller this value is, the smaller is the deviation between the measured and simulated stresses. Considering the performed measurements, there are in total six parameters related to the degeneration network—chemical relaxation times τ_d, one for each ageing temperature θ, and three equilibrium stresses P_{11}^{-}, one for each deformation state λ. The optimisation algorithm minimises the residual value res until it finds the presumed global minimum of the objective function, thereby identifying the first set of parameters. These are then directly used in the next step of the optimisation process—the equilibrium stresses of the degeneration network are represented with the two-parameter Mooney-Rivlin model and the chemical relaxation times with the Arrhenius function. The results of the optimisation process are presented in Fig. 16.5.

The stress in the degeneration network is now characterised with only four parameters, namely the Mooney-Rivlin coefficients C_{10} and C_{01} and the chemical reaction characteristics ν_d and E_d. These are then used to calculate the stress in the degeneration network at different ageing temperatures over the ageing time, and the results are depicted in Fig. 16.6.

It is important to note that the degeneration process parameters were calibrated only for the stresses measured at $t \geq 2 * 10^4$ s to exclude as much of an influence of viscoelastic effects as possible.

The parameters describing the evolution of stress in the regeneration network are identified in a similar manner. The stress measured in intermittent monotonic tests is compared with the total stress equation defined in Sect. 16.2

$$P_{11}(\lambda, t, \theta) = (1 - q_d(t, \theta))\, P_{11}^{-}(\lambda) + (1 - e^{-\frac{t}{\tau_r(\theta)}})\, P_{11}^{+}(\lambda) + res. \tag{16.12}$$

Given that the degeneration process parameters are already known, they are used to calculate the degeneration network contribution to the total stress, leaving the

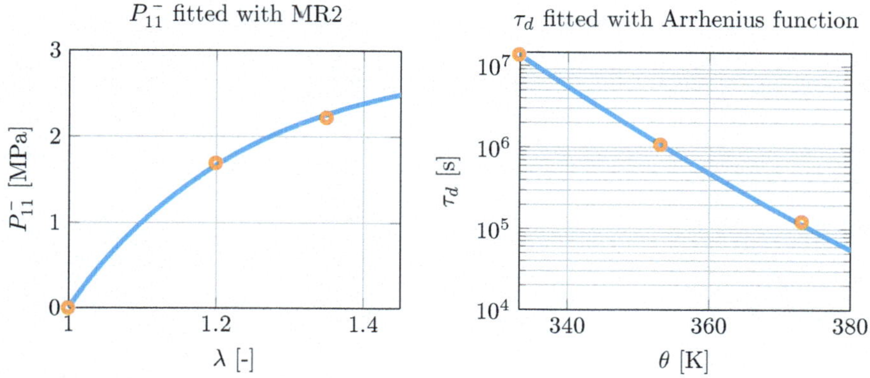

Fig. 16.5 Equilibrium stresses of the degeneration network fitted with the two-parameter Mooney-Rivlin model (left) and chemical relaxation times of the degeneration process represented with the Arrhenius function (right)

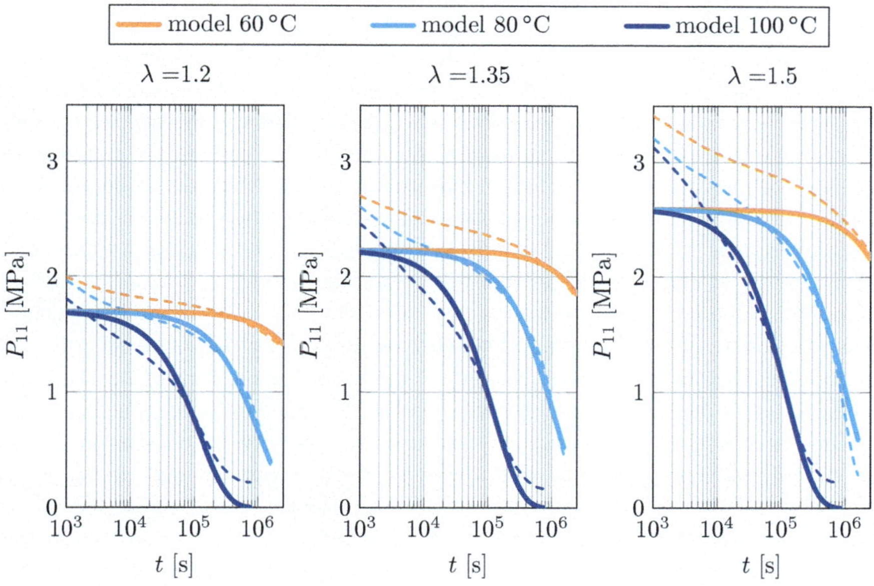

Fig. 16.6 Simulation of the degeneration process for different deformation levels

parameters of the regeneration network to be optimised by the PSO algorithm which reduces the *res* to a minimum. The number of parameters in this optimisation procedure rises to twelve—one chemical relaxation time τ_r for each ageing temperature θ and one equilibrium stress P_{11}^+ for each deformation λ. Since there should be no stress response without any material deformation, the condition $P_{11}^+(\lambda = 1) = 0$ must hold regardless of the ageing state of the material; therefore, the equilibrium stress at zero deformation does not need to be optimised. Analogous to the case of

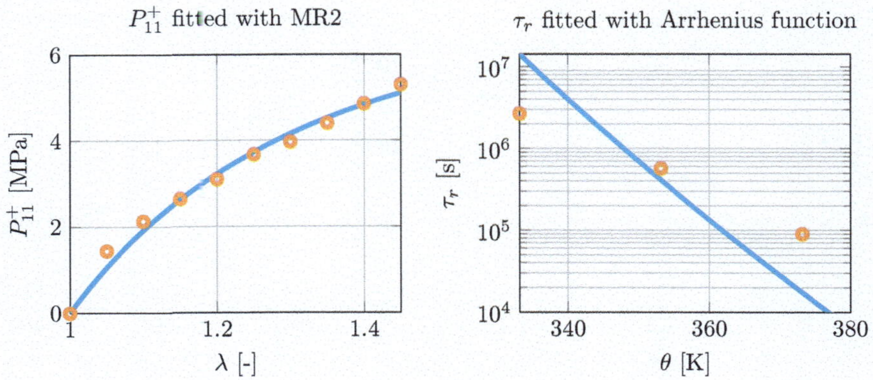

Fig. 16.7 Equilibrium stresses of the regeneration network fitted with two-parameter Mooney-Rivlin model (left) and chemical relaxation times of the regeneration process represented with the Arrhenius function (right)

degeneration network, the optimisation algorithm identifies the equilibrium stresses and the relaxation times, which are then represented with the Mooney-Rivlin model and Arrhenius function, as shown in Fig. 16.7. As one can see, the Mooney-Rivlin hyperelastic function with the parameters D_{10} and D_{01} represents the equilibrium stresses with a satisfactory precision. Unfortunately, the Arrhenius equation does not describe the chemical relaxation times of the regeneration network quite well. It seems to lack the necessary flexibility to approximate the relaxation times well enough if the realistic parameter constraints are considered. This issue is further addressed in the conclusion of this contribution. The relaxation times τ_r identified in the first phase of the regeneration network optimisation process can nevertheless be used together with the base elasticity of the regeneration network to show the performance of the selected ageing model and the fit results are shown in Fig. 16.8.

There are multiple possible reasons for the deviations between the measurements and the calibrated ageing model, one of which is the deformation rate—according to Johlitz et al. (2014), the intermittent tests were performed at a rather high deformation rate of $\dot{\varepsilon} = 0.1\%/\text{s}$. This means that that the viscoelasticity might have influenced the stress measured in the intermittent tests. One can also observe that the ageing model underestimates the stresses in lower ageing times regardless of the ageing temperature. Perhaps it is due to the fact that the degeneration network was calibrated for higher ageing times only. All in all, the ageing model shows a reasonable relation to the experimental data, taking into the account the overall simplicity of the degeneration and regeneration functions and temperature-independent base elasticity of the material.

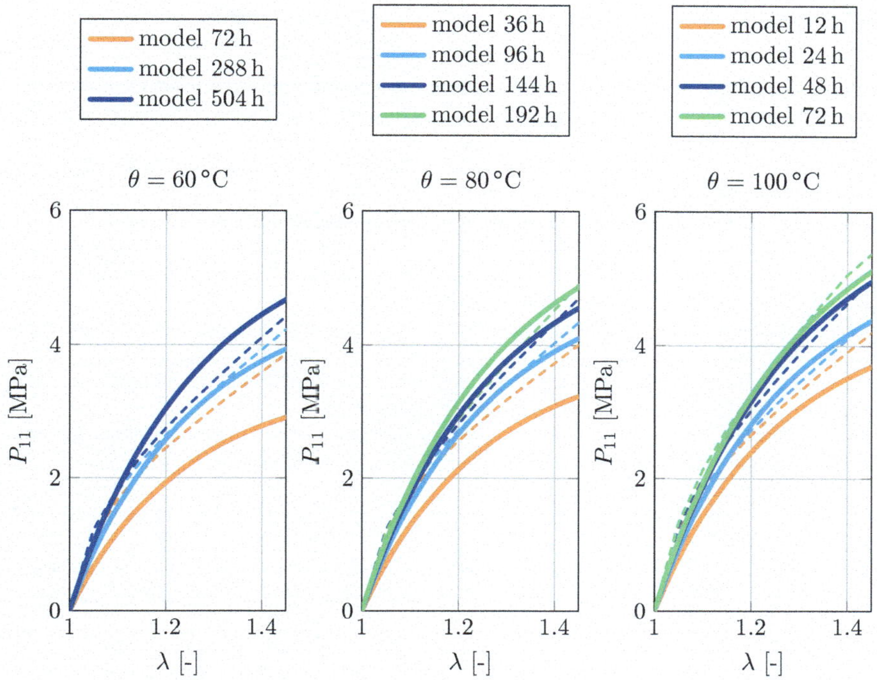

Fig. 16.8 Combined stress of the degeneration and regeneration network calculated using the calibrated ageing model

16.5 Conclusion and Outlook

The herepresented contribution investigates the problem of thermal ageing of elastomers and its very simple way of modelling. The experimental data were used to calibrate the ageing model which represents the degradation of the base elasticity of filled natural rubber as a function of deformation, ageing time, and lastly, ageing temperature.

Of course, the performance of the ageing model can be improved. As the first improvement option, a temperature-dependent hyperelastic model might improve the fit quality and the overall model accuracy in case of the base elasticity being strongly temperature-dependent. In case of being incompatible with the Arrhenius function, a modified Arrhenius function or other temperature-dependent functions could potentially be used to approximate the temperature dependence of chemical relaxation times. Last but not least, the ageing functions $q_{d,r}$ could somehow be connected to the deformation-induced crystallisation phenomena, if the natural rubber is to be investigated for very large deformations.

References

Arrhenius S (1889) Über die Dissociationswärme und den Einfluss der Temperatur auf den Dissociationsgrad der Elektrolyte. Z fur Phys Chem 4U(1):96–116. https://doi.org/10.1515/zpch-1889-0408. Accessed 18 June 2024

Eberhart R, Kennedy J (1995) A new optimizer using particle swarm theory, pp 39–43. https://doi.org/10.1109/MHS.1995.494215

Ehrenstein GW, Pongratz S (2007) Beständigkeit Von Kunststoffen. Hanser, München

Farkas O (2022) Characterisation and simulation of viscoelastic behaviour of NR/BR blend in various loading conditions. Master's thesis, Institute of Applied Mechanics and Mechatronics, Slovak University of Technology, Bratislava, Slovakia

Holzapfel Holzapfel GA (2000) Nonlinear solid mechanics: a continuum approach for engineering. Wiley, New York

Johlitz M (2012) On the representation of ageing phenomena. J Adhes 88(7):620–648. https://doi.org/10.1080/00218464.2012.682905. Accessed 26 Jan 2024

Johlitz M, Diercks N, Lion A (2014) Thermo-oxidative ageing of elastomers: a modelling approach based on a finite strain theory. Int J Plast 63:138–151. https://doi.org/10.1016/j.ijplas.2014.01.012. Accessed 26 Jan 2024

Kennedy J, Eberhart R (1995) Particle swarm optimization, pp 1942–1948. https://doi.org/10.1109/ICNN.1995.488968

Lion A, Johlitz M (2012) On the representation of chemical ageing of rubber in continuum mechanics. Int J Solids Struct 49(10):1227–1240. https://doi.org/10.1016/j.ijsolstr.2012.01.014. Accessed 29 Jan 2024

Petríková I, Marvalova B, Phan Thanh N (2012) Influence of thermal ageing on mechanical properties of styrene-butadiene rubber. In: Constitutive models for rubber VII—proceedings of the 7th European conference on constitutive models for rubber, ECCMR, pp 77–83. https://doi.org/10.1201/b11687-16

Rabanizada N, Lupberger F, Johlitz M, Lion A (2015) Experimental investigation of the dynamic mechanical behaviour of chemically aged elastomers. Arch Appl Mech 85(8):1011–1023. https://doi.org/10.1007/s00419-014-0971-6. Accessed 29 Jan 2024

Tobolsky AV, Prettyman IB, Dillon JH (1944) Stress relaxation of natural and synthetic rubber stocks. J Appl Phys 15(4):380–395. https://doi.org/10.1063/1.1707442. Accessed 05 Feb 2024

Chapter 17
A Constitutive Model of Finite Viscoelasticity for Polyamide 6 Considering Degree of Crystallinity

Sameer Kulkarni, Marie-Christine Reuvers, Tim Brepols, Stefanie Reese, Michael Johlitz, and Alexander Lion

Abstract In this study, the influence of the degree of crystallinity on the mechanical behaviour of semi-crystalline polyamide 6 is investigated. To describe the long-term and time-dependent mechanical response of polyamide 6, a phenomenological constitutive model of finite viscoelasticity is used. This model consists of a parallel arrangement of a nonlinear spring and multiple Maxwell elements, motivated by the 1D rheological generalized Maxwell model. To account for strain rate dependency, a nonlinear viscous evolution equation is proposed. This proposed evolution equation is incorporated into the viscoelasticity model in a thermodynamically consistent manner. On the experimental side, polyamide 6 was melt-mixed with amorphous cyclic olefin copolymer to produce blends representative of polyamide 6 with varying degrees of crystallinity (from 15% to 29%). Various isothermal tests, including monotonic tension, relaxation and cyclic tests at temperatures above the glass transition temperature, were performed on the blends to generate a material database. To obtain the viscoelastic behaviour over an extended period, the time–temperature superposition principle is used. The model parameters are identified through a staggered procedure using a nonlinear optimization method. Initial computations indicated

S. Kulkarni (✉) · M. Johlitz · A. Lion
Institute of Mechanics, University of the Bundeswehr Munich, Werner-Heisenberg-Weg 39, 85577 Neubiberg, Germany
e-mail: sameer.kulkarni@unibw.de

M. Johlitz
e-mail: michael.johlitz@unibw.de

A. Lion
e-mail: alexander.lion@unibw.de

M.-C. Reuvers · T. Brepols · S. Reese
Institute of Applied Mechanics, RWTH Aachen University, Mies-van-der-Rohe-Str. 1, 52074 Aachen, Germany
e-mail: marie.reuvers@ifam.rwth-aachen.de

T. Brepols
e-mail: tim.brepols@rwth-aachen.de

S. Reese
e-mail: stefanie.reese@ifam.rwth-aachen.de

H. Altenbach et al. (eds.), *Lectures Notes on Advanced Structured Materials 3*,
Advanced Structured Materials 226, https://doi.org/10.1007/978-3-031-77403-4_17

that the model could reliably predict the viscoelastic strain rate-dependent loading and relaxation behaviour across different crystallinity levels and temperatures.

Keywords Finite viscoelasticity · Rate dependency · Time–temperature superposition · Semi-crystalline polymer · Polyamide 6 · Degree of crystallinity · Digital image correlation

17.1 Introduction

Thermoplastics offer advantages such as low cost, low density and an excellent balance of mechanical properties. Replacing metals with thermoplastics allows for the redesign of complex parts consisting of assemblies of many components that could not be achieved with metals. As a result, thermoplastics are widely used today in the automotive, textile, packaging, electrical appliance and construction industries (Dorléans et al. 2021; Leaversuch 2004). Polyamide 6 (PA6), commonly known as nylon 6, is a semi-crystalline thermoplastic preferred for most automotive applications among other polyamides. Because of its high mechanical strength, thermal stability, impact resistance, chemical and abrasion resistance, corrosion resistance and insulation qualities, it is increasingly used in the glass fibre-reinforced form in automobiles (Dorléans et al. 2021; Whelan 1994; Qin and Ye 2015; Karak 2024). Its main applications in automobiles include the air intake manifold, the air and cooling systems, the throttle body housing, the cylinder head cover, the acoustic cover, the oil pump and the oil pan and mounting elements for the air duct cover on an oil pan to name a few (Qin and Ye 2015).

Semi-crystalline polymers consist of orderly packed chain sections known as crystals and disordered amorphous regions. All polymers including semi-crystalline polymers are amorphous in the melt state. Crystals are formed during cooling of the molten polymer (Applied Plastics Engineering Handbook 2016). The resulting degree of crystallinity (DOC) can vary depending on cooling conditions like temperature, pressure, cooling rate and the specific orientation of polymer chains, which differ based on the processing technique employed, such as stretching, extrusion or moulding. Additionally, the DOC can be influenced by the presence of nucleating agents, additives or fillers. Investigating the influence of the DOC on the mechanical properties of polymers is crucial for understanding and optimizing the performance of polymer-based materials in a wide range of applications. The DOC directly impacts the material's stiffness, strength, hardening behaviour and toughness (Odrobina et al. 2020; Batista et al. 2016; Galeski 2003; Parenteau et al. 2012). A higher crystallinity generally leads to an increase in stiffness and strength, with reduced toughness and reduced impact resistance. Hence, investigating mechanical behaviour as a function of the DOC has been a topic of interest.

This research aims to characterize and model the mechanical response of PA6 in its solid state above its glass transition temperature (T_g), with a specific focus on simulating the thermoforming process. Thermoforming involves heating a polymer sheet to a pliable forming temperature, shaping it to a mould and simultaneously contact cooling with preheated moulds to retain the shape. Characterizing the viscoelastic behaviour of the material, including strain rate dependency and relaxation, is essential for accurately predicting the residual stresses in the components. One important aspect, thereby, is the consideration of distinct DOCs that evolve in the material depending on the local cooling rates in the laminate. Previous studies (Kulkarni et al. 2022, 2024; Reuvers et al. 2023a, b, 2024a, b) conducted as part of this project have explored different facets of modelling thermo-mechanical PA6 behaviour.

The behaviour of semi-crystalline polymers under finite strains is inherently complex, as it involves a combination of viscoelastic and viscoplastic contributions associated to various deformation mechanisms. Considerable advances have been made in the development of phenomenological models to explain these behaviours. Noteworthy contributions include finite viscoplastic formulations by Barba et al. (2020), Abdul-Hameed et al. (2014) and Drozdov and Gupta (2003). Additionally, some studies focus on small strain viscoelastic formulations (Qi et al. 2019; Lai et al. 2005), while others address small strain viscoelastic–elastoplastic formulations (Popa et al. 2014; Kästner et al. 2012). Few modelling strategies account for the influence of crystallinity on mechanical responses at finite strains using the rule of mixtures. Examples of such viscoplastic formulations can be found in the works of Felder et al. (2020), Ayoub et al. (2011) and Dusunceli and Colak (2008). Additionally, Makradi et al. (2008) proposed a two-phase self-consistent constitutive model that combines elastic–viscoplastic behaviour with microstructure evolution.

In this study, a material framework is developed from a nonlinear viscoelastic standpoint, with the goal of minimizing the number of model parameters and simplifying the identification process by excluding plasticity effects. Over the past few decades, various constitutive frameworks have been proposed based on the multiplicative decomposition of the deformation gradient to model finite strain viscoelasticity, particularly for elastomeric materials (Amin et al. 2006; Dippel et al. 2015; Reese and Govindjee 1997; Reese and Govindjee 1998; Johlitz et al. 2016; Lion 2000; Haupt et al. 2000). The framework employed in this work extends the concept of the Zener model, which incorporates multiple Maxwell elements in parallel, to three-dimensional finite deformations. In this framework, the deformation gradient in each Maxwell-like element is split multiplicatively into elastic and inelastic parts. A primary benefit of this approach is that each decomposition operates independently, which simplifies the extension of the material model by adding additional Maxwell elements (Khan and Farrokh 2006).

Modelling of viscoelasticity-induced strain rate dependency during loading and unloading is normally crucial for the comprehensive constitutive characterization of the material behaviour. The constitutive models of finite viscoelasticity, which rely on the introduction of internal variables, include an evolution equation for the inelastic viscous deformation. A linear viscous flow rule is the simplest evolution equation (e.g., Amin et al. (2002), Johlitz et al. (2007)). However, experimental findings indicate that viscosity under finite strains is also influenced by the applied stress (e.g., Amin et al. (2006)), strain (e.g., Amin et al. (2006), Hoo Fatt and Ouyang (2008)) and strain rate (e.g., Khan and Farrokh (2006), Shim and Mohr (2011)).

Consequently, nonlinear viscous flow rules have been formulated in the literature (e.g., Khan and Zhang (2001), Colak (2005), Amin et al. (2006), Khan and Farrokh (2006), Hoo Fatt and Ouyang (2008), Shim and Mohr (2011)).

In this work, to investigate the influence of DOC, samples with distinct crystallinity levels were produced. The production of polymer samples with varied DOC has historically posed challenges due to the complexities of controlling crystallinity levels, which are primarily influenced by local cooling rates. Achieving different DOC levels has relied on methods such as annealing above T_g (Parenteau et al. 2012; Humbert et al. 2009; Schrauwen et al. 2004; Lee et al. 1995; Felder et al. 2020; Yu et al. 2008; Chen et al. 2013) or controlling cooling rates during solidification (Odrobina et al. 2020; Parodi et al. 2018). However, annealing does not always yield significant variations in crystallinity, particularly in polymers like PA6. Fast-cooled polymers tend to exhibit lower crystallinity levels, yet this crystallinity is often unstable, as evidenced by the experiments detailed in Mileva et al. (2012). Heating the polymer beyond its T_g initiates recrystallization, leading to an increase in overall crystallinity.

In this work, a newer concept has been used to produce polymer samples with varied DOC, which involved the melt-blending of PA6 with amorphous cyclic olefin copolymer (COC) to modify the crystallinity of PA6. The equilibrium crystallinity typically regulates the plateau modulus between the T_g and the melting temperature (T_m). By systematically adjusting the blend composition, the plateau moduli were tuned to produce blends exhibiting mechanical properties similar to those of PA6 with varying DOCs. This method was used by Mather et al. (2003) to produce blends of semi-crystalline polylactic acid (PLA) and amorphous polyvinyl alcohol (PVA) for shape memory applications. In their study, the blends were miscible, meaning that the mixing occurred at a microscopic level, resulting in a homogeneous mixture of the two components, resulting in a single glass transition or melting peak. In this work, on the other hand, the blends of PA6 and COC were immiscible. Typically, such immiscible blends exhibit two glass transitions, or melting peaks, at the respective T_g of the blend constituents. To mitigate this, amorphous COC with a T_g of around $65\,^\circ$C, close to PA6's T_g of about $60\,^\circ$C, was chosen to prevent the occurrence of two T_g's. Given that different crystallinities are not obtained through rapid cooling in this scenario, the resulting crystalline structure in the material is significantly more stable and expected to remain unaltered during testing above T_g and below T_m. The thermal behaviour of the blends is expected to vary depending on their constituents; however, the blends are exclusively intended for evaluating the mechanical properties.

In this study, a thermodynamically consistent phenomenological viscoelastic constitutive framework is developed to model finite strain material behaviour. The temperature and DOC serve as constant input quantities to this framework. Experimental tests, including monotonic tensile tests, relaxation tests and cyclic tests, were conducted (cf. Sect. 17.3). Time–temperature superposition was applied to derive master curves for the stress relaxation behaviour over an extended period of time, which were subsequently utilized in the parameter identification (cf. Sect. 17.4). The model parameters were identified through a staggered procedure using experimental data obtained from PA6-COC blends spanning DOC from 15% to 29% (cf. Sect. 17.5). Initial computations demonstrate that this new model effectively predicts the strain rate dependency and the relaxation behaviour across a broad spectrum of crystallinities and temperatures.

17.2 Material Model Formulation

The experimental findings imply, in general, a nonlinear viscoelastic behaviour. The modelling approach chosen is illustrated by a one-dimensional rheological network diagram (Fig. 17.1a). This diagram serves as a basis model to help understand the upcoming derivation. Isothermal conditions ($\dot{T} = 0$) and no evolution of the crystallinity ($\dot{X}_c = 0$) are assumed, such that X_c and T serve as constant inputs.

17.2.1 Kinematic Considerations

The deformation gradient $\mathbf{F}$ is split multiplicatively into a volumetric $\overline{\mathbf{F}}$ and an isochoric component $\hat{\mathbf{F}}$ with the aid of the determinant $\det(\mathbf{F}) = J$ (Flory 1961):

$$\mathbf{F} = \hat{\mathbf{F}} \cdot \overline{\mathbf{F}} \quad \text{with} \quad \hat{\mathbf{F}} = J^{-\frac{1}{3}}\mathbf{F} \quad \text{and} \quad \overline{\mathbf{F}} = J^{\frac{1}{3}}\mathbf{I}. \tag{17.1}$$

The isochoric component $\hat{\mathbf{F}}$ is further split into elastic and inelastic parts (Lubliner 1985), given by

$$\hat{\mathbf{F}} = \hat{\mathbf{F}}_{\mathrm{e}}^{k} \cdot \hat{\mathbf{F}}_{\mathrm{i}}^{k} \quad \text{for} \quad k = 1, 2, \ldots, n. \tag{17.2}$$

The overall kinematic considerations are illustrated in Fig. 17.1b.

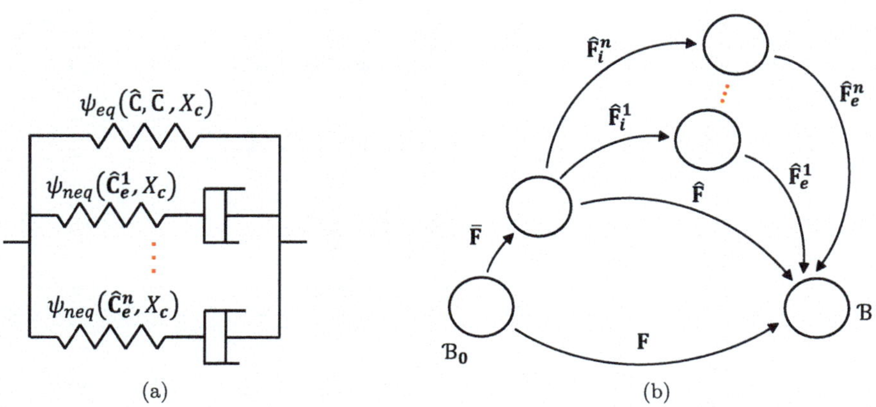

Fig. 17.1 **a** Schematic illustration of the constitutive model. **b** Multiplicative splits of the deformation gradient

17.2.2 Second Law of Thermodynamics: Clausius–Planck Inequality

The free energy density $\rho_0\psi$, serving as the state potential in the model, is formulated using physically meaningful deformation measures. For this purpose, the right Cauchy–Green tensor $\mathbf{C} = \mathbf{F}^{\mathrm{T}} \cdot \mathbf{F}$, the isochoric elastic part of the right Cauchy–Green tensors $\hat{\mathbf{C}}_{\mathrm{e}}^{1} = \hat{\mathbf{F}}_{\mathrm{e}}^{\mathrm{T},1} \cdot \hat{\mathbf{F}}_{\mathrm{e}}^{1}$, $\hat{\mathbf{C}}_{\mathrm{e}}^{2} = \hat{\mathbf{F}}_{\mathrm{e}}^{\mathrm{T},2} \cdot \hat{\mathbf{F}}_{\mathrm{e}}^{2}$, ..., $\hat{\mathbf{C}}_{\mathrm{e}}^{n} = \hat{\mathbf{F}}_{\mathrm{e}}^{\mathrm{T},\mathrm{n}} \cdot \hat{\mathbf{F}}_{\mathrm{e}}^{n}$ are defined as a set of process variables

$$\rho_0\psi = \rho_0\psi\left(\mathbf{C}, \hat{\mathbf{C}}_{\mathrm{e}}^{1}, \hat{\mathbf{C}}_{\mathrm{e}}^{2}, ..., \hat{\mathbf{C}}_{\mathrm{e}}^{n}\right). \tag{17.3}$$

To ensure non-negative internal dissipation, the model equations need to fulfil the second law of thermodynamics. The second law of thermodynamics under isothermal conditions, in the form of the Clausius–Planck inequality, with respect to the reference configuration, is given by

$$-\rho_0\dot{\psi} + \mathbf{S} : \dot{\mathbf{E}} = \rho_0\tilde{\delta} \geq 0. \tag{17.4}$$

The first term indicates the negative temporal change of the free energy density. ρ_0 represents the density associated with the reference configuration. $\mathbf{S} : \dot{\mathbf{E}}$ denotes the volume-related stress power, where $\dot{\mathbf{E}}$ is the time derivative of the Green–Lagrange strain tensor and $\mathbf{S}$ is the second Piola–Kirchhoff stress tensor. In this context, $\tilde{\delta}$ stands for the dissipation per unit mass.

It should be noted that, for a better approximation of the observed viscoelastic material behaviour, typically, k internal variables are required. However, for the sake of clarity, k will be set to 1 in the following derivation. The insertion of the time derivative of the free energy density (17.3) into the dissipation inequality (17.4) leads to

$$\mathbf{S} : \frac{1}{2}\dot{\mathbf{C}} - \rho_0\left(\frac{\partial\psi}{\partial\mathbf{C}} : \dot{\mathbf{C}} + \frac{\partial\psi}{\partial\hat{\mathbf{C}}_{\mathrm{e}}} : \dot{\hat{\mathbf{C}}}_{\mathrm{e}}\right) \geq 0. \tag{17.5}$$

The transformation of $\dot{\hat{\mathbf{C}}}_{\mathrm{e}}$ incorporating kinematic relations yields the following expression:

$$\begin{aligned}&\left[\mathbf{S} - 2\rho_0\left(\frac{\partial\psi}{\partial\mathbf{C}} + (\det\mathbf{C})^{-\frac{1}{3}}\hat{\mathbf{F}}_{\mathrm{i}} \cdot \frac{\partial\psi}{\partial\hat{\mathbf{C}}_{\mathrm{e}}} \cdot \hat{\mathbf{F}}_{\mathrm{i}}^{-\mathrm{T}}\right.\right.\\&\left.\left.-\frac{1}{3}\left(\hat{\mathbf{C}}_{\mathrm{e}}^{\mathrm{T}} \cdot \frac{\partial\psi}{\partial\hat{\mathbf{C}}_{\mathrm{e}}} : \mathbf{I}\right)\mathbf{C}^{-1}\right)\right] : \frac{1}{2}\dot{\mathbf{C}}\\&+2\rho_0\frac{\partial\psi}{\partial\hat{\mathbf{C}}_{\mathrm{e}}} \cdot \hat{\mathbf{C}}_{\mathrm{e}}^{\mathrm{T}} : \frac{1}{2}\left(\hat{\mathbf{L}}_{\mathrm{i}}^{\mathrm{T}} + \hat{\mathbf{L}}_{\mathrm{i}}\right) \geq 0.\end{aligned} \tag{17.6}$$

Here, $\hat{\mathbf{L}}_{\mathrm{i}} = \mathbf{F}_{\mathrm{i}}^{-1}\dot{\mathbf{F}}_{\mathrm{i}}$ represents the inelastic component of the spatial velocity gradient. The evaluation of the inequality according to Coleman and Noll (1963) leads to the constitutive equation

$$\mathbf{S} = 2\rho_0 \left(\frac{\partial \psi}{\partial \mathbf{C}} + (\det \mathbf{C})^{-\frac{1}{3}} \hat{\mathbf{F}}_\mathrm{i}^{-1} \cdot \frac{\partial \psi}{\partial \hat{\mathbf{C}}_\mathrm{e}} \cdot \hat{\mathbf{F}}_\mathrm{i}^{-\mathrm{T}} - \frac{1}{3} \left(\hat{\mathbf{C}}_\mathrm{e}^\mathrm{T} \cdot \frac{\partial \psi}{\partial \hat{\mathbf{C}}_\mathrm{e}} : \mathbf{I} \right) \mathbf{C}^{-1} \right). \tag{17.7}$$

The remaining residual inequality is expressed as

$$\rho_0 \tilde{\delta} = 2\rho_0 \frac{\partial \psi}{\partial \hat{\mathbf{C}}_\mathrm{e}} \cdot \hat{\mathbf{C}}_\mathrm{e}^\mathrm{T} : \frac{1}{2} \left(\hat{\mathbf{L}}_\mathrm{i}^\mathrm{T} + \hat{\mathbf{L}}_\mathrm{i} \right) \geq 0. \tag{17.8}$$

To satisfy the residual inequality $\rho_0 \tilde{\delta} \geq 0$, a proportionality relation is introduced, leading to the following evolution equation for the tensor $\hat{\mathbf{D}}_\mathrm{i}$ and, equivalently, for the inelastic right Cauchy–Green tensor $\hat{\mathbf{C}}_\mathrm{i}$

$$\begin{aligned} \hat{\mathbf{D}}_\mathrm{i} &= \frac{2}{\breve{\eta}} \frac{\partial \psi}{\partial \hat{\mathbf{C}}_\mathrm{e}} \cdot \hat{\mathbf{C}}_\mathrm{e}^\mathrm{T} \\ \Leftrightarrow \dot{\hat{\mathbf{C}}}_\mathrm{i} &= \hat{\mathbf{F}}_\mathrm{i}^\mathrm{T} \cdot \left[\frac{2}{\breve{\eta}} \frac{\partial \psi}{\partial \hat{\mathbf{C}}_\mathrm{e}} \cdot \left[\hat{\mathbf{C}}_\mathrm{e}^\mathrm{T} - \frac{1}{3} \operatorname{tr} \left(\hat{\mathbf{C}}_\mathrm{e}^\mathrm{T} \right) \mathbf{I} \right] \right] \cdot \hat{\mathbf{F}}_\mathrm{i}. \end{aligned} \tag{17.9}$$

Here, $\hat{\mathbf{D}}_\mathrm{i} = \frac{1}{2} \left(\hat{\mathbf{L}}_\mathrm{i} + \hat{\mathbf{L}}_\mathrm{i}^\mathrm{T} \right)$ denotes the rate of inelastic deformation tensor. Additionally, $\breve{\eta} > 0$ is interpreted as the viscosity function (proportional factor).

17.2.3 Specific Choice of Helmholtz Free Energy

The free energy density function $\rho_0 \psi$ is decomposed to capture the equilibrium and non-equilibrium contributions as follows:

$$\begin{aligned} \rho_0 \psi \left(J, I_{\hat{\mathbf{C}}}, I_{\hat{\mathbf{C}}_\mathrm{e}^1}, ..., I_{\hat{\mathbf{C}}_\mathrm{e}^n} \right) &= \rho_0 \psi_\mathrm{eq}^\mathrm{vol}(J) + \rho_0 \psi_\mathrm{eq}^\mathrm{iso} \left(I_{\hat{\mathbf{C}}} \right) \\ &+ \rho_0 \psi_\mathrm{neq} \left(I_{\hat{\mathbf{C}}_\mathrm{e}^1}, ..., I_{\hat{\mathbf{C}}_\mathrm{e}^n} \right). \end{aligned} \tag{17.10}$$

Here, J is the Jacobian determinant representing volumetric changes and $I_{\hat{\mathbf{C}}}$ is the first invariant of the isochoric right Cauchy–Green deformation tensor. The term $\rho_0 \psi_\mathrm{neq}$ accounts for non-equilibrium effects through invariants of the internal variables $\hat{\mathbf{C}}_\mathrm{e}^1, ..., \hat{\mathbf{C}}_\mathrm{e}^n$.
The equilibrium volumetric free energy density is defined as

$$\rho_0 \psi_\mathrm{eq}^\mathrm{vol}(J) = \frac{1}{2} \kappa \left[(J-1)^2 + (\ln J)^2 \right]. \tag{17.11}$$

The equilibrium isochoric free energy density is expressed as

$$\rho_0 \psi_\mathrm{eq}^\mathrm{iso} \left(I_{\hat{\mathbf{C}}} \right) = c_{10} \left(I_{\hat{\mathbf{C}}} - 3 \right). \tag{17.12}$$

The non-equilibrium isochoric free energy density is given by

$$\rho_0 \psi_{\text{neq}}^{\text{iso}} \left(I_{\hat{\mathbf{C}}_{\text{e}}^1}, ..., I_{\hat{\mathbf{C}}_{\text{e}}^n} \right) = \sum_{k=1}^{n} c_{e10}^n \left(I_{\hat{\mathbf{C}}_{\text{e}}^n} - 3 \right). \tag{17.13}$$

In the above equations, $c_{10}(X_c)$ and $c_{e10}(X_c)$ are the crystallinity-dependent shear moduli for the equilibrium and non-equilibrium contributions, respectively, while $\kappa(X_c)$ represents the crystallinity-dependent bulk modulus for the equilibrium contribution.

17.2.4 Calculation of Model Equations

Inserting the free energy densities (17.11)–(17.13) into the stress definition (17.7), the first Piola–Kirchoff stress is obtained as follows:

$$\mathbf{S} = \mathbf{S}_{\text{vol}} + \mathbf{S}_{\text{eq}} + \mathbf{S}_{\text{neq}} = \mathbf{S}_{\text{vol}} + \mathbf{S}_{\text{iso}}. \tag{17.14}$$

The volumetric part of the first Piola–Kirchoff stress is given by

$$\mathbf{S}_{\text{vol}} = \kappa \left[(J - 1) + \frac{\ln J}{J} \right] J \mathbf{C}^{-1}. \tag{17.15}$$

The equilibrium part of the first Piola–Kirchoff stress is defined as

$$\mathbf{S}_{\text{eq}} = 2c_{10} J^{-\frac{2}{3}} \left(\mathbf{I} - \frac{1}{3} (\text{tr}\, \hat{\mathbf{C}}) \hat{\mathbf{C}}^{-1} \right). \tag{17.16}$$

The non-equilibrium part of the first Piola–Kirchoff stress, which is the stress in the Maxwell elements, is given by

$$\mathbf{S}_{\text{neq}} = \sum_{k=1}^{n} 2c_{e10}^k J^{-\frac{2}{3}} \left(\hat{\mathbf{C}}_{\text{i}}^{-1,k} - \frac{1}{3} \text{tr} \left(\hat{\mathbf{C}}_{\text{i}}^{-1,k} \cdot \hat{\mathbf{C}} \right) \hat{\mathbf{C}}^{-1} \right). \tag{17.17}$$

The evolution equation for the internal variable $\hat{\mathbf{C}}_{\text{i}}$ is derived by similarly solving Eq. (17.9)

$$\dot{\hat{\mathbf{C}}}_{\text{i}}^k = \frac{4c_{e10}}{\breve{\eta}^k} \left[\hat{\mathbf{C}} - \frac{1}{3} \text{tr} \left(\hat{\mathbf{C}} \cdot \hat{\mathbf{C}}_{\text{i}}^{-1,k} \right) \hat{\mathbf{C}}_{\text{i}}^k \right]. \tag{17.18}$$

With the definition of relaxation time $\breve{r}^k = \frac{\breve{\eta}^k}{c_{e10}}$,

$$\dot{\hat{\mathbf{C}}}_{\mathrm{i}}^{k} = \frac{4}{\breve{r}^k}\left[\hat{\mathbf{C}} - \frac{1}{3}\operatorname{tr}\left(\hat{\mathbf{C}} \cdot \hat{\mathbf{C}}_{\mathrm{i}}^{-1,k}\right)\hat{\mathbf{C}}_{\mathrm{i}}^{k}\right]. \tag{17.19}$$

The relaxation time $\breve{r}^k$ is defined as a function of temperature T, the isochoric Cauchy–Green tensor $\hat{\mathbf{C}}$ and its rate $\dot{\hat{\mathbf{C}}}$, given by

$$\breve{r}^k(T, \hat{\mathbf{C}}, \dot{\hat{\mathbf{C}}}) = \frac{r^k(T)}{1 + r^k(T)\left|\left(a\|\dot{\hat{\mathbf{C}}}\|^b\right)\right|}. \tag{17.20}$$

Here, $r^k(T)$ represents the temperature-dependent relaxation time, whereas a and b are loading-specific material parameters. In this formulation, when $\dot{\hat{\mathbf{C}}}$ is zero (during relaxation), it ensures $\breve{r}^k$ reduces to $r^k(T)$, indicating that the strain rate does not affect the relaxation behaviour, but rather only influences the loading behaviour.

The temperature influence on relaxation time is described using the standard Williams–Landel–Ferry (WLF) equation (Williams et al. 1955):

$$r^k(T) = r_0^k \exp\left(\frac{C_1(T - T_0)}{C_2 + T - T_0}\right). \tag{17.21}$$

Here, r_0^k denotes the relaxation time at the reference temperature T_0. The parameters C_1 and C_2 represent the WLF constants.

17.3 Experimental Investigations

Before producing the specimens, the desired material blends that represent specific DOCs were prepared. These blends were completely conceived and produced by PSM Merseburg, Germany. In the blending process, PA6-type B granulate provided by Lanxess was compounded with the amorphous copolymer COC, specifically Topas 9506F-500, at four distinct PA6 ratios: 100%, 85/15, 70/30 and 55/45 of PA6 to COC. Subsequently, test specimens of type 1BA from ISO 527-2:2012 were injection moulded from the compounded granulates. This blending methodology aims to achieve varied thermally stable crystallinities across a broad range of DOCs. The alteration in DOC results from adjustments in the PA6 mass content relative to the total compound mass. The DOC was measured using a DSC Q2000 by assessing the enthalpy of melting during the first heating run, using the following formula:

$$X_c = \frac{\Delta H_m}{\Delta H_{100}}. \tag{17.22}$$

Here, the specific fusion enthalpy of $\Delta H_{100} = 190\,\text{J/g}$ for a hypothetical 100% crystalline material is used (Campoy et al. 1998). In this way, the four DOCs X_c = 0.29, 0.24, 0.18 and 0.15 (from highest to lowest PA6 content) were obtained.

Before testing, the specimens were stored in a dry chamber (MP Dry Cabinet IV ST) at 40 °C until their moisture content reached below 0.1%. The tests were conducted using a Zwick/Roell universal testing machine equipped with a 500 N force sensor and a temperature chamber. The tests were performed at the temperatures above T_g (approximately 60 °C). All tests in this investigation were conducted at the Institute of Mechanics, University of the Bundeswehr, Munich.

Initially, uniaxial tensile relaxation tests were performed to characterize the material's relaxation behaviour. For the pure PA6 (with $X_c = 0.29$), the tests were performed at 10 different temperatures ranging from 70 °C to 165 °C. For the three remaining blends (X_c = 0.24, 0.18 and 0.15), tests were limited to temperatures of 70 °C, 80 °C, 105 °C, 120 °C and 150 °C to save time. A strain level of 1% was chosen to ensure measurements remained within the material's linear viscoelastic regime, minimizing the influence of plastic deformation and allowing for the application of the time–temperature superposition principle. The strain rate during loading was set to $\dot{\varepsilon}_x = 0.2\,\%\,\text{s}^{-1}$, followed by relaxation for approx. 600 s. These short-duration tests were conducted to prevent significant changes in crystalline morphology that could occur at temperatures exceeding T_g. In Fig. 17.2, the engineering stress over time relation is presented for the relaxation tests for all four blends.

To gain deeper insights into the material behaviour and to constrain the parameter identification to a narrower range, monotonic uniaxial tension tests were performed, with the lateral contraction measured using the Insta 4D digital image correlation (DIC) system. A slow strain rate of $0.05\,\%\,\text{s}^{-1}$ was selected to minimize viscoelastic effects. The DIC results, which show true (logarithmic) strain fields in both the x and y directions, are presented in Fig. 17.3a and b. To calculate the Poisson's ratio, the average true strains from a central black block on the specimen (refer to Fig. 17.3a and b) were determined. This central region was chosen to avoid strain peaks at the edges caused by boundary effects in the DIC measurements. Poisson's ratio was calculated using the two-dimensional deformation data, defined as the negative ratio of the true longitudinal strain to the true lateral strain. The true strain was chosen because it accounts for the continuous deformation of a material and, hence, more accurately reflects the material's behaviour under large strains.

Figure 17.4a and b illustrate the relationship between instantaneous Poisson's ratio and strain in the x direction. In Fig. 17.4a, data is presented for temperatures of 100 °C and 130 °C, for DOC $X_c = 0.29$. Figure 17.4b shows the same relationship for three DOCs (X_c = 0.29, 0.24 and 0.18) at a fixed temperature of 130 °C. Poisson's ratio data for strains less than $\varepsilon_x = 2\%$ is not displayed due to pronounced fluctuations. Additionally, true stress calculated using the following relationship is plotted in both figures

$$\sigma_x(t) = \frac{F(t)}{\left[1 + \varepsilon_y(t)\right]^2 A_0}. \tag{17.23}$$

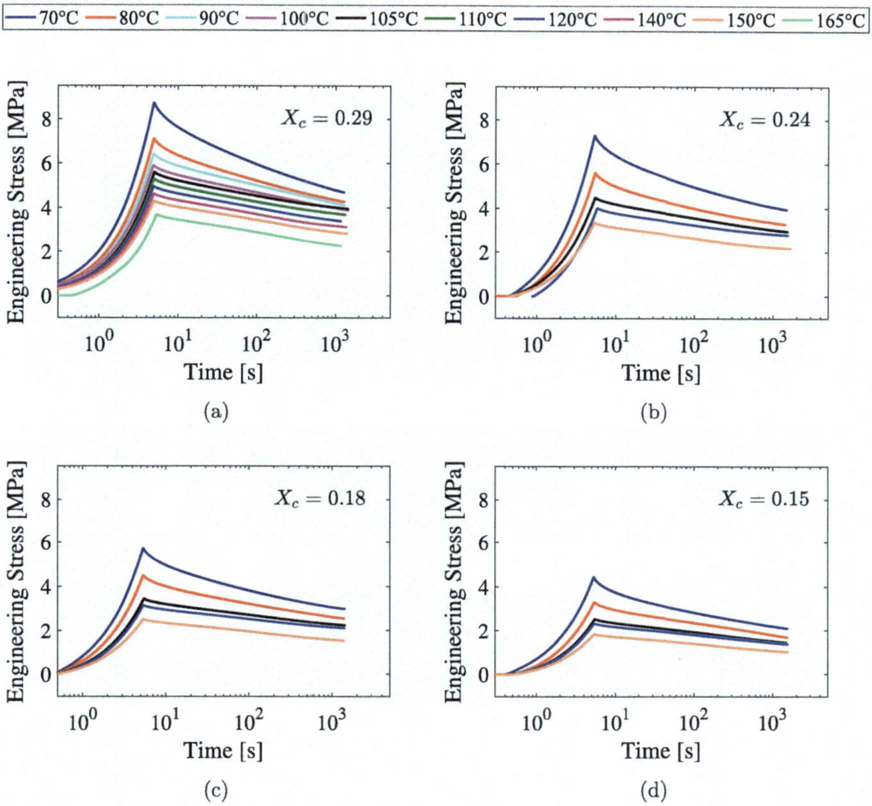

Fig. 17.2 Results of relaxation tests for DOCs $X_c = 0.29$, 0.24, 0.18 and 0.15 for various temperatures above T_g. The maximum strain was $\varepsilon_{\max,x} = 1\%$ and strain rate during loading $\dot{\varepsilon}_x = 0.2\,\%\,\mathrm{s}^{-1}$

Here, ε_y is the engineering strain in the y direction and A_0 is the undeformed cross-sectional area of the specimen.

To explore the material behaviour under varying strain rates, additional monotonic uniaxial tensile tests were performed at 105 °C. Figure 17.5 displays engineering stress versus engineering strain for DOCs $X_c = 0.29$, 0.24 and 0.18 at strain rates of $\dot{\varepsilon}_x = 0.05\,\%\,\mathrm{s}^{-1}$ and $\dot{\varepsilon}_x = 1\,\%\,\mathrm{s}^{-1}$.

Moreover, cyclic loading–unloading–recovery experiments were conducted at 105 ° C. In this procedure, the specimen underwent displacement-controlled loading, followed by unloading until the force reached zero. Subsequently, the force was maintained at a constant level for a time period of around 250 s. In Fig. 17.6, the engineering stress over time (Fig. 17.6a) and the engineering stress versus engineering strain relationship (Fig. 17.6b are presented for this loading procedure at 105 ° C across three DOCs: $X_c = 0.29$, 0.24 and 0.18. The strain rate during loading was $\dot{\varepsilon}_x = 0.25\,\%\,\mathrm{s}^{-1}$ and the maximum strain was $\varepsilon_{\max,x} = 5\%$.

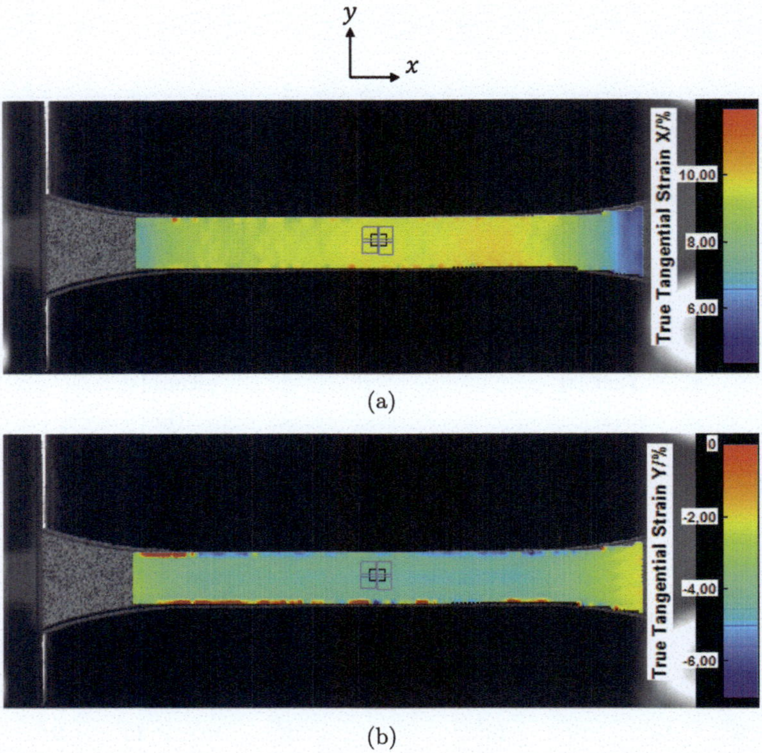

Fig. 17.3 Strain field measured with DIC. **a** True strain in x direction **b** True strain in y direction

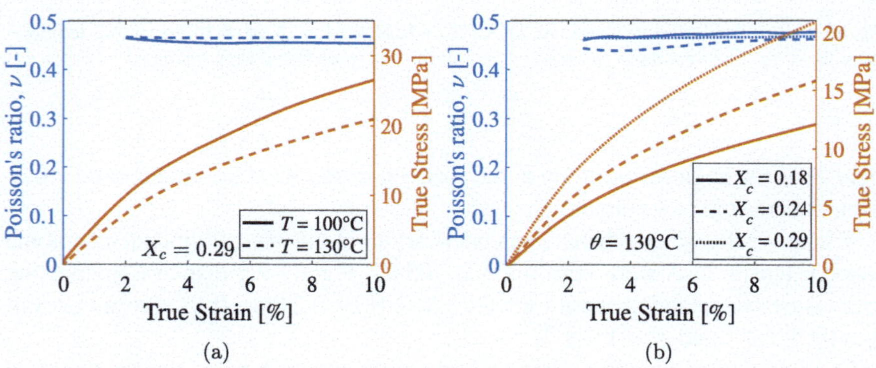

Fig. 17.4 Poisson's ratio ν calculated based on true strains **a** at two temperatures 100 °C and 130 °C measured for the DOC X_c = 0.29 **b** for three DOCs X_c = 0.29, 0.24 and 0.18 at a temperature of 130 °C. The true stress-strain result is additionally plotted to complement the results

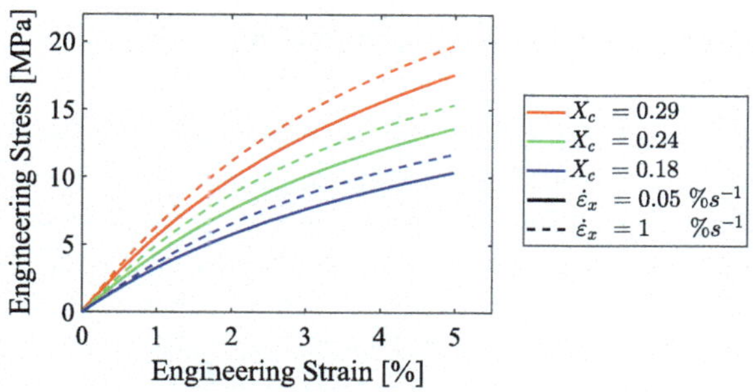

Fig. 17.5 Monotonic, uniaxial tensile tests at 105 °C: Effect of loading rate $\dot{\varepsilon}_x = 0.05\,\%\,\text{s}^{-1}$ and $1\,\%\,\text{s}^{-1}$ and effect of DOC ($X_c = 0.29$, 0.24 and 0.18)

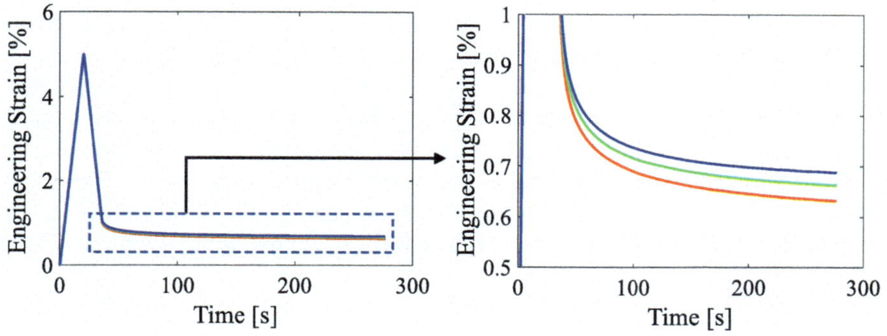

(a)

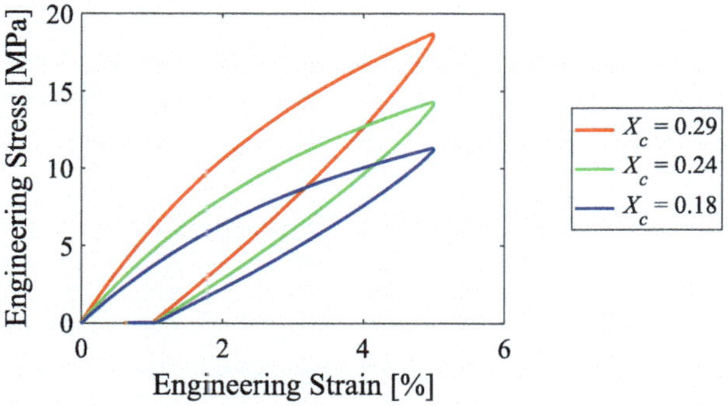

(b)

Fig. 17.6 Results of cyclic loading–unloading–recovery tests conducted at 105° C for multiple DOCs $X_c = 0.29$, 0.24 and 0.18. The maximum strain was $\varepsilon_{\max,x} = 5\%$ and the strain rate during loading and unloading was $\dot{\varepsilon}_x = 0.25\,\%\,\text{s}^{-1}$. **a** Engineering strain over time **b** Engineering stress over engineering strain

17.4 Time–Temperature Superposition

Time–temperature superposition enables the construction of master curves that describe the material's response over a wide range of temperatures and timescales, providing valuable insights into its performance in various applications such as automotive components, electrical housings and industrial machinery (Ferry 1980; Karger-Kocsis 1995).

For the master curve construction, the DOC $X_c = 0.29$ is initially considered. Following the time–temperature principle, the relaxation curves are segmented between time intervals 10 s and 100 s, as shown in Fig. 17.7a. The reference temperature T_0 is arbitrarily set to 105 °C. The curve at T_0 is held fixed while the curves at low temperatures are shifted leftward and those at high temperatures are shifted rightward so that all curves coincide and a smooth curve is obtained. The TTS process is shown for $X_c = 0.29$ in Fig. 17.7b.

Subsequently, master curves are constructed for the remaining blends (with DOCs $X_c = 0.24$, 0.18 and 0.15). Initially, the same horizontal shift factors a_T used at different temperatures for pure PA6 ($X_c = 0.29$) are applied to these blends. The resulting master curves are shown in Fig. 17.7d, f and h, respectively. The same horizontal shift factors for all four DOCs resulted in adequately smooth master curves of all DOCs. The four master curves are combinedly plotted in Fig. 17.8.

17.5 Numerical Implementation and Parameter Identification

The quasi-static form of the balance of linear momentum based on the total Lagrangian formulation, as given by (17.24), is solved to determine the displacement field **u** using COMSOL Multiphysics®

$$\int_{\Omega} \mathbf{S} : \delta \mathbf{E} \, dV = \int_{\partial \Omega} \overline{\mathbf{p}} \cdot \delta \mathbf{u} \, dA \,. \tag{17.24}$$

Here, $\overline{\mathbf{p}}$ represents the external traction or applied loads acting on the boundary $\partial\Omega$. The total second Piola–Kirchoff stress **S**, defined by Eq. (17.14), is incorporated into the formulation.

The evolution equations (17.19) for the inelastic strain in each Maxwell element are also solved using COMSOL. The implicit second-order backward differential method is used within the Finite Element Method framework.

For the identification of model parameters, the iterative inverse FE method is used. A least-squares problem is formulated, where the error is defined as the squared difference between the experimental engineering stress in the loading direction and

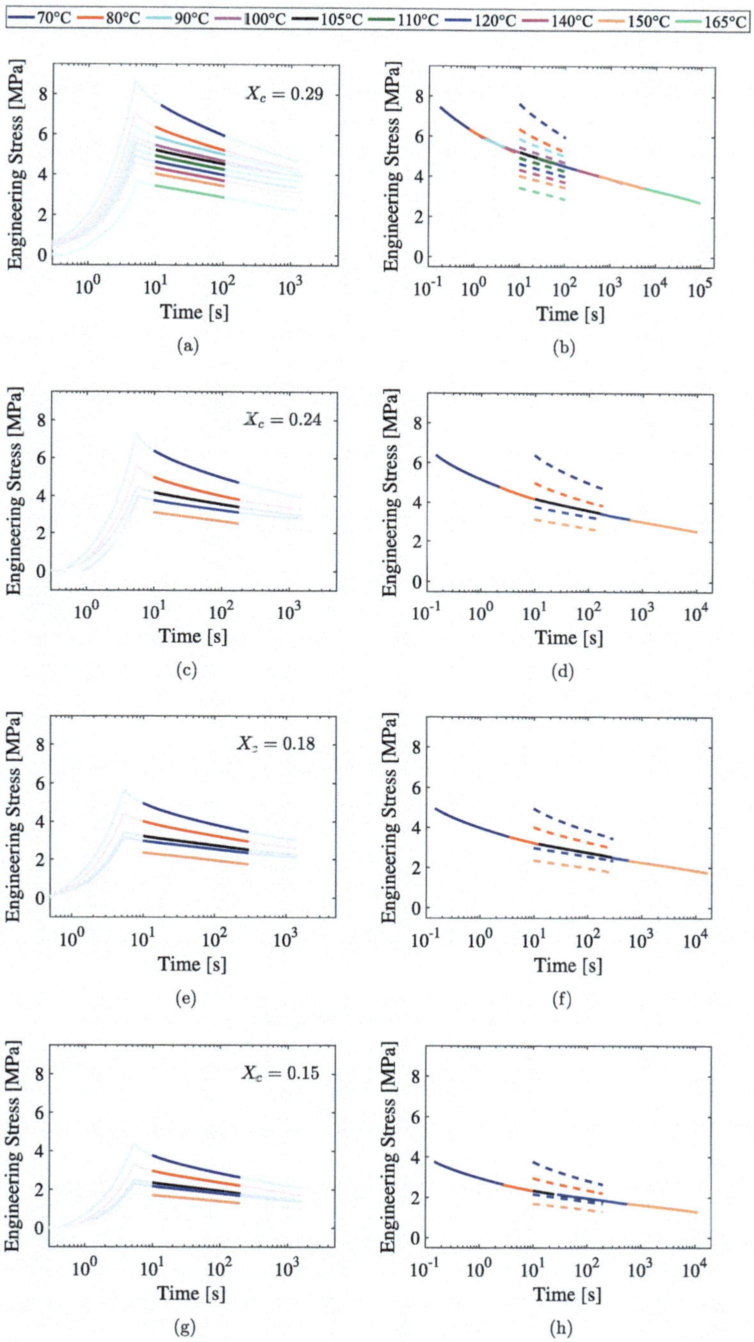

Fig. 17.7 Segmentation of relaxation test results conducted at multiple temperatures above T_g for all DOCs $X_c = 0.29, 0.24, 0.18$ and 0.15 (left). Creation of master curves by shifting the segmented curves, with a reference temperature of $T_0 = 105\,°C$ (right)

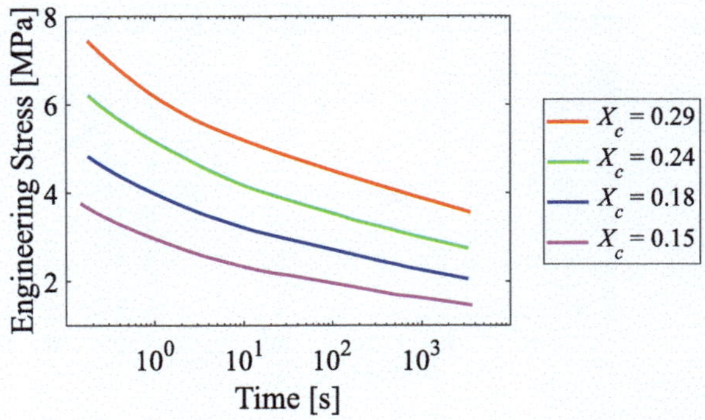

Fig. 17.8 Master curves for all DOCs $X_c = 0.29$, 0.24, 0.18 and 0.15 at 105 °C from Fig. 17.7 are plotted together for clarity

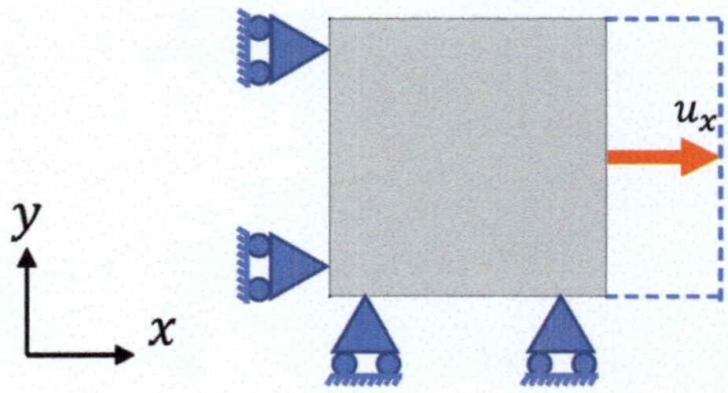

Fig. 17.9 Single-element boundary value problem

the first Piola–Kirchoff stress obtained from a single-element test as illustrated in Fig. 17.9. The error is given by

$$\text{error} = \sum_{t=i} \left(\mathbf{P}_{x,i}^{\text{exp}} - \mathbf{P}_{x,i}\right)^2. \tag{17.25}$$

The boundary value problem of the single-element test (Fig. 17.9) is solved in a 2D plane stress configuration. Model parameters are optimized using the Sparse Nonlinear OPTimizer (SNOPT), a gradient-based optimization algorithm.

17.5.1 Identification of Elastic Parameters

The equilibrium bulk modulus κ and equilibrium shear modulus c_{10} are, in this case, the two unknowns for each DOC for the long-term component of the framework (Neo-Hooke model).

1. As observed in the DIC experiments (in Sect. 17.3), Poisson's ratio ν shows minimal variation and remains close to 0.45 under a measured strain of 10% across different temperatures and DOCs. Consequently, the two unknown parameters are interconnected using the following Lamé relation, assuming a constant $\nu = 0.45$ (Timoshenko and Goodier 1970)

$$\kappa = \frac{2c_{10}(1+\nu)}{3(1-2\nu)}. \tag{17.26}$$

Thus, the identification of elastic parameters is reduced to a single parameter instead of two.

2. During the identification process, parameter c_{10} is determined alongside the viscoelastic parameters, as outlined in Sect. 17.5.2. The dependency of c_{10} on the DOC is assumed to be linear as

$$c_{10}\left(X_c\right) = p^0 X_c + q^0. \tag{17.27}$$

Here, p^0 and q^0 are the parameters to be determined based on the results of multiple DOCs.

17.5.2 Identification of Viscoelastic Parameters

1. As discussed in Sect. 17.4, employing the same horizontal shift factors a_T for all four DOCs resulted in sufficiently smooth master curves. This indicates that the temperature dependence of the viscoelastic properties is consistent across all the DOCs. Therefore, the WLF parameters C_1 and C_2 are optimized once, based on the shift factors for a single DOC $X_c = 0.29$, using the equation

$$\log(a_T) = \log\left(\frac{r_T}{r_0}\right) = \frac{-C_1(T - T_0)}{C_2 + T - T_0}. \tag{17.28}$$

For this fit, the Levenberg–Marquardt algorithm in MATLAB is used. The fit of the WLF equation is shown in Fig. 17.10. The optimized parameters C_1 and C_2 are listed in Table 17.1.

2. In the material framework used (Fig. 17.1), five Maxwell-like elements are used. This selection is based on the observation that the master curves span approximately five decades of relaxation time, allowing each element to cover one decade. The relaxation times r^k are chosen as fixed values: 1 s, 10 s, 100 s, 1000 s and

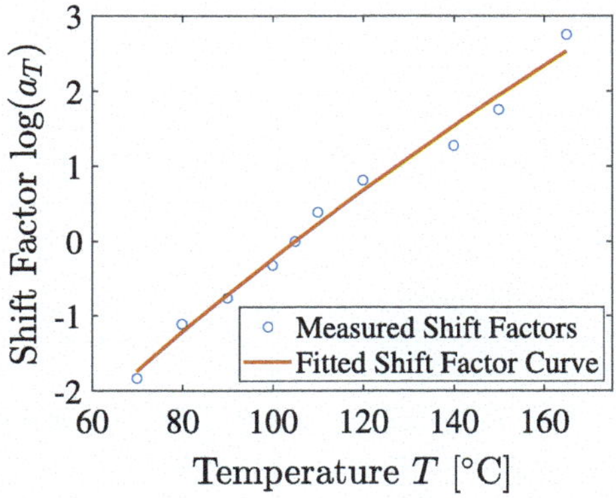

Fig. 17.10 Identification of WLF parameters C_1 and C_2 based on the master curve of DOC $X_c = 0.29$ with $T_0 = 105\,°C$

Table 17.1 Identified material parameters

Material parameters	Symbol	Value
Hyperelasticity: Neo-Hooke model		
Neo-Hooke Parameter 1	$c_{10} = p^0 X_c + q^0$	$1015X_c - 38.2\,\text{MPa}$
Neo-Hooke Parameter 2	$\kappa = \dfrac{2c_{10}(1+\nu)}{3(1-2\nu)}$	$9814X_c - 368.8\,\text{MPa}$
Viscoelasticity		
Stiffness for the Maxwell element 1	$C_{e10}^1 = p^1 X_c + q^1$	$43.68X_c + 4.58\,\text{MPa}$
Stiffness for the Maxwell element 2	$C_{e10}^2 = p^2 X_c + q^2$	$47.70X_c + 4.56\,\text{MPa}$
Stiffness for the Maxwell element 3	$C_{e10}^3 = p^3 X_c + q^3$	$36.00X_c + 1.03\,\text{MPa}$
Stiffness for the Maxwell element 4	$C_{e10}^4 = p^4 X_c + q^4$	$28.35X_c + 1.07\,\text{MPa}$
Stiffness for the Maxwell element 5	$C_{e10}^5 = p^5 X_c + q^5$	$31.11X_c + 1.44\,\text{MPa}$
Relaxation time for the Maxwell element 1	r_0^1	1 s
Relaxation time for the Maxwell element 2	r_0^2	10 s
Relaxation time for the Maxwell element 3	r_0^3	100 s
Relaxation time for the Maxwell element 4	r_0^4	1000 s
Relaxation time for the Maxwell element 5	r_0^5	10000 s
WLF parameters		
Standard parameter 1	C_1	–27.47
Standard parameter 2	C_2	588.97 K
Reference temperature	T_0	378 K
Loading-specific material parameters		
Standard parameter 1	a	3.7092
Standard parameter 2	b	3.7418

10000 s for all DOCs, simplifying the parameter identification process. Here, k denotes the number of Maxwell-like elements ($k = 1, \ldots, 5$).

3. The dependence of c_{e10} on DOC is assumed to be linear, similar to c_{10} and is expressed as

$$c_{e10}^k(X_c) = p^k \cdot X_c + q^k. \tag{17.29}$$

In a staggered parameter identification procedure, the equilibrium stiffness parameters p^0 and q^0, along with the non-equilibrium stiffness parameters p^k (for

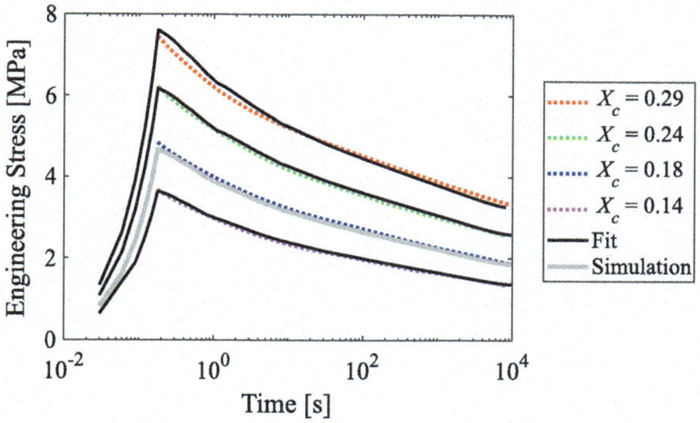

Fig. 17.11 Identification results for fitting of master curves at $T_0 = 105\,°C$ for 3 DOCs $X_c = 0.29$, 0.24 and 0.18, application of identified parameters to $X_c = 0.18$

$k = 1, \ldots, 5$) and q^k (for $k = 1, \ldots, 5$), are identified together. This was followed by identifying rate-dependent parameters a and b while keeping p^0, q^0, p^k and q^k fixed. This approach is adopted due to the decoupled relaxation and loading-specific mechanisms in the model described in Sect. 17.2.4.
The parameters p^0, q^0, p^k and q^k are obtained through a multifit process using the master curves of three X_c values: $X_c = 0.15, 0.24, 0.29$ simultaneously (Fig. 17.8). The master curve for $X_c = 0.18$ is used for verification. During this fitting, the loading-specific parameters a and b in Eq. (17.20) are set to zero. The fitting results are shown in Fig. 17.11. The optimized parameters p^0, q^0, p^k and q^k are listed in Table 17.1.

4. The loading-specific material parameters a and b are assumed to be independent of X_c. They are identified based on the results of two monotonic tensile tests with strain rates $\dot{\varepsilon}_x = 0.05\,\%\,s^{-1}$ and $\dot{\varepsilon}_x = 1\,\%\,s^{-1}$ up to a strain of $\varepsilon_{\max,x} = 5\%$ for a single DOC $X_c = 0.29$ using multifit. The optimized parameters a and b are listed in Table 17.1. The fitting result is shown in Fig. 17.12b. When the loading-specific parameters a and b are set to zero, i.e., the loading part is excluded from the model, the stress results clearly deviate, as observed in Fig. 17.12a, indicating the necessity for the inclusion of loading-specific parameters in the model.

17.6 Model Verification

In order to verify the obtained parameters, the model was evaluated for more experimental cases. The model was evaluated in the same way using simulations on a single element as described in Sect. 17.5.

1. To verify the relaxation behaviour and the TTS over an extended time, a relaxation test at a new temperature $T = 130\,°C$ was simulated. The loading strain during the test was $\varepsilon_{\max,x} = 1\%$ and the loading strain rate was chosen to $\dot{\varepsilon}_x = 0.2\,\%\,s^{-1}$. For the simulation, the relaxation time parameters r^k were updated using shift factors defined by the WLF equation. The engineering stress versus time relation

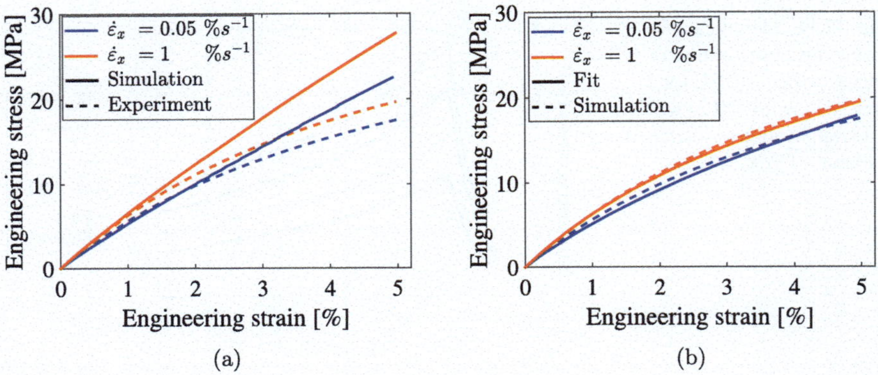

Fig. 17.12 Identification of loading-specific material parameters, through uniaxial, monotonic tests performed for a single DOC, $X_c = 0.29$ at 105 °C. **a** Simulation with loading-specific parameters a and b set to 0, i.e., excluding the loading-specific part of the model. **b** Fitting results after identifying and including the loading-specific parameters a and b

Fig. 17.13 Model verification result: Relaxation test performed at 130 °C for the DOC $X_c = 0.29$, at a loading strain rate $\dot{\varepsilon}_x = 0.2\,\%\,\mathrm{s}^{-1}$ and a maximum strain of $\varepsilon_{\max,x} = 1\%$

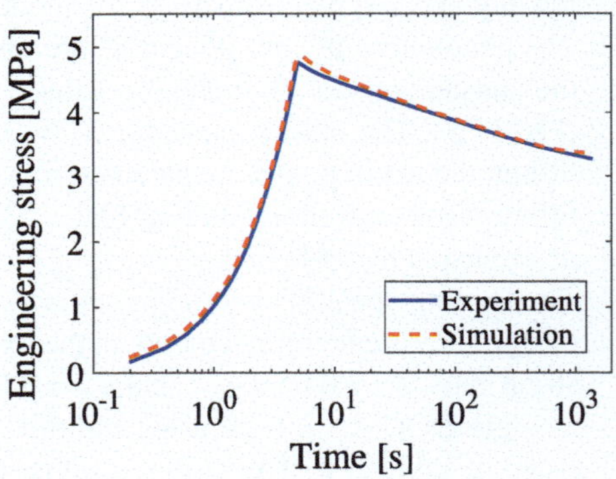

of the evaluated model and the experiment are shown in Fig. 17.13. The model's accuracy was confirmed by its successful prediction of the relaxation behaviour. At higher strains, when nonlinearities are pronounced, TTS is not applicable, which can necessitate further refinement of the model for accurate predictions.

2. The viscoelastic behaviour's dependence on the DOC was approximated by assuming that both the equilibrium and non-equilibrium stiffnesses change linearly with the DOC. This approach proved effective, as the model accurately predicted the relaxation behaviour of PA6 at a crystallinity level of $X_c = 0.18$, as shown in Fig. 17.11.
3. For the verification of the loading-specific part of the model, the monotonic uniaxial test results at two strain rates $\dot{\varepsilon}_x = 0.05\,\%\,\mathrm{s}^{-1}$ and $1\,\%\,\mathrm{s}^{-1}$ conducted on two DOC's $X_c = 0.24$ and 0.18 are used. The engineering stress versus engineering strain relation of the evaluated model and the experiment are shown in Fig. 17.14. The model performed well, correctly predicting the decrease in

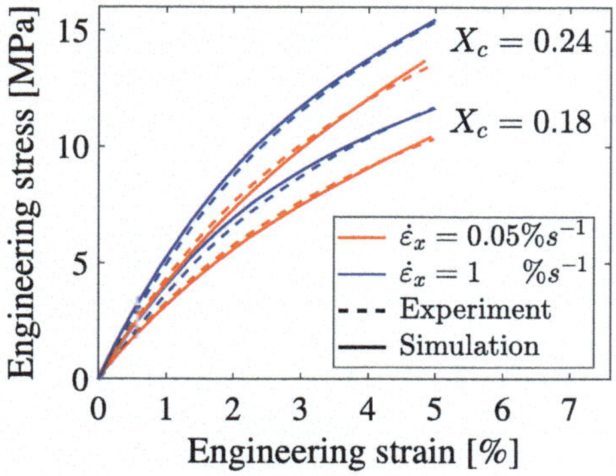

Fig. 17.14 Model verification result: Monotonic, uniaxial tensile tests performed at 105 °C at loading rates $\dot{\varepsilon}_x = 0.05\,\%\,s^{-1}$ and $1\,\%\,s^{-1}$ for DOCs $X_c = 0.24$ and 0.18

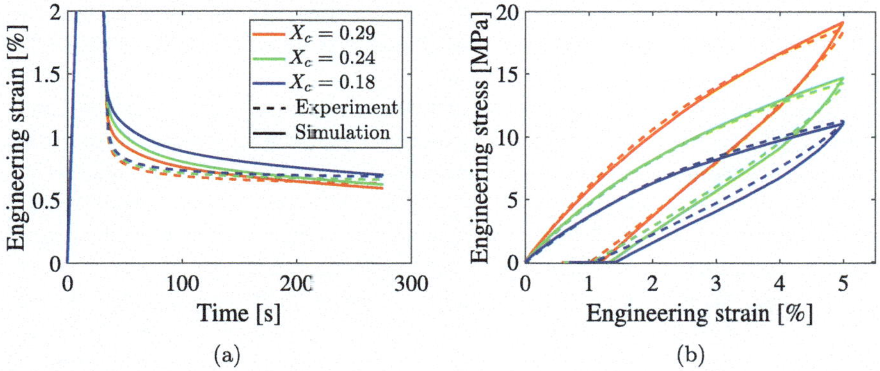

Fig. 17.15 Model verification result: Cyclic loading–unloading–recovery tests performed for three DOC's $X_c = 0.29$, 0.24 and 0.18. The maximum strain was $\varepsilon_{\max,x} = 5\%$ and the strain rate during loading and unloading was $\dot{\varepsilon}_x = 0.25\,\%\,s^{-1}$

stress levels with decreasing DOC. The loading-specific parameters, incorporated through a nonlinear function, were critical in capturing these effects.

4. The model response was evaluated for the cyclic loading–unloading–recovery procedure and compared with the experimental data. The loading strain was 5% and the loading strain rate was $0.25\,\%\,s^{-1}$. In Fig. 17.15a, the engineering strain versus time relation is plotted, whereas in Fig. 17.15b, the engineering stress versus engineering strain relation is plotted. The model effectively predicted the cyclic response, including the unloading phases, across different crystallinity levels. Although the strain after unloading differed slightly, as observed in Fig. 17.15a, the trend of recovery with respect to the DOC was similar to the experimental results.

The verification of the finite viscoelastic model illustrated the model's ability to capture the key aspects of the viscoelastic behaviour, including crystallinity-dependent relaxation and strain rate effects.

17.7 Conclusion

In this work, a phenomenological constitutive model for large isothermal deformations is proposed and applied to the semi-crystalline polymer PA6. The model, developed from a viscoelastic perspective, aimed to minimize the number of parameters and reduce the identification effort by excluding plasticity effects. It employs a 3D generalization of the 1D Zener model, incorporating multiple Maxwell-like elements in parallel. Such formulations typically describe the relaxation behaviour more accurately than the loading behaviour. Therefore, to enhance the material's description during loading, a new nonlinear flow rule is introduced, which incorporates the deviatoric Cauchy–Green strain and its rate.

The framework uses temperature and DOC as fixed input parameters. The time–temperature superposition principle was used to create the stress relaxation master curve for temperatures above T_g (around $60\,^\circ$C), ranging from $80\,^\circ$C to $150\,^\circ$C. Temperature dependence is integrated into the model through the viscosity (or relaxation time) of each Maxwell-like element, while the influence of DOC is introduced through the equilibrium and non-equilibrium stiffness parameters.

For model identification and verification, materials with varying crystallinities were produced using a non-conventional blend methodology to achieve varying crystallinities, spanning around 15%. The model involves two unknown equilibrium stiffness parameters, ten identifiable time-dependent stiffness parameters along with five fixed relaxation time parameters linked to Maxwell-like elements that span a relaxation period of five decades. Additionally, it incorporates two WLF parameters for temperature dependence and two parameters specific to the loading behaviour of the model.

In the staggered parameter identification procedure, the long-term and the time-dependent stiffness parameters are identified based on the stress relaxation master curves. Once these stiffness parameters were determined, they were used to predict the loading-specific parameters based on the monotonic uniaxial tension tests. This approach capitalizes on the fact that the loading-specific part, defined through a nonlinear flow rule, becomes redundant during relaxation (when the strain rate is zero). This redundancy allows for focused determination of the parameters responsible for viscoelastic relaxation.

Model predictions for uniaxial relaxation, monotonic tension tests and cyclic loading–unloading–recovery tests in tension for multiple DOCs at distinct temperatures and loading strain rates showed a reasonable agreement with the experimental observations. The derived model thus effectively captures crystallinity-dependent viscoelastic relaxation and strain rate behaviours.

In future work, the model could be applied to other 3D stress states (compression, shear) beyond uniaxial tension. Additionally, the model could be applied to rubber-like materials where the plastic deformation can be neglected. For other polymeric materials, it could be further developed to include plasticity for a more realistic and comprehensive representation.

Acknowledgements We wish to acknowledge the support provided by the DFG (RE 1057/52-1, project number 454873500) towards this research. Furthermore, we would like to express our gratitude to Dr. André Wutzler from PSM Merseburg, Germany, for his role in conceiving and producing the blends for this project.

References

Abdul-Hameed H, Messager T, Zaïri F, Naït-Abdelaziz M (2014) Large-strain viscoelastic-viscoplastic constitutive modeling of semi-crystalline polymers and model identification by deterministic/evolutionary approach. Comput Mater Sci 90:241–252. https://doi.org/10.1016/j.commatsci.2014.03.043

Amin AFMS, Alam MS, Okui Y (2002) An improved hyperelasticity relation in modeling viscoelasticity response of natural and high damping rubbers in compression: experiments, parameter identification and numerical verification. Mech Mater 34(2):75–95. https://doi.org/10.1016/S0167-6636(01)00102-8

Amin AFMS, Lion A, Sekita S, Okui Y (2006) Nonlinear dependence of viscosity in modeling the rate-dependent response of natural and high damping rubbers in compression and shear: Experimental identification and numerical verification. Int J Plast 22(9):1610–1657. https://doi.org/10.1016/j.ijplas.2005.09.005

Applied plastics engineering handbook: processing and materials. PDL handbook series, 2nd edn. Elsevier, William Andrew, Amsterdam [u.a.] (2016)

Ayoub G, Zaïri F, Fréderix C, Gloaguen JM, Naït-Abdelaziz M, Seguela R, Lefebvre JM (2011) Effects of crystal content on the mechanical behaviour of polyethylene under finite strains: experiments and constitutive modelling. Int J Plast 27(4):492–511. https://doi.org/10.1016/j.ijplas.2010.07.005

Barba D, Arias A, Garcia-Gonzalez D (2020) Temperature and strain rate dependences on hardening and softening behaviours in semi-crystalline polymers: application to peek. Int J Solids Struct 182–183:205–217. https://doi.org/10.1016/j.ijsolstr.2019.08.021

Batista NL, Olivier P, Bernhart G, Rezende MC, Botelho EC (2016) Correlation between degree of crystallinity, morphology and mechanical properties of PPS/carbon fiber laminates. Mater Res 19(1):195–201. https://doi.org/10.1590/1980-5373-MR-2015-0453

Campoy I, Gómez MA, Marco C (1998) Structure and thermal properties of blends of nylon 6 and a liquid crystal copolyester dedicated to the memory of Prof. J. G. Fatou. Polymer 39(25):6279–6288. https://doi.org/10.1016/S0032-3861(98)00181-5

Chen J-W, Dai J, Yang J-H, Huang T, Zhang N, Wang Y (2013) Annealing-induced crystalline structure and mechanical property changes of polypropylene random copolymer. J Mater Res 28(22):3100–3108. https://doi.org/10.1557/jmr.2013.332

Colak O (2005) Modeling deformation behavior of polymers with viscoplasticity theory based on overstress. Int J Plast 21(1):145–160. https://doi.org/10.1016/j.ijplas.2004.04.004

Coleman BD, Noll W (1963) The thermodynamics of elastic materials with heat conduction and viscosity. Arch Ration Mech Anal 13(1):167–178. https://doi.org/10.1007/BF01262690

Dippel B, Johlitz M, Lion A (2015) Thermo-mechanical couplings in elastomers—experiments and modelling. ZAMM 95(11):1117–1128. https://doi.org/10.1002/zamm.201400110

Dorléans V, Delille R, Notta-Cuvier D, Lauro F, Michau E (2021) Time-temperature superposition in viscoelasticity and viscoplasticity for thermoplastics. Polym Test 101:107287. https://doi.org/10.1016/j.polymertesting.2021.107287

Drozdov AD, Gupta RK (2003) Constitutive equations in finite viscoplasticity of semicrystalline polymers. Int J Solids Struct 40(23):6217–6243. https://doi.org/10.1016/S0020-7683(03)00414-1

Dusunceli N, Colak OU (2008) Modelling effects of degree of crystallinity on mechanical behavior of semicrystalline polymers. Int J Plast 24(7):1224–1242. https://doi.org/10.1016/j.ijplas.2007.09.003

Felder S, Holthusen H, Hesseler S, Pohlkemper F, Gries T, Simon J-W, Reese S (2020) Incorporating crystallinity distributions into a thermo-mechanically coupled constitutive model for semi-crystalline polymers. Int J Plast 135:102751. https://doi.org/10.1016/j.ijplas.2020.102751

Ferry JD (1980) Viscoelastic properties of polymers, 3rd edn. Wiley, Chichester. http://www.loc.gov/catdir/bios/wiley041/79028666.html

Flory PJ (1961) Thermodynamic relations for high elastic materials. Trans Faraday Soc 57:829. https://doi.org/10.1039/tf9615700829

Galeski A (2003) Strength and toughness of crystalline polymer systems. Prog Polym Sci 28(12):1643–1699. https://doi.org/10.1016/j.progpolymsci.2003.09.003

Haupt P, Lion A, Backhaus E (2000) On the dynamic behaviour of polymers under finite strains: constitutive modelling and identification of parameters. Int J Solids Struct 37(26):3633–3646. https://doi.org/10.1016/S0020-7683(99)00165-1

Hoo Fatt MS, Ouyang X (2008) Three-dimensional constitutive equations for styrene butadiene rubber at high strain rates. Mech Mater 40(1–2):1–16. https://doi.org/10.1016/j.mechmat.2007.06.002

Humbert S, Lame O, Vigier G (2009) Polyethylene yielding behaviour: What is behind the correlation between yield stress and crystallinity? Polymer 50(15):3755–3761. https://doi.org/10.1016/j.polymer.2009.05.017

Johlitz M, Steeb H, Diebels S, Chatzouridou A, Batal J, Possart W (2007) Experimental and theoretical investigation of nonlinear viscoelastic polyurethane systems. J Mater Sci 42(23):9894–9904. https://doi.org/10.1007/s10853-006-1479-4

Johlitz M, Dippel B, Lion A (2016) Dissipative heating of elastomers: a new modelling approach based on finite and coupled thermomechanics. Contin Mech Thermodyn 28(4):1111–1125. https://doi.org/10.1007/s00161-015-0469-7

Karak N (2024) Advances in biocomposites and their applications. Woodhead Publishing, Cambridge, UK

Karger-Kocsis J (ed) (1995) Polypropylene: structure. Blends and composites. Chapman & Hall, London

Kästner M, Obst M, Brummund J, Thielsch K, Ulbricht V (2012) Inelastic material behavior of polymers—experimental characterization, formulation and implementation of a material model. Mech Mater 52:40–57. https://doi.org/10.1016/j.mechmat.2012.04.011

Khan A, Farrokh B (2006) Thermo-mechanical response of nylon 101 under uniaxial and multi-axial loadings: Part i, experimental results over wide ranges of temperatures and strain rates. Int J Plast 22(8):1506–1529. https://doi.org/10.1016/j.ijplas.2005.10.001

Khan A, Zhang H (2001) Finite deformation of a polymer: experiments and modeling. Int J Plast 17(9):1167–1188. https://doi.org/10.1016/S0749-6419(00)00073-5

Kulkarni S, Loos K, Lion A, Johlitz M (2022) Thermoforming: identification of process-relevant ranges for strain, strain rate, cooling rate, and degree of crystallinity through preliminary simulations. In: Altenbach H, Johlitz M, Merkel M, Öchsner A (eds) Lectures notes on advanced structured materials. Advanced structured materials, vol 153. Springer International Publishing, Cham, pp 303–314. https://doi.org/10.1007/978-3-031-11589-9

Kulkarni S, Reuvers M-C, Brepols T, Reese S, Johlitz M, Lion A (2024) Characterization of crystallization kinetics in polyamide 6 with a focus on modeling the thermoforming process: experiments, modeling, simulations. Contin Mech Thermodyn 36(2):415–431. https://doi.org/10.1007/s00161-023-01268-y

Lai D, Yakimets I, Guigon M (2005) A non-linear viscoelastic model developed for semi-crystalline polymer deformed at small strains with loading and unloading paths. Mater Sci Eng A 405(1–2):266–271. https://doi.org/10.1016/j.msea.2005.06.011

Leaversuch R (2004) Metal replacement accelerates in under-hood parts. Plast Technol 50(3):33

Lee TH, Boey FYC, Khor KA (1995) On the determination of polymer crystallinity for a thermoplastic PPS composite by thermal analysis. Compos Sci Technol 53(3):259–274. https://doi.org/10.1016/0266-3538(94)00070-0

Lion A (2000) Thermomechanik Von Elastomeren: experimente und Materialtheorie: Zugl.: Kassel, Univ., Habil.-Schrift, 2000. Berichte des Instituts für Mechanik, vol 2000, 1. Inst. für Mechanik, Kassel

Liu C, Mather P Thermomechanical characterization of blends of poly (vinyl acetate) with semicrystalline polymers for shape memory applications. In: 61st annual technical conference ANTEC 2003; Conference date: 04-05-2003 Through 08-05-2003

Lubliner J (1985) A model of rubber viscoelasticity. Mech Res Commun 12(2):93–99. https://doi.org/10.1016/0093-6413(85)90075-8

Makradi A, Ahzi S, Belouettar S, Ruch D (2008) Thermoforming process of semicrystalline polymeric sheets: modeling and finite element simulations. Polym Sci Ser A 50(5):550–557. https://doi.org/10.1134/S0965545X0805009X

Mileva D, Kolesov I, Androsch R (2012) Morphology of cold-crystallized polyamide 6. Colloid Polym Sci 290(10):971–978. https://doi.org/10.1007/s00396-012-2657-3

Odrobina M, Deák T, Székely L, Mankovits T, Keresztes RZ, Kalácska G (2020) The effect of crystallinity on the toughness of cast polyamide 6 rods with different diameters. Polymers 12(2). https://doi.org/10.3390/polym12020293

Parenteau T, Ausias G, Grohens Y, Pilvin P (2012) Structure, mechanical properties and modelling of polypropylene for different degrees of crystallinity. Polymer 53(25):5873–5884. https://doi.org/10.1016/j.polymer.2012.09.053

Parodi E, Peters GWM, Govaert LE (2018) Structure¯properties relations for polyamide 6, part 1: influence of the thermal history during compression moulding on deformation and failure kinetics. Polymers 10(7). https://doi.org/10.3390/polym10070710

Popa CM, Fleischhauer R, Schneider K, Kaliske M (2014) Formulation and implementation of a constitutive model for semicrystalline polymers. Int J Plast 61:128–156. https://doi.org/10.1016/j.ijplas.2014.05.010

Qi Z, Hu N, Li G, Zeng D, Su X (2019) Constitutive modeling for the elastic-viscoplastic behavior of high density polyethylene under cyclic loading. Int J Plast 113:125–144. https://doi.org/10.1016/j.ijplas.2018.09.010

Qin Q-H, Ye J (2015) Toughening mechanisms in composite materials. Woodhead Publishing, Cambridge, UK

Reese S, Govindjee S (1997) Mech Time Depend Mater 1(4):357–396. https://doi.org/10.1023/A:1009795431265

Reese S, Govindjee S (1998) A theory of finite viscoelasticity and numerical aspects. Int J Solids Struct 35(26–27):3455–3482. https://doi.org/10.1016/S0020-7683(97)00217-5

Reuvers M-C, Boes B, Felder S, Brepols T, Kulkarni S, Loos K, Johlitz M, Lion A, Reese S (2023) Semi–crystalline polymers at finite strains: a thermo–coupled constitutive model for varying degrees of crystallinity and temperatures. PAMM 23(1). https://doi.org/10.1002/pamm.202200162

Reuvers M-C, Boes B, Felder S, Brepols T, Reese S (2023) A thermo–coupled constitutive model for semi–crystalline polymers at finite strains: application to varying degrees of crystallinity and temperatures. Techn Mech 43(2). https://doi.org/10.24352/UB.OVGU-2023-057

Reuvers M-C, Dannenberg C, Kulkarni S, Loos K, Johlitz M, Lion A, Reese S, Brepols T (2024) Experimental investigation and micromechanical analysis of glass fiber reinforced polyamide 6. https://doi.org/10.2139/ssrn.4862857

Reuvers M-C, Kulkarni S, Boes B, Felder S, Wutzler A, Johlitz M, Lion A, Brepols T, Reese S (2024) A thermo-mechanically coupled constitutive model for semi-crystalline polymers at finite strains: mechanical and thermal characterization of polyamide 6 blends. Contin Mech Thermodyn 36(3):657–698. https://doi.org/10.1007/s00161-024-01288-2

Schrauwen BAG, Janssen RPM, Govaert LE, Meijer HEH (2004) Intrinsic deformation behavior of semicrystalline polymers. Macromolecules 37(16):6069–6078. https://doi.org/10.1021/ma035279t

Shim J, Mohr D (2011) Rate dependent finite strain constitutive model of polyurea. Int J Plast 27(6):868–886. https://doi.org/10.1016/j.ijplas.2010.10.001

Timoshenko S, Goodier JN (1970) Theory of elasticity. Engineering societies monographs, 3rd edn. McGraw-Hill Book Company, New York

Whelan T (1994) Polymer technology dictionary. Springer, Berlin, Heidelberg

Williams ML, Landel RF, Ferry JD (1955) The temperature dependence of relaxation mechanisms in amorphous polymers and other glass-forming liquids. J Am Chem Soc 77(14):3701–3707. https://doi.org/10.1021/ja01619a008

Yu L, Liu H, Xie F, Chen L, Li X (2008) Effect of annealing and orientation on microstructures and mechanical properties of polylactic acid. Polym Eng Sci 48(4):634–641. https://doi.org/10.1002/pen.20970

Chapter 18
Electrical Induced Debonding on Demand of Adhesive Joints for the Automotive Industry

Philip-Marc Barnewitz, Jens Holtmannspoetter, Elisa Arikan, Daniel Schmid, Michael Schüssler, Michael Johlitz, and Alexander Lion

Abstract Structural and semi-structural adhesives are of great importance for the usage in battery systems of electric vehicles. However, a significant challenge is their resistance to loosening. That makes reworking or repairing as well as the reuse of bonded components quite challenging. To address this issue, various methods with different triggers have been explored by several authors to debond adhesive joints. A promising approach is applying an electric field to an adhesive modified with an ionic liquid. Here, the effects are not fully understood yet. Therefore, this work focuses on the debonding mechanism of the mentioned approach using a targeted SEM analysis of the fracture surfaces. In addition, the influence of different bonding conditions and debonding process parameters, e.g. the applied voltage or the aluminium alloy on the reduction of strength is evaluated. It is shown that especially the strength of the induced electric field during the debonding and the layer thickness of the adhesive have a significant influence on the debonding progress.

Keywords Adhesive · Debonding on demand · Automotive industry · Battery systems · Electrical debonding

P.-M. Barnewitz (✉) · J. Holtmannspoetter · E. Arikan
Fuels and Lubricants, Bundeswehr Research Institute for Materials, Institutsweg 1, 85424 Erding, Germany
e-mail: philip-marc.barnewitz@unibw.de

P.-M. Barnewitz · M. Johlitz · A. Lion
University of the Bundeswehr Munich, Werner-Heisenberg-Weg 39, 85579 Neubiberg, Germany

D. Schmid · M. Schüssler
AUDI AG, Auto-Union-Straße 1, 85057 Ingolstadt, Germany

H. Altenbach et al. (eds.), *Lectures Notes on Advanced Structured Materials 3*,
Advanced Structured Materials 226, https://doi.org/10.1007/978-3-031-77403-4_18

18.1 Introduction

Adhesive bonding is one of the most frequently used joining methods for battery systems of modern electric vehicles, as it is able to join different substrates without utilising installation space. Therefore, adhesives with different properties and mechanical characteristics are used for this purpose. Up to 5 kg of different adhesives are used for modern car battery systems, targeting various use cases and requirements, e.g. water-repellence, sealing using soft materials, thermally conductive adhesives as well as structurally strong adhesives. In order to fit the bonded battery systems into circular economy reworking, reusing, or recycling of the bonded battery components must become possible. Today, structural adhesive typically can be debonded only possible by applying high mechanical forces (Habenicht 2008). However, the allowed applicable force to the adhesive layers within automotive battery systems, such as torsional or shear loads, is limited due to the surrounding components. In general, the battery cells within the battery systems must not be damaged to avoid hazard exothermic reactions. Therefore, mechanical debonding is not possible without a controlled weakening of the bond line beforehand.

Multiple options are described in literature for weakening structural adhesive bondings with non-debonding-optimised adhesives. The commonly used method of weakening adhesives is to increase the temperature. If the polymer matrix is heated above the glass transition point (thermoplastics) or above the decomposition temperature, the polymer softens, resulting in a weakening of the adhesive bond (Krumrey 2000). In contrast, a decrease in temperature to the cryothermal range of about minus 80 °C can also lead to a reduction in strength of an adhesive bond due to embrittlement of the polymers (Bartley et al. 2023). However, to prevent damage to the battery cells the narrow operating window of battery cells between minus 40 °C and 60 °C has to be adhered to and forbids applying too high as well as too low temperatures. Another method described in the literature for dissolving adhesive bonds is the interaction with liquid chemical compounds. A distinction must be made between solvents that weaken the bond by dissolving the polymer and bases or acids that cause a degradation reaction of the polymer matrix. Both cases are not applicable for automotive battery systems, as the process times for dissolving the adhesive are too long due to the relatively low diffusion speed (Habenicht 2008). Finally, polymers, including adhesive polymers, can be aged by the absorption of UV radiation. Depending on the polymer composition, it generates polymer degradation reactions, which lead to softening of the material, or polymer crosslinking reactions, which lead to stiffening or embrittlement. Both reactions lead to a reduction in the bond. However, the typical materials used in battery systems are steel and aluminium, and various fibre-reinforced plastics are not transparent to UV radiation. This means that the adhesive bonds can only be weakened when they are accessible from outside, which is practically never the case for most battery systems.

Due to the non-functionality of existing methods to weaken typical structural bonds in battery systems, the adhesive bonds must be designed from the outset for the application of debonding. Numerous methods for debonding bonds on demand

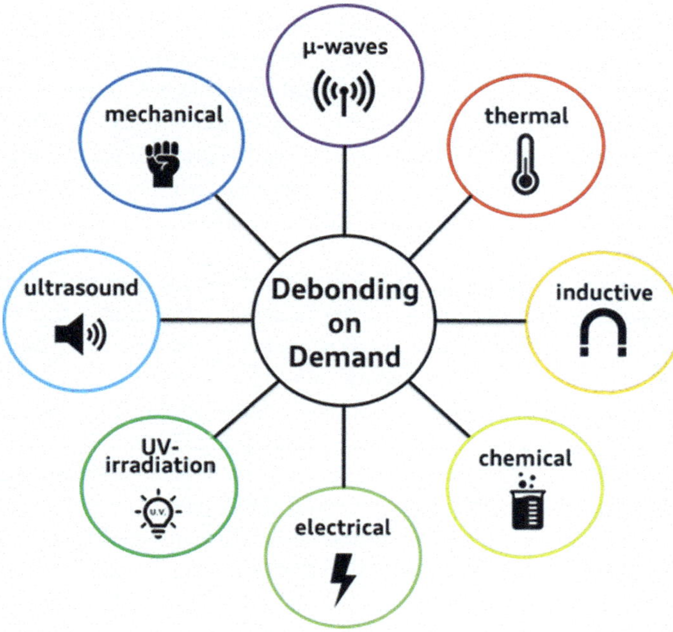

Fig. 18.1 Schematic summary of the different trigger methods for the debonding of adhesive joints

are described in the literature, which have been well summarised by Bandl et al. (2020) and by Mulcahy et al. (2022). A total of eight methods were identified by the authors that can be used as triggers to debond adhesive bondings, which are graphically illustrated in Fig. 18.1.

Debonding-optimised adhesives triggered by a thermal activation are the most frequently described trigger method. A distinction is made between a debonding-optimised polymer matrix, for example through the use of thermoplastics or vitrimers, and debonding through additives contained in the adhesive, such as thermally expandable particles (Banea et al. 2015). The activation of debonding by microwaves and induction relies on very similar processes but can both primarily be used for debonding of non-metallic joining partners. If microwave-absorbing particles, e.g. carbon nanotubes, are introduced into the polymeric adhesive layers, they can be excited by microwaves. This results in a localised heat input and can thus weaken the adhesive bond in a targeted manner. A localised supply of heat can also be induced by magnetic induction through the excitement of ferromagnetic particles contained in the adhesive (Caglara et al. 2024). However, neither of the three debonding-optimised variants are suitable for use in battery systems, as all options would lead to an unlocalised heat input due to the excitement of the metallic joining partners of the bonds, which then could damage the battery cells. In addition, the interaction between the battery cells and the induction fields or microwave radiation is uncertain.

There are also several methods to design an adhesive with a labile backbone towards specific chemicals. For example, Babra et al. (2019) described the usage of an aquatic tetrabutylammonium fluoride solution to degrade the polymer backbone of an silicium-containing adhesive. The reduced molecular weight leads to a reduction in the shear strength of adhesive bonds. In dental medicine, UV-triggered debonding on demand adhesives are used for the debonding of brackets as the adhesive areas are easily accessible by UV-light (Schenzel 2017). Despite that the chemical as well as the UV-triggered debonding on demand optimised adhesives suffer through the slow diffusion and the low transmission as the classical structural adhesive in battery systems.

Ultrasound as a debonding trigger can be used in the waste recycling of lithium ion batteries to separate adhesive from the bonded substrates (Lei et al. 2021). Again, the problem with separating intact battery cells using ultrasonic debonding on demand is that the ultrasound can damage the battery cells. The use of mechanical debondable adhesive, like gecko tapes or stretch release tapes, is not suitable for structural battery applications due to their low shear strength performance.

The use of electrical debonding, however, appears as very promising method, as it is not temperature reliant, acquires no chemical reactants, can be locally applied, and no strong forces are needed. The process of electrical debonding on demand was developed by Gilbert (1999). The author described the usage of an electrochemically debondable composition consisting of a matrix and an electrolyte functionality to adhere to two conductive surfaces. The matrix functionality was used to increase the mechanical strength of the material while enabling adhesion towards the substrates. In contrast, the task of the electrolyte functionality was to make the material conductive, so that included ions can migrate through the adhesive. If a voltage is applied towards the bond line through the substrates, the interfaces between the adhesive and the substrates are weakened. Thus, the substrates can be debonded with low forces afterwards.

Leijonmarck et al. (2012) firstly showed the usage of an ionic liquid (1-ethyl-3-methylimidazoliumethylsulfate) as a conductivity agent. The ionic liquid additionally works as the reactive species to weaken the interfaces of the adhesive bond between two aluminium substrates. When they applied a current to the adhesive joints, debonding occurred at the anode. They postulated that the anodic debonding mechanism is a multiple step process, starting with the oxidation of the anodic substrate to aluminium oxide, followed by the formation of an acidic, patchy sulphur-containing organic film. Whereas at the cathode mainly hydrogen gas was formed. Further, Anduix-Canto et al. (2022) investigated the mechanism of electrically debonding using 1-ethyl-3-methylimidazoliumethylsulfate as the conductive agent using advanced spectroscopy in detail. Figure 18.2 summarises their proposed mechanism of the electrochemical induced debonding on demand. Initially the ionic liquid is dissolved in the polymer matrix with no orientation of the charged ions. The application of an electric potential via the substrates (electrodes) leads to a charge separation in the ionic liquid through the orientation of the anion towards the positively charged anode and the orientation of the cations towards the negatively charged cathode. Different electrochemical reactions take place at both electrodes. At the cathode, a reduction of the cation takes place forming hydrogen gas and a reactive carbene species. The reaction of the carbene species with the polymer matrix leads to

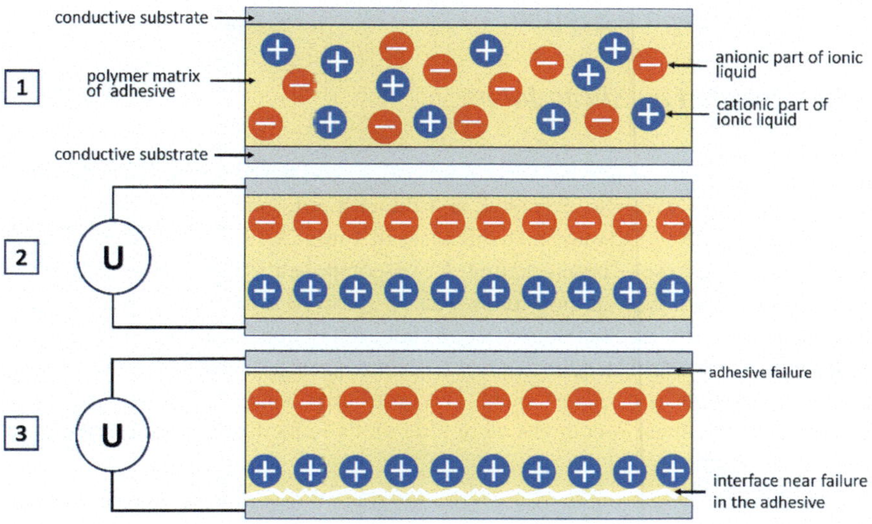

Fig. 18.2 Proposed mechanism of electrical debonding on demand by Anduix-Canto et al. (1) Shows the random distribution of an ionic liquid in an adhesive polymer matrix between two conductive substrates. (2) The orientation of the ions towards the electrode is depicted. (3) Describes the different failure modes through the debonding process

a polymer degradation. The combination of the polymer degradation and the expansion of the formed hydrogen gas leads to a drastic decrease in the interface strength leading to a debondable adhesive joint at the cathode. On the anode they showed similar results to Leijonmarck et al. (2012), showing an oxidation of the aluminium substrate, resulting in the generation of a brittle and weak aluminium oxide layer, which causes the strength decrease of the adhesive joint at the anode.

This work focuses on the influence of different electrochemical debonding parameters as well as the influence of the specimen geometry using an ionic liquid modified epoxy adhesive. The influence on the debonding properties of different aluminium alloys used in the automotive industry as well as the adhesive layer thickness is investigated. Furthermore, the debonding processes are analysed using different electric voltages for different timescales. Finally, investigations regarding the reuse in the sense of rebonding the substrates were made. Therefore, SLS-specimens were prepared using an adhesive which is optimised for electrically induced debonding on demand. Initial and the SLS-strength after the electrochemical debonding process were statically tested. Fracture images were optically characterised. To gain a deeper insight into the mechanism, the topography of the fracture surfaces of the SLS-specimen is analysed with field-emission SEM at low accelerating voltages and the chemical composition via EDX.

18.2 Experimental Section

18.2.1 Single Lap Shear Joints

The preparation of the single lap shear joints (SLS) was based on the DIN EN 1465. The substrates are wrought, extruded, and die-cast automotive aluminium alloys, with dimensions 25 × 120 mm and a thickness of 1 mm. All substrates were cleaned carefully with isopropanol before testing. A two-component epoxy adhesive modified with an ionic liquid was applied to an area of 25 × 10 mm of the top end of the substrate. Afterwards another substrate was adhesively bonded with an overlapping length of 25 × 10 mm. The adhesive layer thickness was adjusted to 0.5 mm, 1 mm, 1.5 mm, and 2 mm using glass beads. The protruding adhesive has been removed carefully. The lap shear specimens were afterwards immediately cured in the oven at 80 °C for 30 min.

18.2.2 Destructive Testing

All tensile shear tests were performed on a Z020 tensile testing machine from Zwick Reoll (Germany). The test speed was set on 10 mm/min and the test was performed at standard conditions. The analysis of the single lap shear strength was determined according to the DIN 1465 as the quotient of the maximal force divided by the measured adhesive area of each specimen. All tests were repeated four times for every parameter with individual samples.

18.2.3 Debonding Process

The debonding process was performed by applying different DC voltages (10 V, 30 V, 50 V) towards the substrates for 2 or 15 min. The positive pole was attached to one substrate and the negative pole to the other substrate of the lap shear joint using tap off terminals. As a voltage source a 1 kW EA-PS 9080-40 T system from Elektro Automatik GmbH (Germany) was used. The currents were measured with an UT803 multimeter from Uni-Trend Technology Co (China).

18.2.4 SEM and EDX-Images

All SEM-images were taken with an Ultra plus scanning electron microscope from Carl Zeiss Microscopy Deutschland GmbH [Germany] using a field emitter source and a secondary electron detector. The acceleration voltage was set on 1 keV to avoid charging effects. For EDX-measurements an Ultim Extreme detector from Oxford instruments [United Kingdom] was used. The acceleration voltage was also set at 1 keV.

18.2.5 *Cleaning and Rebonding*

For the rebonding experiments debonded, but uncleaned single lap shear substrates were reused uncleaned as joining partner. To investigate the influence of the cleaning process, cleaning was done with water or isopropanol. On the uncleaned as well as on the cleaned specimens, the adhesive was applied on the previously bonded area. The other adhesively bonded joining partner in the lap shear specimen was a new aluminium substrate. The adhesively bonded area was set to 10 × 25 mm with an adhesive layer thickness of 1 mm. The lap shear specimens were afterwards immediately cured in the oven at 80 °C for 30 min.

18.3 Analysis and Discussion

18.3.1 *Debonding of Single Lap Shear Joints*

The single lap shear (SLS) strength of different adhesively bonded aluminium alloys in dependence of the debonding process parameters is depicted in Fig. 18.3.

First, the SLS-strength of the reference specimen with an adhesive layer thickness of 0.5 mm was investigated for the different bonded substrates without any debonding

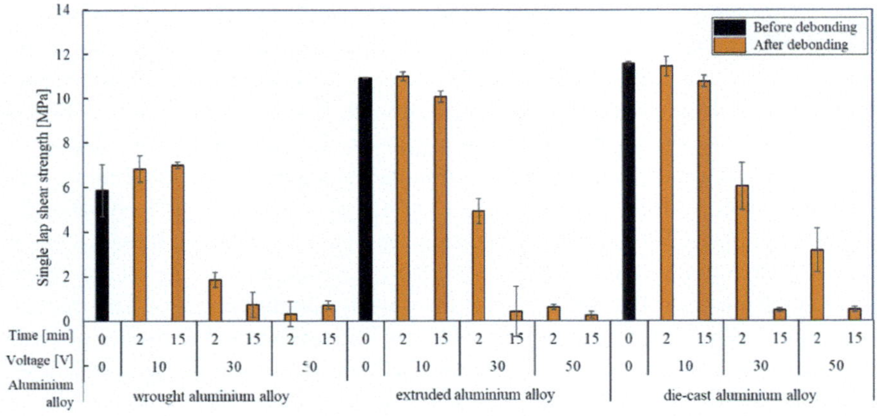

Fig. 18.3 Single lap shear strength of different aluminium alloys in dependence on the timeframe and the voltage of the debonding process

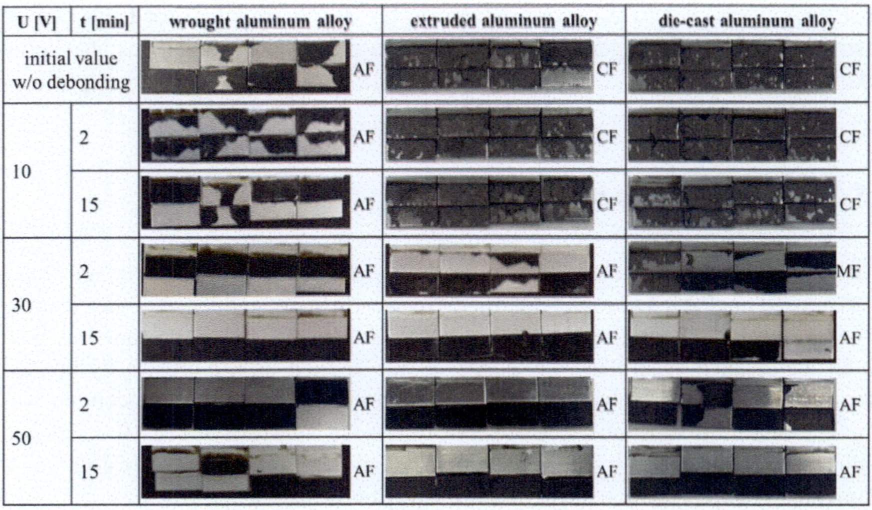

Fig. 18.4 Fracture pattern of the lap shear joints of different aluminium alloys in dependence on the timeframe and the voltage of the debonding process

process (black columns). The wrought aluminium alloy shows a significantly lower SLS-strength (5.9 MPa) than the extruded (10.9 MPa) and the die-cast aluminium alloy (11.5 MPa). An explanation for this could be the lower adhesion of the used adhesive towards the wrought aluminium alloy, as it shows an adhesive fracture in contrast to the cohesive fracture for the extruded and the die-cast aluminium alloy, illustrated in Fig. 18.4.

Furthermore, it is visible that the SLS-strengths decrease with increasing voltages for longer periods. The application of 10 V for 2 or 15 min does not lead to a significant decrease of the SLS-strength or a change in the fracture pattern for any of the investigated alloys in contrast to the initial values. In addition, it can be seen that a differentiation has to be made between the two periods of the measurement series with a voltage of 30 V. The application of the voltage for 2 min decreases the SLS-strength in contrast to the initial strength by approximate 50% whereas the application of the voltage for 15 min leads to decrease of approximate 90%. Also, the fracture images demonstrate different outcomes. While the 2 min series shows partial mixed fracture images, the 15 min series shows only adhesive fracture images. Finally, the 50 V series demonstrate the highest SLS-strength reductions of up to 95% to values under 0,5 MPa, with the only exception of the measurement for an application of 50 V for 2 min on the die-cast aluminium alloy. The fracture images once again illustrate an adhesive failure. The reason for the absence of some adhesive in the fracture images in Fig. 18.4 is that the adhesive bonding is weakened at both the cathode and the anode. This leads to easily detachable adhesive of both substrates.

18.3.2 SEM-Fracture Pattern Analysis

For a more detailed understanding of the fracture pattern and the mechanism a representative cohesively broken non-debonded and an adhesively broken debonded sample was investigated using scanning electron microscopy. The SEM-images are illustrated in Fig. 18.5. The SEM-images of the fraction pattern of the non-debonded illustrate on the left side the aluminium of substrate. On the right side the adhesive consisting of the polymer matrix and the fillers can be seen. The fracture pattern inside the adhesive shows a brittle structure and corresponds to a typical fracture pattern of a stiff epoxy adhesive. Therefore, it can be stated that the SEM-image correlates with the optical image and shows a mostly cohesive failure. In contrast the SEM-image of the debonded sample displays a very different structure. On the light grey aluminium, a dark ductile structure is visible which appears almost like a liquid, agglomerates around the fillers, and forms insulas. This could be the residue of the debonding process formed by an electrochemical induced reaction of the ionic liquid. During this faradic reaction a reactive species is formed, which was described by Anduix-canto et al., and can degrade the polymer structure, leading to a weakening or liquidation of the polymer at the interface (Anduix-Canto et al. 2022). Therefore, the SLS joint fails near the interfaces with a lowered SLS-strength. Due to the small height of the residue the failure pattern was indicated as an adhesive failure by the optical evaluation.

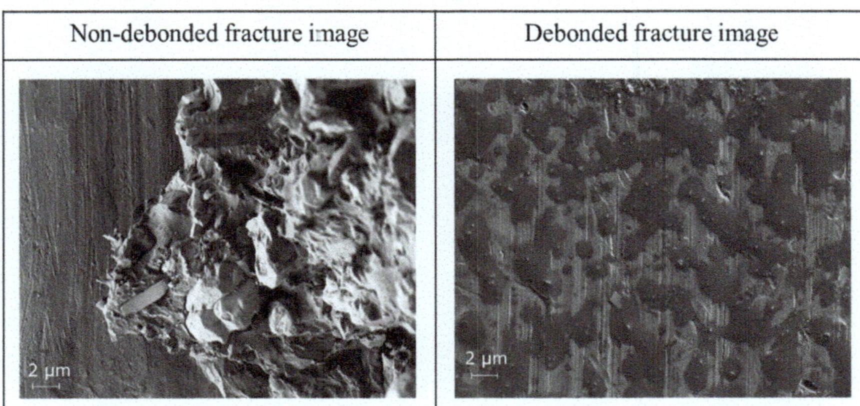

Fig. 18.5 Non-debonded and debonded fracture SEM-images of the single lap shear joints taken with Everhard-Thornley-SE2 detector at 1 keV

To gain a more precise insight into the mechanism, not only the aluminium surface of the fracture surface but also the adhesive surfaces were analysed separately according to the electrodes using SEM. The SEM-images are depicted in Fig. 18.6. The previously analysed SEM-images are depicted in the top left corner as the cathodic aluminium surface. The cathodic adhesive surfaces show the negative structure of the aluminium surface. In addition, the fracture pattern in the adhesive confirms the hypothesis that a polymer degradation occurs, as ductile fractures can also be recognised here. On the aluminium surface of the anode there was no change visible, and no residues of adhesive could be found. On the adhesive surfaces no changes could be found, except the imprints of the aluminium. This leads to the conclusion that on the anode a different mechanism should take place then on the cathode, which doesn't include a degradation of the polymer. Considering the article of Anduix-Canto et al. (2022) regarding the anodic surfaces it is probably a result of a chemical reaction (corrosion) of the aluminium surface on a nanometre scale, which cannot be shown with the used methods.

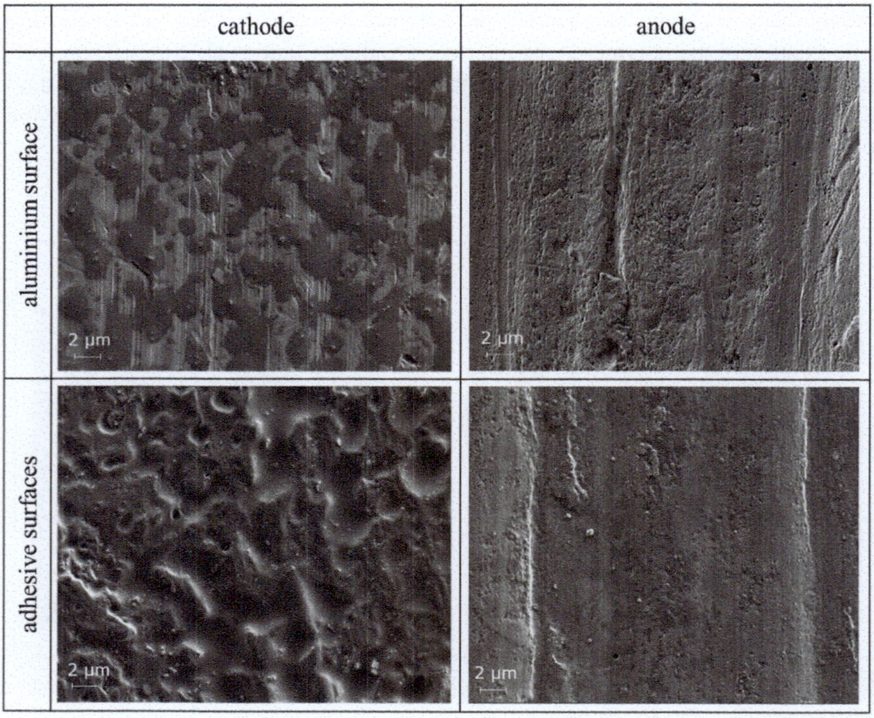

Fig. 18.6 Fracture image of the aluminium and adhesive surfaces for the cathode and anode of the single lap shear joints after the debonding taken with an SEM at 1 keV. Scale is depicted in the images

18.3.3 Investigation of the Debonding on Demand Process

Since we could show that the electrically induced debonding on demand achieves bigger SLS-strength reduction through the input of higher voltages for longer times, the goal was to find a parameter which indicates the status of the debonding. The curve of the electrical current seems to be a suitable indicator for the status of the debonding. In Fig. 18.7 an exemplary current curve over the time is illustrated. The current curve demonstrates an exponential decline over time, which indicates a change in the conductivity of the sample over the time. The decreased current curve is a result of the movement of the ions inside the adhesive to form an ionic double layer at the electrodes. Since the debonding process is dependent on the charge distribution or the presence of the ionic liquid at the electrodes within the adhesive, it can be postulated that after a high percentage drop in the current intensity and a flattening of the current curve, the electrical debonding process will be the most effective. Therefore, we plotted the SLS-strength against the percentage decrease of the current (Fig. 18.8). An almost linear correlation between the SLS-strength and the reduction in the current becomes recognisable. This seems to be promising especially for the high SLS-Strength reduction with high current reduction (>80%), but the scattering increases for samples that have not yet achieved the full current reduction (20–80%), i.e. have a SLS-strength over 1 MPa left. This effect is independent of the aluminium alloy used and the initial SLS-strength. Therefore, the reduction in the electrical current could be used to determine the progress of the debonding.

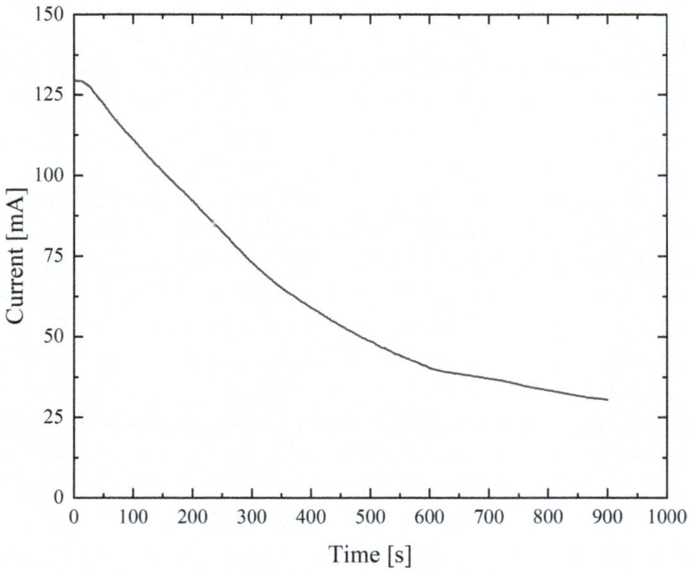

Fig. 18.7 Exemplary current curve over the time for the debonding process of 50 V for 15 min of a 0.5 mm SLS joint

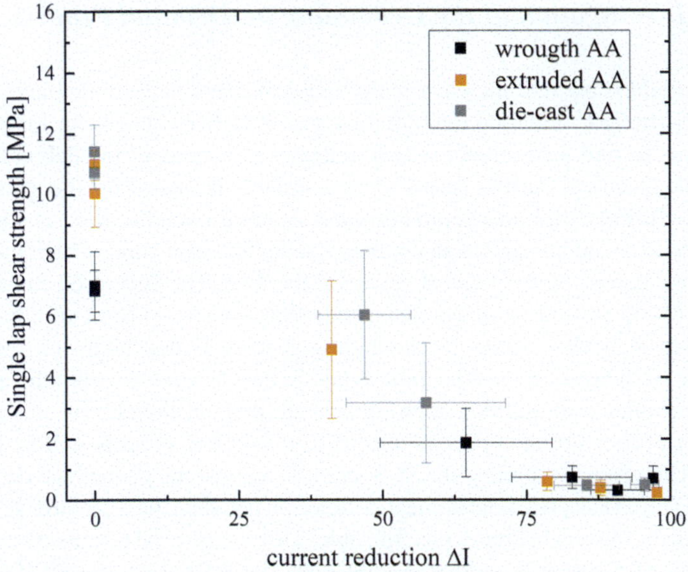

Fig. 18.8 SLS-strength after debonding plotted against the current reduction ΔI during the debonding process

18.3.4 Influence of the Adhesive Layer Thickness on the SLS-Strength

For an industrial process, the robustness towards process deviations, e.g. in the adhesive layer thickness, is important. Therefore, SLS-shear joints were made using die-cast aluminium as substrate with different adhesive layer thicknesses. For debonding voltage, 50 V for 15 min was chosen. The SLS-strength of the initial state and the strength after the debonding are shown in dependence of the adhesive layer thickness in Fig. 18.9. The initial SLS-strengths decrease with increasing adhesive layer thickness from 17 MPa for 0.5 mm to 13 MPa for 2.0 mm This is effect is commonly known in literature and relies on the increased bending moment within the specimen while testing (Habenicht 2008). In contrast, after debonding process low SLS-strengths for the lower adhesive layer thicknesses are shown and a linear increase with increasing layer thickness until nearly no SLS-strength reduction for 2 mm adhesive thickness. In general, it can be stated that the debonding effect decreases with increasing adhesive layer thickness due to the decreasing conductivity of the specimen. This is also expressed by the reduced currents (Fig. 18.10). The thicker adhesive layers reduce the initial current, which inhibits the movement of the ions through the adhesive layer. This results in the absence of the exponential decrease of the current, showing in an unchanged current over the whole time of the debonding process. The decreased ion movement and current prevents the degradation reactions during the debonding resulting in higher residual SLS-strengths.

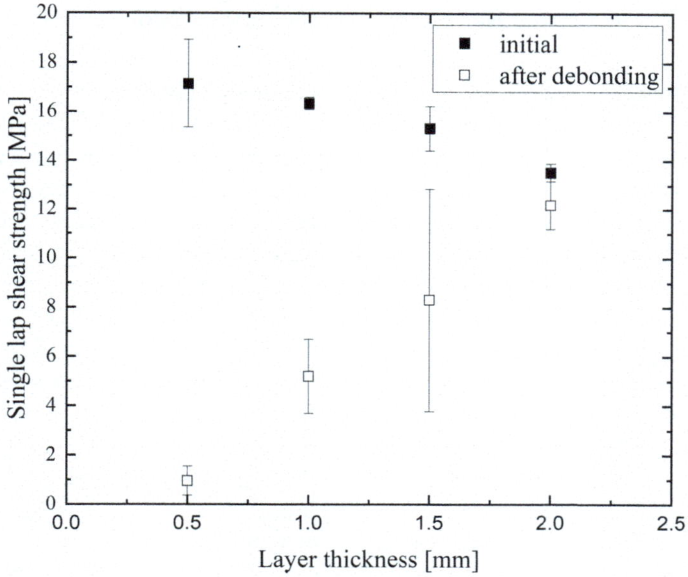

Fig. 18.9 SLS-Strength initial and after the debonding process as a function of the adhesive layer thickness

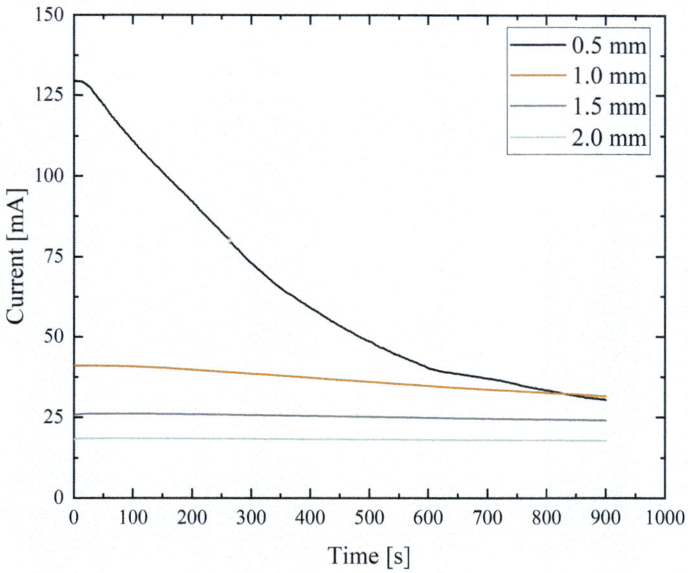

Fig. 18.10 Current curves during the debonding process of SLS joints with different adhesive layer thicknesses

18.3.5 Cleaning and Rebonding of Debonded Substrates

If a debonding process has taken place on automotive components that ought to be reused, a reliable rebonding concept is needed. For this reason, the rebonding of previously debonded substrates is investigated by adhesively bonding the used substrates to new ones. The SLS-strength of the cleaned and uncleaned surfaces is visualised in Fig. 18.11. Adhesive bonding of two new substrates achieved a SLS-strength of 10.4 MPa with a cohesive fracture image. The same results were achieved with used anode surfaces due to the reason that no residues could have been detected in the SEM-images (Fig. 18.12). If the cathode substrates are reused without cleaning, significantly lower SLS-strengths (4.0 MPa) were achieved due to the adhesion-reducing residues. The residue of the debonding process seems to reduce the adhesion of new adhesive towards the surface significantly. Therefore, different solvents as cleaning agents have been used. The water cleaned cathode surfaces show already a SLS-strength improvement (6.1 MPa), and isopropanol cleaned cathode substrates achieve comparable high SLS-strength (10.5 MPa) as new substrates. This can be justified by the SEM-images and EDX-mappings depicted in Fig. 18.12. In the EDX-mappings for uncleaned and water cleaned surfaces many carbon-containing compounds could be found, and the SEM-images show the evidence of the dark residue. After the cleaning with isopropanol no residues can be seen in the SEM-image. Also, the EDX-mapping shows no carbon-containing compounds anymore.

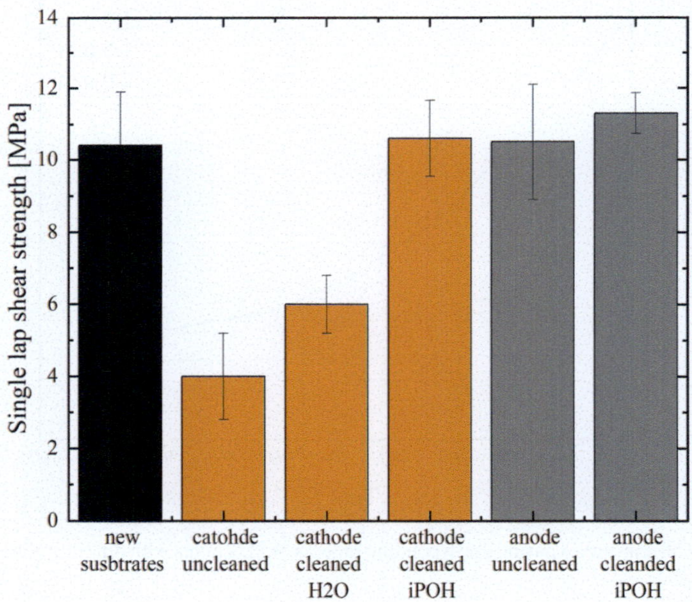

Fig. 18.11 Rebonding SLS-strengths of new, uncleaned, and cleaned aluminium substrates

An additional cleaning of the anode substrates shows no improvement of the SLS-strength. In conclusion, after cleaning with an organic solvent all debonded substrates can be reused.

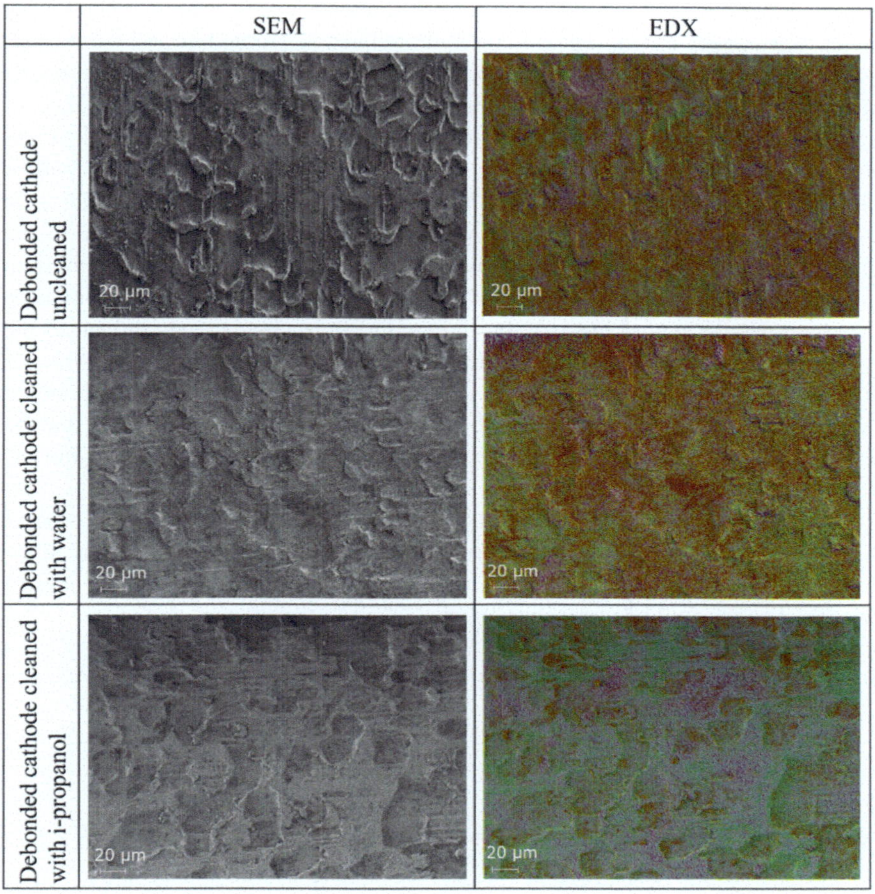

Fig. 18.12 SEM-images and EDX-mappings of uncleaned and cleaned debonded surfaces with different solvents. The measurements were taken with an acceleration speed of 1 kV. In the EDX-mappings the different colours stand for different atoms. Carbon is red, oxygen green, and aluminium violet

18.4 Conclusion

In this study, different debonding on demand methods for adhesive joints were reviewed with regard to a usage in the HV-battery systems of electrical vehicles. It turned out that especially the electrochemical induced debonding on demand has several advantages for this specific field of application. To understand the mechanism in more detail and to define most relevant parameters for the debonding process, several experiments have been performed. The results show that the application of a voltage to an adhesive optimised for debonding leads to a significant decrease in the SLS-strength. The induced currents trigger electrochemical reactions, which weaken the interfaces between the adhesive and the substrates corresponding to reduced adhesive strengths. In line with literature, a polymer degradation at the cathode could be proven by SEM analysis, whereas an oxidation of the aluminium substrate at the anode was on contrary not visible for all three used aluminium substrates. However, the choice of the aluminium alloy substrate only influences the initial bonding strength, with only minor impacts on the debonding process. Longer time scales of the debonding process and higher voltages increase the debonding capabilities. In contrast higher adhesive layer thickness decreases the current and the resulting debonding capabilities. Furthermore, all substrates can be reused after the debonding after cleaning with organic solvents.

Planned investigations on the surface chemistry during the debonding process using X-ray photoelectron spectroscopy or atomic force microscopy should allow deeper insights into the mechanism. In addition, a method to debond higher adhesive layer thicknesses will be under investigation.

Acknowledgements This work was financed by an external Ph.D. project of the AUDI AG.

References

Anduix-Canto C, Peral D, Pérez-Padilla V, Diaz-Rovira AM, Belmez Lledó A, Orme CA, Petrash S, Engels T, Chou KW (2022) Unraveling the mechanism of electrically induced adhesive debonding: a spectro-microscopic study. Adv Mater Inter. https://doi.org/10.1002/admi.202101447

Babra TS, Wood M, Godleman JS, Salimi S, Warriner C, Bazin N, Siviour CR, Hamley IW, Hayes W, Greenland BW (2019) Fluoride-responsive debond on demand adhesives: manipulating polymer crystallinity and hydrogen bonding to optimise adhesion strength at low bonding temperatures. Eur Polym J. https://doi.org/10.1016/j.eurpolymj.2019.07.038

Bandl C, Kern W, Schlögl S (2020) Adhesives for "debonding-on-demand": triggered release mechanisms and typical applications. Int J Adhes Adhes. https://doi.org/10.1016/j.ijadhadh.2020.102585

Banea MD, Da Silva L, Carbas R (2015) Debonding on command of adhesive joints for the automotive industry. I Int J Adhes Adhes. https://doi.org/10.1016/j.ijadhadh.2015.01.014

Bartley A, Chudalla N, Meschut G, Wibbeke TM (2023) Low temperature debonding of toughened structural adhesive joints: a new approach to repairs in the automotive industry. Int J Adhes Adhes. https://doi.org/10.1016/j.ijadhadh.2023.103486

Caglara H, Aksoyc YA, Idapalapatia S, Caglarc B, Sharmab M, Chian KS (2024) Debonding-on-demand Fe3O4-epoxy adhesively bonded dissimilar joints via electromagnetic induction he. J. Adhesion

Gilbert MD (1999) Electrically disbonding materials Patent US 7,968,188 B2, USA

Habenicht G (ed) (2008) Kleben. Grundlagen und Technologie, 6th edn. Springer, Berlin (2008)

Krumrey D (2000) Verfahren zum Entfernen von Klebemitteln von eine Oberfläche eines Fahrzeugkarossiebauteils Patent EP 1 029 602 A2, 9 February 2000

Lei C, Aldous I, Hartley JM, Thompson DL, Scott S, Hanson R, Anderson PA, Kendrick E, Sommerville R, Ryder KS, Abbott AP (2021) Lithium-ion battery recycling using high-intensity ultrasonication. Green Chem. https://doi.org/10.1039/D1GC01623G

Leijonmarck S, Cornell A, Danielsson C-O, Åkermark T, Brandner BD, Lindbergh G (2012) Electrolytically assisted debonding of adhesives: an experimental investigation. Int J Adhes Adhes. https://doi.org/10.1016/j.ijadhadh.2011.09.003

Mulcahy KR, Kilpatrick AFR, arper GDJ, Walton A, Abbott AP (2022) Debondable adhesives and their use in recycling. Green Chem 36–61 (2022)

Schenzel AM (2017) Advanced debonding on demand systems for dental adhesives. Dissertation, KIT (2017)

Chapter 19
Investigation of the Viscosity of Liquid Silicone Rubber with Added Recycled Particles and Impact on the Injection Molding Process

Simon J. Heienbrock, Dennis Mayer, Dennis F. Weißer, Elena Fischer, Alexander Melzer-Bartsch, and Matthias H. Deckert

Abstract Liquid silicone rubber (LSR) has gained considerable importance in various branches of industry due to its unique properties such as high heat resistance, chemical resistance and excellent aging properties. LSR is used in a wide range of applications, from seals to medical devices and food components. The LSR injection molding process enables precise processing with short cycle times, but there is currently no recycling concept for LSR. This study investigates the processing properties of virgin LSR with added mechanically recycled LSR material in the injection molding process. Rheological tests of LSR compounds with varying particle sizes of the recycled material show significant effects on viscosity and processing parameters during injection molding. Nevertheless, processability in the injection molding process is possible. The viscosity of the material increases with a higher proportion of recycling amount, particularly with finer particles, but the surface quality remains very high with these small particles. With larger particles, significant surface defects

S. J. Heienbrock (✉)
Esslingen University of Applied Sciences, Kanalstraße 33, 73728 Esslingen, Germany
e-mail: Simon.Heienbrock@hs-esslingen.de

D. Mayer · D. F. Weißer · E. Fischer · A. Melzer-Bartsch · M. H. Deckert
Institute for Applied Polymer Technologies, University of Applied Sciences Esslingen, Kanalstraße 33, 73728 Esslingen, Germany
e-mail: Dennis.Mayer@hs-esslingen.de

D. F. Weißer
e-mail: Dennis.Weisser@hs-esslingen.de

E. Fischer
e-mail: Elena.Fischer@hs-esslingen.de

A. Melzer-Bartsch
e-mail: Alexander.Melzer-Bartsch@hs-esslingen.de

M. H. Deckert
e-mail: Matthias.Deckert@hs-esslingen.de

H. Altenbach et al. (eds.), *Lectures Notes on Advanced Structured Materials 3*,
Advanced Structured Materials 226, https://doi.org/10.1007/978-3-031-77403-4_19

are noticeable. These results show the potential of LSR recycling for economic and ecological benefits in new products, but at the same time emphasize the need for precise process control and control of the material composition.

19.1 Introduction

Liquid silicone rubber (LSR) has gained importance in various industries due to its exceptional properties and efficient processability. LSR has a number of excellent properties, such as good flexibility at low temperatures and high heat resistance, which makes it suitable for applications in a temperature range from –50 to 200°C. In addition, LSR is resistant to chemicals and resistant to oxidation, aging and weathering. LSR also has high electrical insulating properties and can be used in contact with food and in medical applications due to its biocompatibility. Special customized LSR formulations are also used for optical applications. Due to its unique properties, LSR is used in a variety of industries for seals, vibration dampers, membranes, catheters, insulators, O-rings, or baking molds (Goodship 2017; Henze 1999; Mayer et al. 2022; Momentive Performance Materials Inc 2018).

19.1.1 *Structure of Silicone*

Silicones, also known as polyorganosiloxanes, are synthetic polymers characterized by repeating Si–O bonds (siloxane backbone). As shown in Fig. 19.1, the silicon atoms are additionally linked to organic groups such as methyl groups. Silicones can be classified into liquids, resins and elastomers based on their viscosity. Elastomers, like liquid silicone rubber (LSR), generally consist of vinyl-terminated polydimethylsiloxane (PDMS) (Moretto et al. 2000).

Since its introduction in the late 1970s, LSR has formed its own class of materials and was developed specifically for processing on injection molding machines. Today, LSR is increasingly popular and is usually offered as a two-component system for injection molding applications. To create a three-dimensional network from the two pasty components, an irreversible chemical addition crosslinking is initiated,

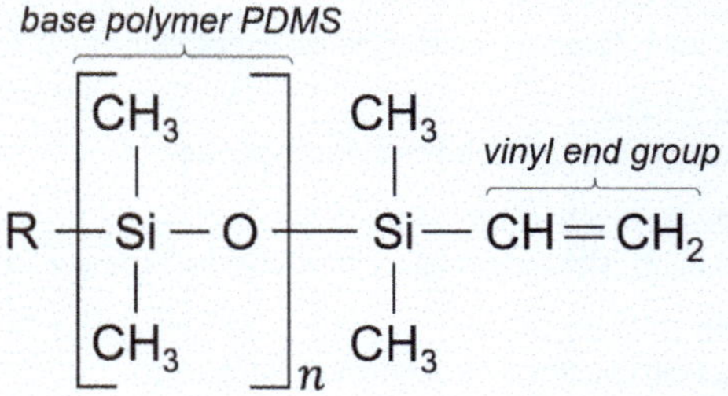

Fig. 19.1 Structural formula of vinyl-terminated polydimethylsiloxane consisting of repeating PDMS units as base polymer and vinyl end groups that are required for the cross-linking reaction (Mayer et al. 2024)

which starts at room temperature and rapidly increases in speed as the temperature rises. An inhibitor is added to slow down the reaction at room temperature and prevent premature cross-linking during processing (Goodship 2017; Moretto et al. 2000). In the initial state, the platinum catalyst is first inactivated by the inhibitor. By increasing the temperature, the platinum-inhibitor complex is deactivated, and the cross-linking begins (Beyer and Wolf 2015). Pyrogenic or precipitated silica is used as reinforcement to ensure the mechanical properties of the LSR (Henze 1999).

Processing LSR in the injection molding process offers numerous advantages: LSR can be precisely processed with short cycle times while maintaining a long pot life at room temperature. Due to its low viscosity, which varies from that of honey to water depending on the type, even complex and thin-walled components with long flow paths can be manufactured with LSR (Weißer et al. 2022).

The rheological properties of LSR are important for the injection molding process and are influenced by several factors, including temperature, shear rate, degree of crosslinking, pressure and reinforcement content. Higher temperatures generally reduce the viscosity until the crosslinking reaction starts, causing the viscosity to increase abruptly. Similarly, a higher shear rate leads to a lower viscosity due to the shear-thinning behavior of LSR (Weißer et al. 2022). One possible method to characterize the viscosity of LSR is to use a rotational rheometer or a high-pressure capillary rheometer (Lauth and Kowalczyk 2015).

19.1.2 LSR Injection Molding Process

The injection molding process for LSR differs from the usual injection molding of thermoplastics and requires special machine technology. The LSR injection molding process can be divided into three main areas: Dosing system, LSR cylinder module and injection mold as shown in Fig. 19.2.

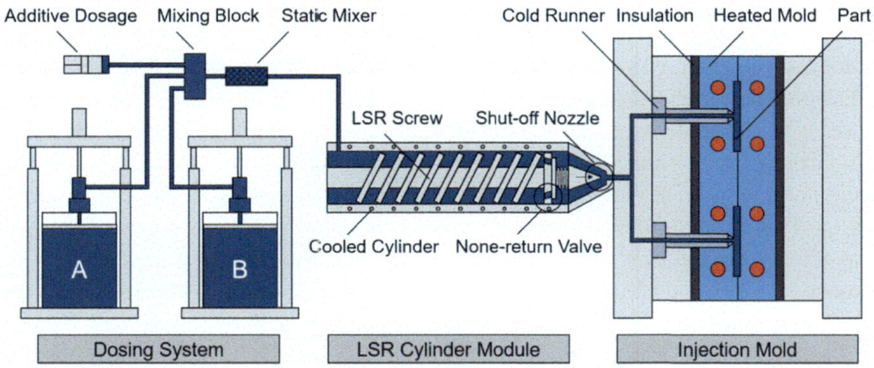

Fig. 19.2 Schematic representation of the liquid silicone rubber injection molding process including its components dosing system, cylinder module and injection mold

The LSR is supplied in two separate containers for components A and B. Piston pumps transfer the material from the 20- or 200-L containers to the injection molding machine through thick hoses. The dosing pressure is typically around 50 bar. Shortly before the injection unit is reached, components A and B are mixed homogeneously in a 1:1 ratio in the static mixer. This is where the first intensive shearing of the material takes place. Additives such as color pigments or pastes can be dosed into the mixing block via an additional additive dosing unit (Mayer et al. 2024; Bont et al. 2021; Weißer et al. 2023).

In the injection unit, the material is conveyed to the mold and metered by the screw, where a second shearing of the LSR material occurs. The shear stress reduces the viscosity by loosening physical interactions. The material is then injected into the hot injection mold where it flows under high pressure due to the narrow sprue geometry. Therefore, the LSR exposes the greatest shear stress during the injection molding process (Beyer and Wolf 2015). During this process, the viscosity of the LSR can drop to values of approx. 1–10 Pa-s at high shear rates for standard silicones (Henze 1999).

The molds for LSR are heated to temperatures between 120 and 220°C to initiate the crosslinking reaction (Bont et al. 2021; Weißer et al. 2021). Often, the mold is divided into cold and hot sections, which are separated by insulation. The cold section includes the sprue and cold runner system, which cools the material until it reaches the cavity (Goodship 2017; Weißer et al. 2023). Components manufactured in this way do not produce any waste from sprues. However, there are also molds that do not have a cold runner system. With these components, a sprue remains on the part, which must be subsequently cut off and disposed of as waste.

19.1.3 Recycling of Silicone Material

The recycling of silicone elastomers offers environmental benefits and technical challenges. One promising method is to grind cross-linked silicone elastomers and reuse the resulting powder as a filler in new polymers. This approach reduces the ecological footprint, reduces dependence on primary raw materials and saves costs.

Despite the beneficial properties of LSR, the structure of silicone poses challenges for recycling. The strong Si–O bond in silicones makes controlled degradation difficult. Grinding cross-linked silicone elastomers and adding the powder as a filler in virgin polymer is an attractive approach to reduce the amount of virgin silicone elastomers. This method offers ecologically and economically relevant potential (Mayer et al. 2024). In addition to reducing the ecological footprint and dependence on primary raw materials, this approach contributes to compliance with regulations on waste prevention and recycling, such as the EU's Circular Economy Action Plan or the Resource Conservation and Recovery Act. The introduction of environmental management systems such as ISO 14001 is also becoming increasingly relevant (Mayer et al. 2024).

In general, polymer recycling, aside from thermal utilization for energy, can be categorized into two main types: chemical recycling, which aims to recover monomers and material reuse (Ikeda et al. 2003). This study focuses exclusively on mechanical recycling. The material reuse of silicone rubber, particularly solvent-free mechanical recycling approaches, is rarely addressed in the literature. Two relevant studies can be considered here (Yamada and Yoshida 2003; Ghosh et al. 2003). In both studies, silicone rubber powder was produced from fully cross-linked components and used as a filler embedded in a silicone rubber matrix.

19.2 Experimental

19.2.1 Material Preparation

Recycled material made out of Silopren LSR 2050 from Momentive Performance Materials Inc. with a hardness of 50 Shore A was used for the investigation in this study. Waste parts and sprues from injection molding tests conducted at Esslingen University of Applied Sciences were used for this purpose. These components were ground at the Chair of Lightweight Structures and Plastics Processing at Chemnitz University of Technology using an ambient grinding system specially developed for the recycling of elastomers (Hoyer 2015; Kroll and Hoyer 2017). Grinding takes place at room temperature and therefore above the glass transition temperature. As a result, the LSR remains flexible during grinding, which leads to particles with an irregular and rough surface. Figure 19.3 shows a microscopic image of these particles, in which the irregular and jagged structures are clearly visible. Due to the rough surface and the resulting opacity, the particles appear black in the microscope image.

In contrast to ambient grinding, cryogenic grinding involves the comminution below the glass transition temperature, which is around -125 °C. This is achieved with the help of liquid nitrogen. Due to the material's brittleness at low temperatures, the ground material produced in this way has a more homogeneous particle shape and a smoother surface. Smaller particle sizes can also be achieved. However, due to the use of liquid nitrogen and complex plant technology, cryogenic grinding is very expensive and more suitable for laboratory scale (Mayer et al. 2024; Baranwal and Klingensmith 2001). For this reason, only ground material from the ambient grinding process is considered in this study.

Milled material normally has a wide particle size distribution, so that it must be sieved to quantify the particle size distribution using a sieve curve. The sieving curve of the produced powder is shown in Fig. 19.4 and indicates that the powder contains a high proportion of large particles. Approximately 30% of the particles are larger than 1600 μm, while only a small fraction is smaller than 600 μm.

Fig. 19.3 Microscope image of particles produced using the ambient milling process. Particles have an irregular and rough surface. Black appearance due to the rough and opaque surface, after grinding

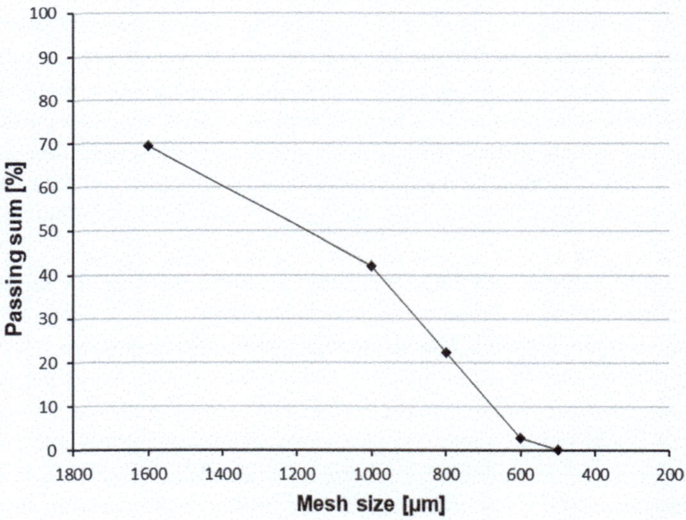

Fig. 19.4 Sieve curve of the LSR powder used within this study. The coarse powder contains 31% particles that are bigger than 1600 μm in size but only 2.8% that are smaller than 600 μm

Accurately determining the particle size of elastomers during sieving is challenging, as the elastic particles deform and larger particles can pass through the sieve meshes. The sieve curve was mainly created for measurement purposes to provide a rough estimate of the particle size distribution. The material obtained from the grinding process of the University of Technology Chemnitz contains the entire particle size distribution that was achieved during ambient grinding process.

One aim of the study is to investigate the influence of different particle sizes on the viscosity of the recycled material and the corresponding processing properties in the injection molding process. The material was therefore sieved into three different particle size fractions using laboratory sieves. Table 19.1 shows the particle size ranges used for the test series. Very large particles were intentionally included in the tests to identify potential processing problems and to examine their effects on component quality. A previous investigation into the processability of unscreened recycled material has already been conducted in Mayer et al. (2024), whereby this study focuses on the mechanical properties of the produced samples.

Figure 19.5 shows the laboratory sieves and the resulting particle fractions. It is noticeable that the finest particles tend to clump together slightly, which is not the case with the coarse and medium particles.

Silopren LSR 2650 silicone was used as the matrix material for the tests. This is also a standard LSR with a specified hardness of 50 Shore A. The measured mechanical properties of both Silopren LSR 2050, from which the recycled material is made, and Silopren LSR 2650 are comparable. The use of different material formulations for matrix and powder is not considered critical, as the materials are very similar.

Table 19.1 Particle sizes of the recyclate LSR material used for the tests to investigate the viscosity and processing properties of the LSR in the injection molding process

Designation of particle size	Coarse	Medium	Fine
Particle size	>1,200μm	600μm – 1,200μm	<600μm

Fig. 19.5 Left: sieves for separating different particle sizes, right: sieved particles in decreasing size. The finest particles stick together a little bit

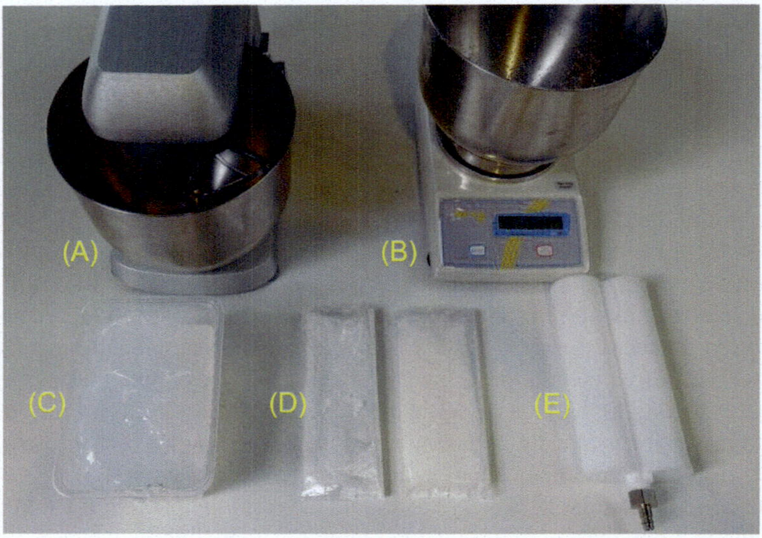

Fig. 19.6 Setup for weighing, mixing and filling the LSR powder compound. (A) Conventional kitchen machine with mixer blade for homogeneous mixing of LSR and powder, (B) balance for weighting the various LSR powder compounds, (C) Pre-mixed LSR out of a standard dosing system laboratory, (D) LSR powder and (E) cartridge and plungers into which the compound is filled

To produce the LSR compounds, different proportions of virgin LSR and powder in the three sieved particle sizes were weighed on a laboratory balance, as shown in Fig. 19.6. The premixed, untreated LSR was fed directly from a standard dosing system. Powder contents of 10% and 20% by weight of the total mass of the mixture were weighed. The matrix and particles were then mixed for 5 minutes using a conventional kitchen machine.

After mixing, the compound was placed in a desiccator under vacuum for an hour to remove trapped air. This is not absolutely necessary, as a large proportion of the air is also pressed out when the material is dispensed, but it reduces the risk of air pockets in the component. The mixtures were then filled into sealable cartridges. These cartridges can be fed to the injection molding cylinder using a cartridge dosing system.

19.2.2 Sample Part and Injection Molding

The test samples were produced on an Arburg 370A-600–170 injection molding machine with an LSR cylinder module. The compound was dispensed using a cartridge dispensing system consisting of a pneumatic dispensing gun, a double cartridge and a static mixer. The setup is shown in Fig. 19.7. The pressure that can be generated with the pneumatic dosing gun can be adjusted manually and is a maximum of 6 bar.

Fig. 19.7 Setup for manual feeding of silicone, (A) pneumatic air gun for generating dosing pressure (max. 6 bar), (B) cartridge with recyclate material, (C) static mixer, (D) injection molding unit

Plates with the dimensions 60 mm × 60 mm and a thickness of 2 mm are produced by injection molding and can be seen in Fig. 19.8. The double plate is injection molded with a 4 mm diameter gate and two film sprues with a width of 60 mm and a height of 1.5 mm. As depicted in Fig. 19.8, a remote pressure sensor is installed to record the cavity pressure during the process. This is a type 6157B pressure sensor from Kistler Instrumente AG.

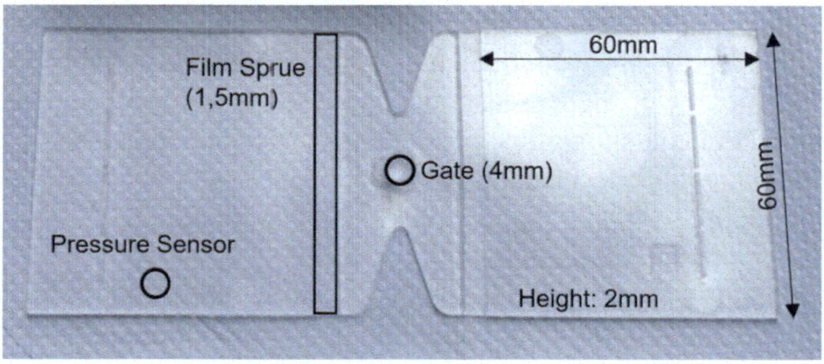

Fig. 19.8 Injection molded plate with the dimensions of 60mm x 60mm and a thickness of 2mm including sprue

The ComoNeo process monitoring system was used to analyze the cavity pressure during injection molding. Both the progression and the pressure level at the end of the injection molding cycle were examined. The cavity pressure was only used to monitor the constant process and was not considered separately in this study.

The mold is heated with oil to a temperature of 160 °C. The test samples were produced with a dosing volume of 18.1 ccm, an injection speed of 15 ccm s^{-1} and a heating time of 40 s. These injection molding parameters were kept constant for all variations of the powder content of 0, 10 and 20 wt.% and for different particle sizes. To evaluate the process, the maximum injection pressure and the required dosing time were recorded in each case.

19.2.3 Rheometer Measurement

The viscosities of the LSR mixtures were measured using an MCR 102e stress-controlled rheometer from Anton Paar GmbH. The measurement setup consisted of a parallel plate geometry with plates that had a diameter of 25 mm. The LSR mixtures were prepared gravimetrically for each measurement using a laboratory precision balance and a stirrer. To accurately measure the viscosity of the mixed material, the measuring gap between the plates had to be larger than the maximum particle size. Therefore, a gap of 1 mm was used for the virgin material and the recycled material with fine particles, 2 mm for the recycled material with medium-sized particles and 3 mm for coarse particles.

To determine the shear-dependent viscosity of the mixtures, a shear rate sweep was carried out over a logarithmic ramp from 0.1 to 100 s^{-1} at a constant temperature of 25 °C. The corresponding measured values are shown in Table 19.2. The step size corresponds to the standard settings for determining the shear-dependent viscosity.

Figure 19.9 shows an example of a measurement of virgin material and a mixture with 20% coarse particles on a logarithmic scale. The viscosity decreases with increasing shear rate for both curves. There is also an overlap between the measurement curves at a shear rate of around 6 s^{-1}. This is due to the fact that material is pushed outwards at higher shear rates and the measuring gap is no longer completely filled with material. This effect is particularly noticeable with large particles. Figure 19.10 shows the particles being pushed outwards at the end of a measurement. Measurements with higher shear rates are therefore no longer comparable with this method. To evaluate the results from the rheometer measurements, a shear rate was selected where this effect had not yet occurred.

Table 19.2 Test program used for measuring the viscosity with 21 steps for shear rate (logarithmic ramp)

Steps [–]	Shear rate [s^{-1}]
1	0,1
2	0,141
3	0,2
4	0,282
5	0,398
6	0,562
7	0,794
8	1,12
9	1,58
10	2,24
11	3,16
12	4,47
13	6,31
14	8,91
15	12,6
16	17,8
17	25,1
18	35,5
19	50,1
20	70,8
21	100

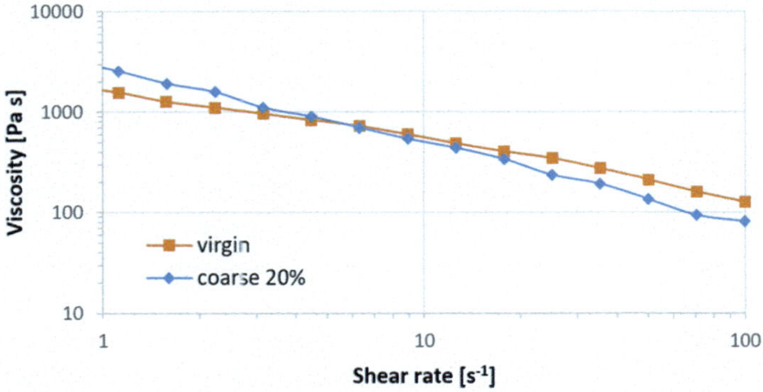

Fig. 19.9 Viscosity measurement of virgin Silopren LSR 2650 and material with 20% coarse powder for shear rates up to 100 s^{-1} (logarithmic scale)

Fig. 19.10 Viscosity measurement for coarse recyclate material with the rheometer MCR 102e from Anton Paar, plate with diameter of 25 mm and a measuring distance of 3 mm

19.3 Experimental Results

19.3.1 Material Viscosity

Investigating the viscosity of liquid silicone rubber combined with recycled material is essential for understanding processing properties and ensuring quality in end products. The aim of these measurements is to evaluate the influence of different particle sizes and recyclate proportions on the viscosity of the material and thus draw conclusions about processing and material properties.

The viscosity measurements were conducted for the virgin material and all recyclate mixtures using the measurement method already described. The results are depicted in Fig. 19.11 for the fine particles, in Fig. 19.12 for the medium particles and in Fig. 19.13 for the coarse particles. The results are shown on a logarithmic scale for all samples up to a shear rate of 10 s^{-1}. Across all graphs, it is evident that viscosity increases with higher proportions of recycled material. For the medium and coarse particles, the curves converge from a certain shear rate, which is due to the expulsion of the silicone particles. This problem has already been discussed in the rheology measurement chapter.

Due to this behavior, a very low shear rate was selected to compare the different mixtures. Although this is not ideal, as very high shear rates are present during injection molding and it is not possible to transfer the results directly, the results nevertheless provide an indication of the change in viscosity as a function of the particle size and the proportion of recycled material.

Figure 19.14 shows the viscosity of all mixtures at a shear rate of 1.12 s^{-1}. It can be seen that the viscosity increases significantly with the fine particles. At 10% recycled content, the viscosity increases by approximately 60%, at 20% recycled content even

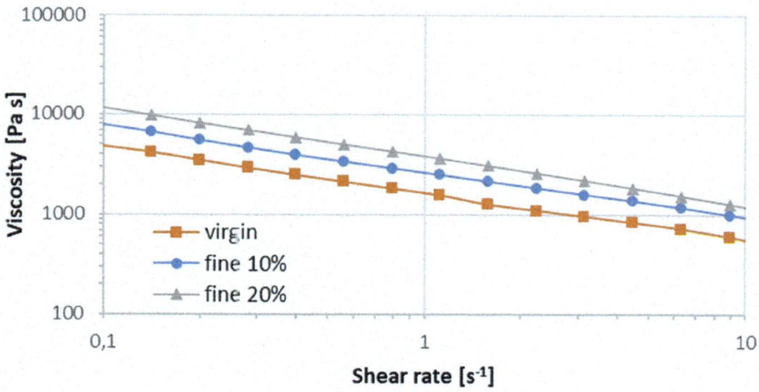

Fig. 19.11 Viscosity over shear rate of virgin Silopren LSR 2650 and recyclate compound with 10% and 20% **fine** powder (logarithmic scale)

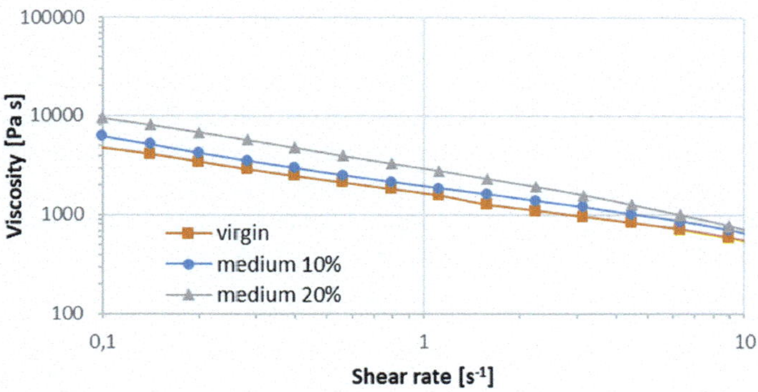

Fig. 19.12 Viscosity over shear rate of virgin Silopren LSR 2650 and recyclate compound with 10% and 20% **medium** powder (logarithmic scale)

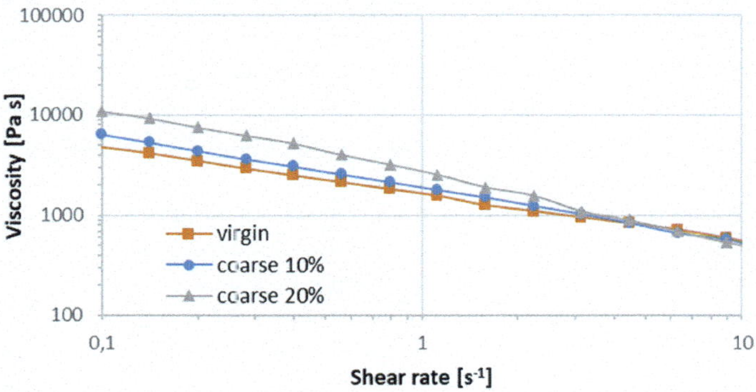

Fig. 19.13 Viscosity over shear rate of virgin Silopren LSR 2650 and recyclate compound with 10% and 20% **coarse** powder (logarithmic scale)

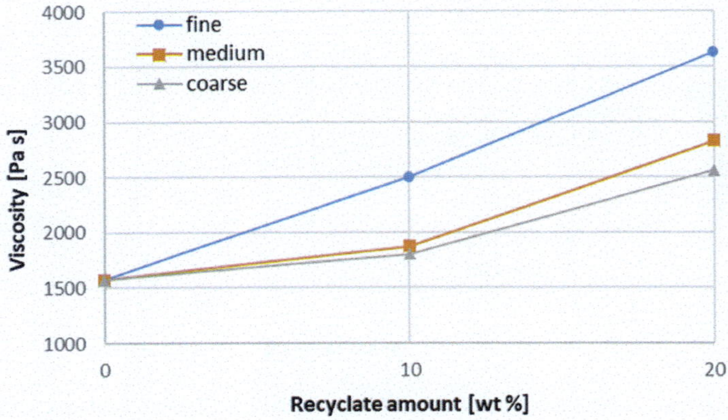

Fig. 19.14 Viscosity of compounds with different amounts of recyclate at a shear rate of 1,12 s^{-1}

by approximately 130%. For the medium and coarse particles, this increase is much smaller and is only approx. 60% for the coarse particles at 20% recyclate content.

The significant increase in viscosity with smaller particles is due to the fact that smaller particles have a larger total surface area in relation to their volume. This larger surface area leads to stronger interactions between the particles and the virgin material, which increases the viscosity (Lauth and Kowalczyk 2015; Ancey and Jorrot 2001; Geng et al. 2017).

19.3.2 Processing

To investigate the influence of the different powder contents on the injection molding process, all compounds were manufactured with the same injection molding parameters. It was observed that the filling of the components did not cause problems for any of the recycled compounds. This indicates that no or very little material was forced back into the cylinder during injection and that the valve gate nozzle of the cylinder module was always sealed.

However, it is not possible to make a clear statement about the function of the non-return valve, as the parts were injected without holding pressure. Furthermore, there was no noticeable drop in the cavity pressure signal, which indicates that the function of the non-return valve was not impaired by the powder particles.

After the tests, the silicone screw was disassembled and the non-return valve was examined. As depicted on the left-hand side in Fig. 19.15, the non-return valve is fully intact, what is affirming a proper functioning. However, it can be seen that there are large silicone particles in the holes of the non-return valve. It is assumed that these are pressed out by the dosing force and had no influence on the production of the silicone parts. For a detailed analysis of the non-return valve's performance, future

Fig. 19.15 Silicone screw (left), non-return valve with visible silicone particles in the holes (right)

studies should involve producing components with holding pressure and analyzing the mass cushion for each component individually.

To evaluate the processability, the dosing time and the injection pressure were analyzed for all mixtures. As can be seen in Fig. 19.16, the injection pressure rises with increasing powder content and is highest for the fine particles. The injection pressure increases from 150 to 215 bar, which corresponds to an increase of about 43%. This increase in pressure is hardly noticeable for the medium and coarse particles.

The significant increase in viscosity observed in the rheology measurements in the previous chapter is not noticeable to the same extent here. A possible reason could be a decrease in viscosity at significantly higher shear rates, which should be investigated further.

The consistent rise in injection pressure with greater powder content can be attributed to the higher solid concentration in the mixture. This leads to higher friction, which means that higher pressures are required to fill the component.

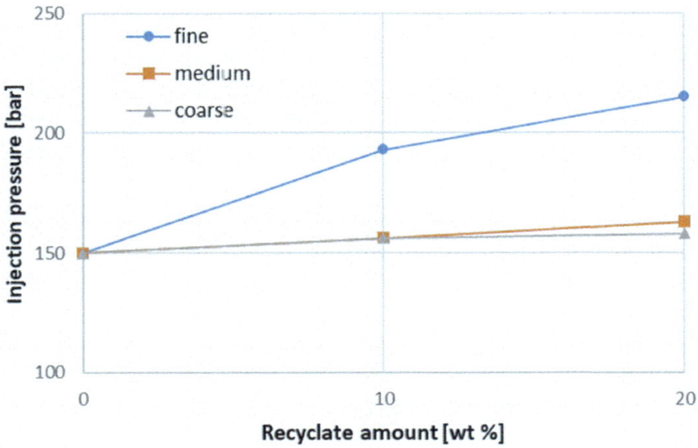

Fig. 19.16 Influence of the powder content and particle size on the injection pressure during injection molding. The injection pressure increases continuously as the powder content increases. The effect with coarse and medium-sized particles is very small

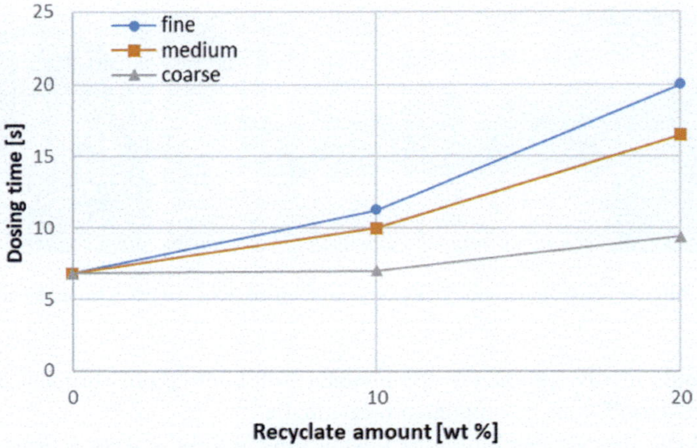

Fig. 19.17 Influence of the powder content and particle size on the dosing time during injection molding. The dosing time increases continuously as the powder content increases. The main reason for the sharp increase is the low pressure in the dosing system of only 6 bar

Figure 19.17 shows the required dosing time for all mixtures. For powder contents of 0–20% by weight, there is a consistent increase in dosing time observed for all particle sizes. For fine recycled material, the dosing time increases from approximately 7 s to approximately 20s. Only a very slight increase can be observed for coarse material.

The constant increase in dosing time has the same reason that prevents the dosing of higher powder contents than 40% by weight. The low pressure of the cartridge dosing system of 6 bar dictates that as viscosity of the mixture increases, material flow from the cartridge to the screw slows down. The increase of the dosing time is therefore mainly due to the lower material pressure in the feed. It can be assumed that such an increase in dosing time would not occur with a conventional dispensing system with a pressure of around 50 bar. There is currently no commercially available solution for feeding silicone particles into the dosing system, for example through an additive dosing device.

The injection pressure and dosing time have already been investigated in Mayer et al. (2024) for a wider range of particle sizes. The results of this study confirm the earlier findings.

19.3.3 Optical Component Analysis

To assess the injection molded parts, the component surface of the silicone sheets was examined for surface defects due to the addition of recycled material using the Leica Olympus SC50 microscope. For this purpose, a sheet without recycled material and sheets with three different particle sizes were analyzed.

Figure 19.18 shows the microscope images at 20 × magnification: (a) shows the sheet with no recycled content, while (b) to (d) show the sheets with a recycled content of 20% and different particle sizes. The injection mold used is not polished, but merely ground. Therefore, the surface of the silicone sheets without recyclate content is characterized by a large number of scratches, which are very well molded due to the low viscosity of the LSR.

The microscope images indicate that the number of surface defects increases with larger particle sizes and higher proportions of recycled material. This is due to the fact that the particles are pressed against the mold wall by the cavity pressure, causing deformation during the molding process. This deformation is reversed after ejection from the mold and the particles attempt to regain their original shape. These areas are recognizable as bumps on the component surface.

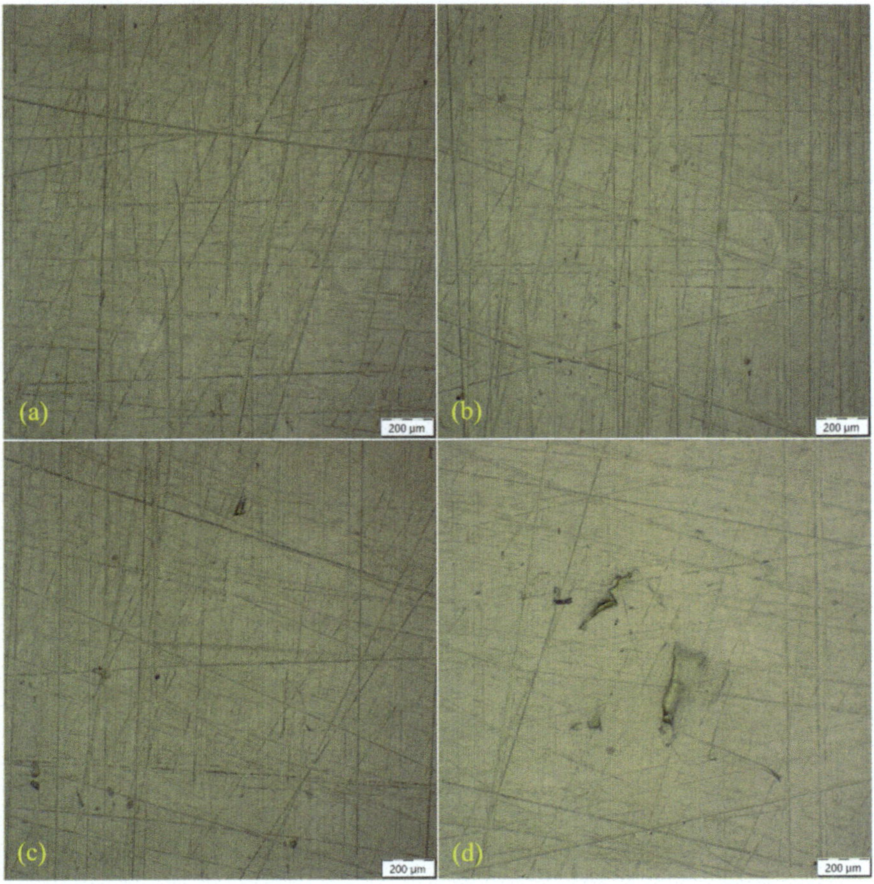

Fig. 19.18 Microscope images of the silicone component surface for different particle sizes in the recyclate **a**: Silopren LSR 2650 natural, **b**: Silopren LSR 2650 with 20% recyclate (particles <600 µm); **c**: Silopren LSR 2650 with 20% recyclate (particles 600 µm –1,200 µm); **d**: Silopren LSR 2650 with 20% recyclate (particles >1,200 µm)

19.4 Conclusion and Outlook

This study examines the use of recycled powder from LSR components as a filler for new LSR. The viscosity of the material mixtures and the processability of the compound with the standard equipment of the liquid injection molding process are analyzed. To investigate the processability, the feeding of the material through the non-return valve, the dosing of the shot volume and the injection pressure were examined. Rheological tests of the material mixtures were conducted to determine material viscosity.

The study demonstrates that the processing of LSR powder compounds with different particle sizes and powder contents of up to 20 wt.% is possible on a standard injection molding machine without any significant process engineering problems. The set and resulting injection molding parameters differ only slightly from the processing of virgin LSR. The injection pressure increased from 150 to 215 bar for the very fine particles (<600 μm) at 20% recyclate content due to the additional interactions caused by the particles. Larger particles showed minimal variation in injection pressure. The dosing system used reaches its limits with the increased viscosities of the material, which leads to a significantly rise of the dosing time. The increase can be explained by the lower material pressure in the dosing system. With conventional dosing systems, which work with significantly more pressure, this effect will not occur to the same extent.

Rheometer measurements also indicated a viscosity increase, notably higher than the increase in injection pressure for all particle sizes. For the fine particles, this increase is even 130% for 20% recycled content, while the injection pressure only increases by 43% for this mixture. The reason for this could be the significantly higher shear rate during injection molding. The rheometer measurements, on the other hand, were carried out at very low shear rates.

The optical examinations of the components show minimal surface defects in the fine particles. The medium and coarse particles show clear surface defects, which indicates that particle sizes above 600 μm are unsuitable for applications requiring high surface quality.

The processability of different particle sizes shown in this study is promising and should form the basis for further, more detailed investigations. In particular, the finding that particle sizes above 600 μm are too large for applications suitable for series production directs the focus to the optimization of smaller particles. This opens up several interesting research questions that should be pursued further.

An important starting point for future studies is the investigation of the adhesion of the particles in the matrix. A better understanding of these interactions could help to improve material properties and processing efficiency. In addition, the development of methods for producing smaller particles is of great interest, as these can potentially lead to a higher quality of the end products.

In addition, future studies should investigate the processability of even higher and finer powder contents. The mechanical characteristics depending on the particle size and surface roughness should also be analyzed in detail to obtain a comprehensive understanding of the material properties. Understanding the relationship between viscosity at low shear rates and injection pressure at high shear rates is crucial for the practical application of rheometer results in injection molding processes and should be further investigated.

Acknowledgements The authors acknowledge the Department of Lightweight Structures and Polymer Technology of the University of Technology Chemnitz for grinding the LSR into powder as well as Momentive Performance Materials Inc. for providing liquid silicone rubber material.

References

Ancey C, Jorrot H (2001) Yield stress for particle suspensions within a clay dispersion. J Rheol 45:297–319. https://doi.org/10.1122/1.1343879

Baranwal K, Klingensmith W (2001) Recycling of rubber, In: Baranwal KC, Stephens HL (eds) Basic elastomer technology. The Rubber Division American Chemical Society

Beyer P, Wolf HP (2015) Effizienz und Qualität durch eine neue Generation rheologieoptimierter Flüssigsiliconelastomere. GAK - Gummi Fasern Kunststoffe 68:676–683

Bont M, Barry C, Johnston S (2021) A review of liquid silicone rubber injection molding: process variables and process modeling. Polym Eng Sci 61(2):331–347. https://doi.org/10.1002/pen.25618

Geng C, Zhang Q, Lei W et al (2017) Simultaneously reduced viscosity and en-hanced strength of liquid silicone rubber/silica composites by silica sur-face modification. J Appl Polym Sci 134:45544

Ghosh A, Rajeev RS, Bhattacharya AK, Bhowmick AK, De SK (2003) Recycling of silicone rubber waste: effect of ground silicone rubber vulcanizate powder on the properties of silicone rubber. Polym Eng Sci 43(2):279–296. https://doi.org/10.1002/pen.10024

Goodship V (2017) Practical guide to injection moulding, 2nd edn. Smithers Rapra

Henze EK (1999) Verarbeitung von Flüssigsilikonkautschuk (LSR) zu technischen Formteilen. Dissertation, Rheinisch-Westfälische Technische Hochschule Aachen

Hoyer S (2015) Neuartige Warmmahltechnologie zum Recycling von Elastomeren und Analyse prozessbedingter Eigenschaften. [Zugl.: Chemnitz, Techn. Univ., Diss., 2014], München: Verl. Dr. Hut. 2015. Ingenieurwissenschaften

Ikeda Y, Huang W, Oku A (2003) Ecycling of monomers and fillers from high-temperature-vulcanized silicone rubber using tetramethylammonium hydroxide. Green Chem 5(5):508–511. https://doi.org/10.1039/B304483C

Kroll L, Hoyer S (2017) Wiederverwertung: alte Reifen fein mahlen., Nachr Chem 65(11):1107–1109. https://doi.org/10.1002/nadc.20174067415

Lauth GJ, Kowalczyk J (2015) Einführung in die Physik und Chemie der Grenzflächen und Kolloide, Springer Spektrum Berlin. Heidelberg. https://doi.org/10.1007/978-3-662-47018-3

Mayer D, Heienbrock SJ, Weißer DF, Deckert MH (2024) Processing recycled silicone powder in liquid silicone rubber injection molding and its influence on mechanical properties. Polym Eng Sci. https://doi.org/10.1002/pen.26847

Mayer D, Heienbrock SJ, Weißer DF et al. (2022) Investigation of adhesion and surface treatments for thermoplastic-liquid silicone rubber joints in multicomponent injection molding. Lect Notes Adv Struct Mater 153:283–301. 10.1007

Momentive Performance Materials Inc (2018) Silicone elastomers

Moretto H-H, Schulze M, Wagner G (2000) Silicones, Ullmann's encyclopedia of industrial. Chemistry. https://doi.org/10.1002/14356007.a24_057

Weißer DF, Walz D, Schmid J, Mayer D, Deckert MH (2021) Calculating the temperature and degree of cross-linking for liquid silicone rubber processing in injection molding. J Adv Manuf Process 3(1):e10072. https://doi.org/10.1002/amp2.10072

Weißer DF, Szántó L, Mayer D, Schmid J, Deckert MH (2022) Investigation of the viscosity of liquid silicone rubber for molding microstructures in the injection molding process. Lect Notes Adv Struct Mater 153:253–282. https://doi.org/10.1007/978-3

Weißer DF, Mayer D, Schmid J, Deckert MH (2023) Novel approach to characterize the cross-linking effect of liquid silicone rubber via cavity pressure analysis during injection molding. J Appl Polym Sci 140(4):e5338. https://doi.org/10.1002/app.53381

Yamada H, Yoshida H (2003) Method for producing recycled silicone rubber product, WO03104328A1; December 18, 2003

Chapter 20
Diffusion Behavior of Pyrolysis Oil from Used Tires in Nitrile Butadiene Rubber

Steffen Seitz, Tobias Förster, Sebastian Eibl, Michael Johlitz, and Alexander Lion

Abstract This study investigates the compatibility of nitrile butadiene rubber (NBR) with pyrolysis oil from used tires. Pyrolysis oils are obtained by thermal treatment of shredded tire waste under oxygen free conditions. Oils are gained from truck, car, and bicycle tires. As a blending component for conventional diesel fuel, the oils are intended to prevent the shortage of fossil raw materials and contribute to ensuring the supply of fuels. When used as fuel, the mixtures come into contact with elastomer materials such as nitrile butadiene rubber. In order to provide a damage-free use of blended fuels in established systems, the compatibility of the elastomers with the fuels must be guaranteed. For this purpose, the diesel fuel and pyrolysis oil compositions was first analyzed using IR spectroscopy. Second the diffusion behavior of diesel-pyrolysis oil fuels in NBR is investigated using sorption tests by immersing them in the respective fuels. After specific time periods, the mass, density and mechanical properties of the NBR are recorded. It turns out that pyrolysis oil has a higher proportion of aromatic compounds than diesel fuel. As one result, the mass of NBR in pyrolysis oil increases more, and the mechanical properties are more strongly influenced than those of NBR stored in diesel fuel.

S. Seitz (✉) · M. Johlitz · A. Lion
Fakultät für Luft- und Raumfahrttechnik, Institut für Mechanik, Universität der Bundeswehr München, Werner-Heisenberg-Weg 39, 85577 Neubiberg, Germany
e-mail: steffen.seitz@unibw.de

M. Johlitz
e-mail: michael.johlitz@unibw.de

A. Lion
e-mail: alexander.lion@unibw.de

S. Seitz · T. Förster · S. Eibl
Wehrwissenschaftliches Institut für Werk- und Betriebsstoffe (WIWeB), Institutsweg 1, 85435 Erding, Germany
e-mail: Tobias1Foerster@bundeswehr.org

S. Eibl
e-mail: SebastianEibl@bundeswehr.org

H. Altenbach et al. (eds.), *Lectures Notes on Advanced Structured Materials 3*,
Advanced Structured Materials 226, https://doi.org/10.1007/978-3-031-77403-4_20

20.1 Introduction

The scarcity of fossil raw materials is a global problem that has far-reaching effects on various aspects of the environment. One consequence is the increasing need for sustainable solutions for renewable energy sources and waste recycling. According to the German Federal Environment Agency, Germany produced 5.7 million tons of plastic waste in 2021 (Umweltbundesamt 2023). In this context, the pyrolysis of plastics is becoming increasingly important as a promising technology.

The pyrolysis of plastic waste has high potential, as it not only recycles waste but also helps to reduce dependence on fossil fuels. Recovering valuable raw materials from waste products supports the transition to a circular economy and boosts resource efficiency. In addition, the oil and gas obtained through pyrolysis can be used as an alternative to conventional fuels. The current challenge is to recycle the various waste fractions efficiently and by type in an industrial process. Waste tires pose a particular challenge due to their complex structure and composition. Pyrum Innovations AG has set itself the task of completely recycling tires (Pyrum Innovations 2024). The oil obtained from the pyrolysis and is intended for a liquid energy source as fuel in vehicles and power generators (Karagöz et al. 2020; Murugan et al. 2008).

The chemical-physical properties must be evaluated to enable the possible use of pyrolysis oils. Since pyrolysis oil is for example to be used as a blending component for diesel fuel for the military vehicles and generators, possible differences and their impact on performance are essential. Pyrolysis oil blends used in established systems, come into contact with built-in elastomers.

Elastomers are widely used in technology as materials for tanks, hoses and seals (Röthemeyer 2013). Fluor silicones, hydrogenated nitrile butadiene rubber (HNBR) or NBR can be used as materials. Due to the properties of NBR and the price, NBR is often used in current vehicles and airplanes. The properties of NBR depend primarily on the acrylonitrile content, which is between 15 and 50 percent by weight. As the acrylonitrile content increases, stronger interactions occur within the material, influencing the properties (Lachat 2008). The flexibility of the material decreases and the resistance to non-polar liquids increases. The high resistance to non-polar liquids such as fuels makes it the ideal material for seals that come into contact with fuels (Starmer 1993).

The aim of this work is to investigate the compatibility of fuels with NBR to use pyrolysis oil as a blending component in diesel fuel. For this purpose, sorption experiments are carried out by immersing NBR to each fuel. After certain periods of time, the mass uptake and mechanical characteristics of the NBR are tested. These data are used to assess whether pyrolysis oil from used tires as a blend component in diesel fuel influences the quality of the sealing properties and can, therefore, potentially be used in current systems.

20.2 Materials

Polymer Service Merseburg produced soot-containing and sulphur-crosslinked NBR sheets with an acrylonitrile content of 18 wt-% and a thickness of 2 mm. The structural formula of NB-copolymer is shown in Fig. 20.1. The formulation of NBR in parts per hundred rubber (phr) is shown in Table 20.1. The elastomer is abbreviated as NBR18 in the following.

The pyrolysis oil was provided by Pyrum Innovations AG Dilligne/Saar, Germany and produced using the Pyrum thermolysis process (Pyrum Innovations 2024). A mixture of shredded used truck, car and bicycle tires was used as the raw material. Before the tests and analyses, the supplied pyrolysis oil was filtered to remove solid residues. Commercially available diesel fuel according to DIN EN 590 was used as a reference. The diesel fuel was stored in an airtight container at a controlled temperature of 23 °C protected from sunlight. Table 20.2 shows the mixtures tested in this work with the volume percentage of pyrolysis oil and the corresponding abbreviations.

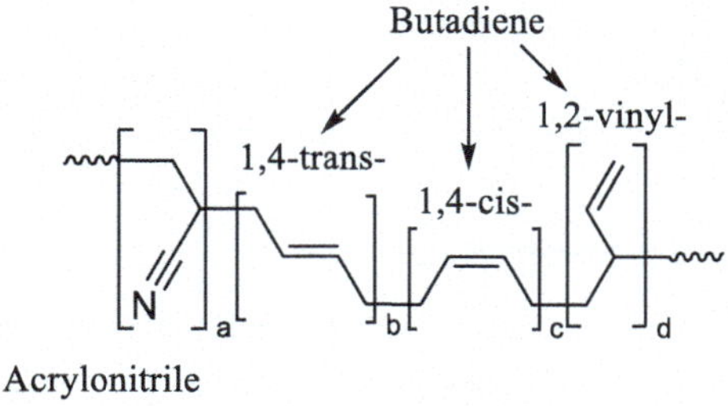

Fig. 20.1 Structural formula of nitril butadiene copolymer

Table 20.1 Composition of the NBR elastomer

Component	Content [phr]
Perbunan 1846	100
Di(2-propylheptyl) phthalate (DPHP)	20
N-(4-Methylpentan-2-yl)-*N*-phenylbenzene-1,4-diamine (6-PPD)	2
Carbon black (Type: N550)	60
Zinc oxide	5
Stearic acid	1
Sulphur	2
N-Cyclohexyl-2-benzothiazolylsulfenamide (CBS)	1, 5
Tetramethyl thiuram monosulfide (TMTM-80)	0, 5

Table 20.2 List of the investigated pyrolysis oil fuel blends

Designation	Volume content of diesel fuel [vol-%]	Volume content of pyrolysis oil [vol-%]
Pyrolysis oil	0	100
DF90_PyOil10	90	10
DF95_PyOil5	95	5
DF99_PyOil1	99	1
Diesel fuel	100	0

20.3 Experimental

20.3.1 Analysis of Fuels

Physical Characteristics

The density and viscosity of the five liquid fuels were recorded using an automatic kinematic viscometer by Anton Parr, model SVM 3001 in accordance with ASTM D7042-21a (viscosity) and ASTM D4052 (density) (ASTM International 2022). The measurements were carried out at 15 °C for the density and 40 °C for the kinematic viscosity.

FTIR-Spectroscopy

FTIR spectra were recorded with a Bruker Tensor 27 spectrometer, which is equipped with a 50 μm zinc-selenid cuvette with a spectral resolution of 4 cm^{-1}. The spectra are averaged over 32 scans in a wavenumber range between 4000 cm^{-1} and 500 cm^{-1}. The spectra were adjusted on their background and normalized to the band's intensity at 2725 cm^{-1}.

20.3.2 Sorption Tests

Sorption tests were carried out to test the compatibility of the diesel fuel, pyrolysis oil and blends with NBR18. First, tensile samples were stamped out of an NBR18 elastomer sheet. Three tensile specimens were stored in the pyrolysis-oil-fuel-blends at 23 °C for 0.5 h, 1 h, 7 h, 24 h, 72 h and 168 h. At the end of the storage period, the samples were dipped in 40/60 petrol to remove the adhering fluid, blotted dry with a paper towel and weighed. Density and the mechanical parameters micro-Shore A Hardness (mShore), uniaxial tensile stress and elongation at break were recorded. A SARTORIUS Secura® 225D-1S analytical balance, equipped with a VF4601 kit from SARTORIUS, was used for mass and density determination using Archimedes' principle. By normalization to the initial weight m_0, the change in mass Δm_g was expressed in weight percent using Eq. (20.1) to ensure the comparability of the samples.

$$\Delta m_{\mathrm{g}} = \left(\frac{m_t - m_0}{m_0}\right) \cdot 100\ \mathrm{wt\%} \tag{20.1}$$

The recorded data points using the solution from Eq. (20.2) of Fick's second law of diffusion. The assumption for the solution is a one-dimensional diffusion in the x-direction.

$$\frac{\Delta m_t}{\Delta m_{\mathrm{eq}}} = 1 - \frac{8}{\pi^2} \sum_{n=0}^{\infty} \left(\frac{1}{(2n+1)^2} \exp\left(\frac{-(2n+1)^2 \pi^2 D t}{h^2}\right)\right) \tag{20.2}$$

The changes in mass after a certain time or when the equilibrium state is reached are Δm_{t} and Δm_{eq}. The samples thickness is h and the mean diffusion coefficient is D, neglecting the change in concentration during the diffusion process (Aminabhavi et al. 1996).

According to DIN ISO 48 4:2021–02, a micro-Shore A hardness sensor was used for a digi test II from Barreis (Deutsches Institut and für Normung e.v. 2021). Tensile tests were performed at 23 °C on a Zwick Roell 1445 testing machine equipped with a 500 N load cell and a ProLine lightXtens 2–100 optical strain sensor. Based on DIN 53,504:2017–03, a prestress of 0.1 MPa and an elongation rate of 200 mm min^{-1} were set at the start.

20.4 Results

20.4.1 Physical Characteristics

Table 20.3 shows the density and viscosity of the fuels. Physical characteristics are standardized for diesel fuel according to DIN ISO 590. For density, this range is between 820 kg m^{-3} and 845 kg m^{-3}, and viscosity is between 2.00 mm^2 s^{-1} and 4.00 mm^2 s^{-1}. With a density of 833 kg m^{-3} and a viscosity of 2.84 mm s^{-1}, the diesel fuel fulfills the target areas. The values obtained for the pyrolysis oil (917 kg m^{-3}, 1.98 mm s^{-1}) are outside the target ranges. The density of the blends increases with an increasing proportion of pyrolysis oil and is within the target areas of the diesel fuel up to 10 vol-% content.

20.4.2 Chemical Analysis

IR spectroscopy was applied to compare differences in chemical functional groups between pyrolysis oil and diesel fuel. Figure 20.2 shows their FTIR spectra of the

Table 20.3 Determined density and viscosity of diesel fuel, pyrolysis oil and their blends

Material	Density[kg m^{-3}]	Viscosity [mm^2 s^{-1}]
Pyrolysis oil	917	1.98
DF90_PyOil10	843	2.85
DF95_PyOil5	838	2.83
DF99_PyOil1	834	2.84
Diesel fuel	833	2.84

two liquids. Alkanes' valence vibrations (C–H, CH_2 and CH_3) can be seen between 3000 cm^{-1} and 2700 cm^{-1}. For CH_2 and CH_3 the typical deformation vibrations are at 1460 cm^{-1} and 1385 cm^{-1} respectively. At 3020 cm^{-1} the valence vibrations of C–H groups in aromatics, whereas in the pyrolysis oil with the shoulder at 3095 cm^{-1} are indicated the (C–H) valence vibrations of alkenes. The valence vibrations of the aromatic C = C bonds can be seen at 1605 cm^{-1} and at 1640 cm^{-1} for C = C bonds in alkenes (Pretsch et al. 2001). It can be noted that the aromatic bonds in the pyrolysis oil spectrum have a higher relative intensity, which indicates a higher proportion of aromatics. At 1710 cm^{-1}, a valence vibration of the C = O bond can be recognized. The broad band at 3395 cm^{-1} for O–H valence vibrations and the bands between 1190 cm^{-1} and 925 cm^{-1}, which result from C–O vibrations, can be seen. This suggests different functional groups, such as alcohols, ketones, esters, or carboxylic acids (Brianc 2018). The comparison between diesel fuel and pyrolysis oil from waste tires shows that both have identical bond classes such as alkanes and aromatics. However, pyrolysis oil is expected to contain a higher content of aromatics and alkenes (Williams and Taylor 1993). In addition, oxygenated molecules can be detected only in the pyrolysis oil. The influence of these differences on the diffusion properties is investigated by sorption tests on NBR.

20.4.3 Sorption Tests

The measured data in Fig. 20.3 show the relative mass change of the immersed NBR18 samples in diesel fuel, pyrolysis oil, and their blends. The mass change is given according to Fick's second law of diffusion following the square root of time. According to Fick's second law, the initial range must be linear, which is the case for all samples. The NBR samples placed in diesel fuel absorbs the least mass with 26 wt-% after $t^{0.5} = 776\ s^{0.5}$ (167 h, see Table 20.4). In contrast, the mass of NBR samples immersed in pyrolysis oil shows the highest increase up to 153 wt-% meaning 5 times higher diffusion into NBR than for diesel fuel. This effect is also reflected in the blends, where the equilibrium uptake increases with increasing pyrolysis oil content 1 vol.% < 5 vol.% < 10 vol.%. There is, therefore, a dependency between the diffusion properties and the volume of pyrolysis oil added to the diesel

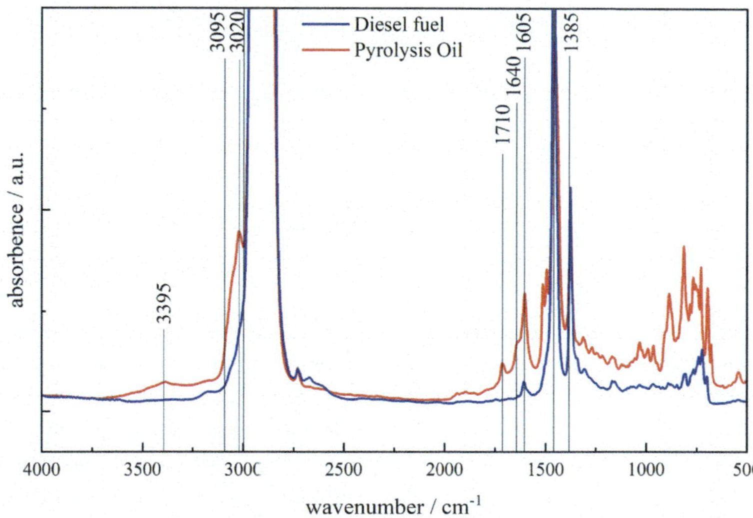

Fig. 20.2 FTIR spectra of diesel fuel and pyrolysis oil

fuel. This observation can be verified by calculating the diffusion coefficient and the mass absorption at equilibrium using Eq. (20.2), shown in Table 20.4.

The blends' diffusion coefficient increases with increasing pyrolysis oil content. With $D = 11.8\ 10^{-6}\ \mathrm{mm}^2\ \mathrm{s}^{-1}$, pyrolysis oil has the highest value and the fastest

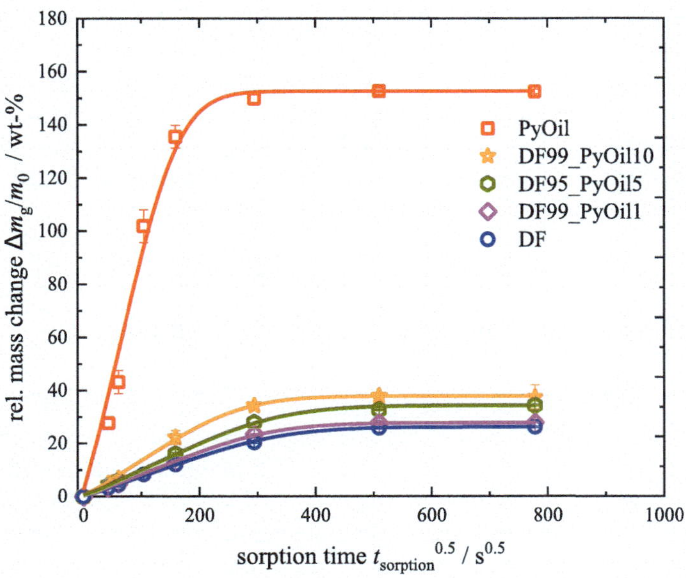

Fig. 20.3 Mass change and sorption curves of NBR in diesel fuel, pyrolysis oil and their blends

Table 20.4 Calculated diffusion coefficients and mass at equilibrium of NBR for different pyrolysis-oil-fuel-blends

Fuel	D x 10^{-6} mm^2 s^{-1}	M_∞ wt-%
Pyrolysis oil	11.8	152.6
DF90_PyOil10	4.3	37.9
DF95_PyOil5	2.8	34.3
DF99_PyOil1	1.8	27.8
Diesel fuel	1.6	26.2

diffusion. Diesel fuel, on the other hand, diffuses at a low rate ($D = 1.6\ 10^{-6}\ mm^2\ s^{-1}$). This means that NBR absorbs significantly more mass and at a higher speed when stored in pyrolysis oil.

The mass of the NBR samples increases as the molecules from the liquids diffuse into the free volume of the material. The diffusion behavior depends on the material's and the fuel's composition (Förster and Blivernitz 2021). The different compositions of diesel fuel and pyrolysis oil can explain the significant differences in the mass absorption and diffusion rate. The higher proportion of aromatics and alkene as well as the oxygen containing compounds results in stronger interactions between the pyrolysis oil and the acrylonitrile group of the NBR (Mathai and Thomas 1996). As a result, a larger mass of the pyrolysis oil diffuses into the NBR samples.

Figure 20.4 shows the density in dependence of the sorption time. At the beginning ($t^{0.5} = 42\ s^{0.5}$, 0.5 h), an increase in density can be seen for the NBR samples immersed in diesel fuels and the blends. As sorption progresses, the density of the NBR samples in the fuels decreases to various levels. The decrease in density depends on the diffused mass and the density of the fluid. The more fuel is adsorbed by NBR, the more the density decreases. When stored in pyrolysis oil, the reduction to a value of 1.015 g cm^{-3} corresponds to the highest decrease. This is due to the high increase in mass of 152 wt-%. The blends decrease in continuous order DF99_PyOil1 < DF95_PyOil5 < DF90_PyOil1 as a function of the blend content of the Pyrolysis oil. At equilibrium, there is no significant difference between the density of NBR stored in diesel fuel and blends.

20.4.4 Mechanical Characteristics While Sorption

The Shore A hardness of NBR samples stored in pyrolysis oil, diesel fuel, and their blends as a function of the root of the sorption time and the fuel used can be seen in Fig. 20.5. NBR samples have a hardness of 76 mShore A and decrease in the process of sorption in all fuels. With a hardness of 45 mShore A and a loss of 41%, NBR samples stored in pyrolysis oil show the highest decrease in hardness after $t^{0.5} = 776\ s^{0.5}$ (167 h). When comparing the diesel fuel and the blend with 10 vol-% pyrolysis oil, the hardness decreases to 54 mShore A and 49 mShore A, respectively,

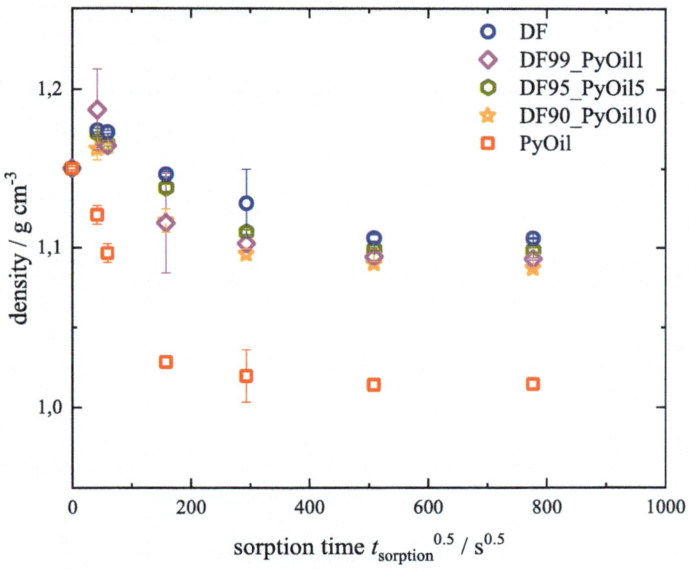

Fig. 20.4 Density evolution of NBR stored in the pyrolysis oil, diesel fuel and their blends in dependence of the sorption time

corresponding to a 7% greater decrease. The fuels act as plasticizers in the NBR samples. The more fuel is adsorbed, the higher the reduction in hardness.

With the uniaxiale tensile stress and the corresponding elongation at break in Fig. 20.6a and b are measured. The tensile stress of 13.5 MPa and an elongation at break of 414% is measured for pristine NBR samples. As with the hardness, a decrease in the properties during sorption into the NBR samples is observed. There is no significant difference in tensile stress and elongation between diesel fuel and the blends. This contrasts with the NBR samples, which was stored in pyrolysis oil. There, the tensile stress decreases by 70% and the tensile elongation at break by 67%.

The reason for the loss of the NBR´s mechanical properties is the fuel that diffuses into the free volume between the polymer chains (Blivernitz 2019). This weakens the interactions between the polymer chains. The material loses its strength, and the chains' cohesion is reduced, making it easier for the chains to slide against each other under a tensile load. Due to the significant higher mass increase during the storage in pyrolysis oil, the chains' interactions are also more strongly influenced, resulting in a higher decrease in mechanical properties compared to the storage in diesel fuel.

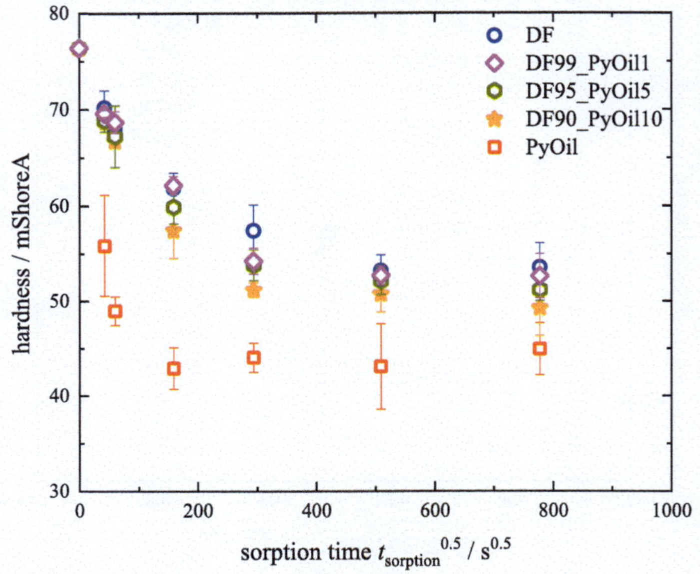

Fig. 20.5 Micro hardness shore A of NBR stored in the different fuels in dependence of the sorption time

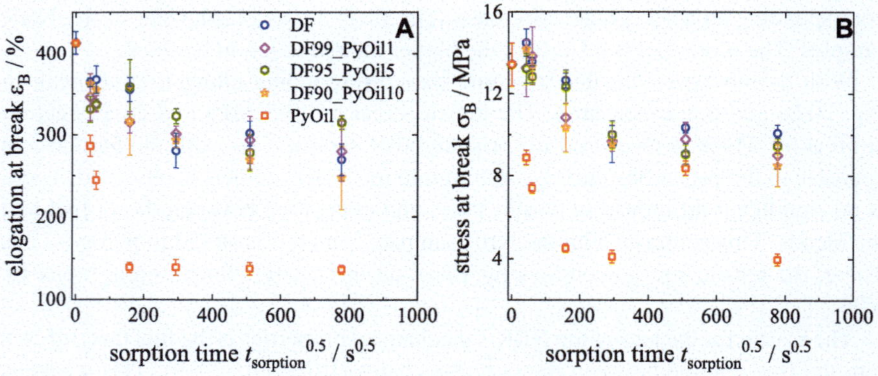

Fig. 20.6 Elongation at break (**a**) and stress at break (**b**) of NBR stored in the different fuels in dependence of the sorption time

20.5 Conclusion

This study investigates the compatibility of NBR with pyrolysis oil from waste tires and blends with conventional diesel fuel thereof, to understand the diffusion behavior of pyrolysis oil towards sealing materials.

Differences in composition between conventional diesel fuel and pyrolysis oil produced from used tires have been determined using FTIR spectroscopy. It can be shown that a higher proportion of aromatic, unsaturated and oxygenated compounds are present in pyrolysis oil. The differences show a strong influence on the diffusion behavior into the NBR. Pyrolysis oil has a considerably higher mass absorption, which significantly degrades the mechanical properties hardness, stress at break and elongation at break. Due to the oxygen containing compounds, the pyrolysis oil has a higher polarity than diesel fuel. This increases the interactions with the nitrile group and results in a higher affinity to NBR. In order to keep the influence of differences as low as possible, the diesel fuel was blended with low amounts of pyrolysis oil (1 vol-%, 5 vol-% and 10 vol-%). When immersed in the blends, slightly higher mass increases were observed than in diesel fuel, whereas mechanical properties did not change significantly.

The results of this study show the relevance of understanding the diffusion properties of pyrolysis oils and fuels in elastomers such as NBR. This will help to prevent possible damage caused by applying pyrolysis oil or blends. Pyrolysis oil from waste tires has the potential to be added to diesel fuel as a blending component. Further research is needed to utilize the full potential.

Acknowledgements The scientific cooperation between the University of the Bundeswehr, Faculty of Aerospace Engineering, Department of Mechanics and the Bundeswehr Research Institute for Materials, Fuels, and Lubricants (WIWeB) is gratefully acknowledged.

References

Aminabhavi TM, Phayde HTS, Ortego JD, Vergnaud JM (1996) A study of sorption/desorption profiles and diffusion anomalies of organic haloalkanes into a polymeric blend of ethylene-propylene random copolymer and isotactic polypropylene. Polymer (Guildf) 37:1677–1684

ASTM International (2022) Standard test method for dynamic viscosity and density of liquids by Stabinger Viscometer (and the Calculation of Kinematic Viscosity) (D7042–21a)

Blivernitz A (2019) Untersuchung der Verträglichkeit von Elastomeren. Dissertation, Universität der Bundeswehr

Brianc S (2018) The C=O bond, part III carboxylic acids. Spektroscopy 33:14–20

DIN Deutsches Institut für Normung e. V. (2021) Elastomere oder thermoplastische Elastomere-Bestimmung der Härte-Teil 4: Eindringhärte durch Durometer-Verfahren (Shore-Härte) (ISO 48–4:2018)

Förster T, Blivernitz A (2021) Migration of mineral oil. J Rubber Res 24:257–269

Karagöz M, Ağbulut Ü, Sarıdemir S (2020) Waste to energy: Production of waste tire pyrolysis oil and comprehensive analysis of its usability in diesel engines. Fuel 275:1–9. https://doi.org/10.1016/j.fuel.2020.117844

Lachat V (2008) Understanding oil resistance of nitrile rubber group interactions at interfaces. Dissertation, University of Akron

Mathai AE, Thomas S (1996) Transport of aromatic hydrocarbons through crosslinked nitrile rubber membranes. J Macromol Sci Phys 35:229–253. https://doi.org/10.1080/00222349608212383

Murugan S, Ramaswamy M, Nagarajan G (2008) The use of tyre pyrolysis oil in diesel engines. Waste Manage 28:2743–2749. https://doi.org/10.1016/j.wasman.2008.03.007

Pretsch E, Bühlmann P, Affolter C, Badertscher M (2001) Spektroskopische Daten zur Strukturaufklärung organischer Verbindungen, 4. Auflage. Springer-Verag Berlin Heidelberg GmbH

Pyrum Innovations AG (2024) Technologie. https://www.pyrum.net/ueber-uns/technologie/. Accessed 10 Jun 2024

Röthemeyer F (2013) Kautschuk Technologie, 3rd edn. Hanser Verlag, München

Starmer P (1993) Swelling of nitrile rubber vulcanizates-part 3: factors affecting maximum swelling. Elastomers an Plastics 25:188–215

Umweltbundesamt (2023) Kunststoffabfälle. https://www.umweltbundesamt.de/daten/ressourcen-abfall/verwertung-entsorgung-ausgewaehlter-abfallarten/kunststoffabfaelle. Accessed 10 Jun 2024

Williams P, Taylor D (1993) Aromatization of tyre pyrolysis oil to yield polycyclic aromatic hydrocarbons. Fuel 72:1469–1474

Chapter 21
Steps Toward Microstructure-Based Modeling of the Influence of Hydrogen on the Fatigue Behavior of Duplex Steels: Non-local Cyclic Polycrystal Plasticity

René Zandomeni and Thomas Seifert

Abstract Wind energy from offshore wind turbines is an important part of the renewable energy revolution. Due to their good combination of strength and ductility and their corrosion resistance, duplex steels, such as 1.4462, are mainly used in these systems. Due to cathodic polarization, the material is exposed to the negative influence of hydrogen, which is known as hydrogen embrittlement. The influence of hydrogen on the material is affected by various aspects, such as the hydrogen activity in the environment and the mechanical loads. In addition, microstructural properties such as grain size, phase distribution, or different grain boundary types also influence hydrogen diffusion and mechanical behavior. Therefore, it is necessary to understand the influence of hydrogen on the fatigue behavior of the duplex steels. This requires modeling of both, diffusion and mechanical behavior, at the microstructural level. In this work, the mechanical behavior as the first step toward microstructure-based modeling of hydrogen-influenced duplex steels is addressed. Since hydrogen traps, such as dislocations, can have a significant influence on diffusion kinetics and hydrogen concentration in the material, a non-local gradient-based single-crystal plasticity model is used accounting for geometrically necessary dislocations. A computational model is developed for the analysis of the micromechanical behavior of the two-phase material and the evolution of the geometrically necessary dislocation densities.

21.1 Introduction

Duplex steels are employed in a wide range of technical applications due to their good combination of strength and ductility and their corrosion resistance. In the context of the energy transition, duplex steels are used in offshore technology, where they are exposed to hydrogen due to cathodic polarization. Such exposure can result in cracks

R. Zandomeni · T. Seifert (✉)
Institute for Digital Engineering and Production (IDEeP), Offenburg University of Applied Sciences, Badstraße 24, 77652 Offenburg, Germany
e-mail: thomas.seifert@hs-offenburg.de

R. Zandomeni
e-mail: rene.zandomeni@hs-offenburg.de

H. Altenbach et al. (eds.), *Lectures Notes on Advanced Structured Materials 3*,
Advanced Structured Materials 226, https://doi.org/10.1007/978-3-031-77403-4_21

due to hydrogen embrittlement (HE), which in turn can lead to the premature failure of components. Understanding the mechanisms of HE and its effect on fatigue of duplex steel is essential for predicting the fatigue life and designing hydrogen-exposed safety-critical offshore components.

Several mechanisms have been proposed to point out the effect of hydrogen on the fatigue life of steels (Lynch 2012). The two main mechanisms of HE are hydrogen-enhanced decohesion (HEDE) and hydrogen-enhanced localized plasticity (HELP) which are supported by several experiments and simulations. Atomic hydrogen diffuses into areas with high localized stresses, as crack tips, inhomogeneities, grain boundaries, interfaces, or stress fields around dislocations. In this context, the HEDE mechanism assumes that hydrogen reduces the strength of the interatomic bonds in areas with high local stresses. Thus, the presence of diffused hydrogen in these areas promotes crack initiation and propagation. Experimental methods for observation of the HEDE mechanism do not exist due to the atomistic nature, but it can be explained with atomistic simulations (Hoagland and Heinisch 1992; Wang et al. 2016). The HELP mechanism assumes that hydrogen enhances the mobility of dislocations, increases nucleation of dislocations, and reduces the internal stress field around dislocations, which collectively leads to a reduction in dislocation interaction stresses (Birnbaum and Sofronis 1994; Robertson 1999). The HELP mechanism can be supported through experimental measurements, including in situ transmission electron microscopy (TEM), micro-deformation experiments, and fractography. Additionally, simulations can be used to investigate this mechanism, employing either atomistic simulations or discrete dislocation dynamics (Lynch 2012; Yu et al. 2019). A summary of the complex HE mechanism and the correlation between microstructure, grain boundary types, hydrogen, and crack formation paths for three different interface types is provided in Li et al. (2024a). There, the influence of hydrogen on different microstructure interfaces and crack formation paths is investigated based on various results from literature using the hydrogen microprinting technique (HMT), atom probe tomography (ATP), and fractography. Furthermore, an overview of the interaction of the different HE mechanisms is given and strategies to improve HE resistance are proposed. The proposed mechanisms indicate that nucleation and movement of dislocations significantly influence diffusion of hydrogen and mechanics and vice versa (Li et al. 2024b; Yuan et al. 2020).

The two-phase austenitic and ferritic microstructure of duplex steels offers the possibility of targeted microstructure optimization with respect to fatigue and HE. On the one hand, austenitic steels generally have high hydrogen resistance due to their crystal structure, while ferritic alloys are susceptible to HE. On the other hand, the different elastic–plastic properties of the two phases promote internal stresses and plasticity at the phase boundaries, which influence hydrogen diffusion and trapping. Microstructure-based computational models can help to gain insights into the complex microstructure–property interactions. To this end, the chemical–mechanical coupling and the cyclic plastic behavior of duplex steels need to be considered in the microstructure-based models, so that the mutual influence of diffusion and mechanics is taken into account. Both aspects are briefly reviewed in the following sections.

21.1.1 Chemical–Mechanical Coupling

The HE mechanisms correlate directly with the microstructure and the mechanical behavior of materials, especially on the microscopic length scale. To gain further insight into the hydrogen-influenced behavior of duplex steels, fully coupled, chemical–mechanical microstructure-based modeling is required.

In order to model fully coupled chemical–mechanical behavior, the majority of works use the analogy between heat transfer and diffusion and implement an Abaqus user-defined material with heat transfer (UMATHT) to solve the diffusion equation, e.g., Abdolvand (2019), Barrera et al. (2016), and Elmukashfi et al. (2020). Therefore, it is usually assumed that hydrogen can diffuse as a dissolved atom in the material at normal interstitial lattice sites (NILS), which are inherent in the crystallographic structure, and that hydrogen can be trapped at trapping sites, which are material inhomogeneities such as vacancies, grain boundaries, or dislocations. The trapping sites have a significant influence on the total amount of hydrogen and the hydrogen diffusion through the NILS. Dislocations act as a relevant trap as they directly influence the mechanical behavior and can transport hydrogen through the material (Dadfarnia et al. 2015).

An early approach for coupling plasticity due to dislocations with hydrogen diffusion and trapping is the empirical equation of Kumnick and Johnson (1980) and Sofronis and McMeeking (1989). It assumes that plastic strain and number of dislocation trapping sites are correlated. This approach is used in microstructure-based models resolving the grain structure of metallic materials to compute local hydrogen concentrations considering local stress and (plastic) strain fields and applying a single-crystal plasticity model. e.g., Hassan et al. (2019) and Tekkaya et al. (2024). Mechanism-based approaches directly consider the dislocation density for the estimation of dislocation trapping sites. They account for the effect of geometrically necessary dislocations (GND), which are generated due to plastic deformations near grain and phase boundaries and, hence, use gradient-based single-crystal plasticity models, e.g., Kumar and Mahajan (2020), Singh et al. (2022), Yuan et al. (2022), and Li et al. (2024b).

21.1.2 Cyclic Plasticity of Duplex Steels and Single-Crystal Plasticity

The cyclic plastic behavior of duplex steels is of particular interest for the design of mechanical components subjected to cyclic loading and in the presence of hydrogen. During cyclic plastic loading, the Bauschinger effect is relevant, which describes the earlier re-plastification of the material after load-reversal. In the work of Kreins et al. (2022a), it is shown that a pronounced Bauschinger effect exists for the studied austenitic-ferritic duplex steels, especially for annealed material conditions. There, the ferritic phase shows higher hardness increasing the difference in elastic–plastic properties of both phases and the local GND density gradient at phase boundaries. According to the theory of Mughrabi (1938), the inhomogeneous dislocation distributions at the phase boundaries result in long-range internal stresses contributing to the Bauschinger effect. Additionally, short-range internal stresses can contribute to the Bauschinger effect (Kreins et al. 2021, 2022b), e.g., due to directional inhibition of dislocation movement and local dislocation pile-ups at obstacles.

The effect of GNDs needs to be considered in microstructure-based models of duplex steels, since they are relevant for the cyclic plastic behavior and for hydrogen diffusion acting as trapping sites. The evolution and the effect of GNDs can be accounted for in second-order gradient-based single-crystal plasticity models. These models use non-local formulations, first, to compute the GND density from the gradient of the plastic deformation (Nye 1953; Demir et al. 2024) and, second, to compute the internal stresses by integration of the stresses induced by GND density gradients over a defined volume controlling the long-range effects (Bayley et al. 2006; Ma 2014).

21.1.3 Aims and Structure

It is the aim of this work to provide a microstructure-based finite-element model for duplex steels that can be used in further steps to assess the influence of hydrogen on their fatigue behavior. To this end, a second-order non-local single-crystal plasticity model is implemented and a computational model is developed for micromechanical analysis of the mechanical behavior including effects from GNDs and the Bauschinger effect.

The structure of this work is as follows: In Sect. 21.2, an overview of the modeling framework of the gradient-based single-crystal plasticity model and the numerical implementation in the finite-element program Abaqus (2023) is given. Section 21.3 shows stress–strain results for cyclic plastic loading and different synthetic microstructures of duplex steel and results regarding the distribution of the GNDs which act as relevant hydrogen traps for a selected microstructure. In Sect. 21.4, the results are discussed before the work is concluded in Sect. 21.5.

21.2 Modeling Framework

In this section, the second-order gradient-based single-crystal plasticity model is introduced and some remarks on the implementation into the finite-element program Abaqus are given.

21.2.1 Gradient-Based Single-Crystal Plasticity Model

For modeling single-crystal plasticity that accounts for GNDs and internal stresses, the second-order gradient model of Ma and Hartmaier (2014) is used. The model employs the concept of super GNDs for the calculation of the internal stresses to avoid the underdetermined system of equations for crystallographic GNDs. The model accounts for kinematic and isotropic hardening due to GNDs.

The total deformation of a single crystal is described by an elastic and a plastic part, where the elastic part defines the elastic deformation and rotation of the lattice and the plastic part defines the purely plastic shearing by crystallographic slip of the lattice. Thus, the total deformation gradient F is decomposed multiplicatively into elastic and plastic parts, F^e and F^p:

$$\mathsf{F} = \mathsf{F}^e \cdot \mathsf{F}^p, \tag{21.1}$$

where F, F^e, and F^p describe the mapping of line elements from the reference configuration to the current deformed configuration, from an intermediate stress-free configuration (with plastic slip referred to the undeformed lattice, later referred to with a superimposed tilde, $\tilde{}$), to the deformed current configuration, and from the reference configuration to the intermediate configuration, respectively.

With the multiplicative decomposition of the deformation gradient, the velocity gradient L can be written as

$$\mathsf{L} = \mathsf{L}^e + \mathsf{F}^e \cdot \mathsf{L}^p \cdot \mathsf{F}^{e-1}, \tag{21.2}$$

where L^e is the elastic part and L^p describes the plastic distortion rate due to dislocation slip in N slip systems:

$$\mathsf{L}^p = \dot{\mathsf{F}}^p \cdot \mathsf{F}^{p-1} = \sum_{\alpha=1}^{N} \dot{\gamma}_\alpha \mathbf{s}_\alpha \otimes \mathbf{m}_\alpha. \tag{21.3}$$

For each slip system α, $\dot{\gamma}_\alpha$ is the slip rate and $\mathbf{s}_\alpha \otimes \mathbf{m}_\alpha$ is the Schmid tensor, where $\mathbf{s}_\alpha$ and $\mathbf{m}_\alpha$ are orthonormal unit vectors in the slip direction and the normal to the slip plane in the undeformed lattice.

The slip rates in Eq. (21.3) are expressed as a power-law function of the resolved shear stress τ_α and considers hardening contributions from GNDs via the resolved back stress $\tau_{\text{kin},\alpha}^{\text{GND}}$ related to kinematic hardening and the passing stress $\tau_{\text{iso},\alpha}^{\text{GND}}$ related to isotropic hardening. Hence, the kinetic law of rate-dependent slip reads (Ma 2014)

$$\dot{\gamma}_\alpha = \dot{\gamma}_0 \left(\frac{|\tau_\alpha + \tau_{\text{kin},\alpha}^{\text{GND}}|}{\tau_0 + \tau_{\text{iso},\alpha}^{\text{GND}}} \right)^n sgn\left(\tau_\alpha + \tau_{\text{kin},\alpha}^{\text{GND}}\right), \tag{21.4}$$

where $\dot{\gamma}_0$ is the reference slip rate, τ_0 is the reference resolved shear stress, and n is the power-law exponent. The resolved shear stress in each slip system is computed from Schmid's law using the second Piola–Kirchhoff stress tensor $\tilde{\mathsf{S}}$ in the intermediate configuration:

$$\tau_\alpha = \tilde{\mathsf{S}} : \mathbf{s}_\alpha \otimes \mathbf{m}_\alpha . \tag{21.5}$$

To describe the elasticity of the single crystal in the intermediate configuration, the crystallographic elasticity tensor C of the single-crystal in the undeformed lattice is used:

$$\tilde{\mathsf{S}} = \mathsf{C} : \frac{1}{2}\left(\mathsf{F}^{\mathrm{e}^{\mathrm{T}}} \cdot \mathsf{F}^{\mathrm{e}} - \mathrm{I}\right), \tag{21.6}$$

where I is the second-order identity tensor and C accounts for the cubic anisotropy including the elastic constants C_{11}, C_{12}, and C_{44}.

The hardening contributions from GNDs are determined by the dislocation density in each slip system $\rho_\alpha^{\mathrm{GND}}$. In accordance with the principles of strain gradient theory, e.g., Nye (1953) and Gurtin and Fried (2010), the GND density is obtained from the dislocation density tensor G and the length of the Burgers vector b and is derived from the curl of the plastic deformation gradient:

$$\frac{1}{b}\mathsf{G} = -\frac{1}{b}\left(\mathsf{F}^{\mathrm{p}} \times \nabla\right) \tag{21.7}$$

$$= \sum_{\alpha=1}^{N} \rho_{\alpha,\perp}^{\mathrm{GND}} \mathbf{l}_\alpha \otimes \mathbf{s}_\alpha + \sum_{\alpha=1}^{N} \rho_{\alpha,\odot}^{\mathrm{GND}} \mathbf{s}_\alpha \otimes \mathbf{s}_\alpha \tag{21.8}$$

$$= \sum_{A=1}^{9} \bar{\rho}_A^{\mathrm{GND}} \bar{\mathbf{l}}_A \otimes \bar{\mathbf{s}}_A . \tag{21.9}$$

The Nabla operator $\nabla = \partial/\partial\mathbf{X}$ specifies the gradient with respect to the undeformed reference configuration. The single-crystal hypothesis allows Eq. (21.7) to be expressed in terms of the GND edge and screw dislocation densities ($\rho_{\alpha,\perp}^{\mathrm{GND}}$ and $\rho_{\alpha,\odot}^{\mathrm{GND}}$) (Gurtin and Fried 2010), the slip direction $\mathbf{s}_\alpha$, and the line direction $\mathbf{l}_\alpha = \mathbf{m}_\alpha \times \mathbf{s}_\alpha$, resulting in Eq. (21.8). The total GND density is then assumed as

$$|\rho_\alpha^{\mathrm{GND}}| = |\rho_{\alpha,\perp}^{\mathrm{GND}}| + |\rho_{\alpha,\odot}^{\mathrm{GND}}|. \tag{21.10}$$

These GND densities are referred to the crystallographic slip systems. However, with nine components of the dislocation density tensor, Eq. (21.8) is an underdetermined system if $N > 9$. Hence, for the calculation of the GND density gradient that determines the internal stresses, the dislocation density tensor is resolved into nine independent dislocation densities $\bar{\rho}_A^{\mathrm{GND}}$ called super GND densities using the nine combinations of the three mutually perpendicular global coordinate axes, represented by $\bar{\mathbf{l}}_A$ and $\bar{\mathbf{s}}_A$ in Eq. (21.9) (Ma 2014).

With the total crystallographic GND densities from Eq. (21.10), the isotropic passing stress is calculated from the Taylor hardening-type equation (Ma 2014; Gottstein 2014)

$$\tau_{\text{iso},\alpha}^{\text{GND}} = c\mu b \sqrt{\sum_{\beta=1}^{N} \chi_{\alpha\beta} |\rho_{\beta}^{\text{GND}}|}, \tag{21.11}$$

where c is the Taylor hardening coefficient, μ is the shear modulus, and $\chi_{\alpha\beta}$ is the cross-hardening matrix to account the self-hardening of forest GND interactions.

The resolved back stress in each slip system resulting from the internal stresses $\tilde{\mathsf{S}}^{\text{GND}}$ due to GND gradients is obtained from Schmid's law (Ma 2014):

$$\tau_{\text{kin},\alpha}^{\text{GND}} = \tilde{\mathsf{S}}^{\text{GND}} : \mathbf{s}_{\alpha} \otimes \mathbf{m}_{\alpha}. \tag{21.12}$$

The calculation of the internal stresses requires the gradient of the super GND densities (Devincre 1995; Ma 2014):

$$\tilde{\mathsf{S}}^{\text{GND}} = \mathsf{F}^{\text{p}} \cdot \left(\sum_{A=1,I=1}^{9,3} \mathsf{C}_{AI} : \frac{1}{2} \frac{\partial \bar{\rho}_{A}^{\text{GND}}}{\partial X_I} L^3 \bar{\mathbf{I}}_A \right) \cdot \mathsf{F}^{\text{p}T}, \tag{21.13}$$

where C_{AI} is a third-order tensor describing the stress field of a dislocation segment and L is a length scale parameter. The internal stress at a material point is computed for a small neighborhood volume L^3 around that point. In the model, latent hardening, i.e., a mutual influence of hardening behavior of different slip systems, is implicitly included via the internal stress. With

$$\frac{\partial \bar{\rho}_{A}^{\text{GND}}}{\partial X_I} = \frac{1}{b} \bar{\mathbf{I}}_A \cdot \frac{\partial \mathsf{G}}{\partial X_I} \cdot \bar{\mathbf{s}}_A \tag{21.14}$$

from Eq. (21.9), this results in a second-order gradient theory, where second-order derivatives of the plastic deformation gradient F^{p} are involved.

21.2.2 Numerical Implementation

The second-order gradient-based single-crystal plasticity model described in Sect. 21.2 is implemented in Abaqus (2023) standard using different user subroutines. The numerical integration of the single-crystal plasticity equations is implemented in a user material subroutine, UMAT. Since the UMAT requires the Cauchy stress tensor, the push forward operation of the second Piola–Kirchhoff stress in the intermediate configuration is used:

$$\sigma = \frac{1}{\det(\mathsf{F}^{\text{e}})} \mathsf{F}^{\text{e}} \cdot \tilde{\mathsf{S}} \cdot \mathsf{F}^{\text{e}T}. \tag{21.15}$$

Moreover, the consistent tangent $\mathrm{d}\sigma/\mathrm{d}\Delta\mathsf{D}$ for Abaqus is supplied, where $\Delta\mathsf{D}$ is the incremental rate of deformation tensor.

The UMAT is coupled with the user subroutine, UEXTERNALDB, for the calculation of the required gradients of F^{p}. In the UMAT, the plastic deformation gradient is defined as a state-dependent variable SDV, which can be accessed via the UEXTERNALDB. In this subroutine, gradients are calculated using a Python script, whereby access is made to the current Abaqus output database following each converged time step. A flow chart of the numerical implementation of this weak-coupling scheme in Abaqus is shown in Fig. 21.1. In the very first step of a simulation, UEXTERNALDB is called to initialize the length of the two global arrays for the first- and second-order gradients (`Fp1grad, Fp2grad`). Following this, UMAT is called to calculate the mechanical response. Once the solution has reached convergence, UEXTERNALDB is called and the required first- and second-order gradients are computed and the results for each element and integration point are stored within the two global arrays. To this end, a Python script computes the nodal values of the plastic deformation gradient by averaging the extrapolated values from adjacent elements and writes them to a binary file. The binary file is then read into a Fortran routine to calculate the first- and second-order gradients. Therefore, it is necessary to use quadratic element types in the simulation, otherwise the second derivatives disappear. In this work, 20-node hexahedral three-dimensional displacement elements are used.

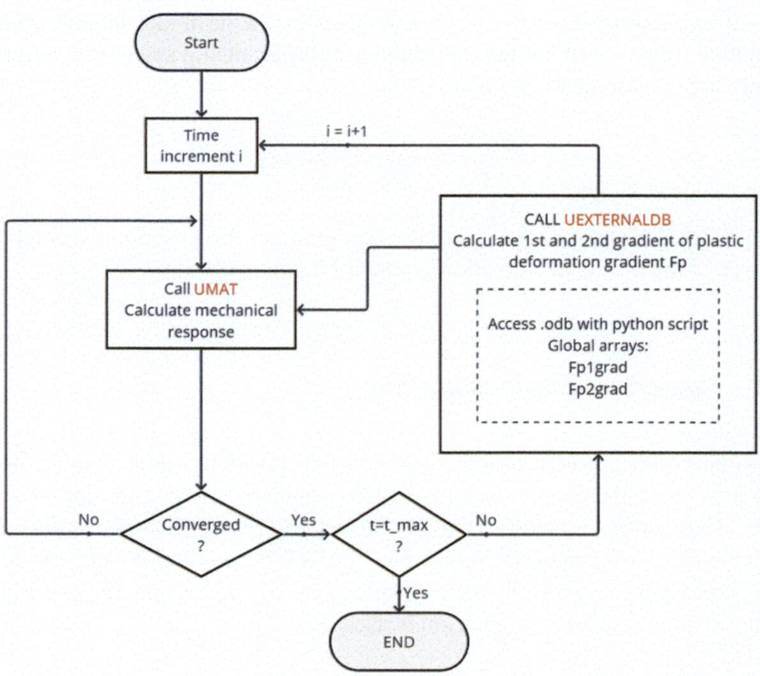

Fig. 21.1 Flow chart of numerical scheme for gradient-based single-crystal plasticity analysis

The first-order gradient of the plastic deformation gradient F^{p} (`Fp1grad`) is calculated from the nodal values $\mathsf{F}^{\mathrm{p}^I}$ and the shape functions N^I of the element, $I = 1, ..., 20$:

$$F^{\mathrm{p}}_{ij,k} = \sum_{I=1}^{20} \frac{\partial N^I}{\partial X_k} F^{\mathrm{p}\,I}_{ij} . \quad (21.16)$$

The second-order gradient (`Fp2grad`) is calculated with second derivative of the element shape functions:

$$F^{\mathrm{p}}_{ij,k,l} = \sum_{I=1}^{20} \frac{\partial^2 N^I}{\partial X_k \partial X_l} F^{\mathrm{p},I}_{ij} . \quad (21.17)$$

These results are transferred to the UMAT at the start of the new time step using a Fortran module block. Due to the weak-coupling scheme, the isotropic passing stress and the resolved back stress due to GNDs from Eqs. (21.11) and (21.12) always lag one time step behind. To avoid significant deviations, small time steps are employed.

21.3 Results

In this section, the modeling framework introduced in Sect. 21.2 is used to investigate the influence of the phase fractions of ferrite and austenite on the mechanical behavior of duplex steels. To this end, polycrystal microstructures of duplex steels with different phase fractions are generated and the results of the macroscopic mechanical behavior for these microstructures under cyclic loading are studied.

21.3.1 Synthetic Microstructure-Based Finite-Element Models

The representative volume elements (RVE) of the synthetic polycrystal microstructure of the duplex steel are generated using data from Kreins et al. (2022a) and the open-source program Dream3D (Groeber and Jackson 2014). Microstructure-based models of the duplex steel polycrystals with varying phase fractions, as given in Table 21.1, are used. Five different models are employed, with the volume fraction of ferrite ranging from 0 to 100%.

Periodic RVEs are generated with equally distributed orientations. For the generation of the orientation, the internal statistical generator of Dream3D is used (Groeber and Jackson 2014). The mean grain size of the two phases is assumed to be 25 μm, estimated from Kreins et al. (2022a). For the generation of the synthetic RVEs, the grains are defined as primary equiaxed with an assumed normal distribution and a standard deviation of 3 μm. In Fig. 21.2, the grain orientations are shown in

Table 21.1 Overview of microstructure-based polycrystal models and phase volume fraction

Model no.	Ferritic fraction in %	RVE size	Element size	Figs.
1	0			21.2b
2	20			21.2c
3	50	100μm	5μm	21.2d
4	80			21.2e
5	100			21.2f

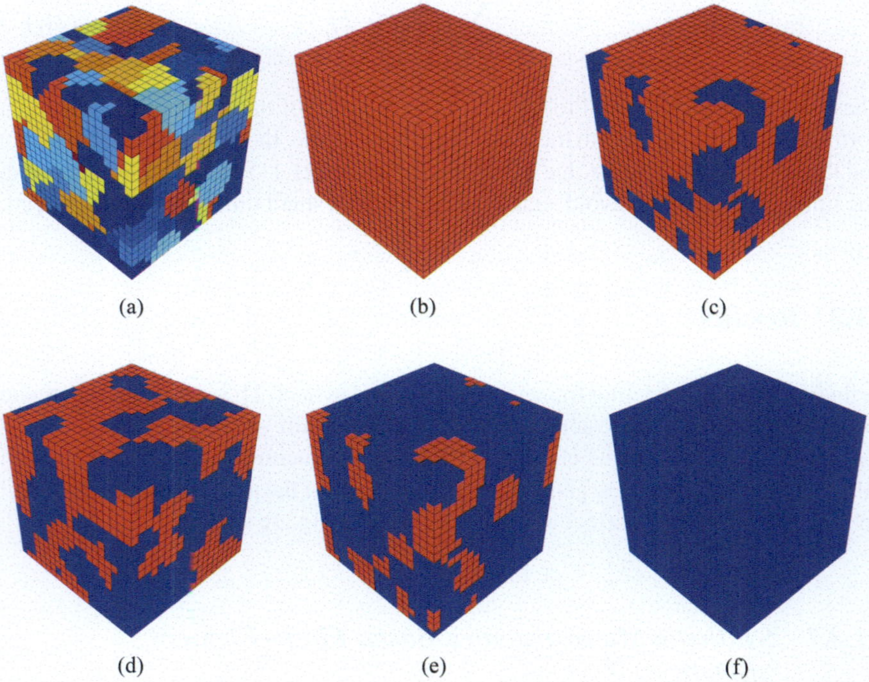

Fig. 21.2 Microstructure-based finite-element models of the duplex steel with different phase volume fractions: **a** Random grain orientation; **b** 100% austenite; **c** 80% austenite (red), 20% ferrite (blue); **d** 50% austenite (red), 50% ferrite (blue); **e** 20% austenite (red), 80% ferrite (blue); **f** 100% ferrite

Fig. 21.2a, where each color stands for different combinations of the three Euler angles. The variation of the volume fraction of the two phases from Table 21.1 is shown in Fig. 21.2b–f.

21.3.2 Polycrystal Cyclic Plasticity

To calculate the macroscopic response of the RVEs, numerical homogenization with periodic boundary conditions is employed as described by Schmidt (2011). Uniaxial cyclic loading conditions are considered where the deformation gradient in one direction is prescribed while stress-free boundaries are applied in the transverse direction and the macroscopic stress in loading direction is computed.

The material properties and model parameters used are summarized in Table 21.2. Values without a reference are chosen reasonably. In this work, it is assumed that slip occurs exclusively in the $\{110\}\langle\bar{1}11\rangle$ slip system family in the ferritic phase. For the austenitic phase, plastic flow in the $\{111\}\langle 1\bar{1}0\rangle$ slip system family is considered.

The macroscopic mechanical behaviors of the five microstructures are shown in Fig. 21.3 as true stress-logarithmic strain curve for a tension-compression test. The stress–strain curves indicate that, as the volume fraction of ferrite increases, the tensile stress also increases due to the higher strength of the ferrite. During unloading and compressive loading, also the pure ferrite has the highest absolute stress for reversed plasticity.

In the context of HE, the amount of GNDs and moreover the areas of high densities are of interest. A qualitative analysis of the GND density is shown in Fig. 21.4 for the

Table 21.2 Mechanical properties and model parameters for austenitic-ferritic duplex steel

Parameter	Ferrite	Austenite	Unit	References
Elastic constants				
C_{11}	236	191	GPa	Zhou et al. (2024)
C_{12}	136	118	GPa	Zhou et al. (2024)
C_{44}	118	139	GPa	Zhou et al. (2024)
Flow rule				
$\dot{\gamma}_0$	1	1	1/s	
τ_0	355	151	MPa	Zhou et al. (2024)
n	10	10	–	
Hardening				
c	0.05	0.05	–	
b	$2.48 \cdot 10^{-7}$	$2.54 \cdot 10^{-7}$	mm	Zhou et al. (2024)
$\chi_{\alpha\beta}$	1	1	–	Ma (2014)
L	10^{-5}	10^{-4}	mm	
Slip system families				
N	12	12	–	
Slip planes	$\{110\}$	$\{111\}$	–	
Slip directions	$\langle\bar{1}11\rangle$	$\langle 1\bar{1}0\rangle$	–	

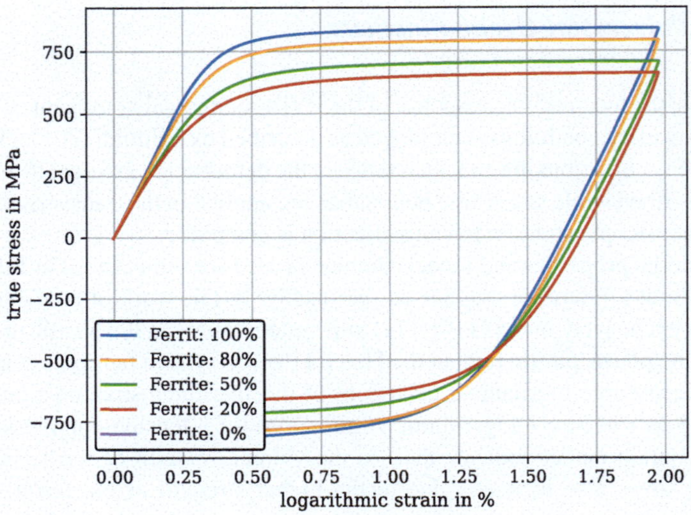

Fig. 21.3 Tension-compression stress–strain curves of the five different microstructures

model No. 3 (50% austenite, 50% ferrite) which is cut in half so that the inside can be seen. The GND density increases as the incompatibility of deformation in each grain increases. This incompatibility leads to a higher first-order gradient of plastic deformation and therefore to a higher dislocation density. The areas of high incompatibility of deformation are, as expected, in the areas of grain and phase boundaries. Comparing Fig. 21.4a with Fig. 21.4b and c the highest amount of GNDs are at grain and phase boundaries. This phenomenon can be attributed to the incompatibility of mechanical properties at phase boundaries and the increased incompatibility at grain boundaries due to the highly misoriented grains. The simulation shows similar results for the distribution of the GND density compared to the experimentally determined results for the density distribution in the work of Kreins et al. (2022a) for an austenitic-ferritic duplex steel.

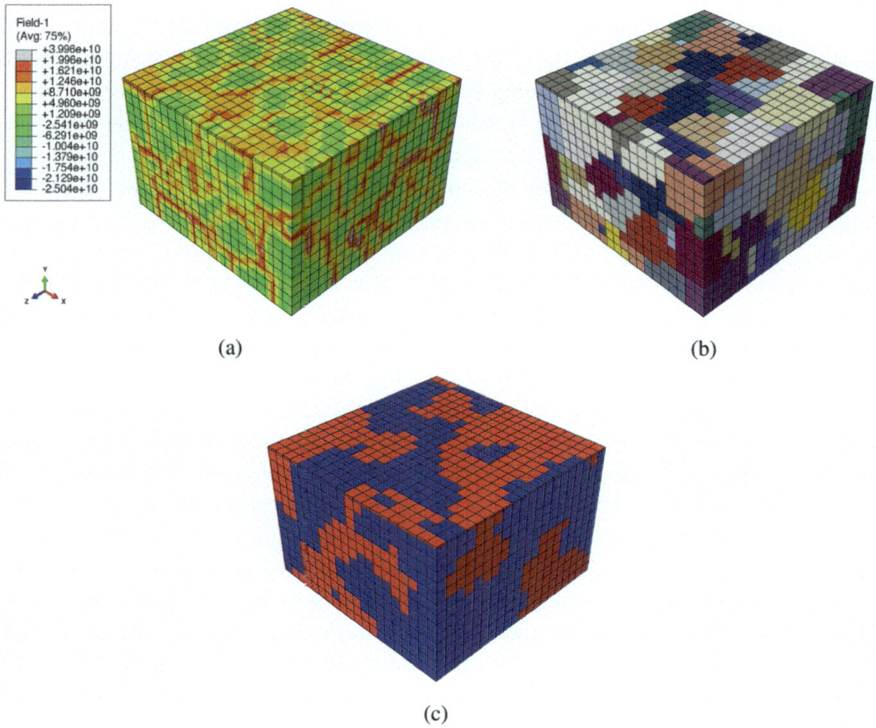

Fig. 21.4 RVE from Fig. 21.2d with half-plane cut in y-direction: **a** Total GND density $\rho_{tot}^{\mathrm{GND}} = \sum_{\alpha} \rho_{\alpha}^{\mathrm{GND}}$ at maximum strain in 1/mm^2; **b** grain structure; **c** phase distribution 50% austenite (red), 50% ferrite (blue)

21.4 Discussion

As a first step toward microstructure-based modeling of the influence of hydrogen on the fatigue behavior of duplex steels, microstructure-based models of duplex steels are generated and a second-order gradient-based single-crystal plasticity model is implemented and applied to assess the macroscopic cyclic mechanical properties as well as local GND densities.

The gradient-based single-crystal plasticity model predicts higher GND densities at grain and especially at phase boundaries. The computed values are reasonable compared to experimental data documented in Kreins et al. (2022a). However, for the synthetic generation of the microstructure with different phase fractions, only very limited information of real microstructures is used. Moreover, the material properties of the single-crystal plasticity model for both phases are only very rough estimates. The models must now be refined with experimental data: on the one hand, statistics of the microstructure are required and, on the other hand, data of the cyclic mechanical behavior of duplex steel with different phase fractions and/or the individual phases must be used to determine the material properties. Challenging is the calibration of the length scale parameter L. Therefore, experimental data with different grain sizes and/or with microspecimens showing size effects is necessary.

What could be shown, however, is that the model and its numerical implementation are robust and can be used for the computational assessment of duplex microstructures. There are some sources for inaccuarcies. On the one hand, averaged values of the plastic deformation gradient at the nodes of the finite-element mesh are used. Hence, a fine mesh is required so that the deviations of the extrapolated values are not too high so that the averaged values are meaningful. However, resolving the gradient of plastic strain and, thus, of GND densities generally requires fine meshes at stress concentrations. Besides the spatial resolution, the temporal resolution is relevant on the other hand. Due to the weak coupling, large time steps result in inaccuracies since the hardening variables are one time step behind. Therefore, mesh as well as time step convergence studies are necessary to find reasonable configurations. Such studies were not performed in detail in this work and are planned as next steps.

The stress–strain curves of the uniaxial tension-compression load suggest a too low Bauschinger effect compared to stress–strain curves measured in experiments. For example, austenitic stainless steels can show stronger hardening and reversed plasticity after unloading from tension at around zero stress (Seifert et al. 2007, 2013). This effect cannot be captured with the current gradient-based single-crystal plasticity model. One reason might be that the internal stresses responsible for the Bauschinger effect in the model are caused by GND gradients only. This effect can be linked to Mughrabi's theory (Mughrabi 1938) and considered a long-range effect. Short-range effects, that are acting quite locally, generating back stresses at obstacles in the grains, are not included in the model, yet. This is why local effects will be included into the model in future works. A suitable extension is the local kinematic hardening model from Schilli et al. (2022), which also takes latent hardening into account and was validated using micro-bending tests (Kreins et al. 2021). In the model, a macroscopic back stress tensor is defined from which the resolved back stresses in the slip systems are obtained via Schmid's law, analogously to Eq. (21.5).

21.5 Conclusion

In this work, a computational model for micromechanical analysis of duplex steels under cyclic loading is developed that uses microstructure-based models of duplex steels and a second-order gradient-based single-crystal plasticity model. The model accounts for internal stresses responsible for the Bauschinger effect and is based on GND densities. This is why the computation model serves as a first step toward microstructure-based modeling of the influence of hydrogen on the fatigue behavior of duplex steels, since GNDs are preferred trapping sites for hydrogen.

The presented results do not include experimental results. Hence, the computational method needs to be calibrated based on experimental data (microstructure statistics, stress–strain data) and validated against experiments.

Acknowledgements The authors would like to thank A. Hartmaier from ICAMS RUB Bochum University (Germany) for valuable discussions and for providing the basic implementation of gradient-based single-crystal plasticity. The authors acknowledge funding of this project by the Deutsche Forschungsgemeinschaft (DFG, German Research Foundation) under project number 504195877. We are grateful for the financial support of the German Research Foundation DFG.

References

Abdolvand H (2019) Progressive modelling and experimentation of hydrogen diffusion and precipitation in anisotropic polycrystals. Int J Plast 116:39–61. https://doi.org/10.1016/j.ijplas.2018.12.005

Barrera O, Tarleton E, Tang H, Cocks A (2016) Modelling the coupling between hydrogen diffusion and the mechanical behaviour of metals. Comput Mater Sci 122:219–228. https://doi.org/10.1016/j.commatsci.2016.05.030

Bayley CJ, Brekelmans WAM, Geers MGD (2006) A comparison of dislocation induced back stress formulations in strain gradient crystal plasticity. Int J Solids Struct 43. https://doi.org/10.1016/j.ijsolstr.2006.05.011

Birnbaum HK, Sofronis P (1994) Hydrogen-enhanced localized plasticity-a mechanism for hydrogen-related fracture. Mater Sci Eng A 176:191–202. https://doi.org/10.1016/0921-5093(94)90975-X

Dadfarnia M, Martin ML, Nagao A, Sofronis P, Robertson IM (2015) Modeling hydrogen transport by dislocations. J Mech Phys Solids 78:511–525. https://doi.org/10.1016/j.jmps.2015.03.002

Demir E, Martinez-Pechero A, Hardie C, Tarleton E (2024) Restraining geometrically-necessary dislocations to the active slip systems in a crystal plasticity-based finite element framework, I. J Plas 178. https://doi.org/10.1016/j.ijplas.2024.104013

Devincre B (1995) Three dimensional stress field expression for straight dislocation segments. Solid State Commun 93:875–878. https://doi.org/10.1016/0038-1098(94)00894-9

Elmukashfi E, Tarleton E, Cocks A (2020) A modelling framework for coupled hydrogen diffusion and mechanical behaviour of engineering components. Comput Mech 66:189–220. https://doi.org/10.1007/s00466-020-01847-9

Gottstein G (2014) Materialwissenschaft und Werkstofftechnik: Physikalische Grundlagen, Berlin Heidelberg

Groeber MA, Jackson MA (2014) DREAM.3D: a digital representation environment for the analysis of microstructure in 3D. Integr Mater Manuf Innov 3:56–72. https://doi.org/10.1186/2193-9772-3-5

Gurtin ME, Fried E, Anand L (2010) The mechanics and thermodynamics of continua. New York

Hassan H, Govind K, Hartmaier A (2019) Micromechanical modelling of coupled crystal plasticity and hydrogen diffusion. Phil Mag 99:92–115. https://doi.org/10.1080/14786435.2018.1530466

Hoagland RG, Heinisch HL (1992) An atomic simulation of the influence of hydrogen on the fracture behavior of nickel. J Mater Res 7:2080–2088. https://doi.org/10.1557/JMR.1992.2080

Kreins M, Wilkes J, Wesselmecking S, Krupp U (2022a) Effect of phase-selective nanoscale precipitates on the Bauschinger effect in Austenitic-Ferritic duplex stainless steels. Metall Mat Trans A 53:3906–3917. https://doi.org/10.1007/s11661-022-06793-0

Kreins M, Büßenschütt K, Wesselmecing S, Krupp U (2022b) On the impact of nanoscale γ" precipitates on the Bauschinger effect in single- and polycrystal Ni-base superalloy 718. Mater Sci Eng A 856. https://doi.org/10.1016/j.msea.2022.144008

Kreins M, Schilli S, Seifert T, Iyer AHS, Colliander MH, Wesselmecking S, Krupp U (2021) Bauschinger effect and latent hardening under cyclic micro-bending of Ni-base Alloy 718 single crystals: Part I. Experimental analysis of single and multi slip plasticity. Mater Sci Eng A 827. https://doi.org/10.1016/j.msea.2021.142027

Kumar R, Mahajan DK (2020) Hydrogen distribution in metallic polycrystals with deformation. J Mech Phys Solids 135. https://doi.org/10.1016/j.jmps.2019.103776

Kumnick AJ, Johnon HH (1980) Deep trapping states for hydrogen in deformed iron. Acta Metall 28:33–39. https://doi.org/10.1016/0001-6160(80)90038-3

Li X, Zhang J, Cui Y, Djukic MB, Feng H, Wang Y (2024a) Review of the hydrogen embrittlement and interactions between hydrogen and microstructural interfaces in metallic alloys: Grain boundary, twin boundary, and nano-precipitate. Int J Hydrog Energy 72:74–109. https://doi.org/10.1016/j.ijhydene.2024.05.257

Li K, Tang B, Zhang M, Dai J, Cao X, Yin B, Zhang Z, Fan J, Li J (2024b) Hydrogen diffusion behavior within microstructures near crack tip: a crystal plasticity study. Mech Mater 195. https://doi.org/10.1016/j.mechmat.2024.105032

Lynch S (2012) Hydrogen embrittlement phenomena and mechanisms. Corros Rev 30:105–123. https://doi.org/10.1515/corrrev-2012-0502

Ma A, Hartmaier A (2014) On the influence of isotropic and kinematic hardening caused by strain gradients on the deformation behaviour of polycrystals. Phil Mag 94:125–140. https://doi.org/10.1080/14786435.2013.847290

Mughrabi H (1938) Dislocation wall and cell structures and long-range internal stresses in deformed metal crystals. Acta Metall 31:1367–1379. https://doi.org/10.1016/0001-6160(83)90007-X

Nye JF (1953) Some geometrical relations in dislocated crystals. Acta Metall 1:153–162

Robertson IM (1999) The effect of hydrogen on dislocation dynamics. Eng Fract Mech 68:671–692. https://doi.org/10.1016/S0013-7944(01)00011-X

Schilli S, Seifert T, Kreins M, Krupp U (2022) Bauschinger effect and latent hardening under cyclic micro-bending of Ni-base Alloy 718 single crystals: Part II. Single crystal plasticity modeling with latent kinematic hardening. Mater Sci Eng A 830. https://doi.org/10.1016/j.msea.2021.142030

Schmidt I (2011) Numerical homogenisation of an elasto-plastic model-material with large elastic strains: macroscopic yield surfaces and the Eulerian normality rule. Comp Mech 48:579–590. https://doi.org/10.1007/s00466-011-0601-x

Seifert T, Schenk T, Schmidt I (2007) Efficient and modular algorithms in modeling finite inelastic deformations: objective integration, parameter identification and sub-stepping techniques. Comput Methods Appl Mech Eng 196:2269–2283. https://doi.org/10.1016/j.cma.2006.12.002

Seifert T, Schweizer C, Schlesinger M, Möser M, Eibl M (2013) Thermomechanical fatigue of 1.4849 cast steel - experiments and life prediction using a fracture mechanics approach. Int J Matre Res 101:942–950. https://doi.org/10.3139/146.110363

Singh V, Kumar R, Charles Y, Mahajan DK (2022) Coupled diffusion-mechanics framework for simulating hydrogen assisted deformation and failure behavior of metals. Int J Plast 157. https://doi.org/10.1016/j.ijplas.2022.103392

Sofronis P, McMeeking RM (1989) Numerical analysis of hydrogen transport near a blunting crack tip. J Mech Phys Solids 37:317–350. https://doi.org/10.1016/0022-5096(89)90002-1

Tekkaya B, Dölz M, Münstermann S (2024) Modeling of local hydrogen concentration on microscopic scale to characterize the influence of stress states and non-metallic inclusions in pipeline steels. Int J Hydrog Energy 50:1274–1287. https://doi.org/10.1016/j.ijhydene.2023.10.158

Wang S, Martin ML, Robertson IM, Sofronis P (2016) Effect of hydrogen environment on the separation of Fe grain boundaries. Acta Mater 107:279–288. http://dx.doi.org/10.1016/j.actamat.2016.01.067

Yu H, Cocks A, Tarleton E (2019) Discrete dislocation plasticity HELPs understand hydrogen effects in bcc materials. J Mech Phys Solids 68:41–60. https://doi.org/10.1016/j.jmps.2018.08.020

Yuan S, Zhu Y, Huang M, Liang S, Li Z (2020) Dislocation-density based crystal plasticity model with hydrogen-enhanced localized plasticity in polycrystalline face-centered cubic metals. Mech Mater 148. https://doi.org/10.1016/j.mechmat.2020.103472

Yuan S, Zhu Y, Zhao L, Liang S, Huang M, Li Z (2022) Key role of plastic strain gradient in hydrogen transport in polycrystalline materials. Int J Plast 158. https://doi.org/10.1016/j.ijplas.2022.103409

Zhou Q, Cong L, Gan L, Ling C, Li D, Busso EP (2024) Effect of thermal ageing on the deformation behaviour of ferrite and austenite in a duplex stainless steel: Micropillar characterisation and dislocation density modelling. Acta Mater 265. https://doi.org/10.1016/j.actamat.2023.119599

Chapter 22
Creep-Fatigue Deformation of the Ferritic-Martensitic Steel P92: Non-unified Viscoplasticity Modeling of Service-Like Material Tests

Markus Schlayer, Gerhard Maier, and Thomas Seifert

Abstract The ferritic-martensitic steel P92 is used for thick-walled components in thermal power plants. Besides continuous operation, the components are exposed to cyclic loadings so that creep-fatigue deformation is relevant for the assessment of the components in finite-element calculations. In this work, a non-unified viscoplasticity model is developed to describe the creep-fatigue deformation behavior of the P92 steel. For the determination of the relevant model parameters of the material model, results of service-like material tests are employed that mimic the material loading conditions in thick-walled components. A good description of the service-like material tests is achieved with the model.

22.1 Introduction

Ferritic-martensitic steels are widely used as material for components in conventional and solar thermal power plants. These components are usually subjected to high pressures and high temperatures around 600 °C due to hot steam (Masuyama 2001). For this temperature range, advantages of ferritic-martensitic steels compared to nickel-based or austenitic steels are, for example, the cost efficiency and a lower thermal expansion (Viswanathan et al. 2008). One frequently used candidate out of the 9–12% chromium heat-resistant steels for power plants is P92.

Nowadays, conventional thermal power plants are operated more flexible with start and stop cycles due to an increasing portion of renewable electricity in the last decades. Also solar thermal power plants are subjected to start and stop cycles

M. Schlayer · T. Seifert (✉)
Offenburg University of Applied Sciences, Institute for Digital Engineering and Production (IDEeP), Badstraße 24, 77652 Offenburg, Germany
e-mail: thomas.seifert@hs-offenburg.de

M. Schlayer
e-mail: markus.schlayer@hs-offenburg.de

G. Maier
Fraunhofer Institute for Mechanics of Materials IWM, Woehlerstrasse 11, 79108 Freiburg, Germany
e-mail: gerhard.maier@iwm.fraunhofer.de

H. Altenbach et al. (eds.), *Lectures Notes on Advanced Structured Materials 3*,
Advanced Structured Materials 226, https://doi.org/10.1007/978-3-031-77403-4_22

because of cloud passages or day and night cycles. The combined stationary and cyclic loading of the components and the resulting creep-fatigue deformation must be taken into account in the service life assessment. For a realistic simulation of the loading conditions of P92 components in the framework of the finite-element method (FEM), there is a need for material models that can describe the creep-fatigue deformation of P92. In the following, works related to creep-fatigue of ferritic-martensitic steels are briefly reviewed, before material models describing creep-fatigue deformation in general are considered.

In Chen et al. (2020), the effect of dwell times at different minimum and maximum stresses on creep-fatigue interaction is investigated for the steel P92. However, the focus is not on the deformation phenomena but on the lifetime behavior. The effect of dwell times in strain-controlled fatigue tests is investigated in Zhang and Xuan (2017) concerning the softening behavior of P92 steel. Due to the strain control, stress relaxation is measured in the strain-controlled dwell times. In this publication, the effect on lifetime of the specimen is investigated as well. Similar results and low-cycle fatigue (LCF) tests with P92 steel are presented in Wang et al. (2018a, b). The creep-fatigue deformation in relation to the damage behavior is investigated in Ji et al. (2015). It is concluded that cyclic loading inhibits creep damage. Investigations on the microstructure and lifetime for creep-fatigue tests with different strain amplitudes are carried out in Mao et al. (2021).

While the creep-fatigue tests are usually performed with fixed trapezoidal waveforms for the controlled strain or stress, real service loadings of power plant components are more complex. Hence, service-like loading conditions mimic the loadings from thermal transients as well as steady-state operations. Service-like creep-fatigue tests are considered for P92 steel in Sonntag et al. (2023b). The service-like material tests contain load reversals and different dwell times in tension and compression. It is assumed that the loading conditions simplify the service conditions even though varying operational pressure and complex stress states are not considered. The material tests are published in Sonntag et al. (2023a). In the previous presented publications, stress-free times were not considered within the creep-fatigue tests. In El-Magd and Shaker (1991), it was demonstrated that unloading of the steel X22CrMoV121 from a peak stress to different lower stresses leads to creep stress. It is shown that negative creep rates occur for lower unloading stresses. From this, it can be concluded that stress-free dwell times after tensile loadings influence the strain rate dependent on the previous loading and should be considered in material tests.

With regard to modeling and simulation of creep-fatigue deformation, mathematical equations and their implementation for usage in the FEM are needed. There are different material models published for ferritic-martensitic and comparable chromium steels. In general, the models are divided into unified and non-unified constitutive models. The non-unified material models include two or more strain tensors for description of inelastic deformations, whereas the unified material models use only one inelastic strain tensor (Chaboche 1993). Non-unified material models are often used for the description of creep-fatigue deformation where the respective strain tensors describe different deformation phenomena, e.g., creep deformation at lower stresses and cyclic deformation at higher stresses individually.

Some non-unified material models use a combination of classic creep and cyclic plasticity models to describe creep-fatigue deformation. In Wang et al. (2021), Ohno et al. (2018), the OHNO–WANG model is used to describe cyclic plastic deformation and the NORTON creep law is used to describe the creep behavior. This material model was adapted for the austenitic-steel 316H in Shimada et al. (2018). Other material models use one or more inelastic strain rates from CHABOCHE type (Chaboche 2008) with FREDERICK–ARMSTRONG-type kinematic hardening laws (Frederick and Armstrong 2007). The models were originally developed in the late 1980s and consequently improved for new materials and deformation phenomena.

Non-unified material models considering an interaction of multiple strain tensors and their respective kinematic hardening laws are proposed in Contesti and Cailletaud (1989) to describe creep-plasticity interaction effects, referred to as multi-mechanism material models. Therefore, two inelastic strain rates are implemented and applied to creep-fatigue deformation modeling of an austenitic stainless steel at the temperature 600 °C. An extension to this multi-mechanism approach is published in Cailletaud and Saï (1995) within a general thermodynamic framework for the additive decomposition of the total strain in an elastic and two or more inelastic strains. There, the inelastic strain rates are of CHABOCHE type with FREDERICK–ARMSTRONG-type kinematic hardening laws (Frederick and Armstrong 2007). In general, no specific deformation mechanism is intended for the inelastic strain rates. That means that the model can be used for creep-fatigue deformation but also may include different deformation mechanisms. However, an interaction of creep and fatigue deformation is implemented in this model but no determination of model parameters on basis of material tests is given. Further developments of multi-mechanism models are published in Saï and Cailletaud (2007), where crystal plasticity modeling and ratcheting behavior for the steel type 316 are addressed. The deformation phenomena which are considered for this austenitic steel are creep, stress relaxation, and cyclic plasticity. Moreover, extensions of the models concerning strain memory effect and non-proportional loading for the austenitic-steel 304L are considered in Taleb and Cailletaud (2010). A review and trends for multi-mechanism models are presented in Saï (2011), where the assessment of multi-mechanism models for different materials like 316 austenitic stainless steel (316SS), the martensitic steel 55NiCrMoV7, or the nickel base superalloys (INCO718) is shown. The mechanisms of strain memory effect and ratchetting are investigated in detail. However, the creep-fatigue deformation is not investigated in this review article. Non-unified material models without interaction are presented for the P92 steel in Bender (2020) and Wang et al. (2015). The publication Bender (2020) uses a modified GRAHAM–WALLES CREEP LAW to predict the creep strain and a CHABOCHE model to predict the cyclic material behavior. In Wang et al. (2015), the decomposition of the total inelastic strain into a viscoplastic strain rate and a creep strain rate in the framework of the BECKER AND HARDENBERG model is presented. The model parameters are determined for the material P92 with isothermal and thermomechanical fatigue tests as well as stress-driven creep tests. In comparison to these publications, this work uses two viscoplastic strain rates of CHABOCHE- TYPE and stress-free times which are considered in the creep-fatigue tests.

In summary, viscoplastic material models for the assessment of creep-fatigue deformation exist where frequently non-unified approaches are used. However, service-like loading conditions for components of thermal power plants are not used for the determination of the model parameters and the validation of the material models. In particular, a material model for creep fatigue of P92 ferritic-martensitic steel for service-like conditions is not available. Furthermore, no dwell times with zero stress (load free times) were considered in these tests. Hence, in this work, a non-unified viscoplasticity model for creep-fatigue deformation of the ferritic-martensitic chromium steel P92 is considered based on service-like creep-fatigue tests so that the operation conditions can be predicted by means of FEM calculations. The aim is that the material model is able to describe the deformation phenomena creep and fatigue individually and in combination as creep-fatigue deformation. Special attention is paid to the anelastic recovery during unloading after previous creep loading.

This work is structured as follows: The material database for which the presented material model is developed is described in Sect. 22.2 where also selected service-like material tests are shown. The non-unified viscoplasticity model and the determination of the corresponding model parameters of the material model are addressed in Sect. 22.3. The results obtained with the developed material model are shown for selected material tests with the material P92 in Sect. 22.4. The discussion of the results in Sect. 22.5 is followed by the summary and an outlook in Sect. 22.6.

22.2 Material Database and Tests

In this work, P92 steel is considered which is referred to as ASTM A335 P92 in accordance to ASME standard or X10CrWMoVNb9-2 in accordance to EN standard (ASTM 2024). The material database which is used in this work for development of the viscoplasticity model consists of material tests carried out within the project "VGB calculation methods" (Oesterlin et al. 2022) and compiled in Table 22.2. The database contains different types of fatigue tests: complex low-cycle fatigue (CLCF, see below), standard low-cycle fatigue (LCF), LCF with dwell times (LCF-H) as well as modified LCF (MLCF, see also below), the latter representing the service-like conditions.

The viscoplasticity model (Sect. 22.3.1) as well as the associated model parameters (Sect. 22.3.2) are determined on basis of all material tests of Table 22.1. However, the focus on this paper is the creep-fatigue deformation and therefore there are shown one complex low-cycle fatigue (CLCF) test and two so-called MLCF tests with complex stress histories and long cycle times exemplary to demonstrate the performance of the non-unified viscoplasticity model regarding the creep-fatigue deformation behavior.

According to Oesterlin et al. (2022), the isothermal CLCF tests are conducted in strain control at the service-like temperature of 600 °C. Figure 22.1 shows exemplary the complex part of a CLCF test. The complex part includes a complex strain history with the different strain rates (10^{-5}, 10^{-4}, 10^{-3} 1/s) and dwell times of 1800 s

Table 22.1 Overview of the material database. CLCF, LCF, and LCF-HT tests conducted in strain control, MLCF tests load controlled

Test	Stress/strain amplitude	Stress/strain dwell	Dwell time	Stress-free time
CLCF	0.2/0.3%		1800 s	–
CLCF	0.4/0.6%	–	1800 s	–
LCF	0.2%	–	–	–
LCF	0.4%	–	–	–
LCF-H	0.4%	0.4%	60 s	–
LCF-H	0.4%	-0.4%	60 s	–
LCF-H	0.6%	0.6%	60 s	–
LCF-H	0.6%	0.6%	600 s	–
LCF-H	0.6%	0.6%	3600 s	–
MLCF	185 MPa	144 MPa	792 s	34 s
MLCF	185 MPa	144 MPa	1600 s	126 s
MLCF	185 MPa	144 MPa	10420 s	306 s

Table 22.2 Model parameters of the non-unified viscoplasticity model

Basic parameters	E in MPa	ν			
Value	156000	0.3			
Parameters ε_1	R_1^e in MPa	K_1 in MPa	n_1	Q_1^∞ in MPa	b_1
Value	0	645	6.37	0	0
Parameters ε_1	C_{11} in MPa	γ_{11}	ϕ_{11}	ω_{11}	R_{11}^{stat} in s^{-1}
Value	250000	1000	0.5	100	$1 \cdot 10^{-9}$
Parameters ε_1	C_{12} in MPa	γ_{12}	ϕ_{12}	ω_{12}	R_{12}^{stat} in s^{-1}
Value	1	1	1	0	$1 \cdot 10^{-9}$
Parameters ε_2	R_2^e in MPa	K_2 in MPa	n_2	Q_2^∞ in MPa	b_2
Value	105	250	9	–58	0.1
Parameters ε_2	C_{21} in MPa	γ_{21}	ϕ_{21}	ω_{21}	R_{21}^{stat} in s^{-1}
Value	1840	23	4	1	$1 \cdot 10^{-9}$
Parameters ε_2	C_{22} in MPa	γ_{22}	ϕ_{22}	ω_{22}	R_{22}^{stat} in s^{-1}
Value	109500	730	2.5	2.5	$6.5 \cdot 10^{-3}$

in tension and compression. The complex part of the fatigue test is followed by a cyclic part with a constant strain amplitude and strain rate (not shown in Fig. 22.1). The strain amplitudes are varied for different CLCF tests as shown in Table 22.1. Based on the results of the CLCF tests, the cyclic viscoplastic material behavior can be measured efficiently in one test with one specimen at the considered temperature of 600 °C and the corresponding model parameters of viscoplasticity models

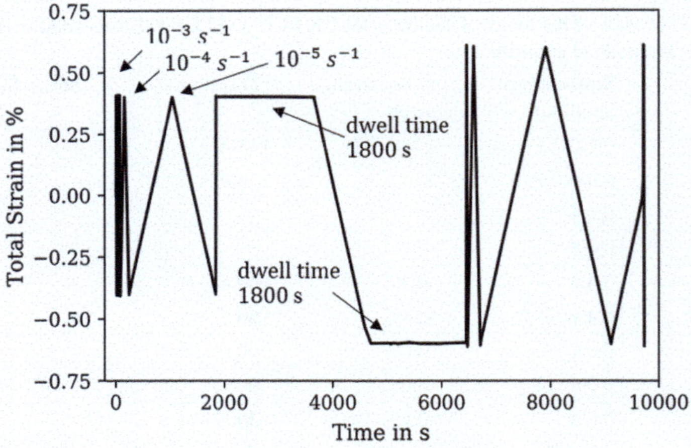

Fig. 22.1 Complex part of the complex low-cycle fatigue deformation test

can be determined efficiently as demonstrated for different materials, e.g., ferritic-martensitic and martensitic steels (Schlayer et al. 2023; Krishnasamy et al. 2010) and nickel-based alloys for power generation (Maier et al. 2011; Fischer et al. 2023). Identical tests on different samples are necessary in order to be able to take into account the scattering from sample to sample. Two CLCF tests are available for this work as shown in Table 22.1.

The MLCF tests are designed for representation of the loading conditions of thick-walled pipes in thermal power plants why they are called service-like tests. Thus, the service-like MLCF tests contain fatigue and creep deformation within one test cycle. One cycle of the MLCF tests is shown in Fig. 22.2 with black markers. The

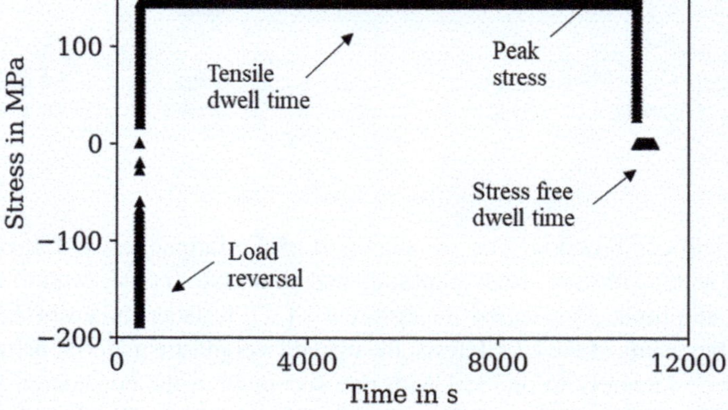

Fig. 22.2 One cycle of the service-like MLCF test at 600 °C

controlled stress is shown on the vertical and the test time on the horizontal axis. The test temperature is 600 °C. Following, one MLCF test cycle is introduced in chronological order, important parts within the shown cycle are noted additionally in the figure: The first test cycle starts from the origin into compression with a minimum stress value of −185 MPa. This represents the start-up phase of a thermal power plant when the inner part of the pipe is heating up faster compared to the outer part of the wall of the pipe, leading to the compressive stress in the inner wall of the pipe. Then, the pipe is fully heated up and the pressure due to the hot steam in the pipe leads to tensile stress within the pipe. The period from the minimum stress to the stress of the dwell time is called load reversal. This load reversal represents fatigue loading. The load reversal is followed by the continuous operation condition of a thermal power plant with the dwell time of 1444 MPa tensile stress. The creep time is varied in the different MLCF tests as shown in Table 22.1. The tensile dwell time is followed by a short tensile stress peak at 185 MPa representing the cooling of the component surface. A dwell time at zero stress represents the normal operation which is maintained for a specific time period until the next cycle follows.

22.3 Non-unified Viscoplasticity Model

In this section, the non-unified viscoplasticity model is presented together with the mathematical equations in Sect. 22.3.1 and the method for the determination of the model parameters in Sect. 22.3.2. The model parameters of the presented material model are given in Table 22.2.

22.3.1 Mathematical Model

The multi-mechanism viscoplasticity model used in this work is a non-unified model and consists of two viscoplastic strain rates according to the works of Chaboche (Chaboche 1989, 1993, 1997). It turned out in this work that an interaction of viscoplastic strains or kinematic hardening laws (as used in the multi-mechanism models) was not necessary in the model for the description of the test data. In the uniaxial formulation, which is used to calculate the test data, the stress can be calculated from the elastic strain and Young's modulus E as

$$\sigma = E \cdot \varepsilon^{\mathrm{el}} = E \cdot \left(\varepsilon - \varepsilon_1^{\mathrm{vp}} - \varepsilon_2^{\mathrm{vp}}\right) \tag{22.1}$$

The elastic strain can be calculated from the total strain ε and the two viscoplastic strains $\varepsilon^{\mathrm{vp}}$ for isothermal conditions. The viscoplastic strains are calculated from the evolution equations:

$$\dot{\varepsilon}_i^{\mathrm{vp}} = \dot{\bar{\varepsilon}}_i^{\mathrm{vp}} \cdot \frac{\sigma - \alpha_i}{|\sigma - \alpha_i|}. \tag{22.2}$$

The index i represents the number of the viscoplastic strain rate and is $i = 1, 2$ since two viscoplastic strains are considered in the model. The flow direction is computed by the second part of Eq. (22.2) from the stress σ and the backstress α, the latter describing kinematic hardening of the material. Furthermore, the equivalent viscoplastic strain rate $\dot{\bar{\varepsilon}}_i^{\text{vp}}$ is calculated by

$$\dot{\bar{\varepsilon}}_i^{\text{vp}} = \dot{\bar{\varepsilon}}_{0,i}^{\text{vp}} \cdot \left\langle \frac{|\sigma - \alpha_i| - R_i^p}{K_i} \right\rangle^{n_i}. \tag{22.3}$$

Again, the index i represents the number of the strain rate. Therefore, the model parameters K, n that describe the viscous material behavior are defined individually for each viscoplastic strain i. The numerator represents the viscous overstress and the denominator K is a model parameter that defines the viscous overstress obtained for the reference strain rate $\dot{\bar{\varepsilon}}_0^{\text{vp}}$. The model parameter R^p is the yield stress and includes isotropic hardening or softening of the material by

$$R_i^p = R_i^e + Q_i^{\infty} \cdot \left(1 - e^{\left(-b_i \cdot \bar{\varepsilon}_i^{\text{vp}}\right)}\right) \tag{22.4}$$

with the model parameters Q^{∞} and b and the yield strength R^e. The isotropic hardening or softening describes obstruction of dislocation movement why it depends on the accumulated viscoplastic strain ε^{vp}.

The Bauschinger effect occurring from internal residual stress fields is considered by kinematic hardening via the backstress α. The backstress for the viscoplastic strain rate i from Eq. (22.3) is computed by the summation of several individual backstresses j from

$$\alpha_i = \alpha_{i1} + \alpha_{i2}. \tag{22.5}$$

Each backstress is described by the FREDERICK–ARMSTRONG-type evolution equation

$$\dot{\alpha}_{ij} = C_{ij} \cdot \dot{\varepsilon}_i^{\text{vp}} - \gamma_{ij} \cdot \varphi_{ij} \cdot \dot{\bar{\varepsilon}}_i^{\text{vp}} \cdot \alpha_{ij} - R_{ij}^{\text{stat}} \cdot \alpha_{ij} \tag{22.6}$$

and cyclic hardening or softening is accounted for in the dynamic recovery term using

$$\varphi_{ij} = \phi_{ij} + \left(1 - \phi_{ij}\right) \cdot e^{-\omega_{ij} \cdot \bar{\varepsilon}_i^{\text{vp}}}. \tag{22.7}$$

The index i represents the strain rate and j represents the number of the backstress within the strain rate. In this model, two backstresses j per viscoplastic strain rate i are used to describe the nonlinear hardening behavior. C, γ, R^{stat}, ϕ, and ω are model parameters. C is the kinematic hardening modulus, γ is the dynamic recovery parameter enabling a nonlinear hardening curve, and R^{stat} quantifies static recovery. The saturation value of kinematic hardening is defined by $C_{\infty} = C/(\gamma \cdot \varphi)$. In Eq. (22.7), ω controls the transition rate to the saturation value ϕ depending on the accumulated viscoplastic strain (Chaboche 2008).

22.3.2 *Determination of Model Parameters*

The determination of model parameters on basis of service-like tests is not yet standard. Therefore, a procedure to determine model parameters is proposed in this section. All model parameters are determined on basis of the material tests of Table 22.1. The compiled model parameter values of the developed material model are given in Table 22.2. Some of the model parameters are determined experience based and some model parameters are determined optimization based by means of numerical methods. The strategy is described in the following itemization. Note that the items are in logical but not mandatory in chronological order. Young's modulus E is determined from the first cycles of LCF tests within the elastic range and verified with hysteresis loops at half lifetime. Poisson ratio is fixed at 0.3. These model parameters could also be taken from a material data sheet.

Step 1: Experience based:

- The yield stress R_1^e of the creep-related strain rate is 0 MPa since creep is a diffusion-based deformation and may occur even for low stresses at high temperatures.
- The static recovery properties R_{1j}^{stat} of the creep-related kinematic hardening are set to low values.
- The yield stress R_2^e of the fatigue-related strain rate is determined for 0.2% plastic strain as mean value of the initial cycles of the LCF tests. The plastic strain is computed by subtracting the elastic strain from the measured total strain. The elastic strain is computed with Young's modulus E.
- For the fatigue-related strain rate, typical values for the model parameters K and n are chosen. The power-law breakdown which represents glide of dislocations is known for $n \geq 8$. Therefore, $n_2 = 9$ is chosen. K_2 is 250 MPa so that the viscous stress in the complex part of the CLCF tests can be described.
- The static recovery properties R_2^{stat} of the fatigue-related strain rate are chosen after definition of the backstresses and the viscous model parameters K_2 and n_2 such that the experimentally observed stress relaxation can be adequately described by the material model.
- For kinematic hardening in the fatigue-related strain rate $\dot{\varepsilon}_2^{\mathrm{vp}}$, two backstresses α_{2j} are used. One backstress develops faster and one backstress slower with the viscoplastic strain rate such that the exponential hardening behavior can be modeled in good agreement with the material tests. The model parameters are determined experience based using the CLCF tests.
- The model parameters Q_2^{∞} and b_2 describing isotropic hardening or softening are determined on basis of the softening observed in LCF tests and in the cyclic part of the CLCF tests. A distinction of isotropic and kinematic softening is possible by comparing the stress–strain hysteresis loops for different cycles. A reduction of the yield strength in the hysteresis loops represents isotropic softening.

- The model parameters ϕ_{2j} and ω_{2j} describing kinematic hardening or softening are determined on basis of the softening observed in LCF tests and in the cyclic part of the CLCF tests as the portions of softening that are not isotropic.
- The validation of the model parameters on 50 cycles of the MLCF tests show overestimation of the creep strain in tensile dwell times. Thus, cyclic hardening of the dynamic recovery term is enabled by the model parameters ϕ_{11} and ω_{11}. The model parameters are determined such that the decreasing creep strain in tensile dwell times is captured.

Step 2: Optimization based:

- The viscous material parameters K_1 and n_1 of the creep-related strain rate $\dot{\varepsilon}_1^{\text{vp}}$ are determined. For this purpose, the strain rates are extracted from the MLCF tests for the tensile dwell times and stress-free times from the measured strain and time. A minimization of the least-square error between the viscoplastic strain rate of the model and the test results is carried out. However, the model includes an effective stress $\sigma - \alpha_1$ (see Eq. (22.3)). Therefore, a mean value of the backstress α_1 is considered which could be varied within the minimization of the least-square error. Following, the model parameters C_{11} and γ_{11} are adapted such that a good description of the optimized mean backstress α_1 is achieved with the viscoplastic strains extracted from the MLCF tests.
- The viscous model parameters K_2 and n_2 can be calculated from the CLCF test by extracting the viscous stresses σ^{v} for the three different strain rates in the complex part. Then two residuals $R_1 = \left(\sigma_1^{\text{v}} - \sigma_2^{\text{v}}\right) + K_2 \cdot \dot{\varepsilon}_1^{(1/n_2)} - K_2 \cdot \dot{\varepsilon}_2^{(1/n_2)}$ and $R_2 = \left(\sigma_2^{\text{v}} - \sigma_3^{\text{v}}\right) + K_2 \cdot \dot{\varepsilon}_2^{(1/n_2)} - K_2 \cdot \dot{\varepsilon}_3^{(1/n_2)}$ are reduced with a numerical solver by variation of K_2 and n_2.
- For further numerical optimizations, the *SciPy* module optimize.minimize is used (SciPy optimize minimize 2024). The aim of the optimization is the minimization of the least-square error between the total computed and measured strain. As input, initial values for all model parameters and their admissible spaces are defined. The admissible space is defined such that the model parameters remain physically reasonable. The initial model parameters are taken from literature data of comparable steels if there are no mathematical approaches for determination of the initial model parameters shown in this section. The variation of the model parameters is done by the *SciPy* optimization tool.
- The material test data have different amounts of data points why the least-square error sum depends on the number of data points. The different deformation phenomena within one test must be weighted as well, e.g., because the dwell times have more data points than the load reversals. Therefore, a weighting of individual data points is implemented. Criterion for the weight is the stress rate which is comparatively low during creep of the material. The stress-free creep reversal is identified by stress rate and a stress range around 0 MPa. In addition, it is possible to weight material tests individually. The weighting is selected from the number of data points, such that the deformation phenomena within one test and the different material tests are equally weighted.

22.4 Results

In this section, the results for the strain-controlled CLCF test with strain amplitudes 0.2/0.3% and the results of two stress-controlled service-like MLCF tests with different tensile dwell times (1600 s and 10420 s) are shown exemplary (see Table 22.1). In the figures in this section, the material tests (black markers) are presented together with the simulation results (red lines). First, the result of the CLCF test is presented followed by the results of the two MLCF tests. Selected LCF-H tests are shown at the end of this section (Sect. 22.4 in Figs. 22.9, 22.10, and 22.11).

Figure 22.3 shows the stress response of the CLCF test with the strain amplitudes 0.2/0.3%. The viscous material behavior in the complex part of the test is described

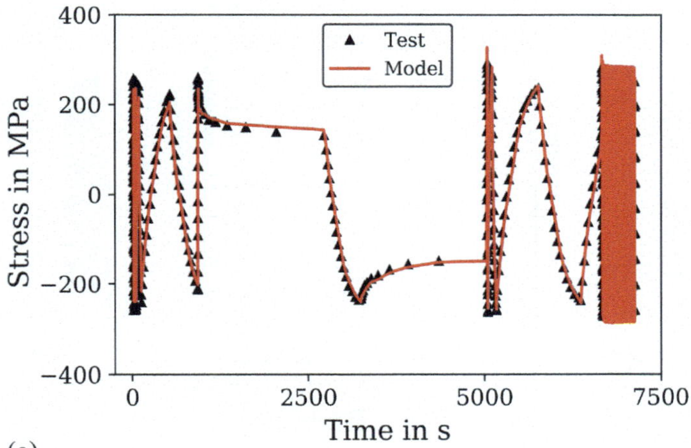

(a)

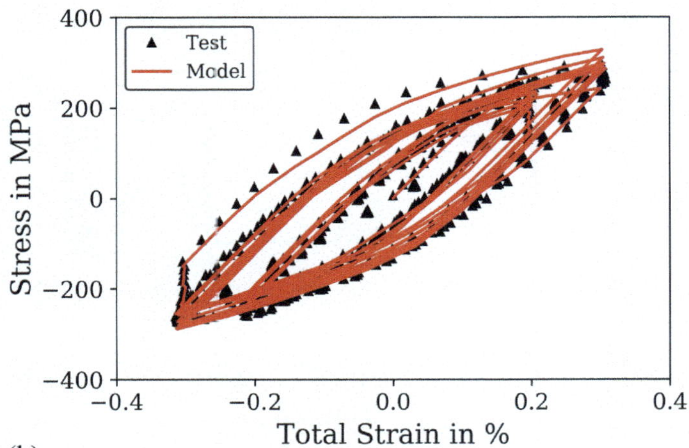

(b)

Fig. 22.3 CLCF test (first 50 cycles) with strain amplitude 0.2/0.3% at 600 °C: **a** stress history; **b** stress–strain hysteresis loops well by the developed material model. The different maximum and minimum stresses which are visible for the different strain rates can be described with the material model. Furthermore, the stress relaxation behavior of the material in the tensile and compressive dwell times is captured by the material model. The stress at load reversal for the higher strain amplitudes of the simulation is in good accordance with the test results. The stress–strain hysteresis loops for 50 cycles show that the hardening behavior is described well by the material model. However, the slight softening behavior which is visible due to smaller maximum stresses in the test results compared with the model results is not fully described by the material model.

The backstresses and viscoplastic strains are presented for the two strain rates of the non-unified model in Fig. 22.4. The backstresses α_{11} and α_{12} are shown with blue

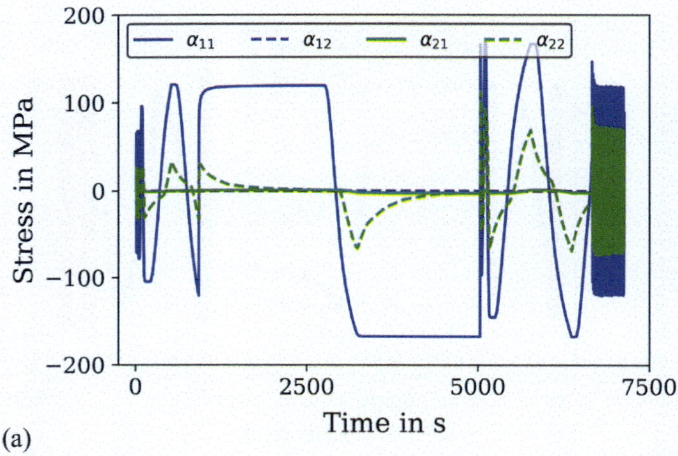

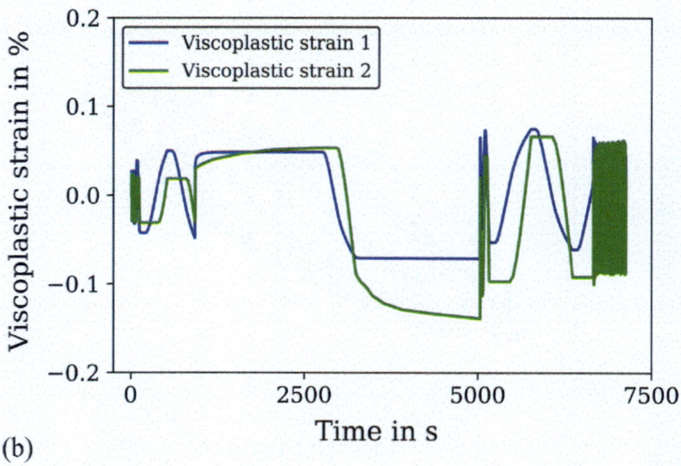

Fig. 22.4 CLCF test (first 50 cycles) with strain amplitude 0.2/0.3% at 600 °C: **a** evolution of the backstresses of the two strain rates; **b** evolution of viscoplastic strains of the two strain rates

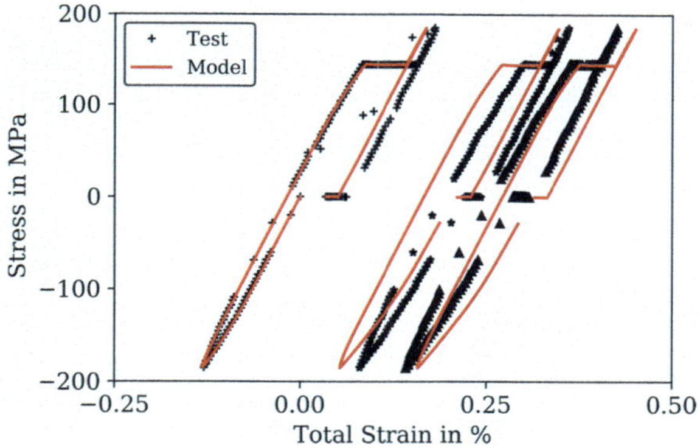

Fig. 22.5 MLCF test (cycle 1, 25, 50) with 1600 s dwell time: stress–strain curve

lines and belong to the first strain rate. The green lines show the backstresses α_{21} and α_{22} of the second strain rate. However, only one backstress of each strain rate is active for this CLCF test. The portion of the first backstress is higher compared to the second one. The stress relaxation behavior in the dwell times is modeled with the second backstress α_{22} of the second strain rate.

The viscoplastic strains show that both viscoplastic strains are active in the CLCF test. The second viscoplastic strain is more dominant for the higher strain amplitudes. A shift to the compression can be observed for the second viscoplastic strain after the compressive dwell time.

For the MLCF test with 1600 s dwell time, the stress–strain curves are shown in Fig. 22.5 for the selected cycles 1, 25, and 50 (left to right) for better visibility. Overall, the total strain increases, therefore the viscoplastic creep strain in the dwell times is higher compared to the viscoplastic compressive load and reverse creep during stress-free loading. The portion of tensile creep decreases with increasing number of cycles. Hence, the distance in total strain is higher between cycles 1 and 25 in comparison to the distance between cycle 25 to cycle 50. The model results show also a decreasing portion of creep strain with increasing number of cycles. However, it is overestimated for cycles 25 and 50.

The creep behavior with negative creep rate during stress-free loading is also described by the material model. Furthermore, the model represents the opened stress–strain hysteresis in the compressive load reversal and the elastic tensile peak stress. Thus, a distinction of the deformation is captured with the material model although the absolute value of the peak stress is the same in tension and compression. The behavior of the backstresses and the two viscoplastic strains is investigated in Fig. 22.6. The backstress consists mainly of the backstress α_{11} of the first strain rate. In chronological order, the backstress decreases in the compressive load reversal and increases in the tensile loading. The maximum value of the backstress within each cycle is reached at the end of the tensile dwell time, respective of the peak stress. The backstress is decreased in the unloading but still remains at a value of around 80 MPa. This remaining backstress is reduced during the stress-free time due to reversed creep. The backstress has a shift to positive values due to the tensile dwell

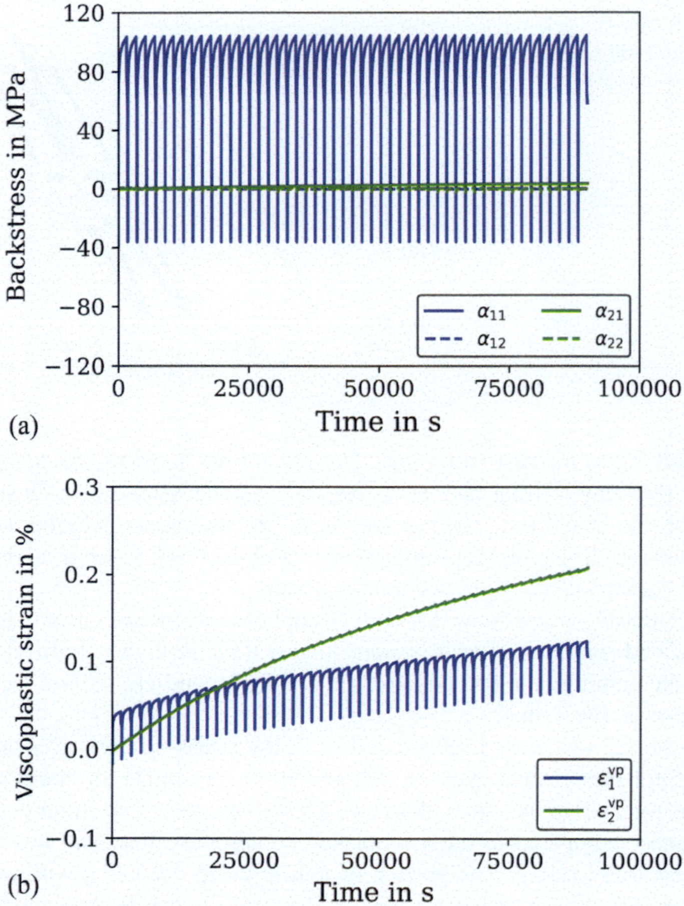

Fig. 22.6 MLCF test (first 50 cycles) with 1600 s dwell time: **a** evolution of the backstresses of the two strain rates; **b** evolution of viscoplastic strains of the two strain rates

times. The backstress α_{21} of the second strain rate slowly increases with increasing number of cycles. It is interesting to note that the second viscoplastic strain ε_2^{vp} is larger than the first viscoplastic strain after a few cycles. Therefore, the total strain is predominantly given by the second strain rate and, thus, the overestimation of the total strain in Fig. 22.5 is particularly from the second strain rate. The first strain rate has a higher strain range within one cycle and describes the viscoplastic behavior in the compressive load reversal as well as the creep reversal during stress-free loading. The results of the second service-like MLCF test, i.e., the MLCF test with 10420 s dwell time, are presented in Figs. 22.7 and 22.8. The difference to the previous shown MLCF test is the longer tensile dwell time. The stress–strain curve for the cycles 1, 25, and 50 in Fig. 22.7 show that the total strain is overestimated in the cycles 25 and 50. However, the viscoplasticity model can describe the initial material behavior.

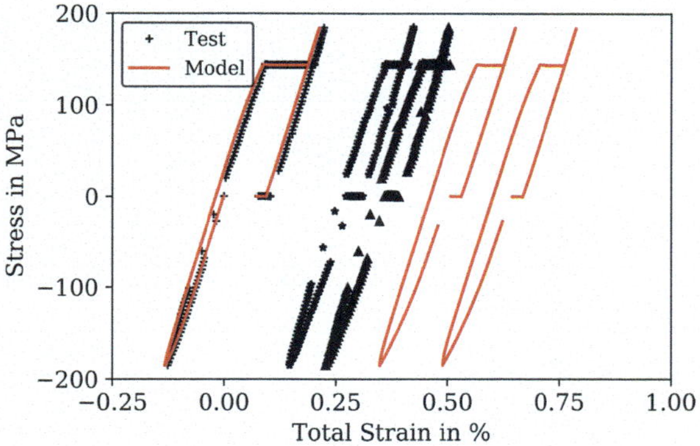

Fig. 22.7 MLCF test (cycles 1, 25, 50) with 10420 s dwell time: stress–strain curve

The creep-fatigue deformation behavior, e.g., the nonlinear hardening during the compressive load reversal, the elastic peak stress, and the reverse creep are described well by the material model, however the total strain is overestimated. As a reason, the second viscoplastic strain $\varepsilon_2^{\mathrm{vp}}$ is higher than the first viscoplastic strain already after the first cycle as shown in Fig. 22.8. Overall, the total strain is higher with increasing tensile dwell time. The reason is the higher portion of creep strain due to the longer dwell time. The evolution of the backstress (Fig. 22.8) is similar to the previous MLCF test with shorter dwell times (Fig. 22.6).

22.5 Discussion

In this work, a non-unified viscoplasticity model is presented for which the model parameters are determined based on different material tests at 600 °C including stress-controlled service-like tests. The model contains a larger set of model parameters. A strategy is proposed to determine the model parameter based on experience and optimization. Due to the applied experience-based approach, the possibility of a different sequence of steps, and a user-defined choice of weights for data points, different persons will obtain different results for the model parameters as documented in this work for the model parameters. This is problematic if different materials are to be compared on the basis of their model parameters and service-like tests. However, the different results of the model parameters might show all a good description of the data (i.e., might all be good local minima of the optimization problem).

The material tests are carried out at the isothermal temperature 600 °C. In service, the temperature varies in start and stop operations of thermal power plants. The transient temperature is not considered in the isothermal tests. In future works, the model parameters could be extended for different isothermal temperatures or thermomechanical fatigue tests could be considered. This requires the proposed strategy for the determination of the model parameters (with all its uncertainties) to be applied for each temperature. A validation of the determined temperature-dependent model

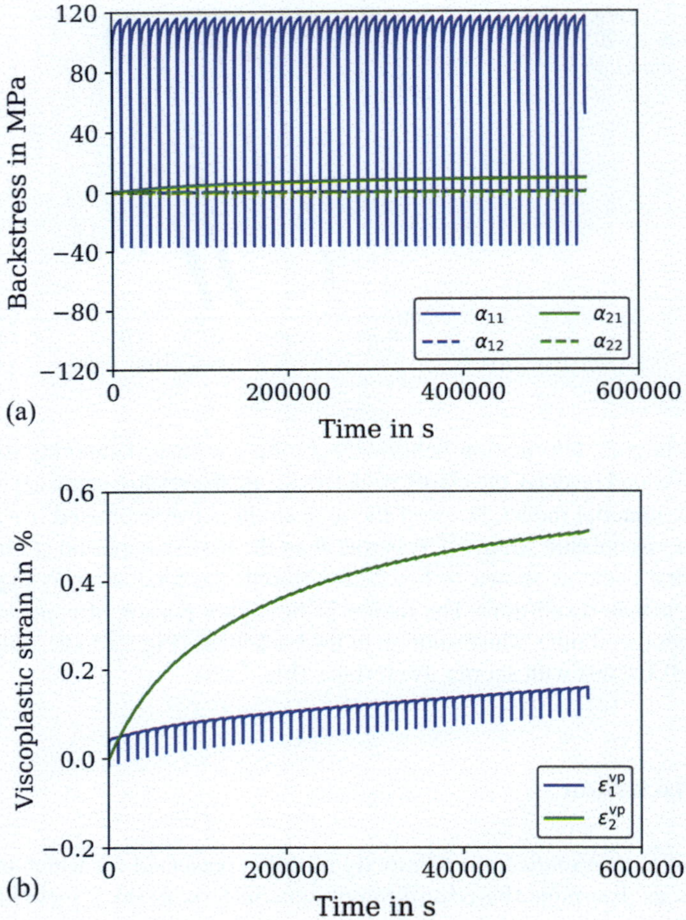

Fig. 22.8 MLCF test (first 50 cycles) with 10420 s dwell time: **a** evolution of the backstresses of the two strain rates; **b** evolution of viscoplastic strains of the two strain rates

parameters based on thermomechanical fatigue tests is recommended then. Moreover, the service-like material tests used in this work for modeling creep-fatigue deformation are stress controlled. An improvement of the service-like behavior could be achieved by changing between stress and strain control within one cycle. The load reversals could be carried out in strain control and the tensile dwell times in stress control. The strain control would fit better to the thermal strains due to transient temperatures which lead to the fatigue loading.

The advantage of the stress-controlled dwell times is the direct measurement of the creep strain in the tensile dwell times. Thus, creep models may be efficiently adapted to the test data. However, modeling of the creep-fatigue tests in this publication with two viscoplastic strain rates of CHABOCHE type has shown that creep and fatigue

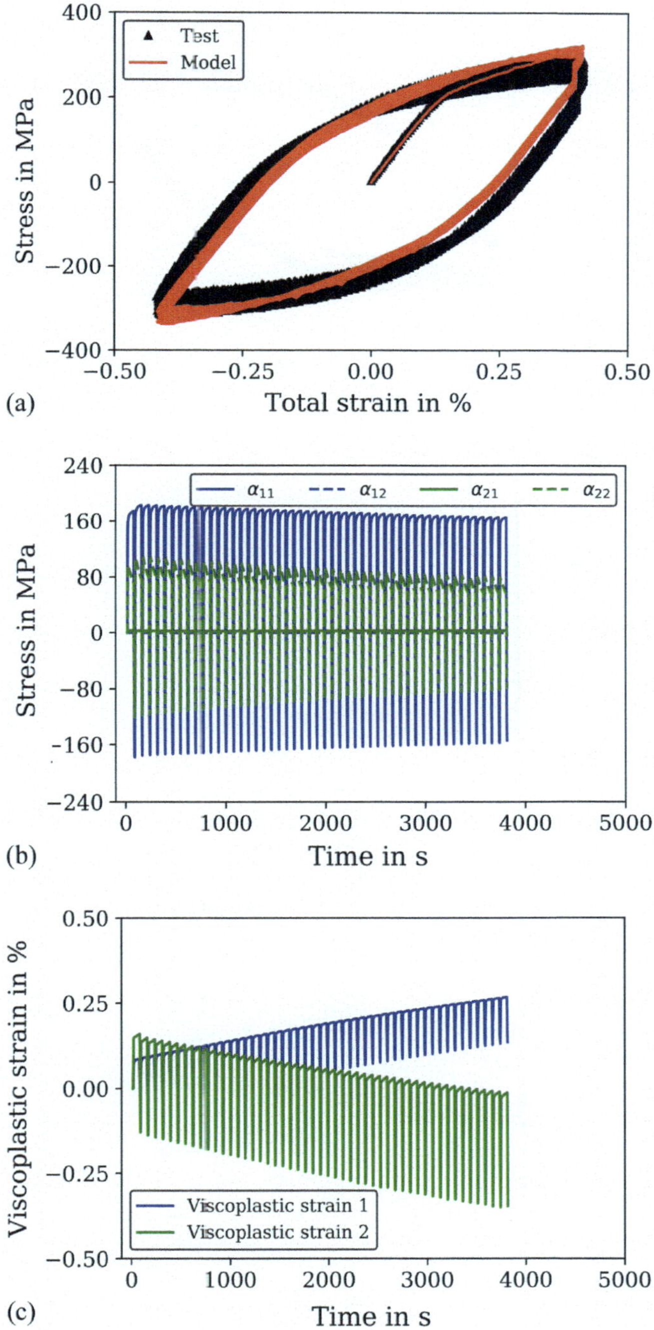

Fig. 22.9 LCF-H test (first 50 cycles) with 60 s dwell time and strain amplitude 0.4% at 600 °C: **a** stress–strain hysteresis loops; **b** evolution of the backstresses of the two strain rates; **c** evolution of viscoplastic strains of the two strain rates

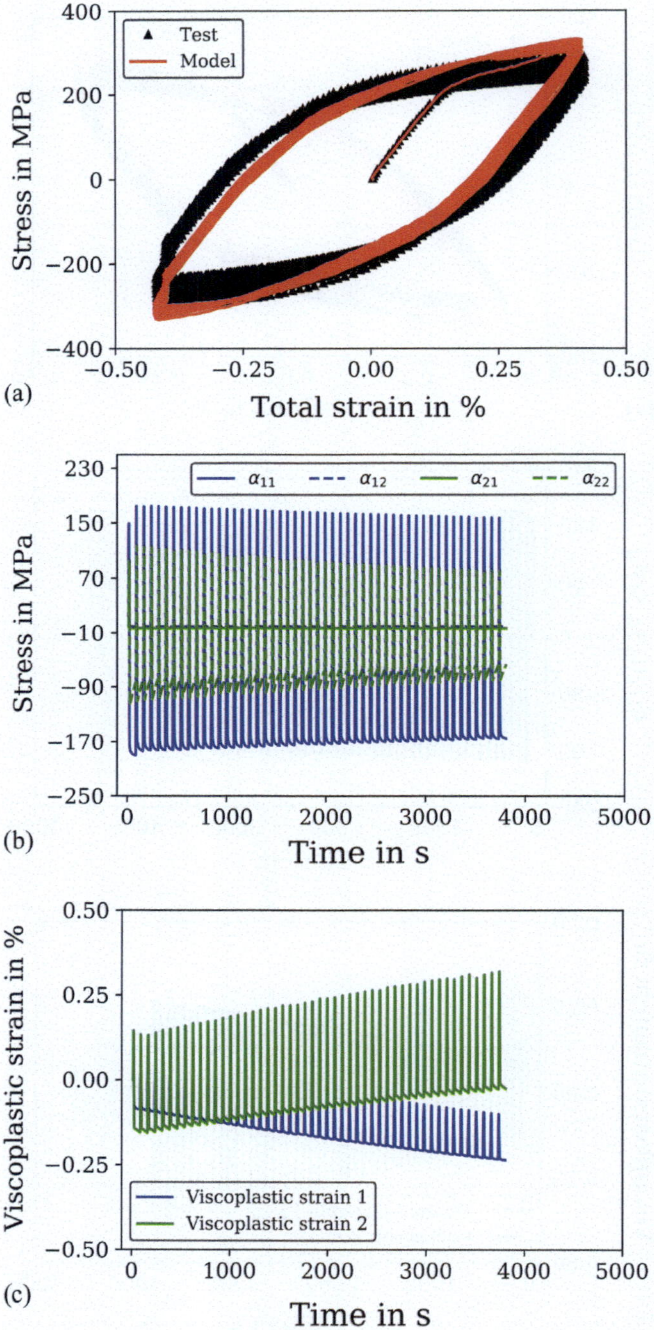

Fig. 22.10 LCF-H test (first 50 cycles) with 60 s dwell time and strain amplitude 0.4% at 600 °C: **a** stress–strain hysteresis loops; **b** evolution of the backstresses of the two strain rates; **c** evolution of viscoplastic strains of the two strain rates

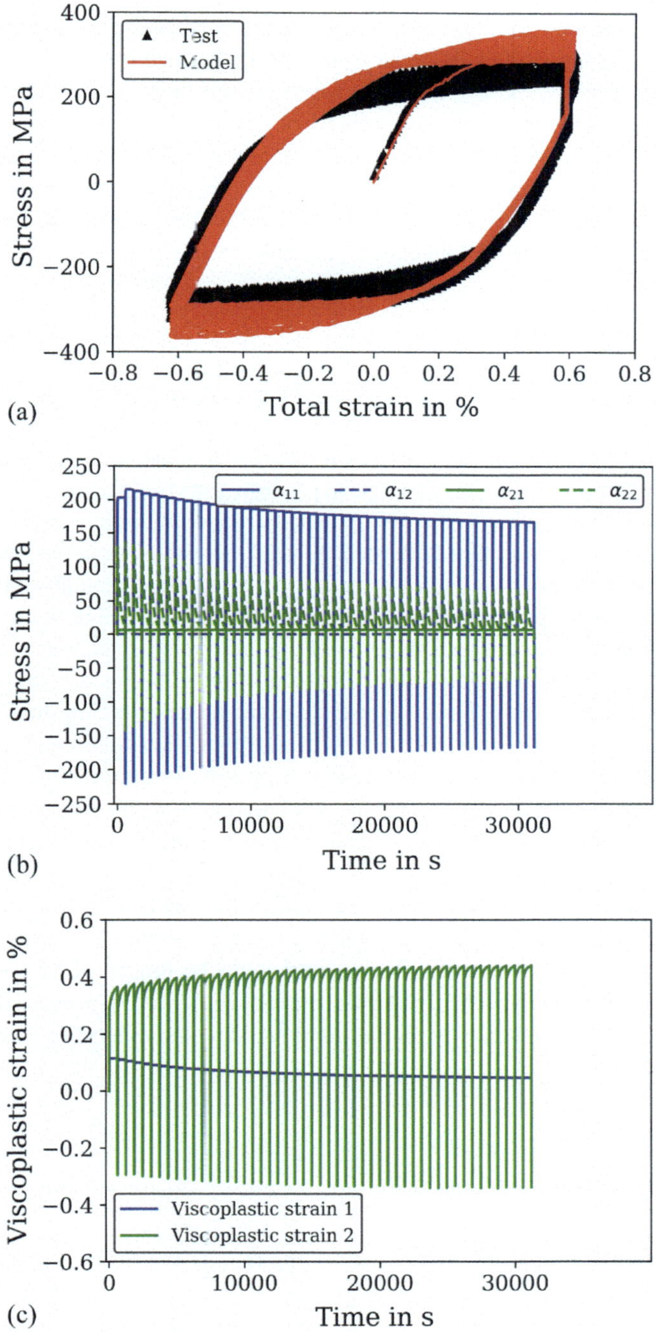

Fig. 22.11 LCF-H test (first 50 cycles) with 600 s dwell time and strain amplitude 0.6% at 600 °C: **a** stress–strain hysteresis loops; **b** evolution of the backstresses of the two strain rates; **c** evolution of viscoplastic strains of the two strain rates

deformation cannot be modeled individually, since both viscoplastic strains ε_1^{vp} and ε_2^{vp} are active during the tensile dwell times and in the load reversals of the MLCF tests. A quick response of the backstress in the load reversals is necessary to describe reverse creep at zero stress after unloading from the creep regime.

The results of the MLCF tests have shown that the maximum peak stress is elastic although the compressive load with the same (negative) absolute stress shows nonlinear hardening. The model results show that this behavior is achieved with the viscoplasticity model. As an explanation, the viscous stress σ_1^{v} of the first strain rate $\dot{\varepsilon}_1^{vp}$ can be computed with

$$\sigma_1^{v} = K_1 \cdot \dot{\varepsilon}_1^{vp\left(\frac{1}{n_1}\right)} \tag{22.8}$$

and gives for the strain rate $\dot{\varepsilon} = 10^{-4}\, s^{-1}$ and the model parameters $K_1 = 645$ MPa and $n_1 = 6.37$ a viscous stress of 152 MPa. Considering Eq. (22.3) the effective stress $|\sigma - \alpha_1|$ gives a higher value in compression than in tension because of the negative backstress in compression. The distinction of elastic or viscoplastic deformations is done by the yield condition in the viscoplasticity model $\Phi_1 = |\sigma - \alpha_1| - \sigma_1^{v}$. If the yield condition Φ_1 is lower than 0, the deformation is elastic. Due to the lower relative stress in tension, the deformation for the tensile peak stress remains elastic. Thus, plasticity of the viscoplastic strain $\varepsilon_1^{vp}$is visible in compression but not in tension.

The fast evolution of the backstress α_1 of the first viscoplastic strain rate $\dot{\varepsilon}_1^{vp}$ was a result of the procedure to determine the model parameters for service-like MLCF tests as well as CLCF tests. Thus, reverse creep can be described in the MLCF tests (see Figs. 22.6 and 22.8). However, the first backstress is also dominant for the CLCF tests in which actually no creep deformation is present. From the point of view that the material model includes one creep and one fatigue strain rate, the first creep strain rate and, thus, the first backstress should not be active contrary to the presented results in Fig. 22.4.

The deviation from the model compared to the test results in the MLCF tests (see Figs. 22.5 and 22.7) with increasing number of cycles is mainly due to the evolution of the second viscoplastic strain. The results could be improved by reducing the evolution of this viscoplastic strain. However, this would lead to a worse description of the CLCF tests, which is why the presented model parameters are determined as compromise.

22.6 Summary and Outlook

In this paper, a non-unified viscoplasticity model is presented to describe creep-fatigue deformation in service-like material tests of steel P92. The model results are presented exemplary for a CLCF test and two service-like MLCF tests together with the material test results. A good overall description of the material tests and the deformation phenomena is achieved with the presented viscoplasticity model. Furthermore, a possible procedure for model parameter adjustment of a multi-mechanism non-unified viscoplasticity model using a combined experience- and optimization-based approach and its application to an experimental database of the steel P92 existing of classical LCF tests with and without dwell times, complex LCF tests, and so-called service-like MLCF tests is presented. As proposed in Sonntag et al. (2023b), the material tests may be more realistic with strain-controlled load reversals and stress-controlled dwell times. To this end, the test procedure could be adapted in future works. The focus in this work is on the deformation phenomena, thus a maximum number of 50 cycles is considered. In future works, the softening behavior of ferritic-martensitic steels at elevated temperatures will be included in the viscoplasticity model. The softening of P92 in fatigue and creep-fatigue tests at high temperatures was comprehensively demonstrated in several material tests, e.g., in Chen et al. (2023), Giroux et al. (2023), Mao et al. (2022), and Zhang et al. (2021).

Acknowledgements The authors acknowledge funding of this project as Industrial Collective Research (IGF) by Federal Ministry for Economic Affairs and Climate Action (BMWK) under project number 22011 N. Furthermore, the technical discussions within the working committee under the direction of Dr. T.-U. Kern, Siemens Energy, are gratefully acknowledged.

References

ASTM A335 P92.: MTS ferrous and nonferrous alloy piping. http://www.metalspiping.com/astm-a335-p92.html. Accessed 24 June 2024

Bender T et al (2019) Developments in assessing welds in high temperature conditions and different stress triaxialities. In: Prakash R, Suresh Kumar R, Nagesha et al. (eds), Structural integrity assessment. Lecture notes in mechanical engineering. Springer, Singapore. https://doi.org/10.1007/978-981-13-8767-8_34

Cailletaud G, Saï K (1995) Study of plastic/viscoplastic models with various inelastic mechanisms. Int J Plast 11(8):991–1005. https://doi.org/10.1016/S0749-6419(95)00040-2

Chaboche JL (1989) Constitutive equations for cyclic plasticity and cyclic viscoplasticity. Int J Plast 5(3):247–302. https://doi.org/10.1016/0749-6419(89)90015-6

Chaboche JL (1993) Cyclic viscoplastic constitutive equations, Part I: a thermodynamically consistent formulation. J Appl Mech 60(4):813–821. https://doi.org/10.1115/1.2900988

Chaboche JL (1997) Thermodynamic formulation of constitutive equations and application to the viscoplasticity and viscoelasticity of metals and polymers. Int J Solids Struct 34(18):2239–2254. https://doi.org/10.1016/S0020-7683(96)00162-X

Chaboche JL (2008) A review of some plasticity and viscoplasticity constitutive theories. Int J Plast 24(10):1642–1693. https://doi.org/10.1016/j.ijplas.2008.03.009

Chen JJ, Ji D, Cai XD et al (2020) Effect of holding duration at maximum and minimum stress on creep fatigue interaction of P92 steel. Mater High Temp 37(1):51–60. https://doi.org/10.1080/09603409.2019.1693141

Chen X, Zhu L, Liu W et al (2023) Numerical predictions of low cyclic fatigue and creep-fatigue behavior of P92 steel under non-isothermal-mechanical loading. Int J Press Vessel Pip 206:105074. https://doi.org/10.1016/j.ijpvp.2023.105074

Contesti E, Cailletaud G (1989) Description of creep-plasticity interaction with non-unified constitutive equations: application to an austenitic stainless steel. Nucl Eng Des 116(3):265–280. https://doi.org/10.1016/0029-5493(89)90087-3

El-Magd E, Shaker C (1991) Ein strukturmechanisches Modell fuer das Kriechverhalten metallischer Werkstoffe bei zeitveraenderlicher Spannung. Materialwiss Werkstofftech 22(2):56–62. https://doi.org/10.1002/mawe.19910220204

Fischer C, Schackert S, Seifert T et al (2023) Experiments and modeling on the stain-controlled time- and temperature-dependent cyclic ratchetting plasticity of the nickel-based superalloy IN100. Materials 16(5):1888. https://doi.org/10.3390/ma16051888

Frederick CO, Armstrong PJ (2007) A mathematical representation of the multiaxial Bauschinger effect. Mater High Temp 24(1):1–26. https://doi.org/10.1179/096034007X207589

Giroux PF, Dalle F, Sauzay M et al (2010) Influence of strain rate on P92 microstructural stability during fatigue tests at high temperature. Procedia Eng 2(1):2141–2150. https://doi.org/10.1016/j.proeng.2010.03.230

Ji D, Zhang LC, Ren J et al (2015) Creep-fatigue interaction and cyclic strain analysis in P92 steel based on test. JMEP. 24(4):1441–1451. https://doi.org/10.1007/s11665-015-1446-9

Krishnasamy RK, Seifert T, Siegele D (2010) A computational approach for thermomechanical fatigue life prediction of dissimilarly welded superheater tubes. Mater Adv Power Eng (2010). https://www.osti.gov/etdeweb/biblio/21588265

Maier R, Riedel H, Seifert T et al (2011) Time and temperature dependent cyclic plasticity and fatigue crack growth of the Nickel-Base Alloy617B - experiments and models. Adv Mater Res 278:369–374. https://doi.org/10.4028/www.scientific.net/AMR.278.369

Mao J, Li X, Wang D et al (2021) Experimental study on creep-fatigue behaviors of chinese P92 steel with consideration of several important factors. Int J Fatigue 142:105900. https://doi.org/10.1016/j.ijfatigue.2020.105900

Mao J, Zhu J, Wang D et al (2022) Influence of 650°C thermal aging on microstructure and creep-fatigue behaviors of P92 steel. J Eng Mater Technol 144(3):031005. https://doi.org/10.1115/1.4053772

Masuyama F (2001) History of Power Plants and Progress in Heat Resistant Steels. ISIJ Int 41(6):612–625. https://doi.org/10.2355/isijinternational.41.612

Oesterlin H, Maier G, Riedel H (2022) VGB-Berechnungsverfahren - Berechnungsverfahren für kritische Kraftwerksbauteile beim flexiblen Betrieb von konventionellen Kraftwerken; Teilvorhaben: Berechnungsverfahren zur Lebensdauer von Bauteilen unter Beruecksichtigung von Spannungsumlagerung und Kriechermuedung : Abschlussbericht V1125/2022. Fraunhofer-Institut fuer Werkstoffmechanik IWM. https://doi.org/10.2314/KXP:1860072291

Ohno N, Sasaki T, Shimata T et al (2018) Effect of cyclic hardening on stress relaxation in SUS316HTP under creep-fatigue loading at 700° C: experiments and simulations. JTAM 56(2):497–510. https://doi.org/10.15632/jtam-pl.56.2.497

Saï K (2011) Multi-mechanism models: present state and future trends. Int J Plast 27(2):250–281. https://doi.org/10.1016/j.ijplas.2010.05.003

Saï K, Cailletaud G (2007) Multi-mechanism models for the description of ratchetting: effect of the scale transition rule and of the coupling between hardening variables. Int J Plast 23(9):1589–1617. https://doi.org/10.1016/j.ijplas.2007.01.011

Schlayer M, Warwas M, Seifert T (2023) A temperature-dependent viscoplasticity model for the hot work steel X38CrMoV5-3, including thermal and cyclic softening under thermomechanical fatigue loading. Materials 16(3):994. https://doi.org/10.3390/ma16030994

SciPy.optimize.minimize, Scientific Python. https://docs.scipy.org/doc/scipy/reference/generated/scipy.optimize.minimize.html. Accessed 24 June 2024

Shimada T, Tokuda K, Yoshida K et al (2018) Creep-fatigue life prediction of 316H stainless steel by utilizing non-unified constitutive model. Procedia Struct Integr 13:1873–1878. https://doi.org/10.1016/j.prostr.2018.12.326

Sonntag N, Juergens M, Uhlemann P et al (2023a) Experimental data from service-like creep-fatigue experiments on grade P92 steel. DIB 49:109333. https://doi.org/10.1016/j.dib.2023.109333

Sonntag N, Juergens M, Skrotzki B et al (2023b) Creep-fatigue of P92 in service-like tests with combined stress- and strain-controlled dwell times. Int J Fatigue 168:107381. https://doi.org/10.1016/j.ijfatigue.2022.107381

Taleb L, Cailletaud G (2010) An updated version of the multimechanism model for cyclic plasticity. Int J Plast 26(6):859–874. https://doi.org/10.1016/j.ijplas.2009.11.002
Viswanathan V, Purgert R, Rawls P (2008) Coal-fired power materials. AM&P 166(8):47–49
Wang J, Steinmann P, Rudolph J et al (2015) Simulation of creep and cyclic viscoplastic strains in high-Cr steel components based on a modified Becker-Hackenberg model. Int J Press Vessel Pip 128:36–47. https://doi.org/10.1016/j.ijpvp.2015.02.003
Wang X, Zhang W, Gong J et al (2018a) Low cycle fatigue and creep fatigue interaction behavior of 9Cr-0.5Mo-1.8W-V-Nb heat-resistant steel at high temperature. J Nucl Mater 505:73–84. https://doi.org/10.1016/j.jnucmat.2018.03.055
Wang X, Zhang W, Gong J et al (2018b) Experimental and numerical characterization of low cycle fatigue and creep fatigue behaviour of P92 steel welded joint. FFEMS. 41(3):611–624. https://doi.org/10.1111/ffe.12722
Wang X, Zhang T, Zhang W et al (2021) An improved unified viscoplastic model for modelling low cycle fatigue and creep fatigue interaction loadings of 9–12%Cr steel. Eur J Mech A-Solid 85:104123. https://doi.org/10.1016/j.euromechsol.2020.104123
Zhang SL, Xuan FZ (2017) Interaction of cyclic softening and stress relaxation of 9–12% Cr steel under strain-controlled fatigue-creep condition: Experimental and modeling. Int J Plast 98:45–64. https://doi.org/10.1016/j.ijplas.2017.06.007
Zhang W, Wang X, Kang Z et al (2021) Characterization of the cyclic softening and remaining creep behaviour of P92 steel weldment. JMR&T 15:1446–1456. https://doi.org/10.1016/j.jmrt.2021.09.013

Chapter 23
Comparative Analysis of Injection Molding and Filling Simulation

Elena Fischer, Sascha Magerle, Lukas Schmidberger, Matthias Deckert, Marius Pflüger, Simon Heienbrock, and Alexander Melzer-Bartsch

Abstract The injection molding industry has to face various challenges, such as the shortage of skilled workers and regulatory requirements like the increasing use of recycled materials. These challenges can be overcome, for example, with artificial intelligence (AI) that predicts component quality and suggests machine control parameters. This could drastically reduce the time and effort required for quality assurance, and it would also make it possible to react more quickly to batch fluctuations caused by recycled materials. Various approaches have already been researched. However, the provision of data for the training of such AI models is very extensive and resource-intensive, as the components have to be produced and then measured. Alternatively, simulation data can be used to train the AI models. This publication examines how close the simulation data is to the real data. This enables a better assessment of how good and reliable an AI model could be with simulation data as a basis. Various process and quality data are compared, for example, the cavity pressure and temperature, various dimensions, and the part weight. In certain cases, very good simulative results were achieved that are very close to reality. In addition,

E. Fischer (✉) · S. Magerle · L. Schmidberger · M. Deckert · M. Pflüger · S. Heienbrock · A. Melzer-Bartsch
Esslingen University of Applied Sciences, Kanalstrasse 33, 73728 Esslingen, Germany
e-mail: Elena.Fischer@hs-esslingen.de

S. Magerle
e-mail: samams00@hs-esslingen.de

L. Schmidberger
e-mail: Lukas.Schmidberger@hs-esslingen.de

M. Deckert
e-mail: Matthias.Deckert@hs-esslingen.de

M. Pflüger
e-mail: Marius.Pflueger@hs-esslingen.de

S. Heienbrock
e-mail: Simon.Heienbrock@hs-esslingen.de

A. Melzer-Bartsch
e-mail: Alexander.Melzer-Bartsch@hs-esslingen.de

H. Altenbach et al. (eds.), *Lectures Notes on Advanced Structured Materials 3*,
Advanced Structured Materials 226, https://doi.org/10.1007/978-3-031-77403-4_23

the material data stored in the simulation program is also examined to control how reliable the data is for the simulations.

Keywords Injection molding · Simulation data · Comparison of reality and simulation · Artificial intelligence

23.1 Introduction

Various production locations, such as the one in Germany, are affected by an increasing relocation of production abroad. High costs are often the reason for this (Spreter 2023). There is also a shortage of skilled workers in Germany. According to a survey conducted by the IFO Institute in 2023, 39% of the companies surveyed in plastics production reported a shortage of skilled workers (). Demographic change is also becoming increasingly problematic. In the industrial manufacturing sector in particular, a decline of 10.6% in potential employees is forecast by 2030 (Ostwald et al. 2030). With the loss of workers comes the loss of a great amount of knowledge and experience. At the same time, demographic change and the general shortage of skilled workers are preventing enough new skilled workers from being trained and further educated.

AI models can be used in injection molding production to compensate for the shortage of skilled workers and the loss of knowledge. Here, for example, quality parameters such as width and length dimensions can be predicted based on the cavity pressure and temperature during the cycle. Many different types of data are required to set up this type of AI. This data can be various pressure and temperature curves that are created experimentally on an injection molding machine using a design of experiments (DoE). The quality characteristics that must be assigned to each injection molding cycle or data curve are also required. However, the execution of such a test plan and the subsequent quality data acquisition are very time-consuming and resource-intensive. A more rapid way to provide data for the AI is the possibility of injection molding or filling simulation.

However, before AI models are developed, this publication draws a comparison between the real and simulated data in order to estimate in advance how good an AI model based on simulation data could be. In the following chapters, the literature review will first discuss how AI models have already been used with simulation data for the quality assurance of the injection molding process. The methodology of real and simulative data collection is then described and the results are discussed. Finally, the work is summarized and further work is pointed out.

23.2 Literature Review

Heinisch researched AI models that optimize the process setup during injection molding. The AI models were built using not only real and simulation data, but also with combined data from simulation and reality. Thin plate test specimens were used, which is why only a 2.5 D simulation was carried out. In addition, the cooling channels were modeled in the simulation. An artificial neural network with the LBFGS algorithm was used for the AI based on real data. In the combined AI model, the simulative quality variables are added to the experimental ones. As the pure simulation models were significantly more affected by errors, an experimental reference cycle and a linear correction of the simulated quality variables were carried out. After the correction, both the component weight and dimensions are significantly closer to the target value. In principle, however, the simulation data should only be used as pre-training and the neuron weights identified in this way should be used for a second training with a reduced number of experimental data (Heinisch 2021).

The SKZ researched Bayesian networks, which were used for process monitoring, error analysis, and control of production processes in injection molding. A simulation model of the molded part including the simulated mold, temperature control channels, and nozzle was also created. However, even after optimization measures, the sim-to-real gap was so large that only experimental data was used for the research (Lambers et al. 2023).

Ketonen and Blech developed an AI model that detects anomalies in the injection molding process during the cycle and identifies the cause of the anomaly. In addition to real data, simulated data was also used. Anomalies, such as a random peak, were randomly added to some of the simulated test data. Each anomaly generated has one to three causes. The F1 score that could be achieved with the simulation data is below average across all models, which indicates that the anomalies in the simulation data could only be captured poorly. Only the DSVAE-AD model (disentangled sequential variational autoencoder-based anomaly detector) achieved a relatively high F1 score of 0.8267 (Ketonen and Blech 2021).

Tercan et al. conducted research into bridging the sim-to-real gap using transfer learning in injection molding simulations. This involves training an AI model with simulation data. The model learns the core relationships between machine control and quality parameters. The model is then adapted to the real process using a few experimental data sets. The component weight was used as the quality criterion to be predicted. There is only an offset in the quality data between the experimental data and the simulation. It was recognized that the more real data is available to the model, the lower the impact of transfer learning on performance. In addition, the sim-to-real gap is sufficiently small even with little real data (Tercan et al. 2018).

Finkeldey et al. developed a machine learning method that predicts the quality characteristics of an injection molding process using simulation data and measurements. The aim was to keep the number of experimental trials small. With combined data from simulation and reality, an XGBoost model performed best. The root mean square error for the part weight without simulation data was 0.04, with 25% simulation data 0.14, with 50% 0.21, and with 75% 0.34 (Finkeldy et al. 2020).

Berger-Weber and Aminabadi investigated the applicability of an experimental design and simulation as a constrained physical model to develop learning methods for predicting component quality. Only simulation data was used. For all quality parameters, an MLP network with backpropagation provided the best results with an R^2 of 0.951 to 0.996. Furthermore, a model with different numbers of neurons and activation functions was developed for each quality feature (Berger-Weber and Aminabadi 2021).

Fernández et al. also developed an artificial neural network with backpropagation using simulation data to make predictions for different quality attributes. For the prediction of measures, the model with eight sigmoid neurons and one linear neuron was the most successful, with an R^2 above 0.99. In addition, a neural network was developed for each quality attribute (Fernández et al. 2023).

Michiels et al. developed a machine learning model based on an LGBM algorithm that was trained and tested using only simulation data. This model predicts which quality class an unknown object is in and how large a certain dimension is. Features were extracted from the simulated data using the Python library "tshfresh". The classification and regression models each achieved an accuracy of 0.994 (Michiels et al. 2022).

Wang et al. used simulations to determine the main factors influencing the warpage and volumetric shrinkage of the part using an analysis of variance. The warpage is most influenced by the packing pressure and the shrinkage by the melt temperature. In addition, the simulation data was used to train and test a kriging and a GEK (gradient-enhanced kriging) model that predicts shrinkage and warpage. The GEK model provided higher accuracy for warpage and shrinkage (Wang et al. 2022).

In the publications mentioned, however, little attention was paid to the quality of simulation data or the direct comparison between reality and simulation data. This publication therefore aims to draw a wide range of comparisons between simulation and real data.

23.3 Methodology

A full-factorial 2^4 test plan with a central point was created for data collection. The injection molding parameters packing pressure, packing time, melt, and mold temperature were varied (Table 23.1). The ABS material Polyman ABS M/MI-A natural was used for the tests. The test part (Fig. 23.1) is designed to be used in the experiments and has both thick- and thin-walled areas.

Table 23.1 DoE ABS Polyman M/MI-A

Setting parameters	Unit	Central point (0)	Variation, full-factorial (±1)
Packing pressure	bar	250	50
Packing time	s	10	3
Melt temperature	°C	230	15
Mold temperature	°C	60	15

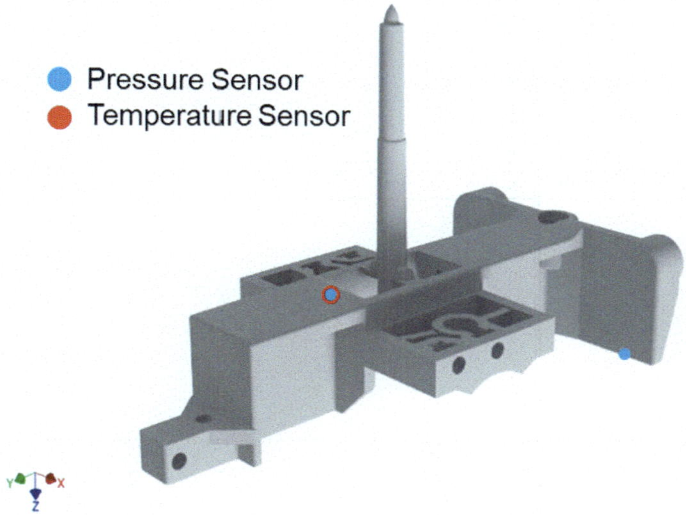

Fig. 23.1 Test part with sensor locations in the cavity

23.3.1 Real Data Collection

The test plan (Table 23.1) was carried out experimentally using an Arburg 370A. Ten components were molded per test series. One pressure sensor is installed in the injection cavity near the sprue and one remote from the sprue, and a temperature sensor is also installed near the sprue (Fig. 23.1). The data from the pressure sensors are processed by a charge amplifier and transmitted to an analog input module from National Instruments. The temperature signal is directly transmitted to a temperature module from National Instruments, which processes the sensor signal. Both signal types undergo further processing in National Instruments' LabVIEW, including timestamping and mapping to a specific production cycle. The data belonging to each production cycle is then written to an external file storage system. The structure of the data acquisition is shown in the following schematic (Fig. 23.2).

A stable process could not be achieved in four different tests. The combination of too high mold temperature and too low packing pressure was most likely the problem. This data was therefore not used for further comparison of reality and simulation.

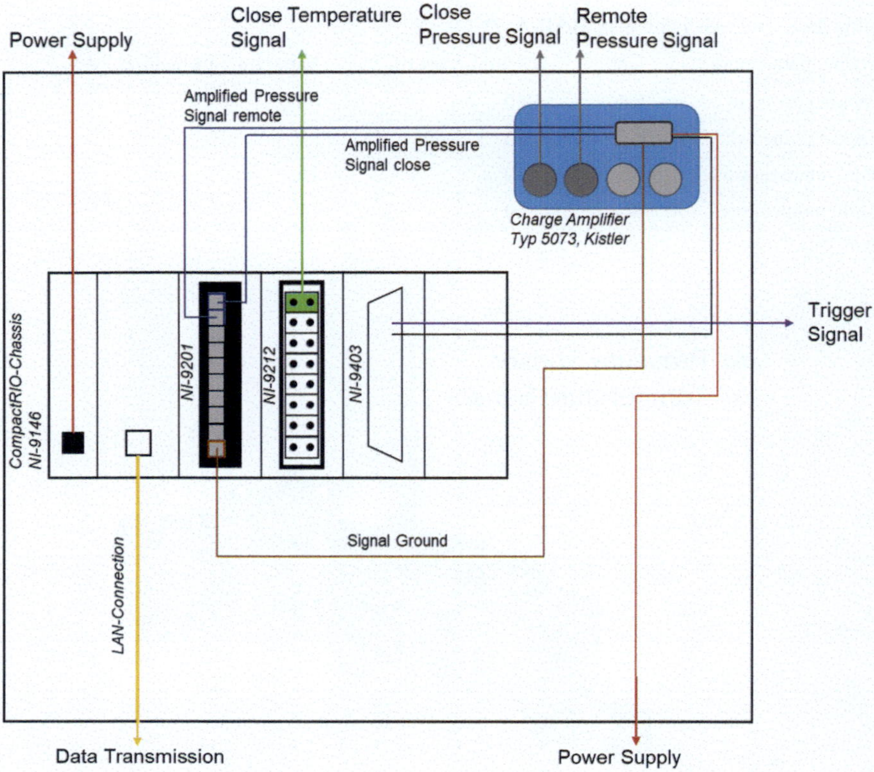

Fig. 23.2 Structure of data acquisition

To determine the quality parameters, two different dimensions (Fig. 23.3) were recorded using a coordinate measuring machine. The first dimension is a tool-related dimension and the second is a tool-unrelated dimension (Kunststoff-Formteilen 2013). Due to post-shrinkage, the real components were initially stored in a measuring room for over a day after production. The components were then measured and the data was assigned to the respective pressure and temperature values.

23.3.2 *Simulative Data Collection*

The simulation of the test plan was carried out using a simulation software. The test component is meshed with prisms in the surface layer and with tetrahedrons in the interior of the component. The sprue is meshed with four inner and five outer layers. These are the mesh arrangements recommended by the simulation program and are suitable for this purpose. A simple mold in the form of a cuboid was recreated around the component. The cooling channels were also reconstructed and placed in

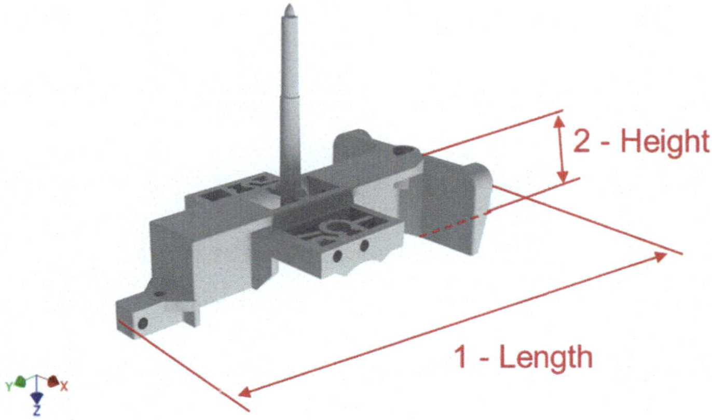

Fig. 23.3 Quality features test part

the mold. The nozzle was also reconstructed, meshed, and simulated. In order to obtain better results, the injection molding machine (Arburg 370A) was also defined in the simulation program. The three real sensors in the cavity were reconstructed in the simulation program at the same locations with sensor nodes (Fig. 23.1).

In order to obtain a complete comparison between reality and simulation, the four tests that were unstable in reality were also simulated. During the simulation of these tests, it was determined that the pressure curves had an unusual appearance and in some cases, the component filling was simulated incompletely. These cycles were therefore also discarded.

The problem with the simulation in general, is that only one cycle can be simulated per setting. Compared to reality, in which ten cycles were carried out per setting, the simulation contains a significantly smaller amount of data. To compensate for this, the variables packing pressure and melt temperature were changed minimally for each test (Table 23.2). This means that five cycles could be simulated per test. Once all simulations had been completed, the pressure and temperature curves and the simulated dimensions of the component and other quality parameters could be read out.

Table 23.2 Parameter variance simulation

Settings for each test series
1. Original test settings
2. Packing pressure is reduced by 5 bar
3. Packing pressure is increased by 5 bar
4. Melt temperature is reduced by 5 °C
5. Melt temperature is increased by 5 °C

23.4 Results and Discussion

23.4.1 Comparison of Simulation and Reality

23.4.1.1 Comparison of the Process Parameters

Comparison of the Cavity Pressure Curves Near the Sprue

The following diagram (Fig. 23.4) shows an example of a DoE test. In each case, five cavity pressure curves close to the sprue from the same DoE test setting are compared from reality and the simulation. It can be seen that the slight changes in the setting parameters within a test series lead to similar but slightly different curves in the simulation. In addition, it can be seen that the real and simulation curves follow a very similar course. Similarly good and realistic simulated pressure curves were also achieved in the other DoE test settings.

Comparison of the Cavity Pressure Curves Remote from the Sprue

The following diagram (Fig. 23.5) illustrates the same experiment from Fig. 23.4, only with a remote pressure sensor and the corresponding simulated, remote cavity pressure curve. Here, a drastic offset in the remote packing pressure level and the time period in which the packing pressure is above zero is immediately apparent between reality and simulation. If the simulation curve from Fig. 23.5 is also compared with the simulation curves from Fig. 23.4, it can be seen that the pressure curve remote from the sprue is almost identical to the curve close to the sprue. Only the time range

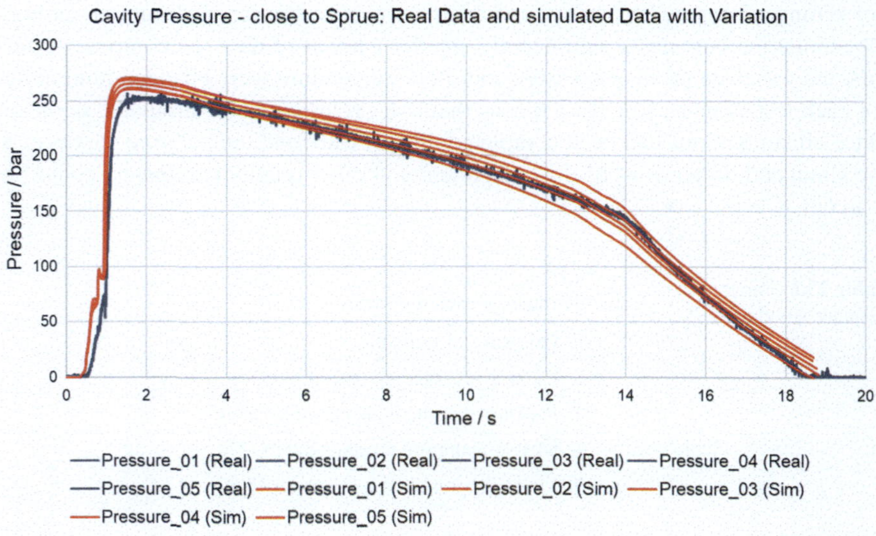

Fig. 23.4 Comparison of the pressure curves close to the sprue from the DoE: Reality and simulation, ABS

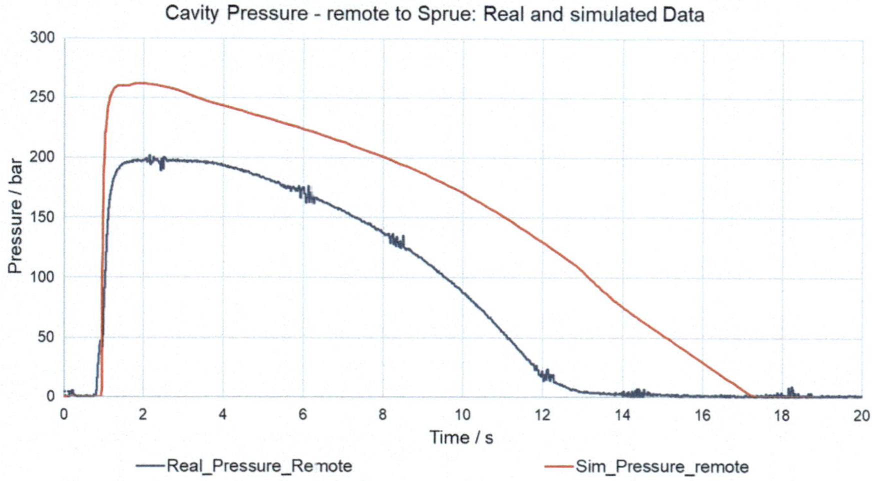

Fig. 23.5 Comparison of the pressure curves remote to the sprue from the DoE: Reality and simulation, ABS

of the measurable packing pressure is slightly shorter in the simulation curve away from the sprue. In reality, the pressure level away from the sprue and the time range in which the pressure is applied is significantly shorter than in the real pressure curves close to the sprue.

Comparison of the Cavity Temperature Close to the Sprue

In addition to the cavity pressure, the cavity temperature can also provide important data. The following diagram (Fig. 23.6) shows that there is an offset of around 15–20 °C between reality and simulation in the rear section of the curves. However, the initial temperature profiles are very different. Although the time of the temperature rise is very close together, the simulation results in a significantly higher peak. This could be due to the fact that the real sensor is slower and therefore cannot measure the drastic rise in temperature that occurs when the plastic melt is injected. In addition, there is a thin barrier between the real sensor in the cavity and the plastic melt, so that in reality the temperature of the plastic melt is not measured directly. The offset after the peak of the simulation curve is due to the fact that more time is needed to cool down from the initial high temperature level to the lower one.

Comparison of the Temperature Distribution

The temperature distribution, without consideration of the temperature level, directly after the component demolding or after the end of the production cycle are similar both in reality and the simulation (Fig. 23.7). In the simulation, the image is created directly after the end of the cycle. In reality, however, the component first had to be demolded and then placed under the thermal imaging camera so that an image could be taken. This process took about 12 s.

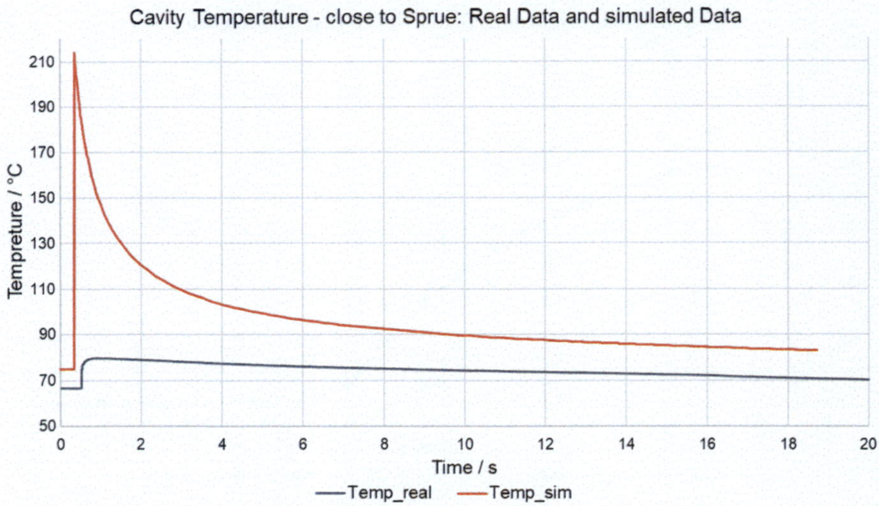

Fig. 23.6 Comparison of temperature curves close to the sprue from the DoE: Reality and simulation, ABS

The temperature on the two side wings and in the rear, wider area are higher in the simulation and in reality, while the temperature along the central bar is lower. With regard to the temperature level, however, there are clear differences between reality and simulation. For example, the maximum temperature in reality is 125 °C, while the scale in the simulation only extends to around 50 °C.

The significant differences in the temperature levels could be related to the fact that the material in the simulation solidified completely more quickly at the end of the cycle and therefore the entire component cools down more quickly. In reality, however, there could still be a hot, molten core in the component during demolding, which would radiate the heat over the component and explain the high temperature. It is also possible that the heat transfer coefficient of the polymer and the mold were assumed to be too high in the simulation and therefore the component cools down more quickly. Another difference to reality is that there are no slides in the reconstructed tool, which is why there is no thermal resistance in the simulation. For comparative measurements to verify the results of the thermal imaging camera, five different parts were measured in reality with a temperature probe directly after production on one of the two side wings in the marked area of the spot of the thermal camera (Fig. 23.7). The mean value for these measurements is 84.18 °C; the median is 85 °C. A disadvantage of this control measurement was that the temperature sensor used was very slow. The measurements were only finished when the temperature dropped slightly. This process took about half a minute, which explains why the temperature level is significantly lower than the approximately 116 °C measured by the thermal camera. This means that the thermal imaging camera is more likely to provide correct results and the simulation cannot reproduce the temperature behavior well.

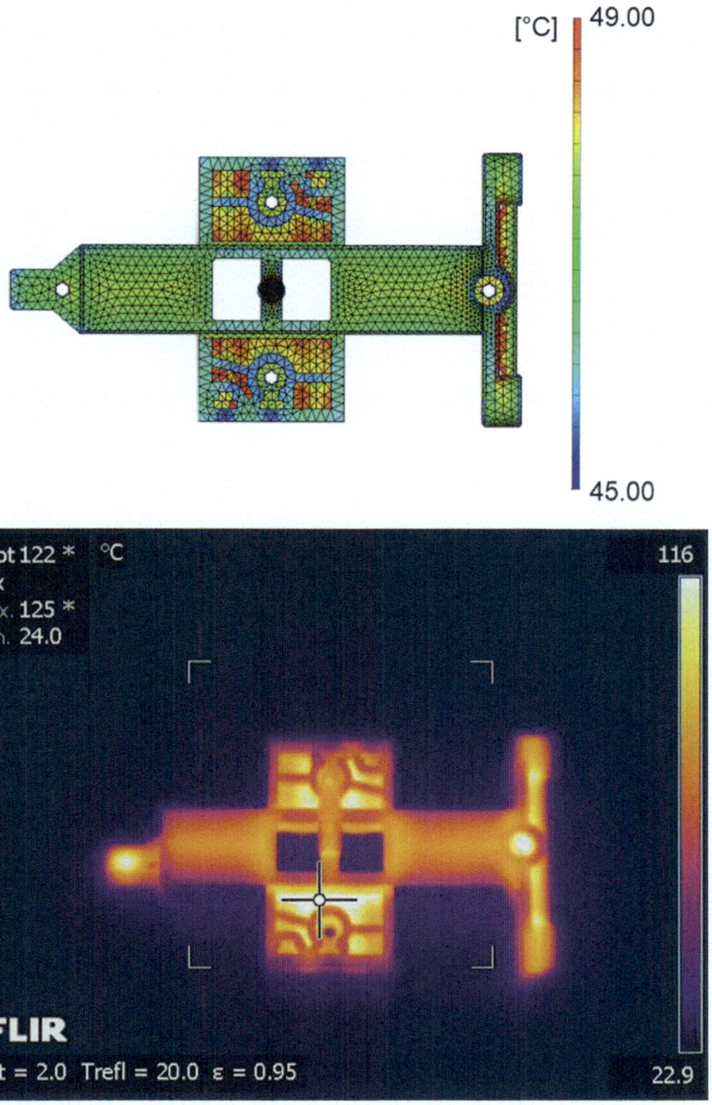

Fig. 23.7 Comparison of the temperature distribution directly after demolding—simulated and real, ABS

23.4.1.2 Comparison of the Quality Parameters

Comparison of the Dimensions

If the dimensions (Fig. 23.3) are compared with each other, it can be seen that the height dimension could be reproduced very well (Fig. 23.9). There is only an offset between simulation and reality, which could be compensated for with one real reference cycle. The length dimension (Fig. 23.8), on the other hand, was simulated less well, as there are large deviations from reality. In general, the fluctuations in the length measurements under the simulated data are only minimal. In the diagrams, only the quality data from the last five cycles were used in the real data in order to ensure a reliable steady-state progression.

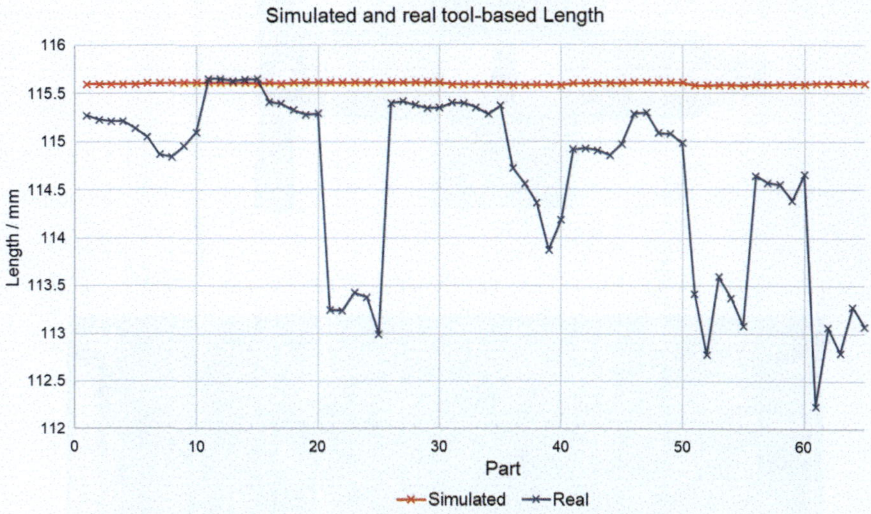

Fig. 23.8 Comparison length-measurement—simulated and real, ABS

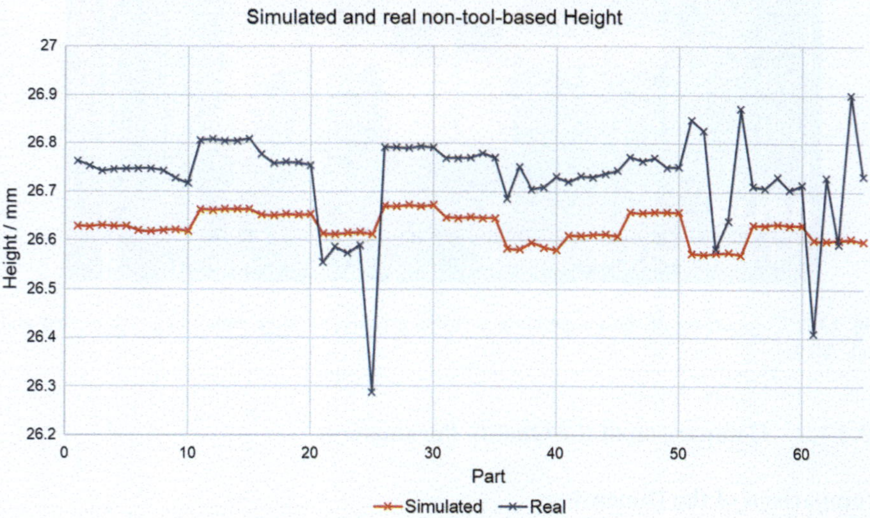

Fig. 23.9 Comparison of height-measurement—simulated and real, ABS

The offset of the height dimension can be explained by the fact that only the component and not the tool was simulated. In reality, the dimensions in the mold cavity are slightly larger so that the component remains dimensionally accurate despite the shrinkage. As the component was simulated instead of the mold in the simulation, this allowance is not present in the cavity. The minimal variance of the length dimension in the simulation could be due to the fact that it is a tool-related dimension, which only allows minimal changes due to the rigid tool. In addition, the simulation does not cover the deformation of the component via the ejectors. This could be observed in the real tests if the component was demolded too early.

Comparison of the Influences of the Machine Parameters

As can be seen from the comparison of the dimensions (Figs. 23.8 and 23.9), the change in the machine control parameters has different influences on the component dimensions. For this comparison, the percentage dimensional change was used when one of the machine control parameters was increased from one test to another. This is summarized in the following bar chart (Fig. 23.10). The two most influential parameters in the simulation are the packing pressure and the packing time. The real dimensions are primarily influenced by the control parameters of mold temperature and packing time. The direction of influence of the control parameters in the simulation and reality are the same except for the melt temperature, only the intensity of influence varies. However, the melt temperature generally has little influence on the two dimensions, both in the simulation and in reality. As with the previous diagrams (Figs. 23.8 and 23.9), the simulated change in height is closer to reality than the simulated change in length. The problem here is the same as in the previous point.

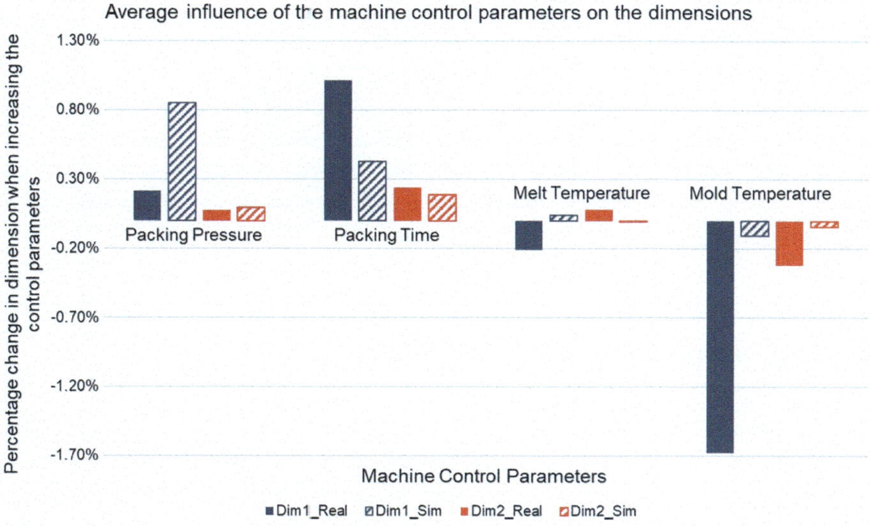

Fig. 23.10 Influence of the control parameters on the quality parameters

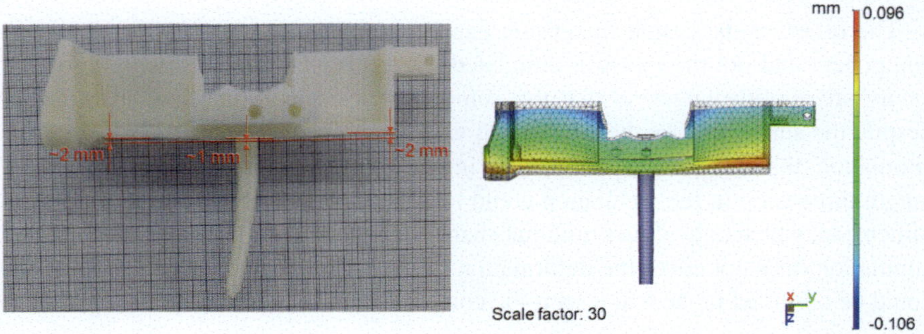

Fig. 23.11 Comparison of the warpage of the Z-axis between reality and simulation

Comparison of the Warpage

The warpage in some tests is most clearly recognizable in reality along the Z-axis. If these real components are compared with the respective simulations, it can be seen that the simulated warpage along the Z-axis (Fig. 23.11) is only minimal. This could also explain the poor representation of the simulated length dimensions (Fig. 23.8). If the warpage is only minimal in the corresponding tests, the length dimension fluctuates less and is, therefore, larger at the same time. In addition, these real components with strong warpage also show greater fluctuations within the same test series. This may be due to the fact that it was difficult to clamp the heavily distorted components in the measuring device for the coordinate measuring machine, resulting in high measurement uncertainties.

Comparison of the Part Weight

In addition to the dimensions and warpage, the component weight can also be simulated as a quality feature. In the following diagram (Fig. 23.12), the simulated component weight of each stable test is used in the original setting. For the real component weights, the average weight was used for each test setting. The weight was used in both cases including the sprue. As with the dimensions, there is only an offset between the real and simulated component weights, which can be compensated for by a real reference cycle.

23.4.2 Comparison of the Material Data

The reason for the deviations between the simulation and reality, in particular the pressure curves far from the sprue, could be the material data stored in the simulation program. The compressibility of the material plays a decisive role here. If this is not correctly or inadequately mapped in the simulation or in the material data, the remote sprue pressure curve is prone to errors. Unfortunately, such compressibility can only be read from the pvT diagram of the material to a limited extent. Nevertheless, the

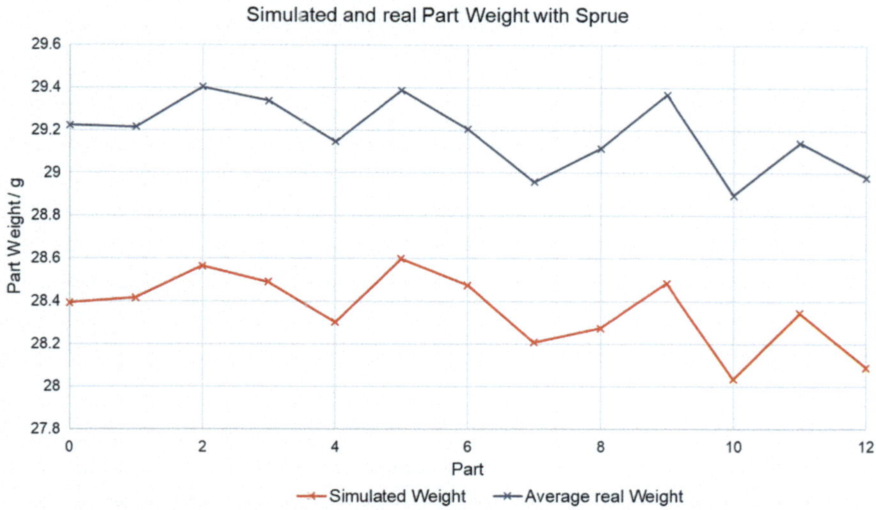

Fig. 23.12 Comparison of part weight—simulated and real, ABS

pvT diagrams from the simulation program and from the material data sheet are compared in the following diagrams (Figs. 23.13 and 23.14). Unfortunately, it was not possible to produce a pvT diagram of the batch of material used, as the necessary measuring instruments were not available.

For the ABS material, more and different pvT curves are specified in the data sheet than in the simulation program. Instead of seven curves, only five are used there and

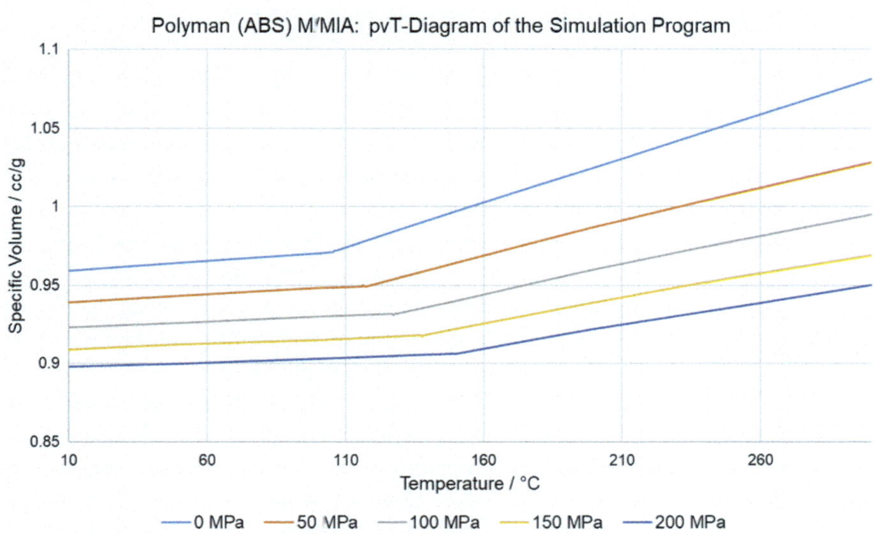

Fig. 23.13 Simulation program: pvT diagram, ABS (CoreTech System Co., Ltd. 2023)

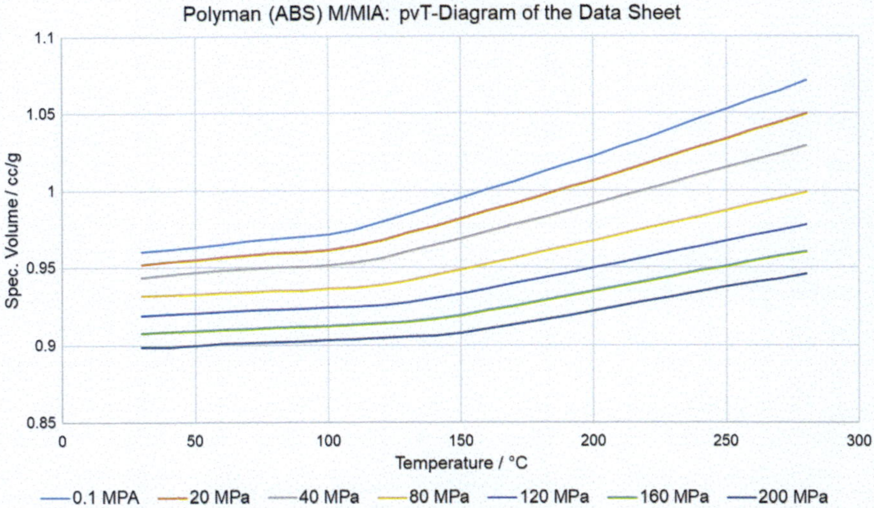

Fig. 23.14 Data sheet: pvT diagram, ABS (Campusplastics, POLYMAN® (ABS) M, MI-A. 2024)

only the 200 MPa curve is used in both. However, these 200 MPa curves appear to be the same. Otherwise, other pressure levels are made available to the simulation. On the one hand, it is questionable where the other curves with different pressures come from and, on the other hand, why less data was used than was actually provided by the manufacturer. It is also noticeable in the pvT diagram of the simulation that the curves were extrapolated and therefore show a temperature range of 10–300 °C instead of 30–280 °C.

In addition to the pvT diagram, the viscosity shear rate diagram is also stored in the simulation program and in the data sheet (Figs. 23.15 and Fig. 23.16). Similar to the pvT diagram, other curves than those specified by the manufacturer were also stored in the simulation program. Only the 240 °C curve is used in both diagrams. However, these have a very similar course in both the simulation program and the data sheet. The curves in the simulation program were also extrapolated so that the sharpness voltage ranges from 10 to 100,000 1/s.

This comparison of the material data also revealed that an incorrect melt volume flow of 23 cm^3/10 min was stored in the simulation program. However, 25 cm^3/10 min is specified in the data sheet. This means that the pressure inside the cavity during injection is slightly higher in the simulation than in reality. This behavior can also be seen in the pressure curve close to the sprue (Fig. 23.4).

Unfortunately, only assumptions can be made as to why different and sometimes incorrect data was stored in the simulation program. In the case of the viscosity shear rate, it can be seen that the deviating temperatures have created a larger information range, which can probably cover a larger range even with extreme settings. For example, if tests are to be simulated with a melt temperature that is too low or too high as recommended. Such an advantage is not apparent in the pvT diagram, as approximately the same pressure range was used for both specifications. One

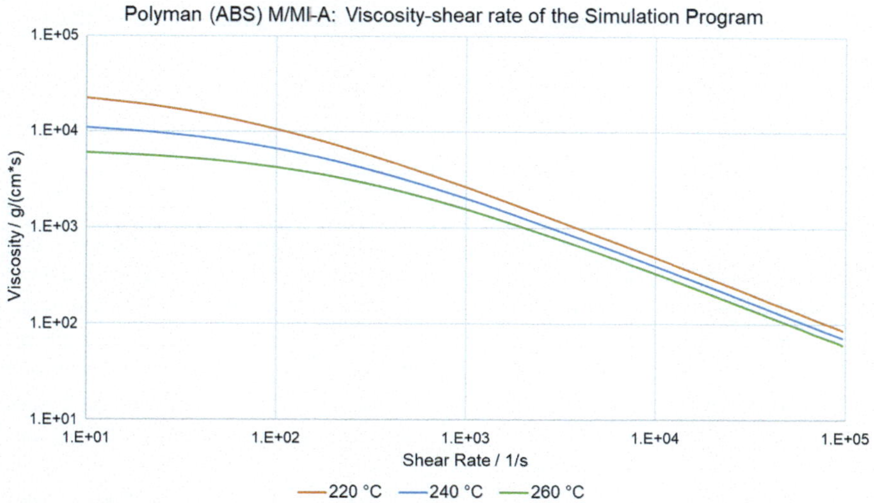

Fig. 23.15 Simulation program: viscosity shear rate, ABS (CoreTech System Co., Ltd. 2023)

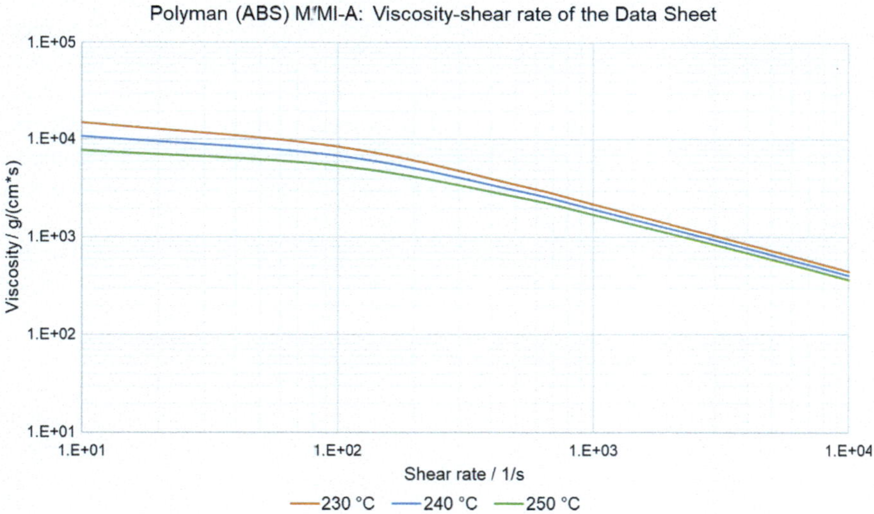

Fig. 23.16 Data sheet: viscosity shear rate, ABS (Campusplastics, POLYMAN® (ABS) M, MI-A. 2024)

possible reason could therefore be that there is too little storage capacity available for all pressure levels and therefore 50 MPa steps were selected. Since the melt volume flow is only a simple value, it is difficult to make assumptions as to why the simulation program uses a different value.

If the deviating information was not determined experimentally, but only derived from the manufacturer's information, this results in higher uncertainties. This could lead to incorrect results being delivered by the simulation program, particularly in areas remote to the sprue.

23.4.3 Comparison of Another Material

The material data in the simulation program was already discussed in the previous subsection. This subsection will show once again how crucial correct material data is in the simulation. A different material (polyethylene—Hostalen GC 7260) was used for this purpose, which contains incorrect material data in the simulation program. The same test component (Fig. 23.1) was produced and simulated as in the previous tests. The following diagram (Fig. 23.17) contains curves for the cavity pressure near and far from the sprue and the cavity temperature near the sprue, both simulated and real.

The temperature curve is mapped very well with the PE from the 15th second onwards. However, as with the ABS material, there is also the problem that a very high temperature peak occurs at the beginning of the simulation. At the beginning of the pressure curves, it is noticeable that the simulated pressure near the sprue and the pressure away from the sprue are initially at the same pressure level. In reality, however, there is a pressure difference of around 50 bar. In addition, both simulated pressure curves drop significantly faster than in reality. This means that the material solidifies much faster in the simulation and the pressure can therefore no longer be applied. There is therefore a clear error in the material data of the PE. Unfortunately, no comparative data could be found for the PE used. It can therefore be assumed that the material data in the simulation program is not reliable.

In the tests with ABS, the height dimension could be reproduced very well in the simulation. Various tests were also carried out with the PE material and the height

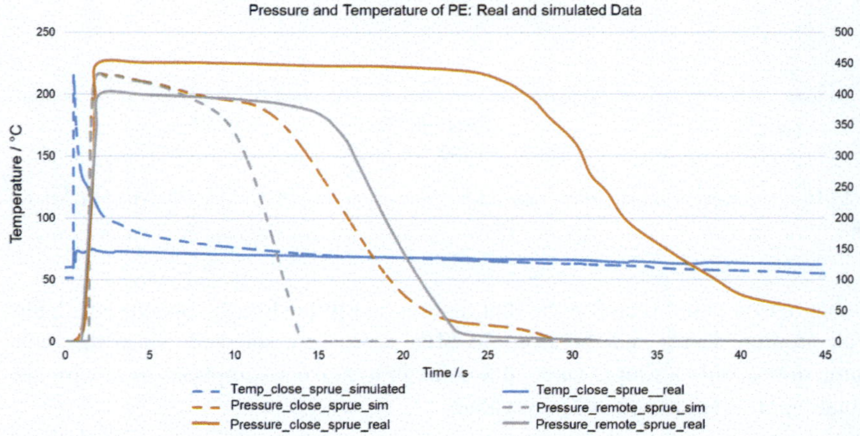

Fig. 23.17 Process data: real and simulated, PE

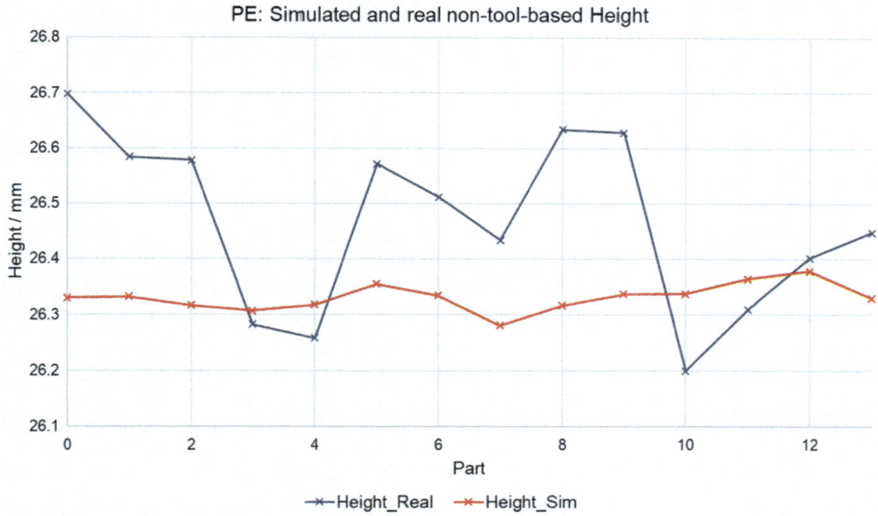

Fig. 23.18 Comparison of height-measurement—simulated and real, PE

was measured. The following diagram (Fig. 23.18) shows a height measurement for each test, both simulated and real. Significant differences are noticeable here, which cannot be compensated for by a simple offset factor. If an AI model were to be trained with this simulated data, the result would be a very unreliable prediction model.

23.5 Summary

In order to check whether simulation data is suitable for training an AI model, the same design of experiments was carried out experimentally and simulatively and compared with each other. An ABS material was used for a test part. It was found that the simulated cavity pressure near the sprue could be reproduced very closely to reality. In contrast, the pressure away from the sprue could be simulated less well, as it is almost identical to the pressure curve near the sprue, which does not correspond to reality. The problem with the simulated temperature curve is that when the melt is injected, the peak is too high and the curve deviates more strongly from the real measurements. However, at the end of the cycle, there is a very similar heat distribution across the component. Unfortunately, the temperature distribution at the end of the cycle is poorly modeled in the simulation. The height dimension could be reproduced very well in the simulation compared to the length dimension. The component weight could also be simulated very well and there is only an offset between reality and simulation. Warpage, on the other hand, was very poorly simulated and is therefore not suitable as a quality parameter. The differences between reality and simulation can most likely be traced back to the data of the materials. If these show errors, the simulation results are unreliable.

23.6 Outlook

The simulation results obtained with the ABS material will be used to train various AI models. These models are intended to predict quality parameters such as component dimensions and weight during the production cycle. For a comparison, the AI models will be also trained with the real data in order to be able to compare how good and robust the AI model is based on simulation data. This will make it easier to see whether training with simulation data is advisable and useful.

References

Berger-Weber GR, Aminabadi SS (2021) The injection mold as a cyber physical system. Using simulation to train its artificial intelligence. https://www.researchgate.net/profile/Saeid-Saeidi-Aminabadi/publication/363116997_The_injection_mold_as_a_cyber_physical_system_Using_simulation_to_train_its_artificial_intelligence/links/630e95121ddd447021213759/The-injection-mold-as-a-cyber-physical-system-Using-simulation-to-train-its-artificial-intelligence.pdf

Campusplastics, POLYMAN® (ABS) M/MI-A (2024). https://www.campusplastics.com/campus/en/datasheet/POLYMAN%C2%AE+(ABS)+MMI-A/LyondellBasell/103/39593403/SI. Accessed 28 May 2024

CoreTech System Co., Ltd. (2023) Moldex3D: SimpaTec GmbH

Heinisch J (2021) Einsatz von maschinellen Lernverfahren zur Einrichtung von Spritzgießprozessen. Dissertation, Rheinisch-Westfälische Technische Hochschule Aachen

Fernández A, Clavería I, Pina C, Elduque D (2023) Predictive methodology for quality assessment in injection molding comparing linear regression and neural networks. https://www.mdpi.com/2073-4360/15/19/3915

Finkeldy F, Volke J, Zarges J-C, Heim H-P, Wiederkehr P (2020) Learning quality characteristics for plastic injection molding processes using a combination of simulated and measured data. https://www.sciencedirect.com/science/article/pii/S1526612520306964

ifo Institut, Herrscht in Ihrem Unternehmen ein Fachkräftemangel?: (Umfrage in Unternehmen des Verarbeitenden Gewerbes; Anteil: "Ja"). https://de.statista.com/statistik/daten/studie/1324209/umfrage/fachkraeftemangel-im-verarbeitenden-gewerbe/. Accessed 17 Jan 2024

Ketonen V, Blech JO (2021) Anomaly detection for injection molding using probabilistic deep learning. https://ieeexplore.ieee.org/document/9468190

Maßhaltigkeit von Kunststoff-Formteilen (2013) DIN 16742, Deutsches Institut für Normung (DIN)

Lambers J et al (2023) Methodik zur Prozessüberwachung, Fehleranalyse und Regelung durch wissensbasierte Modellierung von Produktionsprozessen mittels Bayes'scher Netze: ProBayes. https://www.skz.de/fileadmin/user_upload/Download_PDF/_Forschung/_Projekte/2023/2023_21815N.pdf

Michiels S, de Schryver C, Houthuys L, Vogeler F, Desplentere F (2022) Machine learning for automated quality control in injection moulding manufacturing. https://www.esann.org/sites/default/files/proceedings/2022/ES2022-48.pdf

Dr. Ostwald DA, Dr. Hofmann S, Stohr D, Heieck O (2030) Demografischer Wandel: In Deutschland werden Arbeitskräfte rar: Berufs-, qualifikations-und branchenspezifische Analyse bis zum Jahr. https://www.wifor.com/uploads/2020/05/Ostwald-et-al.-2016-Demografischer-Wandel-In-Deutschland-werden-Arbei.pdf. Accessed 17 Jan 2024

Spreter J (2023) DIHK warnt vor zunehmender Produktionsverlagerung ins Ausland. Zeit online, 2023. https://www.zeit.de/wirtschaft/2023-04/industrie-dihk-produktionsverlagerung-ausland-kosten. Accessed 17 Jan 2024

Tercan H, Guajardo A, Heinisch J, Thiele T, Hopmann C, Meisen T (2018) Transfer-learning: bridging the gap between real and simulation data for machine learning in injection molding. https://www.sciencedirect.com/science/article/pii/S2212827118301914

Wang Z, Bo C, Sun Z, Li J, Gao F (2022) Quality defect analysis of injection molding based on gradient enhanced Kriging model. https://ieeexplore.ieee.org/document/9976740

Chapter 24
Laser Patterning of TiAlN for the Creation of Surfaces with Selective, Anisotropic Wetting

Jens Kortsch-Banzhaf, Fabian Schäfer, Simon Ruck, Florian Köhn, Harald Riegel, Markus Merkel, and Joachim Albrecht

Abstract TiAlN coated tungsten carbide is patterned by ultra-short laser pulses to create modified surface topographies. It turns out that this procedure is applicable to create structures both on the microscale ("large") using a sophisticated beam guiding and on the submicron-scale ("small") owing to the formation of surface modulations by intrinsic laser-TiAlN interaction (laser-induced periodic surface structures, LIPSS). The final goal is the preparation of surfaces with controlled interaction with polar and non-polar liquids. The presented work consists of two steps. First, the laser ablation process itself is studied to extract relevant irradiation parameters for a productive formation of the surface pattern. Second, the corresponding impact on the surface interaction of water and diiodomethane is studied. To separate the effect of "small" and "large" patterns, the LIPSS are created under a particular angle of 60°. Anisotropic drop shapes corresponding to the orientation of the lines are found that can be attributed to different surface topographies. The employed way of looking at the droplets is rather untypical for wetting studies, but indispensable to visualize the anisotropy of the droplet and a possible deflection along the LIPSS.

J. Kortsch-Banzhaf (✉) · M. Merkel
Institute for Virtual Product Development, Aalen University, Beethovenstraße 1, 73430 Aalen, Germany
e-mail: jens.kortsch-banzhaf@hs-aalen.de

M. Merkel
e-mail: markus.merkel@hs-aalen.de

F. Schäfer · S. Ruck · H. Riegel
Laser Application Center LAZ, Aalen University, Beethovenstr. 1, 73430 Aalen, Germany
e-mail: harald.riegel@hs-aalen.de

F. Köhn · J. Albrecht
Research Insitute for Innovative Surfaces FINO, Aalen University, Beethovenstr. 1, 73430 Aalen, Germany
e-mail: florian.koehn@hs-aalen.de

J. Albrecht
e-mail: joachim.albrecht@hs-aalen.de

H. Altenbach et al. (eds.), *Lectures Notes on Advanced Structured Materials 3*,
Advanced Structured Materials 226, https://doi.org/10.1007/978-3-031-77403-4_24

Keywords Laser patterning · Anisotropic surface structures · Laser-induced periodic surface structures (LIPSS) · Ultra-short pulse laser structuring · TiAlN surfaces · Surface-liquid interaction · Anisotropic droplets

24.1 Introduction

Ultra-short pulse (USP) lasers offer tremendous advantages concerning the accuracy of irradiation processes in material and surface processing, respectively. Since thermal exposure is minimized in such a "cold-ablation" process, it is possible to process extremely thin or highly temperature-sensitive materials. Further advantages of ultra-short pulse ablation, in addition to the spatially selective material removal, refer to the lower dependence of material absorption on the laser wavelength. This allows the processing of heat-sensitive materials and structures such as polymers, glasses, thin films, and multi-material stacks. In surface modification and structuring, pulsed lasers play out their advantages in many application areas (Bonse et al. 2000; Meliani et al. 2021; Dumitru et al. 2002; Urbina et al. 2013). The laser material process is an environmentally friendly and wear-free technology that neither requires chemicals nor causes water pollution (Ruck 2021). The production of so-called laser-induced periodic surface structures (LIPSS) plays an eminent role in surface modification (Schille et al. 2013; Wu et al. 2011; Römer et al. 2009; Bonse et al. 2020). The periodicity of the structures can be produced on the length scale of the laser wavelength (Nolte et al. 2016; Straub et al. 2012; Müller et al. 2016; Malshe et al. 2013). For an efficient laser process, the ablation threshold plays a crucial role (Momma et al. 1997; Nolte et al. 2716; Chichkov et al. 1996; Neuenschwander et al. 2012, 2014). Laser surface processing is particularly interesting for wear-resistant coatings like TiAlN. Such hard coatings are often used to increase the wear resistance of cutting tools and can be substantially upgraded by microstructuring or ablation, which leads to an increase in performance or to sustainable postprocessing (Kommer et al. 2017; Denkena et al. 2018). The reuse of cutting tools on WC substrates or decoating has only recently been investigated (Gakovic et al. 2011; Marimuthu et al. 2011; Ragusich et al. 2013). In the use of cutting tools, or more generally, the wettability of solid surfaces is one of the most important properties for many technical applications. Wetting is determined by the structure and chemical composition of the surface (Gleiche et al. 2000; Liu and Jiang 2012; Chen et al. 2013; Milles et al. 2019). The large number of applications ensures a great interest in the targeted interaction of liquids with heterogeneous surfaces. Anisotropic wetting is a typical aspect of surface-liquid interaction with importance in basic research in addition to practical applications (Bonn et al. 2009; Lundgren et al. 2007; Dubey et al. 2017). The interaction of liquids with surfaces can be determined, by an analysis of the drop shape. Determination of the apparent contact angle in the static state based on Young's method is one of the possibilities. Pinning at the surface can be determined by the hysteresis of the advancing and receding contact angles (Li et al. 2007; Zhang and Low 2007; Cao et al. 2007). The interaction with anisotropic surface structures leads to an anisotropic deformation of droplets. Application areas of microscale targeted drop control include microfluidics (transport of droplets) and areas of lab on a chip. On a larger scale, it has been demonstrated that laser treatment can positively influence the wetting properties of battery electrodes (Kleefoot et al. 2022) or can be used as a beneficial step for the preparation of superoleophobic surfaces (Abele et al. 2019).

24.2 Experimental Setup

Laser Patterning: We performed two ablation experiments. For all laser patterning experiments, the number of irradiated pulses per dot and the scan velocity were constant, respectively. A sketch is shown in Fig. 24.1.

First, the ablation threshold was determined for different pulse durations at different fluences. Subsequently, the patterning experiments have been carried out. For this purpose, the ultra-short pulse laser system *Carbide DB3-40W, Light Conversion, Lithuania*, with a pulse duration range of 0.25–10.0 ps, has been used. The center wavelength has been 1030 nm with a repetition rate for single pulses up to 1 MHz with a pulse picker. The laser has been mounted in a high-precision 5-axis processing system that allows 3D laser micro-machining (*microcut, LLT GmbH, Ilmenau, Germany*).

The laser beam is guided into the machine via an optical rail system that goes up to the scanner. An *ExlliSCAN14* (*SCANlab, Puchheim, Germany*) galvanoscanner system and a *telecentric F-Theta* lens guides the laser beam across the surface. To determine the ablation threshold of the TiAlN-thin layer, the pulse duration is varied from 0.34 up to 9.0 ps with a corresponding fluence from 0.3 up to 4.78 J/cm^2.

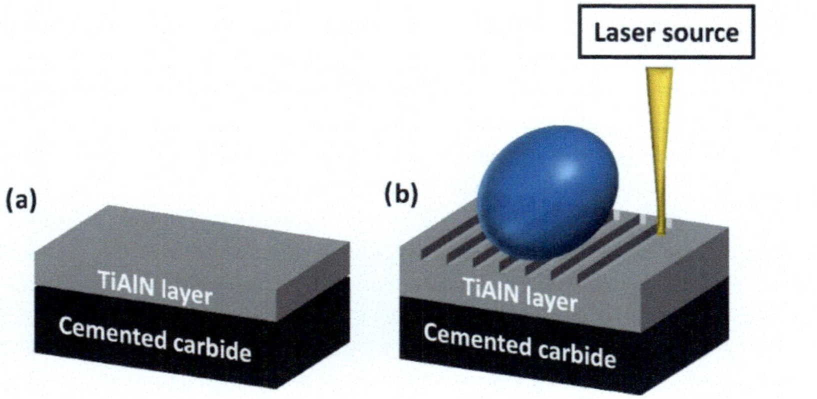

Fig. 24.1 **a** Sketch for a thin TiAlN layer on a cemented carbide substrate in an unstructured state; **b** Sketch for the laser patterning process generating a laser structured TiAlN layer with a deformed liquid droplet on the structured surface

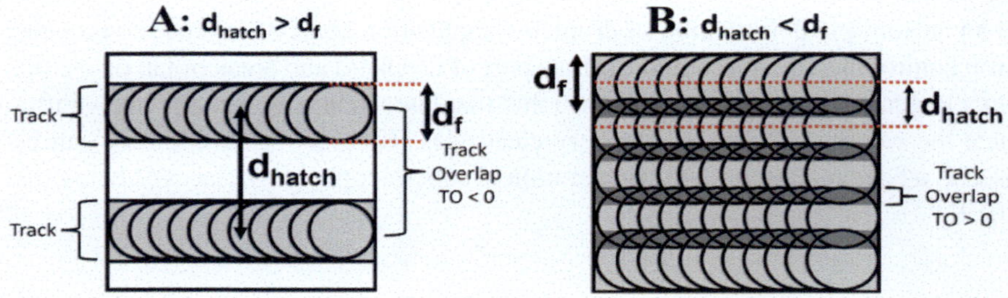

Fig. 24.2 Sketch of surfaces "type A" and "type B". "A" shows the laser structuring for a negative track overlap with $d_{hatch} > d_f$, "B" shows a positive track overlap with $d_{hatch} < d_f$

Subsequently, laser structuring has been executed with a focal diameter $d_f = 17\ \mu m$, a wavelength of $\lambda = 1030$ nm at a peak fluency of $F = 3.5\ J/cm^2$, a pulse overlap of $PO = 80\%$, a repetition rate of $f_{rep} = 200$ kHz and a single pulse duration of $t_p = 600$ fs. A maximum ablation depth of 2.5 μm is reached, which is below the 3 μm thickness of the TiAlN layer. Laser structuring has been performed with five different scan line distances. For each distance, we patterned six fields with a size of 5 × 5 mm on the coated TiAlN surface. Figure 24.2 introduces the two main structure types of patterned surfaces: Structures consisting of distinct individual lines "*type A*" and continuously irradiated surfaces "*type B*".

Related to the scan line distances the track overlap (TO) characterizes the different types. It is determined by:

$$\text{Track Overlap, TO} = 1 - \frac{d_{\text{hatch}}}{d_f} \tag{24.1}$$

The prepared structures are described by five different hatch distances d_{hatch}: $4 \times d_f$ (TO $= -300\%$), $2 \times d_f$ (TO $= -100\%$), $1 \times d_f$ (TO $= 0\%$), $0.75 \times d_f$ (TO $= +25\%$), $0.5 \times d_f$ (TO $= +50\%$).

Equivalent to TO, the structures can be described by the ratio of the irradiated area A_{irr} to the pristine surface A_0. A_{irr} is given a product of the number of laser tracks N, the line length w, and the focus diameter d_f.

The so-created surfaces are analyzed with respect to roughness and morphology using a 3D coherence white light interferometry *Zygo New View 8300, AMETEK, Germany* for measuring the depth and shape of the microablation. In parallel, the topography is investigated with a scanning electron microscope *ZEISS sigma 300vp, ZEISS, Germany* with an *EDAX EDS system, AMETEK, Germany* for element analysis. The interaction of liquids with the surface is investigated using the *Drop Shape Analyzer DSA100, KRÜSS Scientific, Germany.* In contrast to classical contact angle measurements, we perform a top-view drop shape analysis that more precisely allows the influence of the azimuthal angle of the surface pattern. To find an appropriate measure of the influence of the surface on the drop shape we averaged over six measurements each. Test liquids have been deionized water (0.35 μS/cm at 27 °C) and diiodomethane (*Merck, Sigma-Aldrich, CAS 75-11-6, 818,153*) always with a droplet volume of 2 μl.

24.3 Results

Figure 24.3 shows the results of the determination of the fluence-dependent ablation threshold F_{th} for varying pulse durations from 0.34 up to 9.0 ps.

In Fig. 24.3, the ablation depth is plotted against the laser fluence. For short pulses with 0.34 ps and the highest fluence, the highest ablation depth is produced. Long pulses with a duration of 9.0 ps result in the smallest ablation depth at the same fluence. For different pulse durations, regression lines are formed. The intersection points of the individual regression lines with the x-axis result in the ablation thresholds F_{th}.

In Fig. 24.4, the ablation threshold F_{th} is plotted versus pulse duration in the range of 0.34–9.0 ps.

It can be seen that the ablation threshold decreases continuously with increasing pulse duration. Ultra-short laser pulses in the range of 0.34–0.6 ps yield 0.67 J/cm^2 and 0.62 J/cm^2, respectively, while longer pulses in the range of 5.9–9.0 ps yield 0.53 and 0.54 J/cm^2. A reduction of about 20% is observed over the duration range considered, indicating the use of picosecond pulses for effective material removal.

The ablation threshold results of 0.67 and 0.62 J/cm^2 for pulse durations of 0.34 ps (340 fs) and 0.6 ps (600 fs) are comparable to Gakovic et al. (0.66 J/cm^2 for 200 fs, 775 nm titanium-sapphire laser (Gakovic et al. 2011)) and Ragusich et al. (0.63 J/cm^2 for 120 fs, 800 nm titanium-sapphire laser (Ragusich et al. 2013)). Here, the lowering of the ablation threshold for long pulses in the range of 9 ps with 0.54 J/cm^2 does

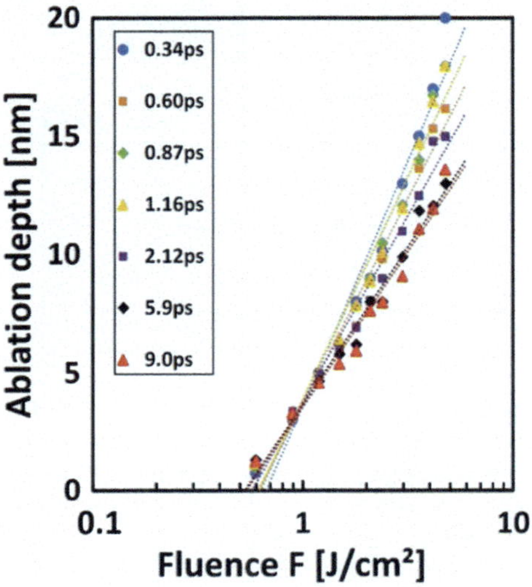

Fig. 24.3 Determination of the ablation threshold Fth for various pulse durations from 0.34 up to 9.0 ps

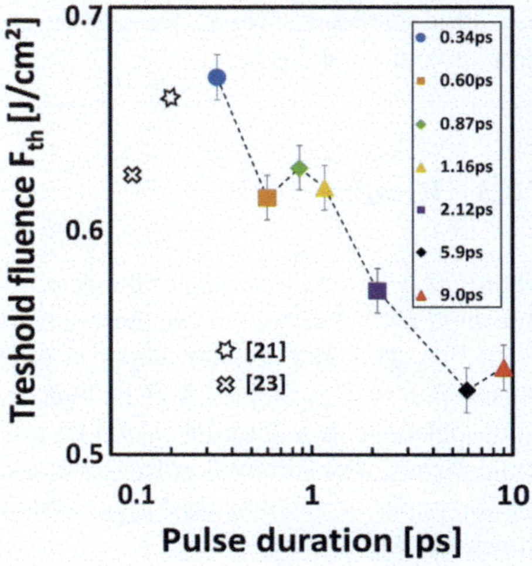

Fig. 24.4 Experimentally obtained threshold of the laser fluence F_{th} dependent on pulse duration. The open symbols refer to literature values (Gakovic et al. 2011; Ragusich et al. 2013)

not seem to apply to the nanosecond range. In the nanosecond range, the ablation threshold increases more than threefold (1.85 J/cm^2 for 18 ns, 248 nm excimer laser (Marimuthu et al. 2011)). EDX analysis also revealed that the composition within the laser lines did not change due to the laser process. The composition remains at a 60:40 ratio of aluminum to titanium.

Figure 24.5 depicts the created surfaces of structures "*type A*" with negative TO and "*type B*" with positive TO, respectively.

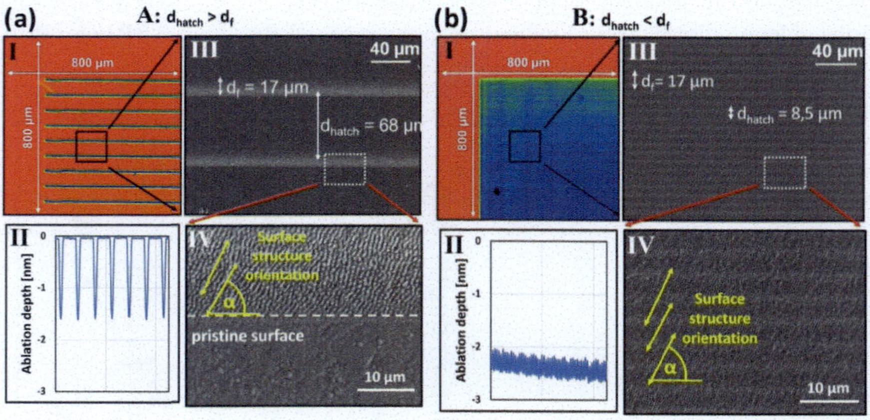

Fig. 24.5 **a** "type A": (I), (II) White light interferometry (WLI) data, (III) and (IV) SEM images describing a surface with TO = −3. (II) shows the obtained height profile extracted vertically from the WLI data. An ablation depth of around 1.5 µm is found in the lines. The SEM image identifies submicron structures oriented under 60° inside the irradiated areas. **b** Present according to data of a surface "type B", TO = +0.5. An increased ablation depth of around 2.2 µm is found. The submicron structures under an angle of 60° are present all over the surface

A negative TO is equivalent to a surface topography consisting of lines, clearly visible in panels 5a) (I)–(IV). (I) shows a white light interferometry image with a field of view of 800 × 800 μm^2. In the image, the created lines can be identified as narrow blue lines on a constant red background. A corresponding depth profile extracted across the lines is given in panel (II). The SEM images (III) and (IV) provide more detailed information on line width and topography. Image (III) shows two tracks having a width of 17 μm and a track distance d_{hatch} of 68 μm (center to center). The tracks can be easily identified as bright horizontal lines in panel (III). In panel (IV) the edge of an individual laser track with the transition to the unprocessed isotropic surface is depicted.

The large magnification of SEM images identifies small and more irregular structures with submicron dimensions in the laser track that exhibit a tilt of about 60° to the line orientation. A sharp edge can be identified at the transition to the pristine surface area that exhibits a granular microstructure.

The large distance between individual tracks characterizes this surface as "*type A*"—in this case, the dominant structure elements have dimensions in the range of several ten microns. In contrast, Fig. 24.5b shows a surface "*type B*", TO = 0.5, exhibiting only submicron surface structures. The white light interferometry image shown in Fig. 24.5b (I) depicts a field of view of 800 × 800 μm^2 with a rectangular-shaped, blue area marking the processed surface. In the SEM image (III) a structure of horizontal lines is still identified throughout the image section referring to individual laser traces. Image (IV) shows a magnification of the rectangular section of (III). Here, continuously distributed, irregular-shaped bright structures can be found with a typical lateral dimension of less than 1 μm. Irrespective of the disorder a predominant orientation of about 60° to the line orientation is found indicated by the yellow arrows in Fig. 24.5b (IV).

When placing a liquid droplet on top of the surfaces "*type A*" and "*type B*", respectively, we expect a modified shape because of the interaction with the anisotropic surface. For a characterization of the surface interaction droplets of deionized water (polar liquid) and diiodomethane (non-polar liquid) with a volume of $V = 2$ μl are used. The droplets are placed on the surface and the occurring drop shape is analyzed from the top view. This method is used to directly access the influence of the different surface patterns that exhibit different orientations. As example, Fig. 24.6 shows two droplets of diiodomethane on top of a "*type A*" and a "*type B*" surface, respectively. The used terminology is given in the figure: droplets are described by a characteristic length "L" and width "W" as well as a macroscopic orientation angle α relative to the laser processing direction. "L" describes the length along the largest extent of the droplet and "W" the maximum width perpendicular to "L".

In case of the "type A" surface, the liquid droplet is found to expand along the laser processing direction (from left to right in the figure). This results in an ellipsoidal droplet with a relative orientation of $a \approx 0°$. In case of "*type B*" an artificial line geometry is absent, only LIPSS cover the whole surface. This means, that the droplet follows the remaining structure that is present on the surface, namely the submicron-scaled LIPSS exhibiting an orientation of $a = 60°$ with respect to the laser processing direction. Accordingly, we find droplets that are rotated by $a \approx 60°$.

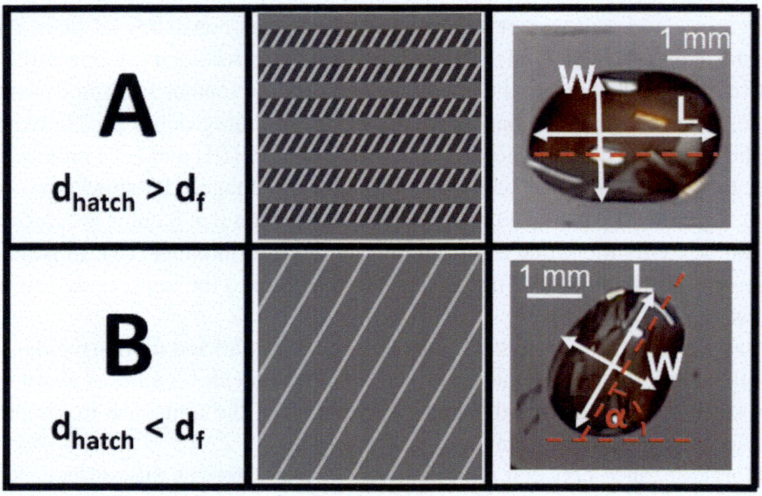

Fig. 24.6 Droplets of diiodomethane with a volume of V = 2 μl are placed on the surface of "type A" (top) and "type B" bottom. "L" and "W" describe the largest and smallest diameters, respectively. In both cases anisotropic droplets are formed; the orientation follows the dominant surface structure

To get a quantitative description we define the anisotropy by $A = d_L/d_W$ as the ratio between the largest and smallest droplet diameter. In case of water and diiodomethane the found values for "A" are plotted in Fig. 24.7 as blue and orange symbols, respectively.

The diagram in Fig. 24.7 shows that for all considered surfaces an anisotropy between 1 and 1.35 occurs. In the region TO < 0 where the dominant structure element is the artificial line geometry given by the orientation of the laser lines we find a drop anisotropy of about 1.2 for both water and diiodomethane. Starting from TO = 0 towards positive values individually written laser lines start to overlap and the dominant structural element of the surface is given by the self-organized LIPSS. Here, a significant difference in anisotropy is found for the two considered liquids. For water we find a reduced anisotropy of 1.0–1.1, in case of diiodomethane the anisotropy increases to values up to 1.35.

These results lead to several interesting points to discuss. First, it was found that the laser produced structures in TiAlN surfaces can be used to influence the wetting of polar and non-polar liquids.

Both artificially generated line patterns and self-organized LIPSS lead to a direct interaction with deposited droplets that exhibit an anisotropic drop shape. If the surface consists of artificially constructed line arrays with typical dimensions of several 10 μm the shape of both water and diiodomethane droplets follows the orientation of this line, irrespective of the fact that the line itself exhibits a more complex microstructure. The found anisotropies for both liquids are basically equivalent.

When reducing the distance of the written laser lines in a way that neighboring lines overlap, the produced surface pattern is governed by structures on a smaller

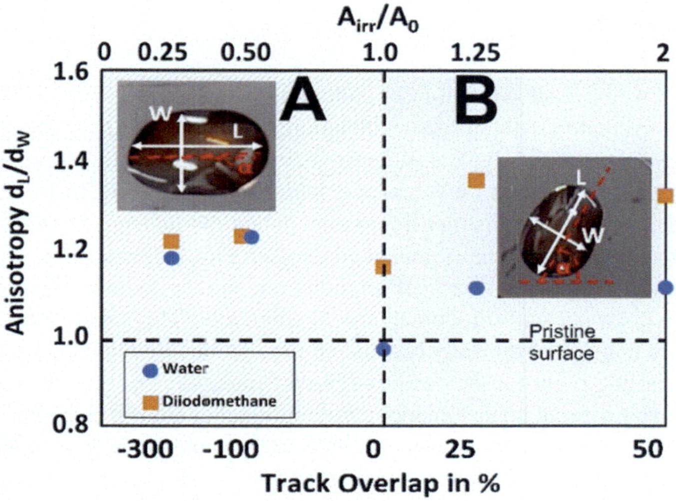

Fig. 24.7 Droplet anisotropy d_L/d_W depending on the track overlap TO from −300% to 50%. Negative TO values refer to "type A" surfaces, and a positive TO indicates surfaces of "type B". The symbols refer to the observed anisotropy in case of water (blue) and diiodome thane (orange). The dashed horizontal line refers to the pristine surface where no anisotropy occurs

length scale. In the absence of individual laser lines self-organized LIPSS oriented under an angle of 60° to the line direction are responsible for the dominant interaction with deposited liquids. Thus, the orientation of the droplets also changes to 60°. However, in this region, a significant difference in anisotropy between polar (water) and non-polar (diiodomethane) liquids is found. The induced anisotropy in diiodomethane increases in the case of LIPSS, for water a strongly reduced anisotropy was found.

This phenomenon can be understood when considering the large surface tension of water that avoids wetting of very narrow surface structures because of the required strong curvature of the solid–liquid interface. Instead, water droplets "sit" on top of the submicron-sized LIPSS, according to the wetting model of Cassie and Baxter (1944). Here, air pockets inside the surface structures are formed that prevent the water droplet from full wetting. Since in this case, the interaction between liquid and surface is drastically reduced a much smaller surface-induced anisotropy is seen. The wetting of diiodomethane with much smaller surface tension can be described by the model of Wenzel (1936) where a full wetting of the LIPSS takes place. This leads to a strong interaction of surface and liquid and, thus to a large anisotropy.

For the purpose of the creation of surfaces that can naturally distinguish between different types of liquids, these LIPSS are highly attractive. Since the interaction can change for different liquids from Cassie-Baxter to Wenzel mechanisms it is in principle possible to control the liquid-surface interaction by surface structures. A necessary prerequisite is the formation of submicron-sized surface structures with high curvature values.

24.4 Summary

In summary, we have shown that laser patterning using ultra-short pulse lasers is a feasible way to create TiAlN surfaces with tailored topography. Artificially generated anisotropic structures with typical dimensions in the range of 10 μm are compared with self-organized patterns on submicron length scale (LIPSS). In particular, the wetting behavior exhibits distinct differences when considering droplets of polar and non-polar liquids. Artificially generated structures lead to anisotropic droplets for all used liquids. Submicron-sized LIPSS, however, show a distinct difference when in contact with water and diiodomethane, respectively. The small, self-organized structures are not wetted by water because of its high surface tension. These results can be the starting point for the design of "smart surfaces" that are able to separate different liquids without further energy consumption. For the full characterization of the surface/liquid interaction, it is required to analyze the wetting process in more detail. Measurements of contact angles and the corresponding hysteresis will provide insight into this topic and have to be the subsequent step.

Acknowledgements The authors would like to acknowledge the invaluable support provided by the ZOT (Center for Optical Technologies, Aalen) for optical surface measurements and from the IMFAA (Materials Research Institute, Aalen) for the material-specific investigations. Furthermore, we would like to thank Wolfgang Engelhart from Walter AG for providing the surfaces.

References

Abele L, Jäger AK, Schulz W, Ruck S, Riegel H, Sörgel T, Albrecht J (2019) Appl Surf Sci 490:56–60. https://doi.org/10.1016/j.apsusc.2019.06.046

Bonn D, Eggers J, Indekeu J, Meunier J, Rolley E (2009) Rev Mod Phys 81:739–805. https://doi.org/10.1103/RevModPhys.81.739

Bonse J, Rudolph P, Krüger J, Baudach S, Kautek W (2000) Appl Surf Sci 154–155:659–663

Bonse J, Kirner S, Krüger J (2020) Laser-induced periodic surface structures. In: Handbook of laser and nano-engineering, pp 1–59

Cao L, Hu H-H, Gao D (2007) Langmuir 23:4310–4314. https://doi.org/10.1021/la063572r

Cassie ABD, Baxter S (1944) Trans Faraday Soc 546:40

Chen F, Zhang D, Yang Q, Yong J, Du G, Si J, Yun F, Hou X (2013) ACS Appl Mater Interfaces 5:6777–6792. https://doi.org/10.1021/am401677z

Chichkov BN, Momma C, Nolte S, Alvensleben F, Tünnermann A (1996) Appl Phys A 63:109. https://doi.org/10.1007/BF01567637

Denkena B, Grove T, Krödel A, Ellersiek L (2018) Procedia CIRP 77:602–605

Dubey PP, Pham QN, Cho H, Kim Y, Won Y (2017) ACS Omega 2:7916–7922. https://doi.org/10.1021/acsomega.7b01178

Dumitru G, Romano V, Weber HP, Sentis M, Marine W (2002) Appl Phys A: Mater Sci Proc 74:729–739. https://doi.org/10.1007/s003390101183

Gakovic B, Radu C, Zamfirescu M, Radak B, Trtica M, Petrovic S, Panjan P, Zupanic F, Ristoscu C, Mihailescu IN (2011) Surf Coat Technol 206:411–416. https://doi.org/10.1016/j.surfcoat.2011.07.042

Gleiche M, Chi LF, Fuchs H (2000) Nature (London) 403:173–175. https://doi.org/10.1038/35003149
Kleefoot M-J, Enderle S, Sandherr J, Bolsinger M, Maischik T, Simon N, Martan J, Ruck S, Knoblauch V, Riegel H (2022) Int J Adv Manuf Technol 118:1987–1997. https://doi.org/10.1007/s00170-021-08004-3
Kommer M, Sube T, Richter A, Fenker M, Schulz W, Hader B, Albrecht J (2017) Surf Coat Technol 333:1–12
Li X-M, Reinhoudt D, Crego-Calama M (2007) Chem Soc Rev 36:1350–1368. https://doi.org/10.1039/B602486F
Liu K, Jiang L (2012) Annu Rev Mater Res 42:231–263. https://doi.org/10.1146/annurev-matsci-070511-155046
Lundgren M, Allan NL, Cosgrove T (2007) Langmuir 23:1187–1194. https://doi.org/10.1021/la060712o
Malshe A, Rajurkar K, Samant A, Hansen HN, Bapat S, Jiang W (2013) CIRP Ann 62:607–628. https://doi.org/10.1016/j.cirp.2013.05.008
Marimuthu S, Kamara AM, Whitehead D, Mativenga P, Li L, Yang S, Cooke K (2011) Proc Inst Mech Eng B 225:1851–1862. https://doi.org/10.1177/0954405411414313
Meliani H, Assoul M, Fontaine M, Malkhasyan V, Gilbin A, Monteil G (2021) Appl Opt 60:621–625. https://doi.org/10.1364/AO.409163
Milles S, Soldera M, Voisiat B, Lasagni AF (2019) Sci Rep 9:13944. https://doi.org/10.1038/s41598-019-49615-x
Momma C, Nolte S, Chichkov BN, von Alvensleben F, Tünnermann A (1997) Appl Surf Sci 109–110:15–19. https://doi.org/10.1016/S0169-4332(96)00613-7
Müller FA, Kunz C, Gräf S (2016) Materials 9:476. https://doi.org/10.3390/ma9060476
Neuenschwander B, Jaeggi B, Schmid M, Hennig G (2014) Phys Proc 56:1047–1058. https://doi.org/10.1016/j.phpro.2014.08.017
Neuenschwander B, Jaeggi B, Schmid M, Rouffiange V, Martin P-E (2012) Optimization of the volume ablation rate for metals at different laser pulse durations from ps to fs. In: Hennig G, Xu X, Gu B, Nakata Y (eds) Laser applications in microelectronic and optoelectronic manufacturing (LAMOM), vol 17, p 824307
Nolte S, Schrempel F, Dausinger F (eds) (2016) Ultrashort pulse laser technology: laser sources and applications, p 195
Nolte S, Momma C, Jacobs H, Tünnermann A, Chichkov BN, Wellegehausen B, Welling H (1997) J Opt Soc Am B, JOSAB 14:2716 https://doi.org/10.1364/JOSAB.14.002716
Ragusich A, Taillon G, Meunier M, Martinu L, Klemberg-Sapieha JE (2013) Surf Coat Technol 232:758–766. https://doi.org/10.1016/j.surfcoat.2013.06.092
Römer G, Huis in't Veld AJ, Meijer J, Groenendijk M (2009) CIRP Annals 58:201–204. https://doi.org/10.1016/j.cirp.2009.03.068
Ruck S (2021) JLMN 16:36–41. https://doi.org/10.2961/jlmn.2021.01.2007
Schille J, Schneider L, Mueller M, Loeschner U, Goddard N, Scully P, Exner H (2013) J. Laser Micro Nanoengin 9:161
Straub M, Afshar M, Feili D, Seidel H, König K (2012) J Appl Phys 111:124315. https://doi.org/10.1063/1.4730381
Urbina JC, Daniel C, Emmelmann C (2013) Phys Proc 41:752–758. https://doi.org/10.1016/j.phpro.2013.03.144
Wenzel RN (1936) Ind Eng Chem 988:28
Wu PH, Cheng CW, Chang CP, Wu TM, Wang JK (2011) J Micromech Microeng 21:1115032. https://doi.org/10.1088/0960-1317/21/11/115032
Zhang F, Low HY (2007) Langmuir 23:7793–7798. https://doi.org/10.1021/la700293y

Chapter 25
Application of the Two-Time-Scale Approach to the Constitutive Equations of High-Temperature Steel

Katharina Knape and Holm Altenbach

Abstract The paper focusses on the derivation of the time-homogenized constitutive equations of the high-temperature steel X20CrMoV12-1 which is mostly used in power plant components. For this purpose, the two-time-scale approach is briefly introduced first explaining the homogenization procedure. The constitutive model considered is described including evolution equations for the inelastic strain and the two inner state variables the backstress and the softening. In the main part, the application of the two-time-scale method to those equations is shown also taking the numerical implementation into account. Since the material is subjected to high mechanical and also thermal loadings leading to highly non-linear behavior, a cyclic test is simulated in the end.

25.1 Introduction

In general, simulating cyclic loading profiles for a component's lifespan is known to be very time consuming due to the time integration procedure with very small time increments when performing a cycle-by-cycle integration (Altenbach et al. 2018; Bhattacharyya et al. 2020). With the overall goal of omitting these drawbacks and still being able to compute the materials response to real-life loading conditions, several methods have been developed over the years, introducing three categories of multi-scale computational strategies in time: multi-time-step methods, the variational multi-scale method in time and methods considering cyclic loadings (Haouala and Doghri 2015). An overview is given in Fig. 25.1.

The LATIN method was originally proposed in Cognard and Ladevèze (1993) and Ladeveze (1985) and uses an elastic solution as a coarse approximation. Two solution subspaces are introduced, with the aim to split the entire interval into sub-intervals

K. Knape (✉) · H. Altenbach
Lehrstuhl für Technische Mechanik, Institut für Mechanik, Fakultät für Maschinenbau, Otto-von-Guericke-Universität Magdeburg, Universitätsplatz 2, 39104 Magdeburg, Germany
e-mail: katharina.knape@ovgu.de

H. Altenbach
e-mail: holm.altenbach@ovgu.de

H. Altenbach et al. (eds.), *Lectures Notes on Advanced Structured Materials 3*,
Advanced Structured Materials 226, https://doi.org/10.1007/978-3-031-77403-4_25

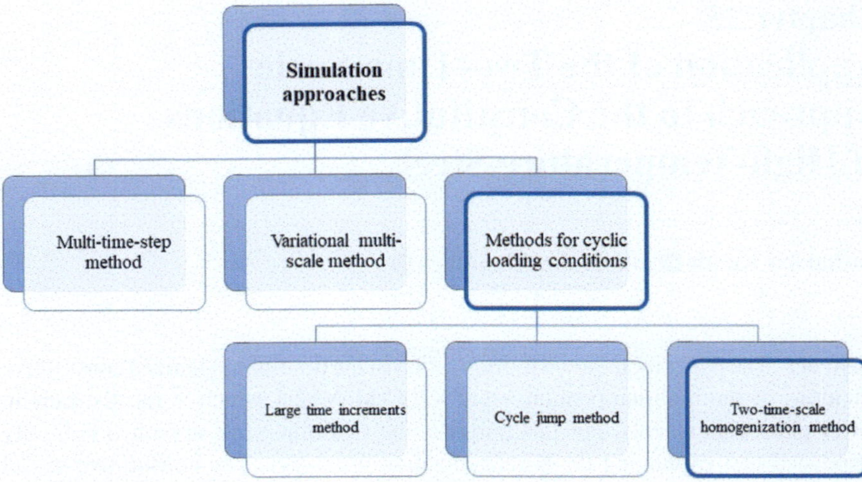

Fig. 25.1 Overview of multi-scale computational strategies in time

where each of them contains tens or hundreds of cycles reducing the numerical calculation time significantly but at the same time, the implementation of the method into a finite element software is rather complex (Haouala 2016). Another possibility to reduce the computational cost is the cycle jumping scheme suggested in Lesne and Savalle (1989) and further applied in Labergere et al. (2015) also modeling the material's response under periodic loading. The basic idea is defining larger time intervals and extrapolating the solution across them. In this paper, considering the application to high-temperature steels in power plant components, the focus will be on applying the two-time-scale approach originally proposed to solve differential equations of dynamical systems and applied, for example, in Altenbach et al. (2018, 2000) and Morachkovskii (1992). Now, it can also be applied to simulate inelastic material behavior (Knape and Altenbach 2022; Oskay and Fish 2004).

The investigation of power plant components and their lifespan has become a large field of research in the past years especially since the energy consumption keeps increasing. Therefore, fluctuations are expected not only to happen more often but also within shorter intervals (Siemens-Energy 2020). Therefore, the highly frequent start-ups and shut-downs of the system result in the need for the material to withstand complex mechanical and thermal loading conditions at operating temperatures ranging up to 900 K. These very demanding operating conditions influence the component's life significantly and worse, could lead to failure earlier than expected. So a reliable simulation of the material's response is mandatory to gain a better understanding of the material's behavior and get the possibility to further adapt the maintenance intervals or even predict the remaining lifetime of the component (Knape and Altenbach 2022).

Using the two-time-scale procedure in a cyclic simulation like that is based on a constitutive model as used in Altenbach et al. (2018). It was firstly introduced by

Armstrong and Frederick (1966) and Frederick and Armstrong (2007), consisting of an equation for the inelastic strain rate tensor and also including an evolution equation for the backstress tensor. In Sect. 25.2, the main background of the method is explained. The practical application to the constitutive equations is then described in Sect. 25.3. They refer to the high-temperature steel X20CrMoV12-1 and were already implemented into the finite element software ABAQUS by the means of a user-defined material subroutine (UMAT) (Eisenträger et al. 2020).

25.2 Two-Time-Scale Approach

The basic idea of the two-time scale approach is the introduction of two different time scales. The first is referred to as t, the slow, macro, or also physical time scale for quasi-static loading, and τ, the fast or micro time scale to capture the influence of the rapidly varying oscillations with the period T (Altenbach et al. 2018; Haouala and Doghri 2015; Sanders and Verhulst 1985). For any set of unknown variables $\boldsymbol{x}$

$$\frac{d\boldsymbol{x}}{dt} = \boldsymbol{X}[t, \boldsymbol{x}(t)], \quad \boldsymbol{x}\,(t = 0) = \boldsymbol{x}_0 \tag{25.1}$$

holds. Both time scales are dependent on each other which is expressed through a scaling parameter μ with respect to the total time t_{end}

$$\mu = \frac{T}{t_{\text{end}}} \ll 1. \tag{25.2}$$

The total time derivative with respect to the two scales can be calculated as

$$\frac{d}{dt} = \frac{\delta}{\delta t} + \frac{1}{\mu}\frac{\delta}{\delta \tau}. \tag{25.3}$$

In the next step, an asymptotic series expansion is applied to this system of partial differential equations with respect to the factor μ (Knape and Altenbach 2022; Oskay and Fish 2004). If these expansions are inserted into Eq. (25.3), the result is a set of differential equations. They can be sorted with respect to the order of μ

$$\mu^{(-1)} : \frac{\delta \boldsymbol{x}^{(0)}}{\delta \tau} = 0, \tag{25.4}$$

$$\mu^{(0)} : \frac{\delta \boldsymbol{x}^{(0)}}{\delta t} + \frac{\delta \boldsymbol{x}^{(1)}}{\delta \tau} = \boldsymbol{X}\left(t, \tau, \boldsymbol{x}^{(0)}\right), \tag{25.5}$$

$$\mu^{(1)} : \frac{\delta \boldsymbol{x}^{(1)}}{\delta t} + \frac{\delta \boldsymbol{x}^{(2)}}{\delta \tau} = \frac{\delta \boldsymbol{X}(t, \tau, \boldsymbol{x}^{(0)})}{\delta \boldsymbol{x}} \boldsymbol{x}^{(1)}. \tag{25.6}$$

Now, the time averaging operator is applied to capture the influence of the micro on the macro scale (Altenbach et al. 2018; Haouala and Doghri 2015)

$$\langle f(t,\tau)\,\rangle = \frac{1}{T}\int_0^T f(t,\tau)\,\mathrm{d}\tau. \tag{25.7}$$

25.3 Constitutive Equations for High-Temperature Steel

When simulating the behavior of a high-temperature steel subjected to cyclic loading conditions, the material model has to not only account for elastic but also inelastic behavior. The constitutive model considered includes a constitutive equation for the inelastic strain rate tensor and a non-linear kinematic hardening rule for the backstress tensor. The material parameters required are successfully identified in Altenbach et al. (2018), Eisenträger (2018), and Eisenträger et al. (2018b) for tempered martensitic steel. The general formulation is based on the additive decomposition of the strain $\boldsymbol{\varepsilon}$ into an elastic $\boldsymbol{\varepsilon}^{\mathrm{el}}$ and inelastic $\boldsymbol{\varepsilon}^{\mathrm{in}}$ part

$$\boldsymbol{\varepsilon} = \boldsymbol{\varepsilon}^{\mathrm{el}} + \boldsymbol{\varepsilon}^{\mathrm{in}}. \tag{25.8}$$

25.3.1 Elastic Behavior

In order to describe the first summand of Eq. (25.8), HOOKE's law is applied

$$\boldsymbol{\sigma} = K\,\mathrm{tr}\left(\boldsymbol{\varepsilon}^{\mathrm{el}}\right)\boldsymbol{I} + 2G\boldsymbol{\varepsilon}^{\mathrm{el}}, \tag{25.9}$$

with

$$K = \frac{E}{3(1-2\nu)}, \quad G = \frac{E}{2(1+\nu)}. \tag{25.10}$$

Here, several material parameters are introduced: Young's modulus E, bulk modulus K, shear modulus G, and Poisson's ratio ν. The stress tensor is described as $\boldsymbol{\sigma}$, tr() defines the trace, and $\boldsymbol{I}$ the unit tensor.

The equation for the elastic strain can now be derived when rearranging Eq. (25.9), and making use of the decomposition of the stress tensor into its spheric and deviatoric part, where $\boldsymbol{\sigma}'$ denotes the stress deviator

$$\boldsymbol{\sigma} = \sigma_m \boldsymbol{I} + \boldsymbol{\sigma}' \tag{25.11}$$

with

$$\sigma_m = \frac{1}{3}\mathrm{tr}\left(\boldsymbol{\sigma}\right), \tag{25.12}$$

and

$$\boldsymbol{\varepsilon}^{\mathrm{el}} = \frac{\sigma_{\mathrm{m}}}{3K}\boldsymbol{I} + \frac{\boldsymbol{\sigma}'}{2G}. \tag{25.13}$$

25.3.2 *Inelastic Behavior*

When describing this kind of highly non-linear behavior, an evolution equation for the inelastic strain $\varepsilon^{\mathrm{in}}$ has to be determined based on the VON MISES stress σ_{vM} and its corresponding strain rate $\dot{\varepsilon}^{\mathrm{in}}_{\mathrm{vM}}$ which yields

$$\dot{\boldsymbol{\varepsilon}}^{\mathrm{in}} = \frac{3}{2}\,\dot{\varepsilon}^{\mathrm{in}}_{\mathrm{vM}}\frac{\boldsymbol{\sigma}'}{\sigma_{\mathrm{vM}}} \tag{25.14}$$

with

$$\sigma_{\mathrm{vM}} = \sqrt{\frac{3}{2}\boldsymbol{\sigma}' : \boldsymbol{\sigma}'}\,, \tag{25.15}$$

and

$$\dot{\varepsilon}^{\mathrm{in}}_{\mathrm{vM}} = \sqrt{\frac{2}{3}\dot{\boldsymbol{\varepsilon}}^{\mathrm{in}} : \dot{\boldsymbol{\varepsilon}}^{\mathrm{in}}} \tag{25.16}$$

including the inelastic strain deviator $\dot{\boldsymbol{\varepsilon}}^{\mathrm{in}}$.

A separation ansatz captures the temperature dependence of the inelastic strain consisting of a stress response function f_σ and a temperature response function $f_\vartheta(\vartheta)$ depending on the absolute temperature ϑ. In literature, e.g., in Eisenträger (2018), several possible ways to describe them can be found.

$$\dot{\varepsilon}^{\mathrm{in}} = f_\sigma(\tilde{\sigma}_{\mathrm{vM}})\; f_\vartheta(\vartheta). \tag{25.17}$$

In Eq. (25.17), $\tilde{\sigma}_{\mathrm{vM}}$ describes the VON MISES active stress, which is defined by decomposing the stress tensor into an active $\tilde{\boldsymbol{\sigma}}$ and a backstress part $\boldsymbol{\beta}$. For steels like X20CrMoV12-1, the backstress as a tensorial variable also has to be calculated with an evolution equation. It depends on the absolute temperature ϑ and the shear modulus G through the function f_G. The influence of the microstructure is included by applying a phase mixture model described with the function f_{m} (Eisenträger et al. 2018b)

$$\boldsymbol{\beta} = \boldsymbol{\sigma}' - \tilde{\boldsymbol{\sigma}}, \tag{25.18}$$

and

$$\dot{\boldsymbol{\beta}} = f_G(G, \vartheta)\,\boldsymbol{\beta} + f_{\mathrm{m}}\left[\dot{\boldsymbol{\varepsilon}}^{\mathrm{in}} - \frac{3}{2}\dot{\varepsilon}^{\mathrm{in}}_{\mathrm{vM}}\frac{\boldsymbol{\beta}}{h(\sigma_{\mathrm{vM}})}\right]. \tag{25.19}$$

Now, the inelastic strain rate is also expressed in terms of the active stress deviator

$$\dot{\boldsymbol{\varepsilon}}^{\text{in}} = \frac{3}{2} f_\sigma(\tilde{\sigma}_{\text{vM}}) f_\vartheta(\vartheta) \frac{\tilde{\boldsymbol{\sigma}}'}{\tilde{\sigma}_{\text{vM}}}. \tag{25.20}$$

In reference to Eisenträger (2018), the softening variable is introduced with a material parameter C and a saturation value for the softening γ_*

$$\dot{\gamma} = C\,(\gamma_* - \gamma)\,\dot{\varepsilon}^{\text{in}}_{\text{vM}}. \tag{25.21}$$

All material parameters included are successfully identified in Eisenträger et al. (2018a).

25.3.3 Application of Two-Time-Scale Procedure to the Constitutive Equations

In case $\boldsymbol{x} = \boldsymbol{\varepsilon}^{\text{in}}$, Eq. (25.3) yields

$$\dot{\boldsymbol{\varepsilon}}^{\textbf{in}} = \frac{d\boldsymbol{\varepsilon}^{\text{in}}}{dt} = \frac{\delta\boldsymbol{\varepsilon}^{\text{in}}}{\delta t} + \frac{1}{\mu}\frac{\delta\boldsymbol{\varepsilon}^{\text{in}}}{\delta\tau}. \tag{25.22}$$

The asymptotic series expansion of the inelastic strain

$$\boldsymbol{\varepsilon}^{\text{in}} = \boldsymbol{\varepsilon}^{\text{in}(0)} + \mu\boldsymbol{\varepsilon}^{\text{in}(1)} + O\left(\mu^2\right) \tag{25.23}$$

with O the Bachmann-Landau notation (or asymptotic notation) (Bachmann 1894; Landau 1909) is now inserted into Eq. (25.22)

$$\dot{\boldsymbol{\varepsilon}}^{\textbf{in}} = \left(\frac{\delta\boldsymbol{\varepsilon}^{\text{in}(0)}}{\delta t} + \frac{1}{\mu}\frac{\delta\boldsymbol{\varepsilon}^{\text{in}(0)}}{\delta\tau}\right) + \mu\left(\frac{\delta\boldsymbol{\varepsilon}^{\text{in}(1)}}{\delta t} + \frac{1}{\mu}\frac{\delta\boldsymbol{\varepsilon}^{\text{in}(1)}}{\delta\tau}\right) + \cdots \tag{25.24}$$

and sorted according to the orders of μ

$$\dot{\boldsymbol{\varepsilon}}^{\textbf{in}} = \frac{1}{\mu}\frac{\delta\boldsymbol{\varepsilon}^{\text{in}(0)}}{\delta\tau} + \left(\frac{\delta\boldsymbol{\varepsilon}^{\text{in}(0)}}{\delta t} + \frac{\delta\boldsymbol{\varepsilon}^{\text{in}(1)}}{\delta\tau}\right) + \mu\left(\frac{\delta\boldsymbol{\varepsilon}^{\text{in}(1)}}{\delta t} + \frac{1}{\mu}\frac{\delta\boldsymbol{\varepsilon}^{\text{in}(2)}}{\delta\tau}\right) + \cdots \tag{25.25}$$

Substituting the right side of Eq. (25.25) with Eq. (25.20) and equating the same orders of μ yields the first part of the $\mu^{(-1)}$-order problem

$$\frac{\delta\boldsymbol{\varepsilon}^{\text{in}(0)}}{\delta\tau} = 0 \tag{25.26}$$

showing that the zero-order inelastic strain does not depend on the fast time scale and is therefore purely elastic. This can also be explained by the physical background that the inelastic strain is a rather slowly evolving material property. Applying the same steps to the evolution equations for the backstress, Eq. (25.19) and the softening variable, Eq. (25.21) results in the following complete $\mu^{(-1)}$-order problem

$$\frac{\delta \boldsymbol{\beta}^{(0)}}{\delta \tau} = 0 \tag{25.27}$$

and

$$\frac{\delta \gamma^{(0)}}{\delta \tau} = 0. \tag{25.28}$$

The $\mu^{(0)}$-order problem can now be derived as follows

$$\frac{3}{2} f_\sigma(\tilde{\sigma}_{\mathrm{vM}}) f_\vartheta(\vartheta) \frac{\tilde{\boldsymbol{\sigma}}'}{\tilde{\sigma}_{\mathrm{vM}}} = \left(\frac{\delta \boldsymbol{\varepsilon}^{\mathrm{in}(0)}}{\delta t} + \frac{\delta \boldsymbol{\varepsilon}^{\mathrm{in}(1)}}{\delta \tau} \right), \tag{25.29}$$

$$f_G(G, \vartheta)\, \boldsymbol{\beta}^{(0)} + f_{\mathrm{m}} \left[\dot{\boldsymbol{\varepsilon}}^{\mathrm{in}(0)} - \frac{3}{2} \dot{\varepsilon}^{\mathrm{in}}_{\mathrm{vM}} \frac{\boldsymbol{\beta}^{(0)}}{h(\sigma_{\mathrm{vM}})} \right] = \left(\frac{\delta \boldsymbol{\beta}^{(0)}}{\delta t} + \frac{\delta \boldsymbol{\beta}^{(1)}}{\delta \tau} \right) \tag{25.30}$$

and

$$C \left(\gamma_* - \gamma^{(0)} \right) \dot{\varepsilon}^{\mathrm{in}}_{\mathrm{vM}} = \left(\frac{\delta \gamma^{(0)}}{\delta t} + \frac{\delta \gamma^{(1)}}{\delta \tau} \right) \tag{25.31}$$

Each of the three variables is now decomposed into a macro and micro part, meaning a mean and an oscillating portion. The time averaging operator introduced in Eq. (25.7) is used where the average of

$$\frac{\delta \boldsymbol{x}^{(1)}}{\delta \tau} \tag{25.32}$$

vanishes (Haouala 2016).

Therefore the total $\mu^{(0)}$-order problem yields

$$\dot{\boldsymbol{\varepsilon}}^{in(0)} = \frac{3}{2} f_\sigma(\hat{\sigma}_{vM}) g(\mathrm{T}) \frac{\tilde{\boldsymbol{\sigma}}'^{(0)}_M}{\hat{\sigma}_{\mathrm{vM}}} \tag{25.33}$$

and

$$\dot{\boldsymbol{\beta}}^{(0)} = f_G(G, \vartheta)\, \boldsymbol{\beta}^{(0)}_M + f_{\mathrm{m}} \left(\dot{\boldsymbol{\varepsilon}}^{in(0)} - \frac{3}{2} \dot{\varepsilon}^{in}_{\mathrm{vM}} \frac{\boldsymbol{\beta}^{(0)}_M}{\beta_*(\sigma_{\mathrm{vM}})} \right) \tag{25.34}$$

$$\dot{\gamma}^{(0)} = C \left(\gamma_*(\sigma_{vM}) - \gamma^{(0)}_M \right) \dot{\varepsilon}^{in}_{\mathrm{vM}} \tag{25.35}$$

where the subscript $\square_M$ implies that the time averaging operator has been applied to the variable and only the terms up to the first order of μ were considered. Higher orders may be calculated according to Sanders and Verhulst (1985).

25.4 Numerical Implementation

The zero-order problem derived in Sect. 25.3.3 can now be implemented numerically. The system of equations can be solved using explicit or implicit integration techniques, here, the implicit EULER method was used. For the beginning of each macro time step, the micro time problem is solved (Haouala and Doghri 2015; Knape and Altenbach 2022). The results serve as input for the calculation of the slow time problem. If convergence with respect to a given tolerance is reached, the variables can be updated and are used again as input for the next micro-scale calculation until the ending time of the simulation is reached. A schematic sketch can be seen in Fig. 25.2 where T denotes the period of the fast time loading, Δt_M defines the macro time step, and the indices n and $n+1$ refer to the times steps respectively. Figure 25.3 shows the implementation of the procedure in terms of a program flowchart.

The first results were already established using the explained numerical implementation. To validate the implementation, a one-element high-temperature creep test was simulated, with a constant load of 50 MPa and another one with a constant load of 100 MPa at a constant surrounding temperature of T= 600 K. The results are depicted in Fig. 25.4 and came to good agreement with the literature (Altenbach et al. 2018; Knape and Altenbach 2022).

After successfully performing the creep test, the model of a pipe was considered, exerted to a cyclic loading profile of the form

$$\varepsilon(t) = A\left(\sin\left(\Omega t - \phi\right)\right) \tag{25.36}$$

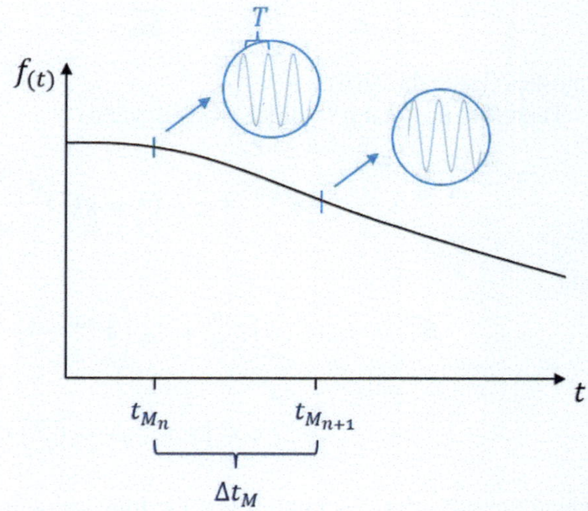

Fig. 25.2 Schematic sketch of the basic principle of the two-time-scale approach

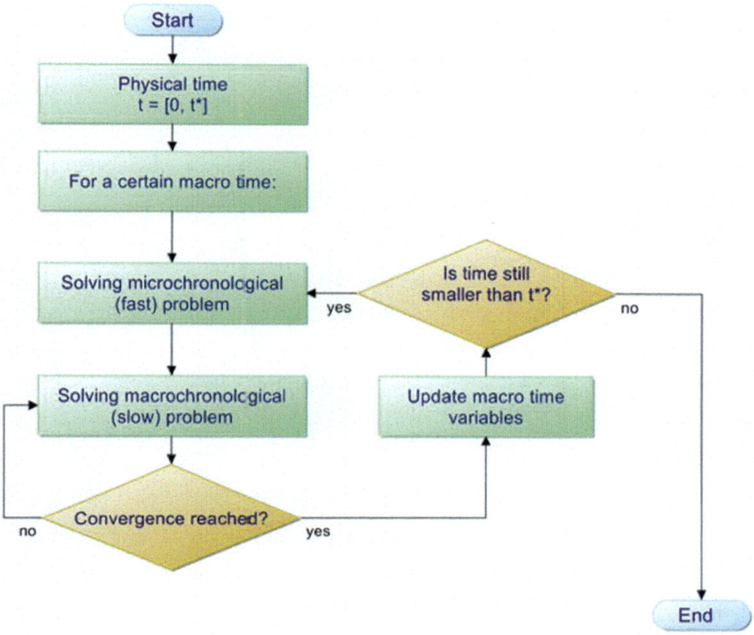

Fig. 25.3 Program flowchart for the implementation

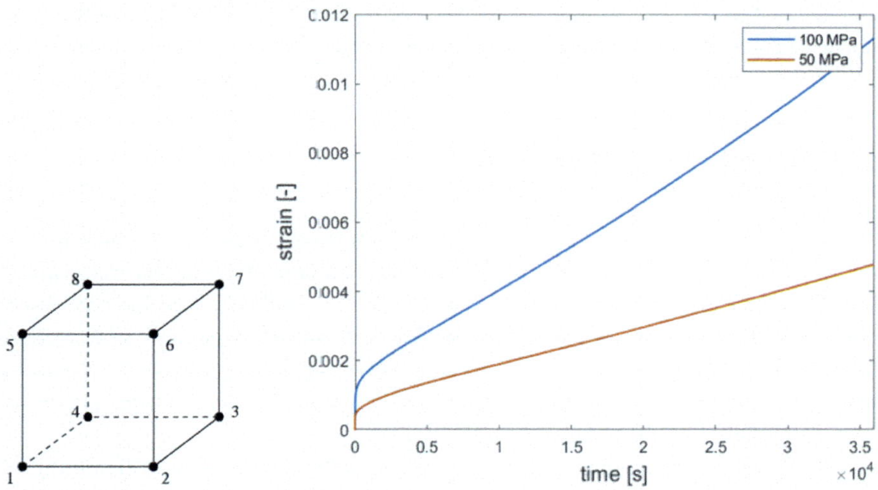

Fig. 25.4 Single element for the simulation (left) and results for the creep test (right)

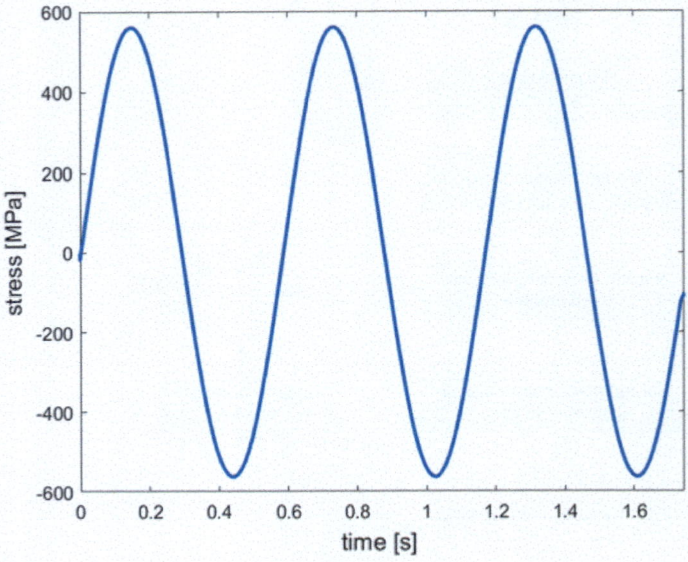

Fig. 25.5 Stress versus time considering profile (25.36)

which was fitted to the experimental data from measuring a hot start with the angular frequency of $\Omega = 10.75$, a phase angle of $\phi = 0.03$ and $A = 0.003$. The result shown in Fig. 25.5 represents the expected cyclic stress response.

25.5 Conclusion and Outlook

The paper shows the derivations of the time-homogenized constitutive equations for high-temperature steel also applying a phase mixture model to capture the microstructure. The considered material model consists of two internal variables; the backstress tensor and the softening which is described as a scalar function, and an evolution equation for the inelastic strain. The model was numerically implemented to perform some cyclic simulations which are supposed to depict the behavior during a hot start with the model of a pipe to serve as a benchmark.

The system of differential equations was implemented as UMAT and their implicit time integration is carried out using the backward Euler method. The results considering a one-element high-temperature creep test come to good agreement with the literature. When moving on to the more realistic model of a steel pipe, the results are plausible and show the expected cyclic stress response.

In the future, thermo-mechanical fatigue should be considered, since the loading case is able to closer represent the real-life loading conditions of a power plant component. Furthermore, it could be of interest to also include a damage parameter defined with its own evolution equation. Due to applying the asymptotic series expansion and the time averaging to that equation as well, the two-time-scale approach allows a numerically efficient way to also consider the damage to adapt the material model even further to reality.

References

Altenbach H, Breslasky D, Naumenko K, Tatarinova O (2018) Two-time-scales and time-averaging approaches for the analysis of cyclic creep based on Armstrong-Frederick type constitutive model. J Mech Eng Sci 233:1690–1700. https://doi.org/10.1177/0954406218772609

Altenbach H, Breslavsky D, Morachkovsky O, Naumenko K (2000) Cyclic creep damage in thin-walled structures. J Strain Anal Eng Des 35(1):1–11. https://doi.org/10.1177/030932470003500101

Armstrong PJ, Frederick CO (1966) A mathematical representation of the multiaxial Bauschinger effect. Cegb report rd/b/n 731, Central Electricity Generating Board

Bachmann P (1894) Analytische Zahlentheorie, vol 2. Teubner, Leipzig

Bhattacharyya M, Dureisseix D, Faverjon B (2020) A unified approach based on temporal homogenisation and cycle jump for thermo-mechanical combined cycle fatigue. Int J Fatigue 131:105320. https://doi.org/10.1016/j.ijfatigue.2019.105320. https://www.sciencedirect.com/science/article/pii/S0142112319304244

Cognard JY, Ladevèze P (1993) A large time increment approach for cyclic viscoplasticity. Int J Plast 9(2):141–157. https://doi.org/10.1016/0749-6419(93)90026-M. https://www.sciencedirect.com/science/article/pii/074964199390026M

Eisenträger J (2018) A framework for modeling the mechanichal behavior of tempered martensitic steels at high temperatures. Phd thesis, Otto-von-Guericke-Universität Magdeburg

Eisenträger J, Naumenko K, Altenbach H (2018a) Calibration of a phase mixture model for hardening and softening regimes in tempered martensitic steel over wide stress and temperature ranges. J Strain Anal Eng 53:156–177. https://doi.org/10.1177/0309324718755956

Eisenträger J, Naumenko K, Altenbach H (2018b) Numerical implementation of a phase mixture model for rate-dependent inelasticity of tempered martensitic steels. Acta Mech 229:3051–3068. https://doi.org/10.1007/s00707-018-2151-1

Eisenträger J, Zhang J, Song C, Eisenträger S (2020) An SBFEM approach for rate-dependent inelasticity with application to image-based analysis. Int J Mech Sci 182:105778. https://doi.org/10.1016/j.ijmecsci.2020.105778

Frederick CO, Armstrong PJ (2007) A mathematical representation of the multiaxial Bauschinger effect. Mater High Temp 24(1):1–26. https://doi.org/10.1179/096034007X207589

Haouala S (2016) Time and space (4D) homogenization for viscoelastic-viscoplastic solids under large numbers of cycles. https://api.semanticscholar.org/CorpusID:124211350

Haouala S, Doghri I (2015) Modeling and algorithms for two-scale time homogenization of viscoelastic-viscoplastic solids under large number of cycles. Int J Plast 70:98–120. https://doi.org/10.1016/j.ijplas.2015.03.005

Knape K, Altenbach H (2022) Modeling the inelastic behavior of high-temperature steels using a two-time-scale approach. In: Altenbach H, Johlitz M, Merkel M, Öchsner A (eds.) Lectures notes on advanced structured materials, pp 357–367. Springer International Publishing, Cham. https://doi.org/10.1007/978-3-031-11589-9_23

Labergere C, Saanouni K, Sun ZD, Dhifallah MH, Li Y, Duval YL (2015) Prediction of low cycle fatigue life using cycles jumping integration scheme. Appl Mech Mater 784:308–316. https://doi.org/10.4028/www.scientific.net/AMM.784.308

Ladeveze J (1985) Methodes constructives pour la simulation numerique d'ecoulement de fluides non newtoniens. Phd thesis, Universite Paris 6

Landau E (1909) Handbuch der Lehre von der Verteilung der Primzahlen. Teubner, Leipzig

Lesne PM, Savalle S (1989) An efficient cycles jump technique for viscoplastic structure calculations involving large number of cycles. ONERA, TP no. 1989-138, p 13

Morachkovskii O (1992) Nonlinear creep problems of bodies under the action of fast field oscillations. Int Appl Mech 28(8):489–495. https://doi.org/10.1007/BF00847065

Oskay C, Fish J (2004) Fatigue life prediction using 2-scale temporal asymptotic homogenization. Int J Numer Meth Eng 61:326–359. https://doi.org/10.1002/nme.1069

Sanders JA, Verhulst F (1985) Averaging methods in nonlinear dynamical systems. Springer, New York, Applied Mathematical Sciences. https://doi.org/10.1007/978-1-4757-4575-7

Siemens-Energy: Innovative power plant solutions (2020). https://www.siemens-energy.com/global/en.html. Accessed 02 March 2023

Chapter 26
A System for Evaluating the Health Status of Traction Battery Systems in Electric-Powered Vehicles

Andreas Herkommer, Dirk Schuhmann, and Markus Merkel

Abstract Mobility exists in a dynamic and constantly evolving environment. The shift towards electrified vehicles is accelerating, particularly within the passenger car and commercial vehicle segments. A significant driver of this change is the impending ban on new registrations of vehicles powered solely by combustion engines, set to take effect in Europe in 2035. However, partially and fully electrified vehicles remain highly price-sensitive due to the cost of traction batteries. The service life of these batteries diminishes over time and is largely dependent on the specific usage profile. Currently, it is not possible to determine key parameters of traction batteries, such as the state of health or service life, in an independent and standardized manner. This article presents a methodological approach for independently determining traction battery data. An online-based evaluation tool processes and analyzes batteries life date and historical data using an AI-based evaluation method to draw conclusions about battery longevity. Based on past and anticipated user profiles, this tool can forecast the primary service life of the battery in the dynamic environment of the vehicle, as well as its secondary service life in a static environment as an energy storage device.

A. Herkommer (✉) · M. Merkel
Institute for Virtual Product Development, Aalen University of Applied Science, Beethovenstr. 1, 73430 Aalen, Germany
e-mail: andreas.herkommer@hs-aalen.de

M. Merkel
e-mail: markus.merkel@hs-aalen.de

D. Schuhmann
Ingenieursbüro Weigel, Hauptstr. 48, 74595 Langenburg, Germany
e-mail: dirk.schuhmann@ib-weigel.de

H. Altenbach et al. (eds.), *Lectures Notes on Advanced Structured Materials 3*,
Advanced Structured Materials 226, https://doi.org/10.1007/978-3-031-77403-4_26

26.1 Introduction

Battery technology is advancing rapidly, with enhanced energy densities, shorter charging times, and the growing significance of electrified mobility driving the emergence of increasingly diverse and complex battery topologies. The advancements are not only differentiated in terms of performance but also reflects the complexity of modern battery systems. The trend towards increased usage of electrically powered vehicles is evident in various forecast models, which predict continuous and exponential growth (Thielmann et al. 2020; Statista 2030). This development is largely driven by European Union regulations mandating a significant transition to low- or zero-emission drive technologies to meet carbon dioxide reduction targets (Bundesumweltministerium 2024). By 2035, Europe will ban new registrations of purely combustion-powered vehicles. In this context, traction batteries have become the dominant technology, enabling emission-free mobility through the use of renewable energy sources (Bundesumweltministerium 2024). Additionally, the integration of degraded batteries in second-life applications is becoming increasingly important, offering opportunities to extend battery service life, enhance resource efficiency, and promote the sustainability of the entire battery system (Europäisches Parlament 2024).

The evaluation of traction batteries, which serve as both energy storage units and drive systems for electric vehicles, is thus gaining increasing importance. Currently, lithium-ion batteries dominate the market for traction batteries in electrically powered vehicles (Thielmann et al. 2020). These batteries consists of multiple interconnected cells that store energy on an electrochemical basis (Bauer 2017; Korthauer 2013). The interaction of all cell components and their material properties is crucial for determining the characteristics of traction batteries (Korthauer 2013; Forouzan et al. 2016). Experience from the mobile electronics industry provides valuable insights and measures for optimizing service life, but these are only a secondary priority in the case of new procurement due to low financial costs (Thielmann et al. 2020). Given the high acquisition costs of electrically powered vehicles and the expense of battery replacement, it is essential to test traction batteries and accurately predict their service life and durability (Sonnenholzner 2024). Recording and monitoring battery parameters, such as the state of health (SoH) and the state of charge (SoC), are therefore central aspects of ongoing research.

Battery health typically describes the usable capacity in relation to the total capacity, as represented by the equation

$$(1)\boldsymbol{SoH} = \frac{\boldsymbol{C_{rem}}}{\boldsymbol{C_{bat}}}$$

where C_{rem} denotes the usable capacity and C_{bat} represents the total capacity (Korthauer 2013; Sonnenholzner 2024).

The state of charge (SoC) indicates the current charge level of a battery and can be described by the following formula:

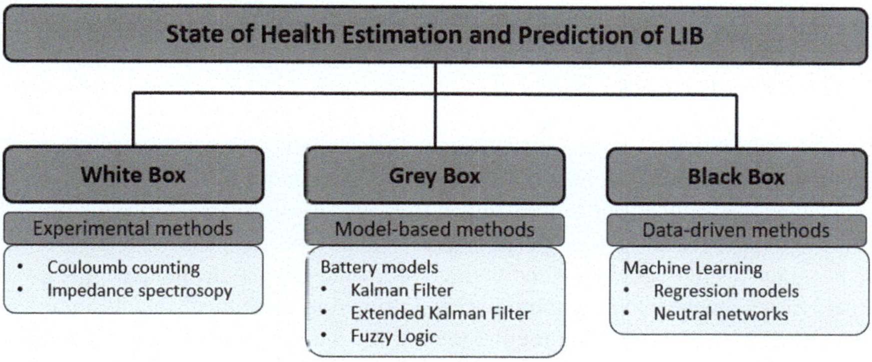

Fig. 26.1 State of health estimaticn an prediction of lithium-ion-batteries

$$(2)\boldsymbol{SoC}(\boldsymbol{t}) = \boldsymbol{SoC}(\boldsymbol{t} - 1) + \int_{0}^{t} \frac{\boldsymbol{I}}{\boldsymbol{C_{rem}}}$$

In this equation, *t* represents time, *SoC(t)* is the state of charge at time t, *SoC(t*–1) is the initial state of charge, and *I* is the current flowing into or out of the battery. The usable battery capacity is denoted by C_{rem}.

In the context of batteries, degradation is differentiated between calendar and cyclical life (Korthauer 2013). Calendar life refers to the degradation process that occurs over time, independent of usage, resulting from various chemical reactions within the cell that deteriorate performance and capacity (Korthauer 2013). In contrast, cyclical life pertains to the degradation associated with the number of charge and discharge cycles, which directly correlates with capacity reduction. Besides capacity reduction, battery degradation is also indicated by increasing internal resistance (Newe et al. 2023). The assessment of a battery's state of health encompasses a variety of methods and technologies, as illustrated in Fig. 26.1, and these can typically be classified into three distinct groups (Yao et al. 2021).

Electrochemical tests, such as impedance spectroscopy and Coulomb counting, have demonstrated exceptional accuracy for evaluating battery cells under controlled laboratory conditions (Haußmann and Melbert 2017; Käbitz 2016; Kohs 2022). Impedance spectroscopy determines the internal resistance of the battery by applying a variable test signal, which necessitates direct access to individual battery cells for precise measurement and analysis of electrical signals. However, this method is economically impractical for in-situ applications due to the limited accessibility without complete disassembly of the battery system. Coulomb counting, on the other hand, measures the remaining battery capacity by integrating the electrical current over time. Nevertheless, in practical applications, traction batteries are neither fully charged nor fully discharged to ensure their protection, which significantly impacts the accuracy and reliability of this methodology (Sonnenholzner 2024). Alternative approaches for determining the state of health of batteries focus on various modeling techniques (Yao et al. 2021). These include model-based approaches that integrate physical or chemical processes within the batteries to understand degradation mechanisms at the cell and system levels. Another option is empirical models, which rely on experimental data to establish mathematical relationships between operating parameters and the SoH. Additionally, data-driven approaches utilize machine learning algorithms to capture complex non-linear relationships between measurement data and the SoH (Kohs 2022; Birkl 2017; Deng et al. 2022). Although these models are generally easier to implement, they provide less insight into the underlying degradation mechanisms. Despite significant advancements in battery research, the mechanisms of degradation and their complex interplay at the microscopic level remain largely unclear. Various approaches to studying these mechanisms are time-consuming and costly, as they require testing the aging effects of specific battery cells under all relevant combinations of operating conditions.

Current methods for determining SoH have predominantly been tested at the cell level and are challenging to apply to traction batteries (Newe et al. 2023). These methods are typically based on full cycles under controlled laboratory conditions and do not adequately account for all relevant factors such as ambient temperature and usage conditions. This limits their application for the dynamic use of traction batteries under real conditions as well as for their static use in the second-life area (Pires Mimoso 2023; Zhang et al. 2023). In addition to the difficulties in recording and monitoring battery parameters, determining the SoH during operation is challenging. Vehicle and battery manufacturers provide limited information, often only considering the gross and net capacity of the battery as relevant factors (Hubik 2021). At the same time, monitoring capacity loss and the resulting reduction in health is further complicated by dynamic loads such as recuperation and other operational influences. Electric vehicles are subject to various aging processes that depend on usage intensity and other general conditions (Hubik 2021). Extreme driver behaviors, such as prolonged driving under full load or extended parking under unfavorable conditions, accelerate the aging of traction batteries. Additionally, fast charging at rates exceeding 50 kW imposes considerable strain on the battery (Sonnenholzner 2024). Over time, traction batteries experience a decline in performance, range, and value. Degradation or potential faulty battery cells can also pose safety risks (Hubik 2021).

The demand for a reliable evaluation of battery systems, considering their unique characteristics and complex structures, is increasingly evident. However, standardized procedures for assessing the SoH of traction batteries are currently lacking. Consequently, manufacturers employ various time-consuming methods to evaluate battery conditions, leading to disparate estimations (Thielmann et al. 2020; Sonnenholzner 2024). To mitigate this uncertainty, car manufacturers offer warranties for traction batteries, typically covering up to eight years or 160,000 km. During this period, the battery is guaranteed for replacement if its remaining capacity falls below 70–80% of its original capacity. Some manufacturers extend these warranties up to ten years or one million kilometers (Sonnenholzner 2024). A comprehensive analysis of the factors' influencing SoH is crucial for a thorough understanding and evaluation of battery degradation in operation (Yao et al. 2021). When a battery's capacity drops to 80% of its original value, it is no longer suitable for dynamic use in vehicles and must be replaced. Systematic classification of traction batteries can facilitate their integration into second-life applications, optimizing economic and ecological benefits (Hubik 2021). This paper presents a methodical approach for the independent determination of traction battery data. An online-based evaluation tool processes and analyzes battery service life data and historical data using an AI-supported method, enabling reliable predictions about battery longevity. By leveraging past and future user profiles, the tool can forecast service life in both dynamic vehicle environments and second-life applications.

26.2 Methods

The aim of the research project is to develop a battery test system that enables the autonomous and manufacturer-independent recording and transparent interpretation of operating data from traction batteries in electric vehicles. The intended system should allow a reliable and fast evaluation of the batteries within a few minutes in order to ensure their suitability for short and medium-term applications in electromobility. In addition, a forecast of the utilization potential in second-life applications of the traction battery is made possible with the highest possible accuracy. The test system provides a new way of quickly and accurately determining original data via the OBD interface of the vehicle during operation. Figure 26.2 illustrates the principle and the different elements of the battery test system.

Compact laboratory cycles and test cycles on the road are developed for data recording, which enable a quick and precise assessment of the battery condition. Different traction batteries as well as the influences of different usage profiles are considered. A data-driven approach with AI is used to investigate and interpret the charging and discharging curves, which enables rapid condition analysis. This approach focuses on selected load points and makes it possible to obtain precise information about the condition of the battery more quickly and efficiently by using a limited number of measurement points instead of complete cycles. Interdependencies between a few measuring points and the charging and discharging characteristics are to be investigated in order to enable an evaluation of the traction battery based on individual characteristics of the charging and discharging characteristics using AI methods. In addition, the investigation of the correlation between the SoH and the SoC is crucial for the development of a reliable evaluation method.

The data obtained during operation will be used to create an evaluation database with AI algorithms that includes various charging and discharging characteristics and their significant points. This database is used not only to detect ageing effects, but also to predict performance data. The system aims not only to evaluate the current condition of the battery, but also to make reliable forecasts about its future usage potential, particularly with regard to second-life applications.

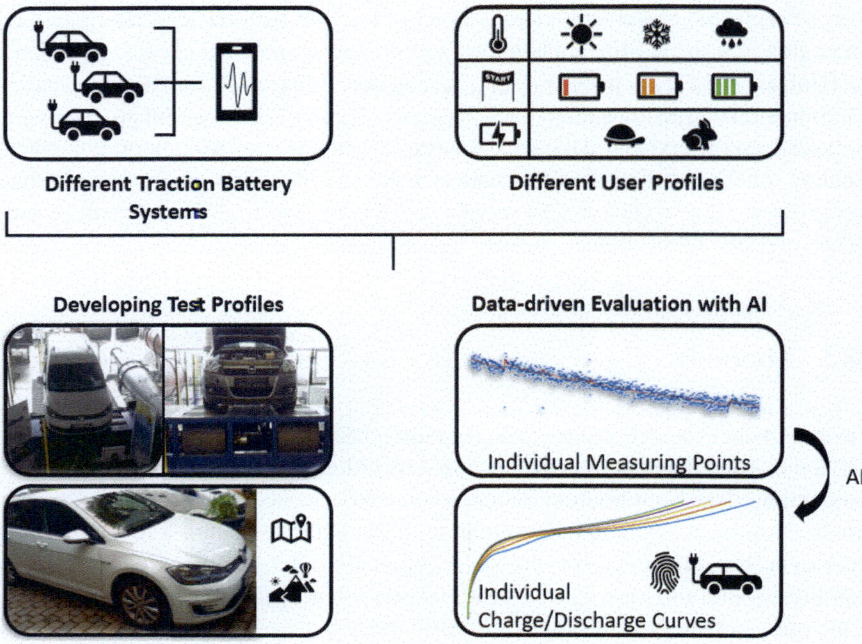

Fig. 26.2 Overview of the planned battery test system

26.3 Developing Test Profiles and Interpretation of the Battery Data

During the detailed tests and measurements, a chassis dynamometer manufactured by MAHA Maschinenbau Haldenwang GmbH is used to assess electric vehicles under controlled conditions. At the same time, the traction batteries are evaluated through tests under real road conditions on the road. A key challenge is the development of a manufacturer-independent measurement procedure that ensures transparent and reproducible data acquisition. It is essential to create similar thermal, physical, electrical and metrological conditions for precise recording of the characteristic curves. The relevant measured variables include battery current, battery voltage, SoC and battery temperature in particular, in order to be able to accurately classify the measurement results. In order to ensure a high level of accuracy regarding the battery condition, consistent conditions must be maintained during the test drives or influencing factors must be integrated into a calculation model.

A defined real drive test cycle is required in order to be able to precisely record the discharge current during a test drive. A diagnostic device is required to record the relevant battery parameters during the test drives. The diagnostic tool VCDS from the manufacturer Ross-Tech, which is compatible with all Volkswagen Group models, is used for this purpose. The device is connected via the vehicle's OBD interface and accesses various control units in order to read the measured values from the sensors. The primary data of the traction battery is recorded via various data paths and evaluated individually so in order to visualize basic data and usage profiles of the traction battery that are difficult to access. When developing a test cycle under real conditions and to detect ageing effects, various criteria must be taken into account. The reproducibility of the measurement conditions is of central importance in order to ensure comparability of the measurements. Deviations, such as fluctuations in the ambient temperature, must either be controlled or taken into account in the evaluation to ensure good results. These measures guarantee that measurement deviations are not due to external influences, but to the actual characteristics of the traction battery. Additional measuring devices can also be used, e.g. to record the outside temperature during the test trials. A separate measuring device is also required to record GPS data as well as speed and altitude profiles. Based on the cycle on the road, these test cycles can be simulated on a chassis dynamometer. The speed profile as a function of time and the height profile are used to simulate the original route profiles.

The recording of battery data and the creation of an evaluation database for various characteristic curves and critical points enables the identification of ageing effects and sub-optimal use, allowing a precise evaluation of the traction batteries in terms of performance and service life. Various vehicles are used for the investigations in order to systematically analyze the topologies and charging and discharging curves of the traction batteries. As the current SoH of the vehicles cannot be read out directly, the test vehicles are selected based on their mileage in order to cover a wide range of mileages and ageing conditions. Three different types of electric vehicle are currently included in the study (refer to Table 26.1). By procuring two identical vehicles with differently aged traction batteries, the relationship between ageing effects and changes in the characteristic curves can be investigated. A vehicle with a lower mileage and healthier battery is the VW e-up!, while the comparison vehicle with a lower SoH is characterized by a higher mileage.

The transition of traction batteries to second-life applications is also being examined in detail in order to enable predictions to be made about their potential use (see Fig. 26.3). The system developed and implemented for the transparent assessment of battery condition is based on original data and is aimed at commercial users, fleet operators and testing and monitoring organizations. The goal is to achieve a quick evaluation, providing immediate solutions for assessing rented electric vehicles. In

Table 26.1 Types of electric vehicles

Characteristics	VW e-up! – 2022	VW e-up! – 2021	VW e-golf
Year of manufacture	2022	2021	2020
Engine power [kW]	61		100
Engine torque [Nm]	212		290
Capacity [kWh]/Range [Km]	36.8/260		35.8/200
Unladen weight [Kg]	1248		1615
Mileage [Km]	2001	32,539	20,550

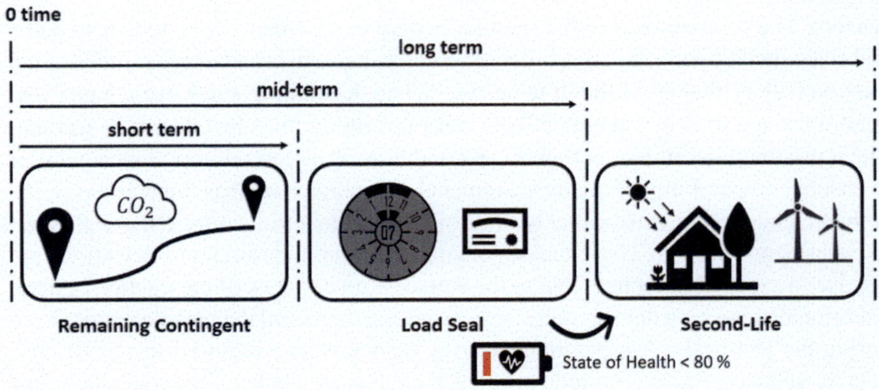

Fig. 26.3 Evaluation and application potential of traction batteries

addition, the system allows more precise predictions regarding range and maintenance requirements. A detailed analysis provides medium-term insights into battery performance as well as long-term forecasts on the potential for secondary use of the batteries. The introduction of an official label ensures that the performance of the traction batteries is guaranteed in continuous use and that the road safety of the vehicles is maintained in the long term. In the context of second-life scenarios, the residual use potential of traction batteries is considered. A defined limit is set above which the battery is no longer suitable for primary use. The remaining utilization potential is then recommended for various second-life applications. The extrapolation of condition data to predict the remaining service life represents a further challenge for which suitable physical-technical models must be developed.

26.4 Summary

The necessity of a reliable evaluation of the traction battery by determining its state of health based on its individual characteristics and complex structures becomes clear. The present study has provided significant insights into a new method for determining the state of health of electrically powered vehicles and highlighted important implications. Despite this progress, some questions remain that should be clarified in future studies and measurements on electric vehicles, whether on the chassis dynamometer or on the road. By specifying the different ageing mechanisms and their influence on the charging and discharging cycles, correlations can be interpreted more precisely and the state of health of the traction battery can be determined. By comparing the recorded discharge and charge characteristics of batteries that are in different stages of ageing, the relationships between ageing effects and changes in the characteristics can be better understood. Systematic classification of traction batteries can enable targeted and simpler integration into second-life applications, thereby achieving optimum economic and ecological benefits. In order to train the data-driven approach using artificial intelligence and achieve comprehensive results, a sophisticated algorithm is required that interpolates measurement results from different user profiles to a comparable level. Furthermore, the transformation of the data must not compromise the accuracy of the measurement results.

References

Bauer M (2017) Elektrische und mechanische Verfahren zur Detektion von Alterungseffekten in Lithium-Ionen-Batterien. Dissertation, Technical University of Munich (2017)

Birkl C (2017) Diagnosis and prognosis of degradation in lithium-ion batteries. Dissertation, University of Oxford

Bundesumweltministerium: Warum überhaupt Elektromobilität? https://www.bmuv.de/themen/verkehr/elektromobilitaet/ueberblick-elektromobilitaet. Accessed 14 June 2024

Bundesumweltministerium: Die EU-Verordnungen zur Verminderung der CO2-Emissionen von Straßenfahrzeugen – Gesetze und Verordnungen https://www.bmuv.de/gesetz/die-eu-verordnungen-zur-verminderung-der-co2-emissionen-von-strassenfahrzeugen. Accessed 14 June 2024

Deng Z, Hu X, Li P (2022) Data-driven battery state of health estimation based on random partial charging data. IEEE T Power Electr 37:5021–5031. https://doi.org/10.1109/TPEL.2021.3134701

Europäisches Parlament: Neue EU-Vorschriften für nachhaltigere und ethisch bedenklose Batterien. https://www.europarl.europa.eu/topics/de/article/20220228STO24218/neue-eu-vorschriften-fur nachhaltigere-und-ethisch-bedenkenlose-batterien. Accessed 14 June 2024

Forouzan MM, Chao C-W, Bustamante D (2016) Experiment and simulation of the fabrication process of lithium-ion battery cathodes for determining microstructure and mechanical properties. J Power Sourc 312:172–183. https://doi.org/10.1016/j.jpowsour.2016.02.014

Haußmann P, Melbert J (2017) Spannungsgeregelte Impedanzspektroskopie mit beitbrandigen Anregungssignalen für Lithium-Ionen-Zellen in Kfz-Anwendungen. TECH MESS 84:411–425. https://doi.org/10.1515/teme-2017-0018

Hubik F (2021) Hauptuntersuchungen bei Elektroautos lückenhaft – Experten schlagen Alarm. Preprint at https://www.handelsblatt.com/mobilitaet/elektromobilitaet/tuev-und-dekra-hauptuntersuchung-bei-elektroautos-lueckenhaft-experten-schlagen-alarm/27864550.html (2021)

Käbitz SR (2016) Untersuchung der Alterung von Lithium-Ionen-Batterien mittels Elektroanalytik und elektrochemischer Impedanzspektroskopie. Dissertation, RWTH Aachen University

Kohs A (2022) Batteriemodell zur Prädiktion des Gesundheitszustands von Lithium-Ionen-Batterien. Springer Fachmedien Wiesbaden, Wiesbaden

Korthauer R (2013) Handbuch Lithium-Ionen-Batterien. Springer, Berlin

Newe H, Schmuck N, Annuth R (2023) Vergleich von Konzepten zur datenbasierten Ermittlung der Alterung von Fahrzeugbatterien. Helmut-Schmidt-Universität, Universität der Bundeswehr Hamburg

Pires Mimoso R (2023) Maschinelles Lernen in der Batteriedatenanalyse. Master thesis, FH Vorarlberg

Sonnenholzner J (2024) Warum der Akkuzustand beim E-Auto oft für Rätsel sorgt. Tagesschau.de

Statista: Elektroautos – Anteil am Pkw-Bestand 2030. https://de.statista.com/statistik/daten/studie/1202904/umfrage/anteil-der-elektroautos-am-pkw-bestand-in-deutschland/. Accessed 14 Jan 2024

Thielmann A, Wietschel M, Funke S (2020) Batterien für Elektroautos – Faktencheck und Handlungsbedarf – Sind Batterien für Elektroautos der Schlüssel für eine nachhaltige Mobilität der Zukunft?, Frauenhofer ISI

Yao L, Xu S, Tang A (2021) A review of lithium-ion battery state of health estimation and prediction methods. WEVJ 12:113. https://doi.org/10.3390/wevj12030113

Zhang M, Yang D, Du J (2023) A review of SOH predictions of Li-Ion batteries based on data-driven algorithms. Energies 16. https://doi.org/10.3390/en16073167

Chapter 27
Energetic Analysis of CO_2 Conversion with MW-Plasmalysis

Tanja Hasenjäger and Walter Czarnetzki

Abstract In terms of the urgent global challenges of climate change and the increasing concentration of greenhouse gases in the atmosphere (Friedlingstein in Global Carbon Budget, 2023; Lee and Romero in Climate Change 2023: Synthesis Report, 2023), the reduction of CO_2 emissions is the focus of efforts by scientists, politicians and industry representatives worldwide (Davis and Lewis in Net-zero emissions energy systems, 2018). The decomposition of CO_2 by using plasmalysis, in particular microwave assisted plasmalysis, is an innovative process that has attracted significant interest in recent years as it is potentially more efficient and environmentally friendly than traditional methods (Silva et al. in Study of CO_2 Decomposition in Microwave Discharges by Optical Diagnostic Methods, 2016; Ong et al. in Journal of cleaner production 336, 2022; Zhang and Zhang in Environmental Science and Technology, 2021). For the evaluation of the plasmalysis process, the specific energy input *SEI* and the energy efficiency η among other key values are typically used (Fridman in Plasma Chemistry, 2008). This paper introduces two new key numbers to these indicators, the global energy input *GEI* and the global energy efficiency γ. With the help of these two indicators, parameter dependencies of the plasmalysis process can be better understood in the future, and optimization potentials can be better uncovered. This also makes it easier to compare different plasmalysis processes with each other. Various parameters such as microwave power, gas flow rate, and pressure are systematically varied in order to investigate their influence on the *GEI*.

T. Hasenjäger (✉) · W. Czarnetzki
Faculty Mechanical and Systems Engineering, University of Applied Sciences Esslingen, Kanalstraße 33, 73734 Esslingen, Germany
e-mail: Tanja.Hasenjaeger@hs-esslingen.de

W. Czarnetzki
e-mail: Walter.Czarnetzki@hs-esslingen.de

H. Altenbach et al. (eds.), *Lectures Notes on Advanced Structured Materials 3*, Advanced Structured Materials 226, https://doi.org/10.1007/978-3-031-77403-4_27

27.1 Introduction

The splitting of CO_2 offers a promising solution for reducing the concentration of greenhouse gases in the atmosphere (Britun et al. 2017). Innovative technologies such as plasmalysis can be used to convert CO_2 into products such as carbon monoxide and oxygen, see Formula (27.1) (Fridman 2008).

$$CO_2 \leftrightarrows CO + \frac{1}{2}O_2 \tag{27.1}$$

The microwave (MW)-plasmalysis process is carried out in a reactor specially designed to generate and maintain plasmas using microwave radiation. Plasma is an ionized gas that can convert stable molecules into value-added chemicals. For the optimization of the plasmalysis process, three factors are of crucial importance, which have so far been used in the analysis and evaluation of the process. The degree of conversion χ, the specific energy input *SEI,* and the energy efficiency η.

In chemical processes, the degree of conversion χ describes the proportion of the reactant that is converted into a desired product. This is often expressed as a percentage and provides information on how efficient and effective a chemical process is. The gas expansion, which is often disregarded, must be taken into account here (Belov et al. 2018). This results in the degree of conversion χ, where y stands for the molar/volume fraction, see Formula (27.2) (Wanten et al. 2023).

$$\chi = \frac{1 - y_{CO_2}^{out}}{1 + \frac{y_{CO_2}^{out}}{2}} \cdot 100\% \tag{27.2}$$

The specific energy input *SEI* is the energy used per converted unit and is an important indicator of the efficiency of the process. A low *SEI* with the same degree of conversion χ means, that less energy is required to achieve the same result. The *SEI* is formed with the input power P and the flow rate $\dot{V}_{\mathrm{tot}}^{\mathrm{in}}$, shown in Formula (27.3) (Wanten et al. 2023).

$$SEI\left(\frac{kJ}{L}\right) = \frac{P(kW)}{\dot{V}_{\mathrm{tot}}^{\mathrm{in}}\left(\frac{L}{\mathrm{min}}\right)} \cdot 60\left(\frac{s}{\mathrm{min}}\right) \tag{27.3}$$

where the *SEI* is usually expressed in kJ/mol, see formula (27.4) (Wanten et al. 2023). Here V_m is the molar volume of the ideal gas.

$$SEI\left(\frac{\mathrm{kJ}}{\mathrm{mol}}\right) = SEI\left(\frac{kJ}{L}\right) \cdot V_m\left(\frac{L}{\mathrm{mol}}\right) \tag{27.4}$$

V_m depends on the calibration of the mass flow controller (MFC). As the MFC used, has been calibrated to 'EU normal', $V_m = 22.41L/\mathrm{mol}$ applies (Wanten et al. 2023). The *SEI* and the degree of conversion χ are used to determine the energy efficiency η, see Formula (27.5) (Wanten et al. 2023). ΔH° is the standard reaction enthalpy, which is + 283 kJ/mol for CO_2 splitting (Snoeckx and Bogaerts 2017).

$$\eta(\%) = \frac{\chi(\%) \cdot \Delta H^\circ\left(\frac{\mathrm{kJ}}{\mathrm{mol}}\right)}{SEI\left(\frac{\mathrm{kJ}}{\mathrm{mol}}\right)} \tag{27.5}$$

For comparability within a process and between different processes, however, it is also relevant to consider not only the power input P, which was used in Formula (27.3), but also to evaluate the total power applied. The efficiency of the magnetron generator plays a crucial factor in this. The energy required for the peripherals, which is necessary for running the process, is also of major importance. Only when these factors are taken into account, the process can be evaluated and compared in detail and an optimum operating point identified. Energy consumption is a critical factor in the scalability of an industrial process. For this reason, two further factors are defined in this context.

Firstly, the global energy input *GEI* is introduced, see formula (27.6), which is formed in a similar way to *SEI* in formula (27.3), but takes into account the total power required. P_{tot} takes into account the efficiency of the magnetron generator, the cooling power used in the process, and the energy applied by the vacuum pump for the vacuum. The power for the 3-stub tuner and the MFC is minimal and independent of the operating point, which is why it is not taken into account here.

$$GEI\left(\frac{kJ}{L}\right) = \frac{P_{\text{tot}}(kW)}{\dot{V}_{\text{tot}}^{\text{in}}\left(\frac{L}{\text{min}}\right)} \cdot 60\left(\frac{s}{\text{min}}\right) \tag{27.6}$$

Analogously, the *GEI* can also be expressed in kJ/mol like the *SEI*, see formula (27.7).

$$GEI\left(\frac{\mathrm{kJ}}{\mathrm{mol}}\right) = GEI\left(\frac{kJ}{L}\right) \cdot V_m\left(\frac{L}{\mathrm{mol}}\right) \tag{27.7}$$

Furthermore, the factor γ is introduced, see formula 27.8, which expresses the global energy efficiency.

$$\gamma(\%) = \frac{\chi(\%) \cdot \Delta H^\circ\left(\frac{\mathrm{kJ}}{\mathrm{mol}}\right)}{GEI\left(\frac{\mathrm{kJ}}{\mathrm{mol}}\right)} \tag{27.8}$$

In the following sections, the *GEI* is measured and analyzed for various operating points of the MW-plasmalysis system and correlations between the individual parameters are shown. In this analysis, the degree of conversion is initially disregarded.

27.2 Experimental Part

27.2.1 Laboratory Setup

The plasmalysis setup of the University of Esslingen is schematically shown in Fig. 27.1. The power supply for the magnetron generator (B) is provided by the control cabinet (A), which can supply several kW of power. Here the total required power P_{tot} of the magnetron generator is measured. The magnetron generator (B) generates electromagnetic waves in the microwave range, which is setup for a frequency of 915 MHz and is currently tuned for splitting CO_2. The plasmalysis system is equipped with two recirculating chillers (C1)(C2), which extract the generated process heat. (C1) is for cooling the magnetron generator and is setup for 18 degrees. (C2) is for cooling the reaction chamber and is setup for 14°. For both cooling towers the required power was measured. Section (D) of the system consists of a fine tuning of the microwave energy, as well as a 3-stub tuner that adjusts the impedance of the microwaves to minimize the reflective power in the system and therefore destructive interference (Mühleisen 2001). Because the required power of the 3-stub tuner is relatively small and constant at different operating points, it was not incluced in the *GEI*. The microwaves are guided through the surfaguide and reflected at the short circuit. The short circuit is movable and must be adjusted, so that the interference between the outgoing and return microwaves creates a standing wave, which produces a maximum in the reaction chamber. The waveguide walls of the surfaguide converge in the region of the tube to increase the field strength (Spencer and Gallimore 2013). At the narrowest diameter of the surfaguide, it crosses a quartz tube (G) which is transparent to microwave radiation—here the discharge is initiated. In the laboratory setup, microwaves are used with surface wave discharge. The microwaves spread along the interface between the quartz tube and the plasma column. The plasma then absorbs the wave energy. The controller (I) can be used to control the flow rate of the gas being used, which is provided by (F) which symbolizes the storage of the initial product, so in this case CO_2. The mass flow controller also requires very low power and is consistent at different operating points, so like the 3-stub tuner, it wasn't considered in the *GEI*. The gas chromatograph (J) can analyze the composition of the gas products by type and amount of substance, which is necessary for analyzing the conversion rate χ.

27.2.2 Experiments

In order to find out the dependencies of the individual parameters on the required power and ultimately the *GEI*, systematic measurements were carried out, in which initially only one parameter was varied at a time. The following operating point applies to the measurements in the Sects. 3.1 to 3.5: MW-Power $P = 1000W$, flow rate $\dot{V}_{tot}^{in} = 1000sccm$, rotational speed of the vacuum pump $n = 4800\frac{1}{\text{min}}$, pulse

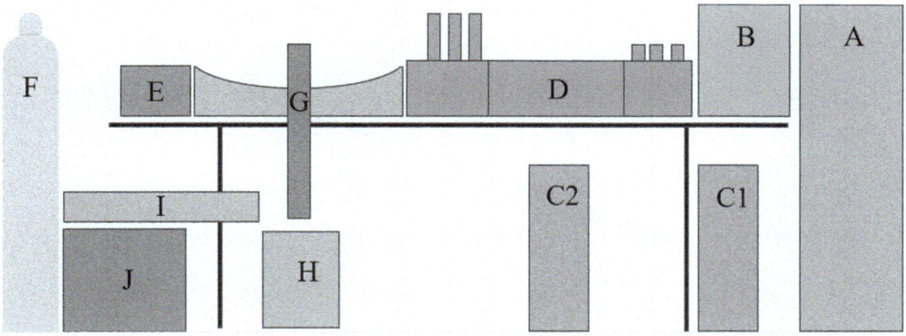

Fig. 27.1 Schematically laboratory setup of the MW-plasmalysis plant

time on $t_{on} = 180\mu s$ and pulse time off $t_{off} = 240\mu s$. Only the parameter discussed in the corresponding chapter is varied. The energy required by the individual units is measured over a period of four minutes for each measurement and an average is calculated. This compensates for possible temporal fluctuations and minimizes potential errors. In addition, all units were measured simultaneously at one operating point in order to reduce possible deviations. In Sect. 3.6, several parameters were then varied at one operating point in order to show interactions between the parameters and to show extreme values of the individual key values at different operating points.

27.3 Results and Discussion

27.3.1 Variation of the MW-Power

In this chapter, the provided MW-power varies between $P = 1000$W and $P = 5000$W. The reflected power of the MW output is always 0. Figure 27.2 shows a linear relationship between the power input and the effective power of the magnetron generator.

Furthermore, it can be seen, that the electrical losses decrease with increasing power and thus the efficiency of microwave generation improves. The explanation for this could be, that there are basic losses that are independent of the power level but become more significant in relation to the power generated (Radoiu and Mello 2022). Microwave generators are also usually designed for higher power levels. In other terms, the circuit design and the electronic components have improved energy transfer and lower internal resistance at high power levels.

In Fig. 3a it can be seen, that higher power requires more energy to cool the reaction chamber and the magnetron generator. This means, that the total power required, as well as the *SEI* and the *GEI*, as shown in Fig. 3b, increases.

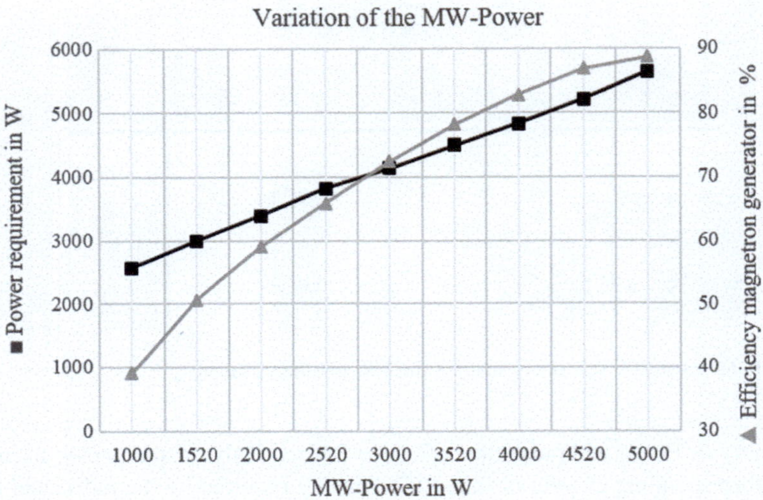

Fig. 27.2 Influence of the produced MW-Power on the required power of the magnetron generator and the efficiency of the magnetron generator

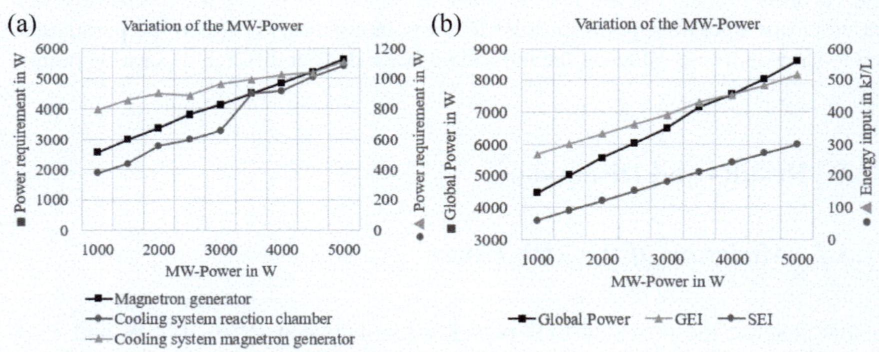

Fig. 27.3 Influence of the generated MW-power on **a** the required power of the magnetron generator and the cooling systems and on **b** the global power and the energy input

27.3.2 Variation of the Flow Rate

In this section, the flow rate was varied from $\dot{V}_{\mathrm{tot}}^{\mathrm{in}} = 100sccm$ up to $\dot{V}_{\mathrm{tot}}^{\mathrm{in}} = 5000sccm$ in steps of $500sccm$.

It can be seen, that the energy requirement of the three components in Fig. 4a remains almost constant. This means that the global power required in Fig. 4b also remains approximately constant. However, the *SEI* and *GEI* change as a result of the change in the flow rate. Due to the initially very low flow rate, the change in *SEI* and *GEI* is very large at the beginning.

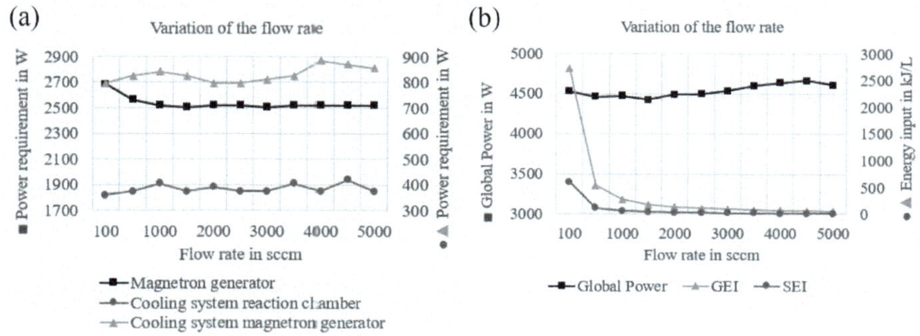

Fig. 27.4 Influence of the flow rate on **a** the required power of the magnetron generator and the cooling systems and on **b** the global power and the energy input

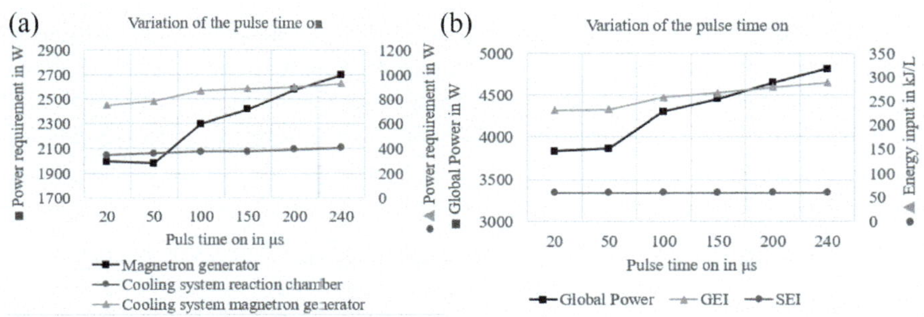

Fig. 27.5 Influence of the pulse time on **a** the required power of the magnetron generator and the cooling systems and on **b** the global power and the energy input

27.3.3 Variation of the Pulse Time On

In this chapter, the pulse time at which microwaves are transmitted is varied from $t_{on} = 20\mu s$ up to $t_{on} = 240\mu s$.

In Fig. 5a can be seen, that the greater the pulse time on, the higher the energy input per time and therefore, as with the increase in power, the greater the cooling requirement and the greater the power required by the magnetron generator. This also increases the total power required in Fig. 5b. The *SEI* does not change, as the pulse time is not taken into account in the calculation, whereas the *GEI* changes.

27.3.4 Variation of the Pulse Time Off

In this section, the pulse time off, during which no microwaves are transmitted, is varied from $t_{off} = 180\mu s$ up to $t_{off} = 1000\mu s$.

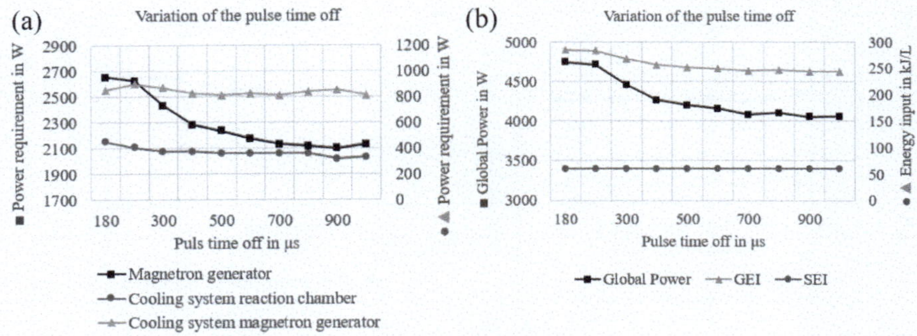

Fig. 27.6 Influence of the pulse time off on **a** the required power of the magnetron generator and the cooling systems and on **b** the global power and the energy input

Analogous to Sect. 3.3, the required power of the magnetron generator decreases with greater pulse time off, see Fig. 6a. A slight drop in the required cooling tower power can be seen. As a result, the total energy requirement decreases accordingly, as can be seen in Fig. 6b. In turn, the *SEI* remains constant and the *GEI* decreases.

27.3.5 Variation of the Rotational Speed of the Vacuum Pump

In the following section, the speed of the vacuum pump is set from $n = 2500\frac{1}{\text{min}}$ up to $n = 4800\frac{1}{\text{min}}$.

Figure 7a shows that changing the speed of the vacuum pump or changing the pressure in the reaction chamber does not change the energy requirement of the three components. However, the *GEI* increases due to the increasing energy requirement of the vacuum pump, see Fig. 7b. Since the power of the vacuum pump is not taken into account in the *SEI*, it remains constant. It can be seen in Fig. 8a that a higher vacuum is generated with a higher speed. This effect can also be reproduced with a lower flow rate. With regard to the power required by the vacuum pump shown in Fig. 8b, this not only increases with a higher speed but also with a higher flow rate. With a higher gas flow rate, the pump has to compress and move more gas volume per unit of time, which requires more energy. Higher flow resistance in the system and higher friction losses can also be a cause.

27.3.6 Variation of Different Parameters

In order to cover more than just the relationship between the energy requirements of individual parameters, the parameters were systematically varied in combination. For this purpose, a low value '–' and a high value '+' were defined for each parameter,

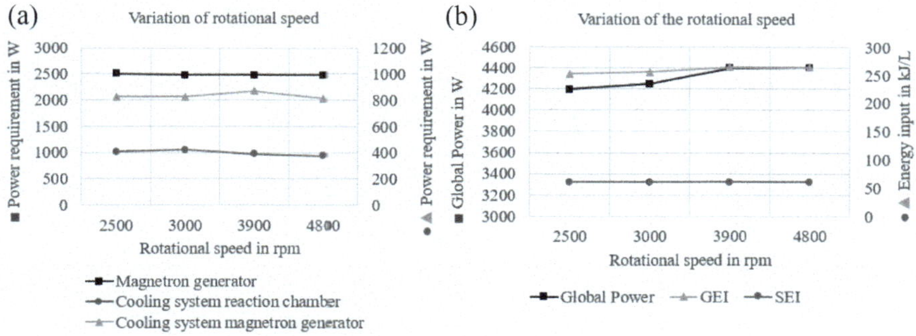

Fig. 27.7 Influence of the rotational speed of the vacuum pump on **a** the required power of the magnetron generator and the cooling systems and on **b** the global power and the energy input

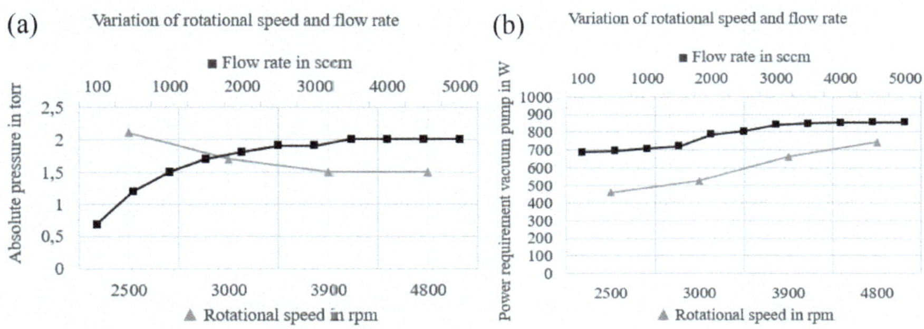

Fig. 27.8 Influence of the flow rate and the speed of the vacuum pump on **a** the generated vacuum and **b** the required power of the vacuum pump

which are listed in Table 27.1. The measurement matrix contains 32 measurements, which are shown in Table 27.1. As in the previous section, each parameter series was measured over four minutes and an average was calculated.

The results with the highest and lowest values of the individual key values are shown in Table 27.2, which however relate to the 32 measurements. The correlation between a high power input of the magnetron generator and the high required cooling power is clearly recognizable, which consequently results in a high total power requirement and a high *GEI*. Experiment no. 3 in the table is an example of the lowest *SEI* value, but experiments no. 8; 14; 15; 18; 25; 28; also had the same *SEI*. Experiment no. 2 is an example of the largest *SEI* value, but experiment no. 2; 11; 12; 13; 20; 24; 26; 29 also had the same *SEI*. This also shows that a much finer differentiation of different measurements is possible with the *GEI* and therefore enables a better evaluation of the process.

Table 27.1 Combined variation of the parameters power, flow rate, rotational speed, pulse time on, and pulse time off

Parameter	Unit	−	+
Power	W	1000	4000
Flow rate	sccm	100	3000
Rotational speed	1/min	2500	4800
Pulse time on	μs	100	180
Pulse time off	μs	180	500

Parameter	Unit	1	2	3	4	5	6	7	8	9	10	11	12	13	14	15	16	17	18	19	20	21	22	23	24	25	26	27	28	29	30	31	32
Power	W	−	+	−	−	−	−	+	−	−	−	+	+	+	−	−	−	+	−	−	+	+	+	+	+	−	+	+	−	+	+	+	+
Flow rate	sccm	−	−	+	−	−	−	+	+	−	−	−	−	−	+	+	−	+	+	−	−	+	+	−	−	+	−	+	+	−	+	+	+
Rotational speed	1/min	−	−	−	+	−	−	−	+	+	−	+	−	−	−	−	+	+	+	+	−	−	−	+	−	−	+	+	+	+	−	+	+
Pulse time on	μs	−	−	−	−	+	−	−	−	+	+	−	+	−	+	−	−	−	+	+	+	+	−	+	+	+	−	+	+	+	+	−	+
Pulse time off	μs	−	−	−	−	−	+	−	−	−	+	−	−	+	−	+	+	−	−	+	+	−	+	−	+	+	+	−	+	+	+	+	+

Table 27.2 Extreme values of various key values of the experimental series

Parameter	Unit	Smallest value		Highest value	
		Nr	Value	Nr	Value
Magnetron generator	W	16	1935	12	6300
Cooling system magnetron generator	W	8	735	12	1215
Cooling system reaction chamber	W	3	300	12	1290
Vakuum pump	W	30	435	32	840
Global Power	W	6	3624	12	9252
GEI	kJ/L	15	72,9	12	5551
SEI	kJ/L	3	20	2	2400

27.4 Conclusions and Remarks

A lower energy consumption has an impact on the economic profitability of the process, reduces the ecological footprint (Luo and Sun 2024), and is crucial for the sustainable development of the plasmalysis process. This applies to the use of renewable as well as non-renewable energies. Several measurements have shown that a consideration of the energy required globally, as well as the individual components and their interaction, is important for optimizing the process. The newly introduced key value *GEI* now enables a global view of the system and better comparability of different operating points as well as the comparability between different plasmalysis systems. All in all, the degree of conversion and the global energy efficiency must also be considered for the final evaluation of the operating points, which then allows a detailed statement to be made about the process. The system can then be optimized more effectively based on these findings.

References

Belov I, Vermeiren V, Paulussen S, Bogaerts A (2018) Carbon dioxide dissociation in a microwave plasma reactor operating in a wide pressure range and different gas inlet configurations. J CO_2 Utilization. https://doi.org/10.1016/J.JCOU.2017.12.009

Britun N, Chen G, Silva T, Godfroid T, Delplancke-Ogletree M, Snyders R (2017)Enhancing the greenhouse gas conversion efficiency in microwave discharges by power modulation. https://doi.org/10.5772/67875

Davis SJ, Lewis NS (2018) Net-zero emissions energy systems 360(6396). https://doi.org/10.1126/science.aas979

Fridman A (2008) Plasma chemistry.https://doi.org/10.1017/CBO9780511546075

Friedlingstein P (2023) Global Carbon Budget 2023. https://doi.org/10.5194/essd-15-5301-2023

Lee H, Romero J (2023) Climate Change 2023: Synthesis Report. https://doi.org/10.59327/IPCC/AR6-9789291691647

Luo H, Sun Y (2024) The impact of energy efficiency on ecological footprint in the presence of EKC: evidence from G20 countries. Energy 304. https://doi.org/10.1016/j.energy.2024.132081

Mühleisen R (2001) A guide for microwave application

Ong MY, Nomanbhay S, Kusumo F, Show PL (2022) Application of microwave plasma technology to convert carbon dioxide (CO_2) into high value products. J Clean Prod 336.https://doi.org/10.1016/j.jclepro.2022.130447

Radoiu M, Mello A (2022) Technical advances, barriers, and solutions in microwave-assisted technology for industrial processing. Chem Eng Res Des 181.https://doi.org/10.1016/j.cherd.2022.03.029

Silva T, Britun N, Godfroid T, Snyders R (2016) Study of CO_2 decomposition in microwave discharges by optical diagnostic methods. https://doi.org/10.5772/60692

Snoeckx R, Bogaerts A (2017) Plasma technology—a novel solution for CO_2 conversion?. https://doi.org/10.1039/c6cs00066e

Spencer LF, Gallimore AD (2013) CO_2 dissociation in an atmospheric pressure plasma/catalyst system: a study of efficiency. Plasma Sources Sci Technol 22

Wanten B, Vertongen R, de Meyer R, Bogaerts A (2023) Plasma-based CO_2 conversion: how to correctly analyze the performance?. J Energy Chem. https://doi.org/10.1016/J.JECHEM.2023.07.005

Zhang S, Zhang Y, et al (2021) Microwave plasma systems for advanced environmental applications. Environ Sci Technol. https://doi.org/10.1021/acs.est.1c04739

Chapter 28
A Review of Renewable Energy Production and Storage Systems for Households

Leonardo Rodrigues, João M. P. Q. Delgado, Adélio Mendes, and Ana S. Guimarães

Abstract The construction industry is increasingly focusing on sustainable and environmentally friendly solutions to reduce its carbon emissions and material consumption. The emergence of economically viable renewable technologies has sparked hope for decarbonizing the sector. In the past decade, discussions regarding building energy efficiency have been conducted in governmental bodies such as the EU Commission, leading to the proposal and approval of protocols like nearly zero energy buildings. A thorough literature review on the application of renewable technologies in households and operational management was performed. Due to the intermittent nature of solar energy, integrating photovoltaic panels with storage technologies offers users greater energy independence from the grid, leading to monthly savings on energy costs. In addition, these systems help to mitigate peak loads on the electricity grid. However, these technologies have a high initial investment, which is unattractive to many users. As such, the correct sizing of the equipment for a household is essential to make it attractive to the end-user. Simulation tools are non-expensive fundamental tools for insight into the topic. Advancements in these and home energy management systems are covered in this work.

L. Rodrigues (✉) · A. Mendes
LEPABE—Laboratory for Process Engineering, Environment, Biotechnology and Energy, Faculty of Engineering, University of Porto, Rua Dr. Roberto Frias, 4200-465 Porto, Portugal
e-mail: lrodrigues@fe.up.pt

ALiCE—Associate Laboratory in Chemical Engineering, Faculty of Engineering, University of Porto, Rua Dr. Roberto Frias, 4200-465 Porto, Portugal

A. Mendes
e-mail: mendes@fe.up.pt

L. Rodrigues · J. M. P. Q. Delgado · A. S. Guimarães
CONSTRUCT-LFC—Building Physics Laboratory, Faculty of Engineering, Universidade Do Porto, Rua Dr. Roberto Frias, S/N, 4200-465 Porto, Portugal
e-mail: jdelgado@fe.up.pt

A. S. Guimarães
e-mail: anasofia@fe.up.pt

H. Altenbach et al. (eds.), *Lectures Notes on Advanced Structured Materials 3*,
Advanced Structured Materials 226, https://doi.org/10.1007/978-3-031-77403-4_28

28.1 Introduction

The soaring new 20s introduced several challenges to humankind. First came the COVID-19 pandemic, which completely broke the *status quo* and dynamics of the world. Then came the current energy crisis in Europe. Adding to this, ongoing issues such as climate change and the observed temperature increase around the world call for action (Allen et al. 2022).

In the face of challenges, changes are made. Regarding climate change, the Intergovernmental Panel on Climate Change (IPCC) has made an assessment on the topic and proposed different net emission mitigation options. Across the different sectors, the theme of adopting renewable and cleaner energy sources as well as carbon capture is mentioned (Summary for Policymakers 2022). In fact, the shift in the energy paradigm towards renewable sources has been around for at least half a century (Brandon and Kurban 2017). For that, huge efforts have been made in researching new technologies and develop new solutions as replacements for the well-established fossil fuels. Treaties like the Paris Agreement show the necessity of this change as well as its urgency (Nations 2015). The European Green Deal aims to have "no net emissions of greenhouse gases in 2050 (2019). The strategy was initially proposed in the communication "A Clean Planet for all" (Commission). Paired with these actions, the discussion around sustainability became prevalent in 1992 when the United Nations declared it as the guiding principle for the twenty-first century (Walter Klöpffer 2014). The report "Our Common Future" discussed the future of mankind and how sustainable development would be fundamental across all sectors (Our Common Future 1985).

As the world population keeps increasing, there is a continuous need for the construction of new buildings. It is known that the construction sector is one of the largest users of energy and material resources, as well as one of the big contributors to pollution (Akadiri et al. 2012). Regarding CO_2 emissions, reports show that, in 2016, energy use in buildings was responsible for 17.5% of the global greenhouse gas emissions, with 10.9% being done by residential buildings (Ritchie et al. 2020). In 2020, households were responsible for 28% of the final energy consumed, compared to 26.5% in 2019. This change can be explained by the lockdown measures implemented during this period, further supported by the decline of energy consumed in the transportation sector, going from 30.8% in 2019 to 28.4% in 2020 (Eurostat 2024).

As such, the sector has evolved with solutions to mitigate the total energy needed, decarbonization as well as reducing the material wastes during construction. This has been supported by governmental organizations that keep creating new policies. The European Directive 2010/31/EU stated that all new public buildings must be nearly zero energy buildings (NZEBs) after the 31st of December 2018, expanding the requirements to all buildings by the end of 2020 (European Parliament CotEU 2010). In fact, the European Commission goes further, proposing that by 2030 all new buildings must be zero-emission (ZEB).

A nearly zero energy building is characterized as having extremely high energy efficiency, with almost zero or very low energy requirements being met as much as feasible by renewable energy sources generated on-site or locally (D'Agostino et al. 2021). This can be achieved through passive solutions such as thick insulation and highly efficient heating, ventilation, and air conditioning (HVAC) systems and pairing photovoltaic panels with energy storage systems (Attia et al. 2017; Becchio et al. 2015; Chastas et al. 2016).

Given the different weather observed in the different Member States of the European Commission, there must be different specifications for NZEBs in different locations. As such, there is a lack of uniformity in the quantification of values for primary energy use in NZEBs.

Another major hurdle in the decarburization of the building industry is the fact that most of the building stock was built in the last century when there was no legislation regarding energy efficiency. For example, looking at Portugal, the data from Fig. 28.1 highlights that roughly two-thirds of the current buildings were built before 1990, which are expected to not be energy efficient.

Adding to this, a study conducted by Statistics Portugal revealed that 35.8% of Portugal's buildings need some sort of repairs, where 9.4% need medium repairs and 4.6% deep repairs (Estatística and Buildings (No.) by Geographic location at Census date 2013).

As time went by, the evolution of the concept of NZEBs has changed to incorporate several aspects. Originally only encompassing energy efficiency and renewable energy production, nowadays NZEBs also take into consideration important topics like being economically viable, incorporation of electric vehicles into the household ecosystem, and social causes like occupant health and comfort as well as fighting energy poverty (D'Agostino et al. 2021).

As such, a significant push for renewable technologies in buildings has been made. Given the intermittent nature of solar energy, energy storage solutions must also be

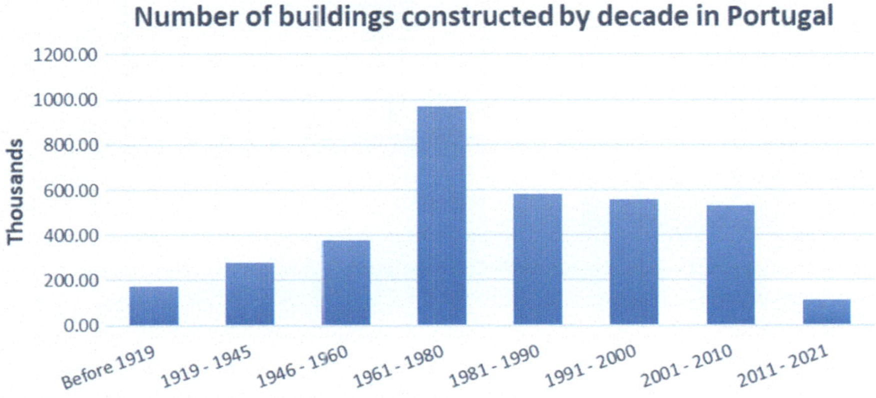

Fig. 28.1 Number of buildings constructed by decade in Portugal. Data taken from Estatística and Buildings (No.) by Geographic location at Census date (2013)

considered. Pairing these technologies with an efficient management system can reduce the need for energy from the grid, therefore leading to cost savings.

This work aims to review possible solutions based on renewable technologies for retrofitting or designing new buildings.

28.2 Literature Review

28.2.1 Self-produced Energy

There are several options for self-production of energy in a household. Photovoltaic panels (PV) are the most common technology, as they are easy to install and can provide clean energy. The typical installation has other components, such as inverters and electricity meters. Figure 28.2 shows a diagram of a PV system.

As presented in Fig. 28.2, a PV system needs a solar panel to convert sunlight into electrical energy, an inverter to convert DC electricity (represented in yellow) into AC electricity (represented in blue), and an electricity meter to keep track of energy flows into and out of the grid as well as the energy consumption.

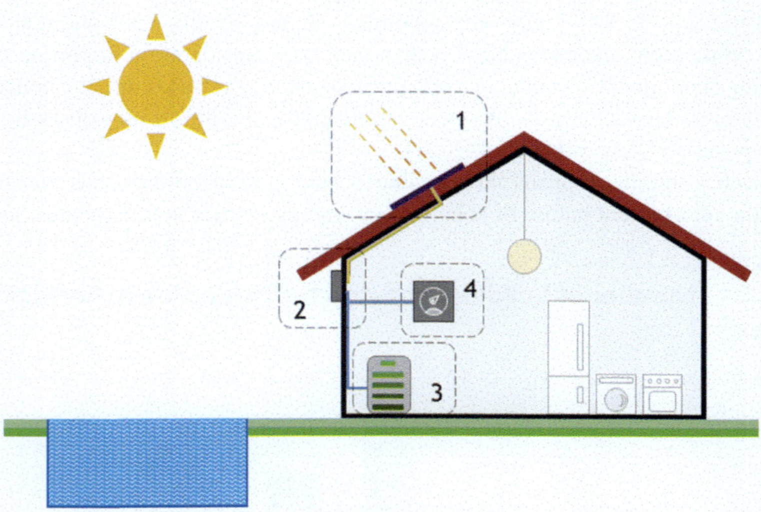

Fig. 28.2 Solar panel system scheme. 1- Photovoltaic panel; 2- inverter; 3- energy storage system; 4- electricity meter

28.2.2 Photovoltaic Panels

Photovoltaic panels are devices that convert sunlight into electricity. They are made up of individual cells that contain semiconductor materials. When sunlight hits the PV cells, it excites the electrons in the semiconductor material and causes them to flow, generating an electric current. This process is called the photovoltaic effect and was first discovered in 1839 by Alexandre-Edmond Becquerel (1841). PV panels are often used in solar energy systems to generate electricity for homes, businesses, and other facilities as they generate clean, renewable energy since they do not emit any greenhouse gases or other pollutants during operation. PV panels are also relatively low maintenance, as they do not require any fuel or moving parts.

It took almost five decades after the discovery of the photovoltaic effect for the first solar cell to be created by Charles Fritts. The cell was made out of selenium and gold and had less than 1% efficiency (Fraas 2014). The first silicon-based solar cells were developed in the 1950s. Efficiencies rose from single to double digits in that decade, and soon after they started being used for space ventures. By the turn of the century, both single and polycrystalline cells reached efficiency values superior to 20%, with single-crystal cells performing better than polycrystalline ones (Green 2009).

The evolution of solar cells is grouped into different generations. The first generation is based on crystalline silicon wafers. It is characterized by having high costs and high efficiencies, as obtaining pure silicon crystals is an expensive process. Despite this, as of 2021, most of the market share of PV panels is made out of these cells (ISE 2023). Given the high cost of manufacture, PV modules made out of these cells tend to have a long payback period. The second generation of cells aimed to reduce costs by using thin layers (i.e., 1 μm). Other materials, such as copper indium gallium selenide (CIGS) or cadmium telluride (CdTe)/cadmium sulfide (CdS), were also developed. Thin layer cells tend to be cheaper than the first-generation counterparts but also have lower efficiency values. In addition, some of the materials used to manufacture non-silicon-based cells are rare metals, which add to the cost of materials. Finally, the emergent technologies of solar cells encompass the third generation. These include the dye-sensitized solar cells (DSSC) as well as perovskite solar cells (PSC), which will discussed further below. Most of these technologies are still under research and development, although some of them have already started to hit the market (Muñoz-García et al. 2021). The advantages these technologies bring are namely due to the cheaper materials and their manufacturing process.

The photovoltaic market is rapidly growing. Figure 28.3 shows the evolution of the cumulative installed solar PV capacity worldwide over the last two decades.

As shown in Fig. 28.3, the growth has been significant in the last decade and it is expected to hit 2.3 TW by 2025 (Europe 2002). One of the principal motives for said growth can be attributed to the cost of a PV module significantly decreasing in the last few years, as shown in Fig. 28.4.

Since 2020, both solar and wind energy have become cheaper than fossil fuels, as shown below.

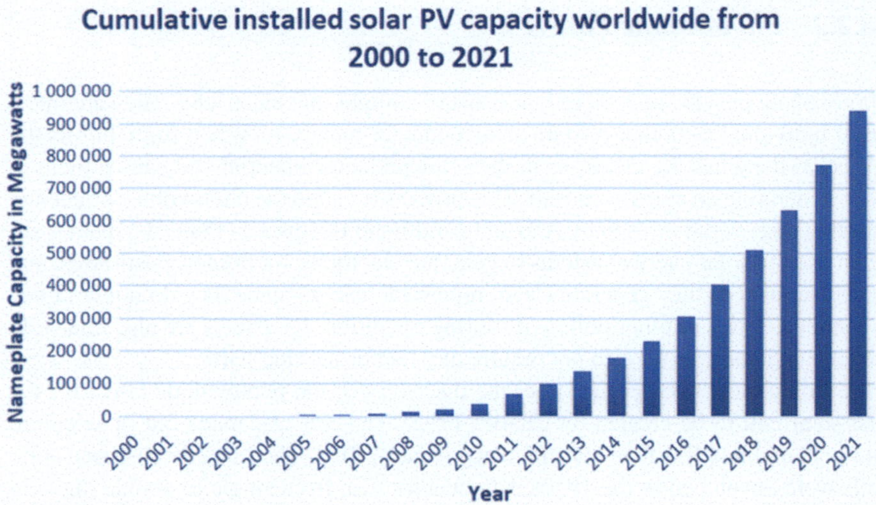

Fig. 28.3 Cumulative installed solar PV capacity worldwide from 2000 to 2021. Data taken from Statista (2000)

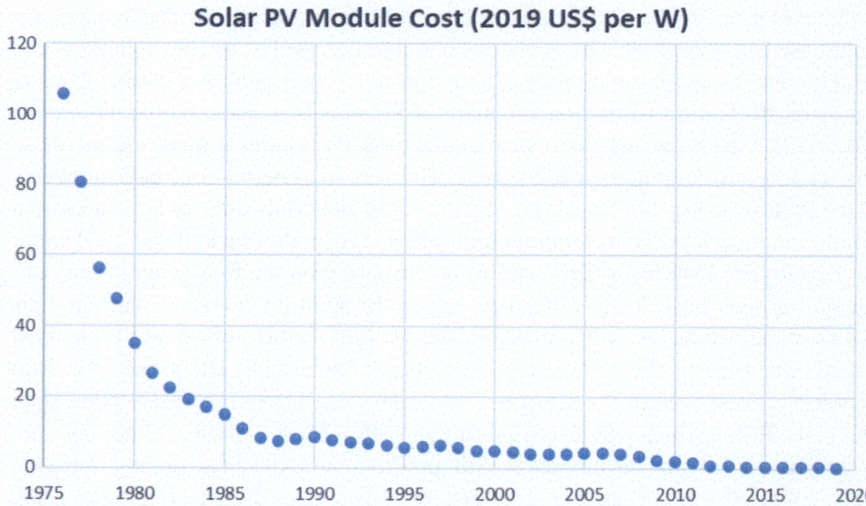

Fig. 28.4 Solar PV module cost (in dollars per watt). Data taken from (IRENA NFLIREA 2023)

According to the data shown in Fig. 28.5, solar energy is now a cost-competitive technology, and as such, it is expected to have a higher market share of energy production in the future. Research and development of other types of cells and promising technologies are emerging in the market. As of 2021, silicon-based cells had 95% of the PV market share (ISE 2023). This is expected to change as other technologies such as DSSC and PSC mature and become commercially viable.

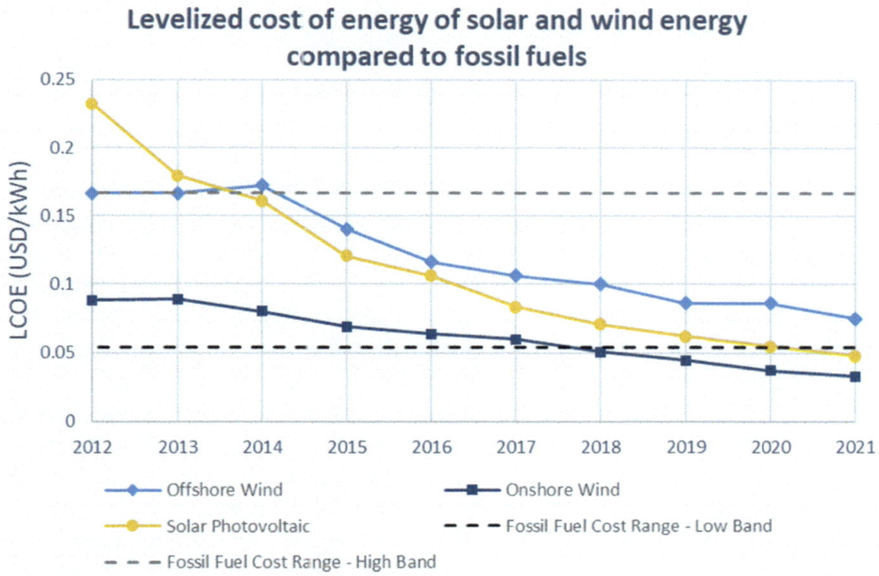

Fig. 28.5 Levelized cost of energy of solar and wind energy compared to fossil fuels. Data taken from Agency et al. (2021)

One of the main disadvantages of PV cells is that it is an intermittent source of energy, as it relies on sunlight. This property makes PV panels by themselves undesirable for household applications, as people typically spend most of their sunlight hours away from home for work or leisure. This calls for the coupling of PV panels and an energy storage system, such as batteries. Other disadvantages include the high initial cost of installation and equipment which can be unfeasible for poorer households. Another disadvantage lies in the fact that most PV technologies rely on direct irradiation, which means there is a loss of efficiency if the module does not follow the sunlight path/ is not in open air facing the sun. Some PV technologies mentioned below can absorb diffuse irradiation, making them viable for other non-rooftop applications such as horizontal façades, windows, or indoor applications (Mathews et al. 2019; Hagfeldt et al. 2010; Li et al. 2021).

For PV to become a major player in the energy sector, large PV parks must be built to generate enough energy to power cities. This can be seen as a disadvantage as it would require large areas of terrain to be dedicated for PV panels. However, a counterargument to this is that the world needs an area of 1375.28 km^2 of solar panels to produce all the energy needed in 2017, roughly a bit less than the size of Arizona in the US (Power 2023).

One last disadvantage is the small argument that PV panels are not aesthetic. However, some PV technologies like DSSC are colorful and can ally their practicality with art.

28.2.2.1 Dye-Sensitized Solar Cells

With a ground-breaking proposal in 1991 by Brian O'Regan and Micheal Grätzel, dye-sensitized solar cells have become a heavily researched topic as an emergent PV technology. The authors created a cell made out of titanium dioxide, TiO_2, a semiconductor material well known for its photoelectrochemical applications, coated with a charge-transfer dye responsible for light absorption and charge separation (O'Regan and Grätzel 1991). The efficiency reported was between 7 and 8%, which can seem low when compared to the values observed for silicon cells, but it completely changed the paradigm of this technology, as before efficiencies were below 1% (Hagfeldt et al. 2010). The low cost of materials and easiness of preparation of the cells made it a viable subject for further exploration. Adding to this, the fact that DSSCs absorb diffusive irradiation and have semi-transparent electrolytes makes them viable for indoor applications or windows (Muñoz-García et al. 2021). Multiple configurations have been tested, as most parts of the cell can be changed, such as the dye used, as well as the electrolyte and morphology of the films.

This technology has matured after 30 years of research and development. As it currently stands, devices have been proposed with power conversion efficiencies (PCE) up to 13.5%, with stability tests lasting more than 1000 h with minor loss of efficiency (Muñoz-García et al. 2021). This makes DSSCs not competitive when compared with Si-based solutions. However, the colorful characteristics of the dyes make them interesting in building integrated photovoltaic (BIPV) products. Another potential application of DSSCs is in low-power devices for the Internet of Things (IoT). For example, Fujikura, a Japanese company, has started implementing DSSCs in sensors for both indoor and outdoor applications (Muñoz-García et al. 2021).

28.2.2.2 Perovskite Solar Cells

Over the last decade or so, perovskite solar cells (PSC) have gained attention from industry and scientists as a promising next-generation photovoltaic technology. They were first proposed in 2009 by Kojima et al. (2009), who used a perovskite as a light absorber in a DSSC. The material demonstrated astoundingly high efficiency as a light absorber while also exhibiting extremely high hole and electron conductivities. This sparked research on this new thin-film PV technology.

The term "perovskite" is used to describe the absorption material, which uses the ABX_3 crystal structure (Snaith 2013). The aforementioned work used $CH_3NH_3PbI_3$ and $CH_3NH_3PbBr_3$ and reported 3.8% PCE. By now, single junction PSCs have already reached PCE values of 25.8% (Min et al. 2021). Tandem perovskite-silicon systems have reached PCE values of 32.5% (HZB 2022). However, efficiencies are not the only thing that matters in these types of devices. One of the major issues with PSCs is their sensitivity to moisture and oxygen (Mesquita et al. 2018). These quickly degrade the perovskite absorber, making the device stable for a short period of time. For that, studies are being conducted on the process of cell encapsulation using laser-assisted glass frit bonding of soda-lime glass substrates (Emami et al.

2017). Another disadvantage of most PSCs is the inclusion of the toxic element lead in its composition. Lead-free perovskite is also under research; however, the PCE obtained using these compositions is much lower than that of lead-based PSC (Duan et al. 2021).

28.2.3 Heat Pumps

As the name suggests, the heat pump takes heat from a source and transfers it to another area (Jouhara et al. 2018). Their working principle is very similar to that of a refrigerator or an air conditioner. Since they rely on heat transfer, they are more efficient than heat-generating technologies. Their coefficient of performance (COP) is usually higher than 3, meaning that the energy output is at least 3 times higher than the energy needed for operation.

A heat pump is comprised of a connected system of a compressor, two heat exchangers, and an expansion valve. This closed loop has a refrigerant inside, which is a fluid that undergoes through phase changes during the cycle. A diagram of a heat pump is shown in Fig. 28.6.

The compressor pressurizes the refrigerant and heats it, turning it into a high-pressure, high-temperature fluid. The accumulated heat is transferred to the exterior via a heat exchanger, where the fluid condenses, releasing energy as latent heat. The new liquid is still kept at high pressure, which is reduced at the expansion valve, lowering both temperature and pressure. It then goes through another heat exchanger, where heat from the exterior is transferred into the fluid, vaporizing it. The cycle can be reversed if the opposite effect is intended.

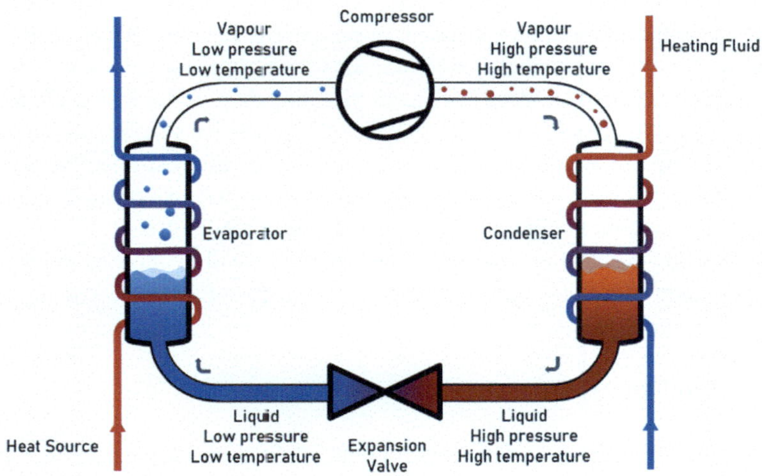

Fig. 28.6 Diagram of the working principle of a heat pump

The refrigerant is chosen due to its low boiling point, condensation under low pressure, and low vapor volume. Traditional systems had F-refrigerant gases that have large global warming potentials, but in the past years, there has been a regulation and a change into other alternatives that make heat pumps reduce greenhouse gas emissions by at least 20% compared with a gas boiler when running on emissions-intensive electricity (IEA 2022).

28.2.4 *Energy Storage*

Energy production and storage are two symbiotic agents in energy management, as most renewable sources of energy are intermittent or unpredictable. Energy storage systems (ESSs) can lower the cost of energy production, and reduce fuel consumption and therefore the amount of greenhouse gases emitted to the atmosphere (Mahlia et al. 2014). Additionally, energy storage is crucial for electrical systems as it enables load balancing and peak shaving, frequency regulation, dampens energy fluctuations, and boosts power quality and dependability (Koohi-Fayegh and Rosen 2020). It has been proven that ESSs are able to reduce household electricity costs (Emmett and Roberts 2021). Considering the scope of this work, most household energy storage relies on batteries (Koohi-Fayegh and Rosen 2020; IRENA 2030).

A battery uses electrochemical cells to store electricity as chemical energy. The positive and negative electrodes that make up each cell are where the redox reactions happen. An electrolyte, often a solution containing dissociated salts, separates the electrodes and allows ions to move between them. When these electrodes are externally linked, chemical processes take place simultaneously at both electrodes, releasing electrons and producing the current that the user may access (Dunn et al. 2011). Numerous types of batteries have been heavily researched, including lead, lithium-ion, sodium-based, nickel-based, or redox flow batteries (Divya and Østergaard 2009). Limitations such as storage capacity, cost, response time, and efficiency are the major hurdles of an ideal ESS, as to this date, there is no solution that satisfies all the aforementioned properties (Worku 2022). Studies have been done on the economic viability of batteries, and they were found to be not cost-effective (Pandžić 2018). Despite this, the cost of battery storage systems is expected to decrease by 2030, making them more attractive to consumers (IRENA 2030).

As lithium-based batteries and redox flow batteries will be studied in the scope of this work, a more detailed description of each technology is given below.

28.2.4.1 Lithium-Ion Batteries

These batteries use lithium-intercalation compounds as the electrodes of the cell (see Fig. 28.7). While charging, Li^+ ions are transported across the electrolyte from the cathode to the anode, where they accumulate between the carbon layers of the electrode. The opposite occurs when the battery is discharged (Divya and Østergaard

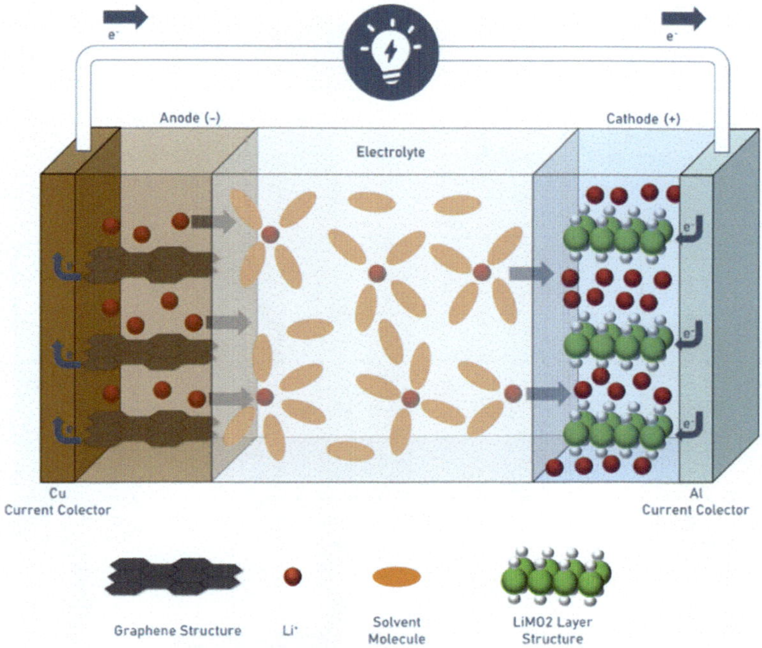

Fig. 28.7 Scheme of a Li-ion battery during discharge

2009; Goodenough and Park 2013). Lithium-ion (Li-ion) batteries are known to have long life cycles, low molecular weight (which makes them lightweight), high charge densities, voltage and power compared to other metal-based technologies, high response time, high efficiency, low maintenance, are made out of abundant materials and most of its metal components can be recycled (Mahlia et al. 2014; Dunn et al. 2011; Divya and Østergaard 2009; Zhang et al. 2021; Nitta et al. 2015; Yang et al. 2011; Harper et al. 2019). The main drawbacks of Li-on batteries rely on the high cost of the packing and protection against overcharging, high manufacturing CO_2 emissions, and the risk of heat generation and potential fire (Mahlia et al. 2014; Zhang et al. 2021; Yang et al. 2011; Larcher and Tarascon 2015). Despite this, Li-ion batteries have overtaken the portable devices market and have enabled the rise of the electric vehicle market (Li et al. 2018). Research is being done on potential alternatives to Li-ion batteries such as Li-air batteries or the use of other metals like sodium (Zhang et al. 2021; Hu 2016).

28.2.4.2 Redox Flow Batteries

A redox flow battery (RFB) is a type of electrochemical energy storage system that uses the principles of reversible electrochemical reactions to store and release energy. This mechanism is governed by species' oxidation and reduction (redox) reactions,

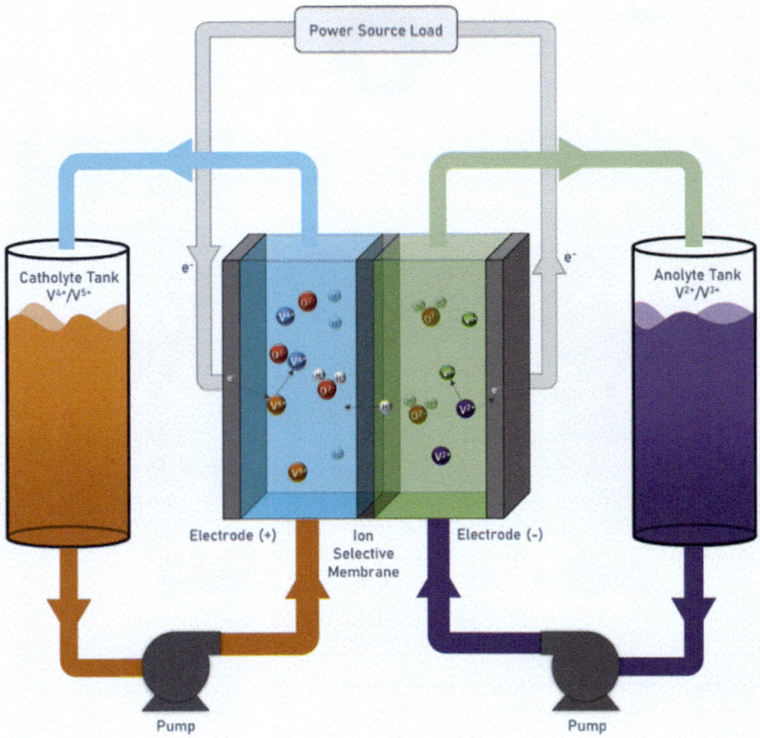

Fig. 28.8 Schematic of the various components of a redox-flow battery

which are contained within separate electrolyte solutions (Weber et al. 2011). These electrolytes are usually made up of transition metal salts with multiple oxidation states. They are kept from mixing using an ion-exchange membrane. A scheme of an RFB is shown in Fig. 28.8.

When discharging, the anolyte (electrolyte in the anode side of the cell) is pumped to the cell, where it is reduced to a lower oxidation state. The free electrons flow to an external circuit to the cathode, promoting the oxidation reaction, which charges the catholyte (electrolyte in the cathode side of the cell) to a higher oxidation state. This process is reversed when charging the battery (Dunn et al. 2011).

RFBs are suitable for various energy storage requirements since their energy density and power output are decoupled and can be individually tailored. Power output is related to the design of the electrochemical cell stack while the energy density depends on the volume of electrolytes (Iwakiri et al. 2021). This modularity is one of the technology's numerous advantages. Other advantages include the ability to be fully discharged (which typically harms other battery technologies), high energy efficiencies (up to 90% in large applications), long life cycles, operation at low-temperature levels, no self-discharge (the electrolytes are kept separately), fast response time due to quick redox reaction rates and quick swap between charging/discharging modes. However, RFBs also have disadvantages, namely the added capital, operation and maintenance costs of pumps, low energy density which demands higher volume, lower energy efficiency when compared to other ESSs, electrolyte contamination due to ion cross-over in the membrane as well as high cost of precious metals/ redox pairs (Emmett and Roberts 2021; Dunn et al. 2011; Divya and Østergaard 2009; Weber et al. 2011; Al-Fetlawi et al. 2009; Clemente and Costa-Castelló 2020). A complete review of each component of a redox flow battery is done by Arévalo-Cid et al. (2021). RFBs can have aqueous organic or non-organic aqueous electrolytes, as well as non-aqueous ones. Different redox pairs are considered for this technology. The most mature and common one is the vanadium redox flow battery (VRFB), where electrolyte contamination is nullified since both electrolytes have vanadium in different oxidation states. This is also commonly known as an all-vanadium redox flow battery. Other types of redox flow batteries include zinc-based chemistries, iron-based chemistries, vanadium-bromine, polyoxometalates (POMs), semi-solid (slurry flow batteries), among others (Emmett and Roberts 2021; Weber et al. 2011; Iwakiri et al. 2021; Wang et al. 2013; Shin et al. 2013; Soloveichik 2015; Park et al. 2016; Arenas et al. 2018; Gentil et al. 2020). The redox reactions of all-vanadium redox flow batteries can be expressed as the following (Tokuda et al. 1998):

$$\text{Positive electrode:} VO^{2+} + H_2O - e^- \underset{\text{Discharge}}{\overset{\text{Charge}}{\leftrightarrows}} VO_2^+ + 2H^+ \tag{28.1}$$

$$\text{Negative electrode:} V^{3+} + e^- \underset{\text{Discharge}}{\overset{\text{Charge}}{\leftrightarrows}} V^{2+} \tag{28.2}$$

The electrolytes found in VRFB are made out of sulphuric acid (usually 3 M) and vanadium. The electrodes are usually carbon or graphite felt, and membranes are commonly made out of nafion (Iwakiri et al. 2021). Current research focuses on achieving higher power outputs at lower prices, which is usually accomplished by lowering the ohmic resistances, which are the main cause of power loss (Aaron et al. 2011). The technology has already been commercialized, as can be seen in Bryans et al. (2018).

The solar redox flow battery (SRFB) is a similar emergent technology worth mentioning. It combines the working principle of photoelectrochemistry (PEC) with RFBs, where the energy is stored in the redox pairs that are photocharged by a semiconductor-based photoelectrode (Wedege et al. 2018; Cao et al. 2018). The technology is still under lab scale, where the hurdles of upscaling are being tackled (Silva et al. 2022). Modeling is also being used to help undertake challenges (Delgado et al. 2021).

28.2.5 Building Modeling Tools

Different modeling tools can be applied in the study and design of buildings. Building information modeling (BIM) is a digital representation of a building, with all the fundamental data and parameters, acting as a database with all the important information (Hensen and Lamberts 2019). It has grown in popularity with the rise of digital technology as it provided benefits in construction planning and design phases. Another tool that has become prevalent in the construction industry is building energy modeling (BEM). By taking into account the geometry of a building, materials used, as well as local weather and temperature conditions, BEM can predict the thermal loads and energy needs (Coakley et al. 2014). One of the most used BEM software is called EnergyPlus® (EP), whose development started in 1996 (Crawley et al. 2001). The software is being updated to this date and is used as a base tool for other software with added features like OpenStudio or DesignBuilder (NREL: OpenStudio 2023; Designbuilder 2023). Examples of other popular software are TRNSYS (University of Wisconsin--Madison 2017), eQuest (DOE 2018), and ESP-r (Unit ESR: ESP-r. 2018). Crawley et al. have done a review of 20 different BEM software available, later complemented by other works (Coakley et al. 2014; Crawley et al. 2008; Gao et al. 2019; Harish and Kumar 2016).

BEM modeling can be achieved through three different types of models: white box, black box, and gray box models (Li and Wen 2014). White box models are systems that are entirely defined by physics laws, which can be translated using equations (Loyola-González 2019). They are the most accurate out of the three types of models and are usually preferred. The main drawback is the extensive detail and information needed to represent all the phenomena occurring in the system fully. On the other hand, black box models are data-driven models (Loyola-González 2019). They use statistics to correlate the operation data with the energy consumption. While easier to build when compared to white box models, they are not as versatile and their results are mostly dependent on the quality of the training data. The compromise between these two models is found in gray box models, where simplified physical models are used to save computational time, and unknown parts of the model are complemented with experimental data (Sohlberg and Jacobsen 2008; Braun and Chaturvedi 2002).

EnergyPlus has three main components: a simulation manager, a heat and mass balance simulation module, and a building systems simulation module. The simulation manager is responsible for coordinating the entire simulation process, successfully feeding results from both simulation modules to each other for a more integrated result (Crawley et al. 2001). EP was created so that users can write and easily incorporate their own modules. Published works on EP cover topics such as building renovation (Zhu 2014), green roofs (Sailor 2008), building integrated solar panels (Vuong et al. 2015), heat pumps (Priarone et al. 2020), phase-changing materials (Tardieu et al. 2011), urban building energy modeling (Reinhart and Cerezo 2016), NZEBs (Attia et al. 2013), among others (Brown et al. 2014; Monfet et al. 2007; Yu et al. 2015a; Ahn et al. 2014; Shabunko et al. 2018; Ramos and Ghisi 2010; Delgado et al. 2023).

However, BEM has inherent limitations. First, to faithfully represent the building, an extensive amount of information regarding all the components involved in the simulation is needed. This demanding task can be hard to tackle, as not every single piece of information might not be available to the modeler, such as material properties, occupant behavior, or real specifications of the equipment. This usually translates into variance between the simulated and real data, which can be found in studies using BEM (Karlsson et al. 2007; Turner et al. 2008). Adding to this, each module of the software has some inherent uncertainty which adds up to the final results being different from the real measured data. Most of the validations of BEM software are done only taking into account the building load (that means the energy needed by all the equipment in the household) (Coakley et al. 2014). This is usually done by following the guidelines published by ASHRAE (Air-Conditioning and Guideline 14 2014), IMPVP (EVO 2007), or M&V y (Doe 2008). Other useful tools are the Commercial Reference Building project done by the U.S. Department of Energy and the material guidelines available in ASHRAE Standard 90 (Deru et al. 2011; Air-Conditioning 2022).

One of the largest factors of uncertainty relies on occupant behavior (Clevenger and Haymaker 2006; Guerra Santin et al. 2009; Zhang et al. 2014). While usually modeled as a static schedule (Mahdavi and Tahmasebi 2019), it is known that, in reality, the behavior is dynamic (Mahdavi and Tahmasebi 2019). Yan et al. did a complete review of occupant behavior and strategies to tackle this problem (Yan et al. 2015). Model calibration is often sought out as a solution. A complete review of calibration methods was carried out by Coakley et al. (Coakley et al. 2014).

Given the nature of most BEM software taking input files in text format, the geometry already drawn for BIM is usually required to be redrawn. To address this, BIM-based BEM technologies have become popular in recent years (Gao et al. 2019). Software like SketchUp (Trimble: SketchUp. 2023) or Revit (Autodesk: Autodesk Revit: BIM software to design and make anything. (2023). Accessed 21–08-2023 2023. 2023) is used for 3D modeling, and BIM can then transpose information so that it can be easily incorporated into BEM (Bahar et al. 2013). This is achieved through different file formats like gbXML (Garwood et al. 2018; Xu et al. 2019) or IFC (Ivanova et al. 2015; Bazjanac and Crawley 1997).

One final limitation of BEM is the complexity of the model. The need for know-how and the lack of user-friendly interfaces make it less attractive for non-simulation experts, which restricts the usage of BEM (Ahn et al. 2014).

The incorporation of third-party tools with EP makes it a powerful asset for BEM. Studies have created "script-editing-based" optimization models to automate the modeling process and develop control systems (Li and Wen 2014). One such topic is Model Predictive Control (MPC), where EnergyPlus is paired with optimization tools such as MATLAB (Inc and version, 2022) or Python (2009). Roth et al. describe four useful tools for enhancing EP through scripting (Amir et al. 2018). One such tool is known as Eppy, a Python library that allows communications with EP through the input and output files (Santos P: EPPY. 2020). Eppy has been used in the literature (Lee et al. 2018; Kubwimana and Najafi 2023; Jason 2016), as well as MATLAB (Lee et al. 2018; Delgarm et al. 2016). Other examples of work using EP and other tools can be found in Benincá et al. (2023), among others (Bre and Fachinotti 2017; Chiesa et al. 2021, 2023; Yang et al. 2019; Xu et al. 2016; Matti et al. 2013; Taoning et al. 2021; Yu et al. 2015b; Méndez Echenagucia et al. 2015; Carlucci et al. 2015; Mokhtarian Mobarakeh and Sayyaadi 2023; García Kerdan et al. 2016).

Regarding the optimization algorithms for BEM, it has been found that the Non-Dominated Sorting Genetic Algorithm II (NSGA-II) (Deb et al. 2002) is preferred (Delgarm et al. 2016; Bre and Fachinotti 2017; Yu et al. 2015b; Méndez Echenagucia et al. 2015; Brownlee and Wright 2015). Hamdy et al. compared the performance of seven algorithms and recommended guidelines for optimization models (Hamdy et al. 2016). Finally, one library with various multi-objective optimization algorithms is Pymoo (Blank and Deb 2020).

28.2.6 Home Energy Management System

A smart home allows the monitoring of the appliances so that, through the use of management software, one can optimize the electric consumption to lower costs and/or help mitigate electric grid strains due to heavy electricity use (Zandi et al. 2018). This concept became a reality with the introduction of wireless communication protocols such as Wi-Fi and Zigbee in the past 20 years as well as the emergence of smart grids (Li 2013; Peretto 2010). As shown in Fig. 28.9, the term Internet of Things has become a hot topic of research and development, as well as the plethora of "smart" gadgets that have become standard in the past decade.

As Beaudin and Zareipour found in 2015 (Beaudin and Zareipour 2015), there is no agreed nomenclature for a system to manage the energy component in a household. A mixture of terms including "home", "energy", "management", "controller", "system", "unit", and "assistant", among others, are often found in the literature. To simplify, going forward all the systems will be described as home energy management system, or HEMS.

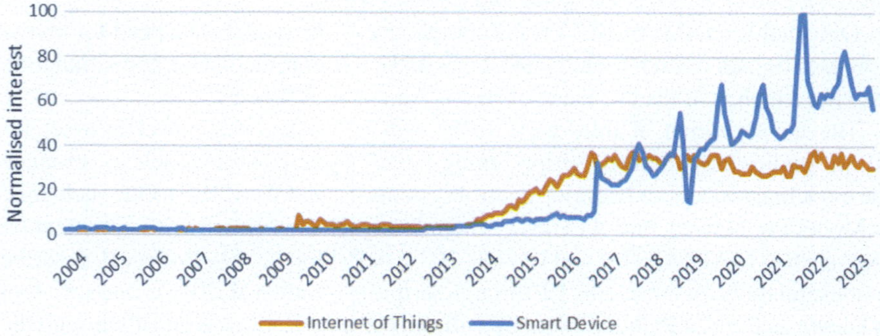

Fig. 28.9 Google trends since 2004 for terms "Internet of Things" and "Smart Device". Data Source: Google Trends

A HEMS should efficiently monitor the electricity ecosystem of a household, from its generation through technologies like solar panels as well as its storage and consumption (Zhou et al. 2016). HEMS can use predictive modeling of the weather forecast or energy market prices to schedule the devices to guarantee the user's comfort and/or reduce electricity costs. The concept of demand response is often paired with HEMS: changing typical electricity consumption schedules for the economic benefit of the user as well as helping with grid balancing (Palensky and Dietrich 2011). Typically, demand response can be divided into two types: Price-based and incentive-based. The former has the user adjust his appliance used according to the electricity price variation during the day, while the latter is based on pre-established measures done between the user and the power companies where monetary compensation or electricity price reductions are expected in trade of reducing power consumption to help out during situations of power shortage or high grid load (Ma et al. 2021).

The system has multi-tasking features: monitors the building load and the energy generation and storage system; records and compiles the data in a database that ideally is user-friendly and has easy visualization; controls the appliances and the electricity flow in the household; manages the devices and allows for manual input and finally should have an alarm system set up in case of failure or if user decisions are needed (Zhou et al. 2016; Ma et al. 2021).

Challenges to HEMS are the diversity of home appliances, the uncertainty of future consumption tied with weather and price predictability, as well as infrastructure complications and preferred optimization parameters. As such, HEMS should be robust enough to be able to address all of these issues. Despite these challenges, research on HEMS has been done regarding the optimization of energy consumption [157, 158] and the different components and structures of HEMS (Zhou et al. 2016; Ma et al. 2021; Risteska Stojkoska and Trivodaliev 2017).

As the concept of smart homes is rather recent, most of the household stock is not ready for retrofitting. Smart plugs with consumption metering are not currently deemed necessary; therefore, most dwellings do not have them. Workarounds are proposed by companies like Cleanwatts, whose product Kiome, formerly known as Cloogy, is an energy management solution that allows the measurement of the energy consumption of a house (Santos 2018).

The Kiome ecosystem has multiple components. The power plug allows the monitoring and controlling of equipment. The clamp can be applied to the switchboard and measure the energetic consumption. The transmitter will then send the data collected by the clamp to the Kiome hub, which then sends the data to the database, which can be accessed using a smartphone app. Ethernet signal is necessary to connect the hub to the internet. The Kiome ecosystem communicates with itself using the Zigbee protocol.

One mentioned limitation of this software is the different appliances. Most of them do not have communication features so scheduling cannot be done fully automatically. Moreover, some appliances have their own communication protocols that are not compatible with other brands. A case-by-case approach is then needed to be done. Zigbee has advantages for smart home applications given its low power consumption, high durability, decent data transfer speeds, can support a large number of network nodes, is low cost, and wireless (Zhou et al. 2016).

One of the main components of a HEMS is acquisition devices. These measure the energy consumption of household appliances and the energy generation and storage systems. They should be able to turn on and off by the HEMS for an easier automation process. These devices communicate with a control center, which is the "brain" of the system. Here, the software runs and takes into account the real-time information of the acquisition devices. It also needs to have access to the internet for the weather prediction tools as well as remote access via a desktop or a smartphone. Although being barely mentioned so far, security is also a very important component of this system, and should not be glanced over.

28.3 Conclusions

Facing our planet's demographic and climate evolutions, humanity keeps evolving and adapting toward sustainable solutions. Regarding buildings, new concepts like NZEBs have become standards for new constructions. The fast advancement of renewable technology, such as photovoltaic panels and batteries, enables cleaner energy generation and wiser energy consumption. While the price of solar energy has become cost-competitive compared to traditional fossil fuel alternatives, the equipment cost for both generation and storage remains expensive, making it unappealing to the average household. Advancements are being made on both fronts to reduce the cost and increase equipment's efficiencies and life expectancy. As such, the correct sizing of the equipment for a household is essential to make it attractive to the end-user.

Given the worldwide call to action regarding the decarbonization of the energy sector, significant progress has been made in developing renewable energy and storage solutions for residential use. This enables clean energy production and, through the use of management software, improves the energy efficiency of the household. The variety of household appliances, the unpredictability of future usage related to weather and price fluctuations, infrastructural difficulties, and desired optimization parameters provide challenges for HEMS. However, it is expected that new constructions will be more compatible with the concept of smart homes, and therefore, a complete integration of automation and equipment management is expected.

Acknowledgements This work was financially supported by national funds through the FCT/MCTES (PIDDAC), under the projects R2UTechnologies – Modular Systems with a reference C644876810-00000019 and 2022.06841.PTDC-BlueHouseSim: Development of numerical simulation tools and methodologies for high-efficiency off-the-electrical-grid houses, with https://doi.org/10.54499/2022.06841.PTDC. In addition, this work was financially supported by Base Funding-UIDB/04708/2020, with https://doi.org/10.54499/UIDB/04708/2020; Programmatic Funding- UIDP/04708/2020, with a https://doi.org/10.54499/UIDP/04708/2020 of the CONSTRUCT funded by national funds through the FCT/MCTES (PIDDAC); and FCT through the individual Scientific Employment Stimulus 2020.00828.CEECIND/CP1590/CT0004, https://

doi.org/10.54499/2020.00828.CEECIND/CP1590/CT0004. Adding to this, the authors thank the Agenda "H2Driven Green Agenda", nr. C644923817-00000037, investment project nr. 50, financed by the Recovery and Resilience Plan (PRR) and by European Union—Next Generation EU. Finally, this work was also financially supported by national funds through FCT/MCTES (PIDDAC): LEPABE, UIDB/00511/2020 (https://doi.org/10.54499/UIDB/00511/2020) and UIDP/00511/2020 (https://doi.org/10.54499/UIDP/00511/2020) and ALiCE, LA/P/0045/2020 (https://doi.org/10.54499/LA/P/0045/2020).

References

Aaron D, Tang Z, Papandrew AB, Zawodzinski TA (2011) Polarization curve analysis of all-vanadium redox flow batteries. J Appl Electrochem 41(10):1175–1182. https://doi.org/10.1007/s10800-011-0335-7

Agency IRE (2022) Renewable power generation costs in 2021 Abu Dhabi: IRENA

Ahn B-L, Jang C-Y, Leigh S-B, Jeong H (2014) Analysis of the effect of artificial lighting on heating and cooling energy in commercial buildings. Energy Procedia 61:928–932. https://doi.org/10.1016/j.egypro.2014.11.997

Ahn K-U, Kim Y-J, Park C-S, Kim I, Lee K (2014) BIM interface for full vs. semi-automated building energy simulation. Energy Build 68:671–8. https://doi.org/10.1016/j.enbuild.2013.08.063

American Society of Heating R (2022) Air-Conditioning E. ASHRAE Standard 90.1–2022—Energy Standard for Sites and Buildings Except Low-Rise Residential Buildings

American Society of Heating R (2014) Air-Conditioning E. ASHRAE Guideline 14: Measurement of Energy and Demand Savings

Akadiri PO, Chinyio EA, Olomolaiye PO (2012) Design of a sustainable building: a conceptual framework for implementing sustainability in the building sector. Buildings 2(2):126–152

Al-Fetlawi H, Shah AA, Walsh FC (2009) Non-isothermal modelling of the all-vanadium redox flow battery. Electrochim Acta 55(1):78–89. https://doi.org/10.1016/j.electacta.2009.08.009

Allen MR, Dube OP, Solecki W, Aragón-Durand F, Cramer W, Humphreys S, et al (2022) Framing and context. In: Intergovernmental panel on climate C, editor. Global Warming of 15 °C: IPCC Special Report on Impacts of Global Warming of 15 °C above Pre-industrial Levels in Context of Strengthening Response to Climate Change, Sustainable Development, and Efforts to Eradicate Poverty. Cambridge University Press, Cambridge, pp 49–92

Amir R, Jamie B, Scott C, Peter E, Jason G, David G, et al (2018) Scripting frameworks for enhancing EnergyPlus modeling productivity. In: 2018 building performance analysis conference and simbuild co-organized by ASHRAE and IBPSA-USA: ASHRAE/IBPSA-USA, pp 312–9

Arenas LF, Loh A, Trudgeon DP, Li X, Ponce de León C, Walsh FC (2018) The characteristics and performance of hybrid redox flow batteries with zinc negative electrodes for energy storage. Renew Sustain Energy Rev 90:992–1016. https://doi.org/10.1016/j.rser.2018.03.016

Arévalo-Cid P, Dias P, Mendes A, Azevedo J (2021) Redox flow batteries: a new frontier on energy storage. Sustain Energy Fuels 5(21):5366–5419. https://doi.org/10.1039/D1SE00839K

Attia S, De Herde A, Gratia E, Hensen JLM (2013) Achieving informed decision-making for net zero energy buildings design using building performance simulation tools. Build Simul 6(1):3–21. https://doi.org/10.1007/s12273-013-0105-z

Attia S, Eleftheriou P, Xeni F, Morlot R, Ménézo C, Kostopoulos V et al (2017) Overview and future challenges of nearly zero energy buildings (nZEB) design in Southern Europe. Energy Build 155:439–458. https://doi.org/10.1016/j.enbuild.2017.09.043

Autodesk (2023) Autodesk Revit: BIM software to design and make anything. https://www.autodesk.eu/products/revit/. Accessed 21 Aug 2023

Bahar YN, Pere C, Landrieu J, Nicolle C (2013) A thermal simulation tool for building and its interoperability through the building information modeling (BIM) platform. Buildings. https://doi.org/10.3390/buildings3020380

Bazjanac V, Crawley DB (1997) The implementation of industry foundation classes in simulation tools for the building industry

Beaudin M, Zareipour H (2015) Home energy management systems: a review of modelling and complexity. Renew Sustain Energy Rev 45:318–335. https://doi.org/10.1016/j.rser.2015.01.046

Becchio C, Dabbene P, Fabrizio E, Monetti V, Filippi M (2015) Cost optimality assessment of a single family house: building and technical systems solutions for the nZEB target. Energy Build 90:173–187. https://doi.org/10.1016/j.enbuild.2014.12.050

Becquerel AE, Memoire sur les effects d´electriques produits sous l´influence des rayons solaires. Annalen der Physick und Chemie. 1841;54

Benincá L, Crespo Sánchez E, Passuello A, Karini Leitzke R, Grala da Cunha E, Maria González Barroso J (2023) Multi-objective optimization of the solar orientation of two residential multi-family buildings in south Brazil. Energy Build 285:112838. https://doi.org/10.1016/j.enbuild.2023.112838

Blank J, Deb K (2020) Pymoo: multi-objective optimization in python. IEEE Access 8:89497–89509. https://doi.org/10.1109/ACCESS.2020.2990567

Brandon NP, Kurban Z (2017) Clean energy and the hydrogen economy. Philos Transact A Math Phys Eng Sci 375(2098):20160400. https://doi.org/10.1098/rsta.2016.0400

Braun JE, Chaturvedi N (2002) An inverse gray-box model for transient building load prediction. HVAC&R Res 8(1):73–99. https://doi.org/10.1080/10789669.2002.10391290

Bre F, Fachinotti VD (2017) A computational multi-objective optimization method to improve energy efficiency and thermal comfort in dwellings. Energy Build 154:283–294. https://doi.org/10.1016/j.enbuild.2017.08.002

Brown N, Ubbelohde MS, Loisos G, Philip S (2014) Quick design analysis for improving building energy performance. Energy Procedia 57:3041–3050. https://doi.org/10.1016/j.egypro.2014.10.340

Brownlee AEI, Wright JA (2015) Constrained, mixed-integer and multi-objective optimisation of building designs by NSGA-II with fitness approximation. Appl Soft Comput 33:114–126. https://doi.org/10.1016/j.asoc.2015.04.010

Bryans D, Amstutz V, Girault HH, Berlouis LEA (2018) Characterisation of a 200 kW/400 kWh vanadium redox flow battery. Batteries. https://doi.org/10.3390/batteries4040054

Cao L, Skyllas-Kazacos M, Wang D-W (2018) Solar redox flow batteries: mechanism, design, and measurement. Adv Sustain Syst 2(8–9):1800031. https://doi.org/10.1002/adsu.201800031

Carlucci S, Cattarin G, Causone F, Pagliano L (2015) Multi-objective optimization of a nearly zero-energy building based on thermal and visual discomfort minimization using a non-dominated sorting genetic algorithm (NSGA-II). Energy Build 104:378–394. https://doi.org/10.1016/j.enbuild.2015.06.064

Chastas P, Theodosiou T, Bikas D (2016) Embodied energy in residential buildings-towards the nearly zero energy building: a literature review. Build Environ 105:267–282. https://doi.org/10.1016/j.buildenv.2016.05.040

Chiesa G, Fasano F, Grasso P (2021) A new tool for building energy optimization: first round of successful dynamic model simulations. Energies. https://doi.org/10.3390/en14196429

Chiesa G, Fasano F, Grasso P (2023) Simulated versus monitored building behaviours: sample demo applications of a perfomance gap detection tool in a northern italian climate. In: Sayigh A (ed) Towards net zero carbon emissions in the building industry. Springer International Publishing, Cham, pp 109–133

Clemente A, Costa-Castelló R (2020) Redox flow batteries: a literature review oriented to automatic control. Energies. https://doi.org/10.3390/en13174514

Clevenger CM, Haymaker J (2006) The impact of the building occupant on energy modeling simulations. In: Joint international conference on computing and decision making in civil and building engineering. Montreal, Canada, pp 1–10

Coakley D, Raftery P, Keane M (2014) A review of methods to match building energy simulation models to measured data. Renew Sustain Energy Rev 37:123–141. https://doi.org/10.1016/j.rser.2014.05.007

Comission E (2018) Communication From The Commission To The European Parliament, The Council, The European Economic And Social Committee, The Committee Of The Regions and the european investment bank—A Clean Planet for all. A European strategic long-term vision for a prosperous, modern, competitive and climate neutral economy. Brussels

Comission E (2019) Communication From The Commission To The European Parliament, The Council, The European Economic And Social Committee And The Committee Of The Regions—The European Green Deal. Brussels

Crawley DB, Lawrie LK, Winkelmann FC, Buhl WF, Huang YJ, Pedersen CO et al (2001) EnergyPlus: creating a new-generation building energy simulation program. Energy Build 33(4):319–331. https://doi.org/10.1016/S0378-7788(00)00114-6

Crawley DB, Hand JW, Kummert M, Griffith BT (2008) Contrasting the capabilities of building energy performance simulation programs. Build Environ 43(4):661–673. https://doi.org/10.1016/j.buildenv.2006.10.027

da Silva LT, Dias P, Monteiro R, Vilanova A, Ivanou D, Mendes A (2022) A 25 cm^2 solar redox flow cell: facing the engineering challenges of upscaling. Adv Energy Mater 12(5):2102893. https://doi.org/10.1002/aenm.202102893

D'Agostino D, Tzeiranaki ST, Zangheri P, Bertoldi P (2021) Assessing nearly zero energy buildings (NZEBs) development in Europe. Energ Strat Rev 36:100680. https://doi.org/10.1016/j.esr.2021.100680

Deb K, Pratap A, Agarwal S, Meyarivan T (2002) A fast and elitist multiobjective genetic algorithm: NSGA-II. IEEE Trans Evol Comput 6(2):182–197. https://doi.org/10.1109/4235.996017

Delgado NM, Monteiro R, Mendes A (2021) The first approach to dynamic modeling of a solar vanadium redox flow cell. Nano Energy 89:106372. https://doi.org/10.1016/j.nanoen.2021.106372

Delgado JMPQ, Guimarães AS, Poças Martins J, Parracho DFR, Freitas SS, Lima AGB et al (2023) BIM and BEM interoperability—evaluation of a case study in modular wooden housing. Energies. https://doi.org/10.3390/en16041579

Delgarm N, Sajadi B, Delgarm S, Kowsary F (2016) A novel approach for the simulation-based optimization of the buildings energy consumption using NSGA-II: case study in Iran. Energy Build 127:552–560. https://doi.org/10.1016/j.enbuild.2016.05.052

Deru M, Field K, Studer D, Benne K, Griffith B, Torcellini P, et al (2011) US Department of energy commercial reference building models of the national building stock

Designbuilder (2023) DesignBuilder: EnergyPlus Simulation. https://designbuilder.co.uk/simulation/. Accessed 17 Aug 2023

Divya KC, Østergaard J (2009) Battery energy storage technology for power systems—an overview. Electric Power Syst Res 79(4):511–520. https://doi.org/10.1016/j.epsr.2008.09.017

DOE U (2008) Measurement and Verification (M&V) Guidelines for Federal Energy Projects, V3.0. Office of Scientific and Technical Information (OSTI)

DOE (2018) eQuest: the Quick Energy Simulation Tool. https://doe2.com/equest/index.html. Accessed 19 Aug 2023

Duan C, Zhao Z, Yuan L (2021) Lead-free cesium-containing halide perovskite and its application in solar cells. IEEE J Photovoltaics 11(5):1126–1135. https://doi.org/10.1109/JPHOTOV.2021.3095457

Dunn B, Kamath H, Tarascon J-M (2011) Electrical energy storage for the grid: a battery of choices. Science 334(6058):928–935. https://doi.org/10.1126/science.1212741

Emami S, Martins J, Andrade L, Mendes J, Mendes A (2017) Low temperature hermetic laser-assisted glass frit encapsulation of soda-lime glass substrates. Opt Lasers Eng 96:107–116. https://doi.org/10.1016/j.optlaseng.2017.04.006

Emmett RK, Roberts ME (2021) Recent developments in alternative aqueous redox flow batteries for grid-scale energy storage. J Power Sources 506:230087. https://doi.org/10.1016/j.jpowsour.2021.230087

University of Wisconsin--Madison (2017) Solar Energy L: TRNSYS, a transient simulation program. https://sel.me.wisc.edu/trnsys/. Accessed 19 Aug 2023

Estatística INd. Buildings (No.) by Geographic location at Census date [2021] (NUTS - 2013) (2021) Time period of construction and Dimension of the repairing; Decennial—Statistics Portugal, Population and housing census. Lisboa: INE

Europe S (2002) Global Market Outlook for Solar Power 2022–2026

European Parliament CotEU. Directive 2010/31/EU of the European Parliament and of the Council of 19 May 2010 on the energy performance of buildings. https://eur-lex.europa.eu/legal-content/EN/ALL/?uri=CELEX:32010L00312010

Eurostat (2024) Final Energy Consumption by Sector. In: Eurostat, editor

EVO (2007) International performance measurement & verification protocol. Efficiency Valuation Organisation

Gao H, Koch C, Wu Y (2019) Building information modelling based building energy modelling: a review. Appl Energy 238:320–343. https://doi.org/10.1016/j.apenergy.2019.01.032

García Kerdan I, Raslan R, Ruyssevelt P (2016) An exergy-based multi-objective optimisation model for energy retrofit strategies in non-domestic buildings. Energy 117:506–522. https://doi.org/10.1016/j.energy.2016.06.041

Garwood TL, Hughes BR, O'Connor D, Calautit JK, Oates MR, Hodgson T (2018) A framework for producing gbXML building geometry from point clouds for accurate and efficient building energy modelling. Appl Energy 224:527–537. https://doi.org/10.1016/j.apenergy.2018.04.046

Gentil S, Reynard D, Girault HH (2020) Aqueous organic and redox-mediated redox flow batteries: a review. Curr Opin Electrochem 21:7–13. https://doi.org/10.1016/j.coelec.2019.12.006

Goodenough JB, Park K-S (2013) The Li-ion rechargeable battery: a perspective. J Am Chem Soc 135(4):1167–1176. https://doi.org/10.1021/ja3091438

Green MA (2009) The path to 25% silicon solar cell efficiency: history of silicon cell evolution. Prog Photovoltaics Res Appl 17(3):183–189. https://doi.org/10.1002/pip.892

Guerra Santin O, Itard L, Visscher H (2009) The effect of occupancy and building characteristics on energy use for space and water heating in Dutch residential stock. Energy Build 41(11):1223–1232. https://doi.org/10.1016/j.enbuild.2009.07.002

Hagfeldt A, Boschloo G, Sun L, Kloo L, Pettersson H (2010) Dye-sensitized solar cells. Chem Rev 110(11):6595–6663. https://doi.org/10.1021/cr900356p

Hamdy M, Nguyen A-T, Hensen JLM (2016) A performance comparison of multi-objective optimization algorithms for solving nearly-zero-energy-building design problems. Energy Build 121:57–71. https://doi.org/10.1016/j.enbuild.2016.03.035

Harish VSKV, Kumar A (2016) A review on modeling and simulation of building energy systems. Renew Sustain Energy Rev 56:1272–1292. https://doi.org/10.1016/j.rser.2015.12.040

Harper G, Sommerville R, Kendrick E, Driscoll L, Slater P, Stolkin R et al (2019) Recycling lithium-ion batteries from electric vehicles. Nature 575(7781):75–86. https://doi.org/10.1038/s41586-019-1682-5

Hensen JLM, Lamberts R (2019) Building performance simulation—challenges and opportunities. Routledge, Building performance simulation for design and operation, pp 1–10

Hu Y-S (2016) Development of Sodium-Ion Batteries for Grid-Scale Energy Storage. ECS Meeting Abstracts.;MA2016–03(1):71. https://doi.org/10.1149/MA2016-03/1/71

HZB (2022) World record back at HZB: Tandem solar cell achieves 32.5% efficiency. https://www.helmholtz-berlin.de/pubbin/news_seite?nid=24348;sprache=en. Accessed 15 March 2023

IEA (2022) The future of heat pumps. Paris

Inc. TM (2022) MATLAB version: 9.13.0 (R2022b). The MathWorks Inc.

IRENA (2017) Electricity storage and renewables: costs and markets to 2030. International Renewable Energy Agency, Abu Dhabi

(IRENA) NFLIREA (2023) Solar (Photovoltaic) panel prices. In: Data OWi, editor

ISE F (2023) Photovoltaics Report
Ivanova I, Kiesel K, Mahdavi A (2015) BIM-generated data models for EnergyPlus: a comparison of gbXML and IFC Formats
Iwakiri I, Antunes T, Almeida H, Sousa JP, Figueira RB, Mendes A (2021) Redox flow batteries: materials, design and prospects. Energies. https://doi.org/10.3390/en14185643
Jason G (2016) Using Python and Eppy for a Large National Simulation Study. ASHRAE & IBPSA-USA SimBuild 2016: Building performance modeling conference: ASHRAE/IBPSA-USA, pp 230–7
Jouhara H, Khordehgah N, Almahmoud S, Delpech B, Chauhan A, Tassou SA (2018) Waste heat recovery technologies and applications. Thermal Sci Eng Progress 6:268–289. https://doi.org/10.1016/j.tsep.2018.04.017
Karlsson F, Rohdin P, Persson ML (2007) Measured and predicted energy demand of a low energy building: important aspects when using building energy simulation. Build Serv Eng Res Technol 28(3):223–235. https://doi.org/10.1177/0143624407077393
Kojima A, Teshima K, Shirai Y, Miyasaka T (2009) Organometal halide perovskites as visible-light sensitizers for photovoltaic cells. J Am Chem Soc 131(17):6050–6051. https://doi.org/10.1021/ja809598r
Koohi-Fayegh S, Rosen MA (2020) A review of energy storage types, applications and recent developments. J Energy Storage 27:101047. https://doi.org/10.1016/j.est.2019.101047
Kubwimana B, Najafi H (2023) A novel approach for optimizing building energy models using machine learning algorithms. Energies. https://doi.org/10.3390/en16031033
Larcher D, Tarascon JM (2015) Towards greener and more sustainable batteries for electrical energy storage. Nat Chem 7(1):19–29. https://doi.org/10.1038/nchem.2085
Lee D, Lim MC, Negash L, Choi HL (2018) EPPY based building co-simulation for model predictive control of HVAC optimization. In: 2018 18th international conference on control, automation and systems (ICCAS), pp 1051–5
Li Y (2013) Design of a key establishment protocol for smart home energy management system. In: 2013 fifth international conference on computational intelligence, communication systems and networks, pp 88–93
Li X, Wen J (2014) Review of building energy modeling for control and operation. Renew Sustain Energy Rev 37:517–537. https://doi.org/10.1016/j.rser.2014.05.056
Li M, Lu J, Chen Z, Amine K (2018) 30 years of lithium-ion batteries. Adv Mater 30(33):1800561. https://doi.org/10.1002/adma.201800561
Li B, Hou B, Amaratunga GAJ (2021) Indoor photovoltaics, the next big trend in solution-processed solar cells. InfoMat 3(5):445–459. https://doi.org/10.1002/inf2.12180
Fraas L (2014) Low cost solar electric power
Loyola-González O (2019) Black-box versus white-box: understanding their advantages and weaknesses from a practical point of view. IEEE Access 7:154096–113. https://doi.org/10.1109/ACCESS.2019.2949286
Ma Y, Chen X, Wang L, Yang J (2021) Investigation of smart home energy management system for demand response application. Front Energy Res 9
Mahdavi A, Tahmasebi F (2019) People in building performance simulation. Building performance simulation for design and operation. Routledge, pp 117–45
Mahlia TMI, Saktisahdan TJ, Jannifar A, Hasan MH, Matseelar HSC (2014) A review of available methods and development on energy storage; technology update. Renew Sustain Energy Rev 33:532–545. https://doi.org/10.1016/j.rser.2014.01.068
Mathews I, Kantareddy SN, Buonassisi T, Peters IM (2019) Technology and market perspective for indoor photovoltaic cells. Joule 3(6):1415–1426. https://doi.org/10.1016/j.joule.2019.03.026
Matti OP, Mohamed H, Ala H. Mobo A (2013) New software for multi-objective building performance optimization. In: Proceedings of building simulation 2013: 13th conference of IBPSA: IBPSA, pp 2567–74
Méndez Echenagucia T, Capozzoli A, Cascone Y, Sassone M (2015) The early design stage of a building envelope: multi-objective search through heating, cooling and lighting energy

performance analysis. Appl Energy 154:577–591. https://doi.org/10.1016/j.apenergy.2015.04.090

Mesquita I, Andrade L, Mendes A (2018) Perovskite solar cells: materials, configurations and stability. Renew Sustain Energy Rev 82:2471–2489. https://doi.org/10.1016/j.rser.2017.09.011

Min H, Lee DY, Kim J, Kim G, Lee KS, Kim J et al (2021) Perovskite solar cells with atomically coherent interlayers on SnO_2 electrodes. Nature 598(7881):444–450. https://doi.org/10.1038/s41586-021-03964-8

Mokhtarian Mobarakeh J, Sayyaadi H (2023) A novel methodology based on artificial intelligence to achieve the formost Buildings' heating system. Energy Convers Manage 286:116958. https://doi.org/10.1016/j.enconman.2023.116958

Monfet D, Zmeureanu R, Charneux R, Lemire N (2007) Computer model of a university building using the EnergyPlus program

Muñoz-García AB, Benesperi I, Boschloo G, Concepcion JJ, Delcamp JH, Gibson EA et al (2021) Dye-sensitized solar cells strike back. Chem Soc Rev 50(22):12450–12550. https://doi.org/10.1039/D0CS01336F

Nations U (2015) Paris agreement to the United Nations framework convention on climate change

Nitta N, Wu F, Lee JT, Yushin G (2015) Li-ion battery materials: present and future. Mater Today 18(5):252–264. https://doi.org/10.1016/j.mattod.2014.10.040

NREL (2023) OpenStudio. https://openstudio.net/. Accessed 17 Aug 2023

O'Regan B, Grätzel M (1991) A low-cost, high-efficiency solar cell based on dye-sensitized colloidal TiO_2 films. Nature 353(6346):737–740. https://doi.org/10.1038/353737a0

Our Common Future—Report of the World Commission on Environment and Development. Brundtland: United Nations

Palensky P, Dietrich D (2011) Demand side management: demand response, intelligent energy systems, and smart loads. IEEE Trans Industr Inf 7(3):381–388. https://doi.org/10.1109/TII.2011.2158841

Pandžić H (2018) Optimal battery energy storage investment in buildings. Energy Build 175:189–198. https://doi.org/10.1016/j.enbuild.2018.07.032

Park M, Ryu J, Wang W, Cho J (2016) Material design and engineering of next-generation flow-battery technologies. Nat Rev Mater 2(1):16080. https://doi.org/10.1038/natrevmats.2016.80

Peretto L (2010) The role of measurements in the smart grid era. IEEE Instrum Meas Mag 13(3):22–25. https://doi.org/10.1109/MIM.2010.5475163

Power A (2023) Powering the entire world with solar: surface area and panel requirements. https://www.axionpower.com/knowledge/power-world-with-solar/. Accessed 08 Aug 2023

Priarone A, Silenzi F, Fossa M (2020) Modelling heat pumps with variable EER and COP in EnergyPlus: a case study applied to ground source and heat recovery heat pump systems. Energies. https://doi.org/10.3390/en13040794

Van Rossum GaD, Fred L (2009) Python 3 Reference Manual. CreateSpace

Ramos G, Ghisi E (2010) Analysis of daylight calculated using the EnergyPlus programme. Renew Sustain Energy Rev 14(7):1948–1958. https://doi.org/10.1016/j.rser.2010.03.040

Reinhart CF, Cerezo DC (2016) Urban building energy modeling—a review of a nascent field. Build Environ 97:196–202. https://doi.org/10.1016/j.buildenv.2015.12.001

Risteska Stojkoska BL, Trivodaliev KV (2017) A review of internet of things for smart home: challenges and solutions. J Clean Prod 140:1454–1464. https://doi.org/10.1016/j.jclepro.2016.10.006

Ritchie H, Roser M, Rosado P, CO_2 and greenhouse gas emissions. Our World in Data

Sailor DJ (2008) A green roof model for building energy simulation programs. Energy Build 40(8):1466–1478. https://doi.org/10.1016/j.enbuild.2008.02.001

Santos D (2018) Caracterização e análise crítica do perfil de consumos de energia para aquecimento. Universidade do Porto, Departamento de Engenharia Civil

Santos P: EPPY (2020). https://eppy.readthedocs.io/en/latest/readme.html. Accessed 11 Aug 2023

Shabunko V, Lim CM, Mathew S (2018) EnergyPlus models for the benchmarking of residential buildings in Brunei Darussalam. Energy Build 169:507–516. https://doi.org/10.1016/j.enbuild.2016.03.039

Shin S-H, Yun S-H, Moon S-H (2013) A review of current developments in non-aqueous redox flow batteries: characterization of their membranes for design perspective. RSC Adv 3(24):9095–9116. https://doi.org/10.1039/C3RA00115F

Snaith HJ (2013) Perovskites: the emergence of a new era for low-cost, high-efficiency solar cells. The J Phys Chem Lett 4(21):3623–3630. https://doi.org/10.1021/jz4020162

Sohlberg B, Jacobsen EW (2008) Grey box modelling—branches and experiences. IFAC Proc Vol 41(2):11415–11420. https://doi.org/10.3182/20080706-5-KR-1001.01934

Soloveichik GL (2015) Flow batteries: current status and trends. Chem Rev 115(20):11533–11558. https://doi.org/10.1021/cr500720t

Statista (2023) Cumulative installed solar PV capacity worldwide from 2000 to 2021 (in megawatts)

Summary for Policymakers (2023) Climate Change 2022—Mitigation of Climate Change: Cambridge University Press, pp 3–48

Taoning W, Greg W, Eleanor SL (2021) A Python library for Radiance matrix-based simulation control and EnergyPlus integration. In: Proceedings of building simulation 2021: 17th conference of IBPSA: IBPSA, pp 1604–10

Tardieu A, Behzadi S, Chen J, Farid M (2011) Computer simulation and experimental measurments for an experimental PCM-impregnated office building. In: Proceedings of building simulation 2011: 12th conference of international building performance simulation association

Tokuda N, Kumamoto T, Shigematsu T, Deguchi H, Ito T, Yoshikawa N, et al (1998) Development of a redox flow battery system. Sumitomo electric technical review-english edition, pp. 88–94

Trimble: SketchUp (2023). https://www.sketchup.com/plans-and-pricing/sketchup-pro. Accessed 21 Aug 2023

Turner C, Frankel M, Council U (2008) Energy performance of LEED for new construction buildings. New Build Inst 4(4):1–42

Unit ESR: ESP-r (2018). https://www.esru.strath.ac.uk/Courseware/ESP-r/tour/. Accessed 19 Aug 2023

Vuong E, Kamel RS, Fung AS (2015) Modelling and simulation of BIPV/T in EnergyPlus and TRNSYS. Energy Procedia 78:1883–1888. https://doi.org/10.1016/j.egypro.2015.11.354

Walter Klöpffer BG (2014) From LCA to sustainability assessment. In: Wiley, editor. Life Cycle Assessment (LCA), pp 357–74

Wang W, Luo Q, Li B, Wei X, Li L, Yang Z (2013) Recent progress in redox flow battery research and development. Adv Func Mater 23(8):970–986. https://doi.org/10.1002/adfm.201200694

Weber AZ, Mench MM, Meyers JP, Ross PN, Gostick JT, Liu Q (2011) Redox flow batteries: a review. J Appl Electrochem 41(10):1137–1164. https://doi.org/10.1007/s10800-011-0348-2

Wedege K, Bae D, Smith WA, Mendes A, Bentien A (2018) Solar redox flow batteries with organic redox couples in aqueous electrolytes: a minireview. The J Phys Chem C 122(45):25729–25740. https://doi.org/10.1021/acs.jpcc.8b04914

Worku MY (2022) Recent advances in energy storage systems for renewable source grid integration: a comprehensive review. Sustainability. https://doi.org/10.3390/su14105985

Xu W, Chong A, Karaguzel OT, Lam KP (2016) Improving evolutionary algorithm performance for integer type multi-objective building system design optimization. Energy Build 127:714–729. https://doi.org/10.1016/j.enbuild.2016.06.043

Xu W, Chong A, Lam K, Wang H (2019) A new BIM to BEM framework: the development and verification of an open-source gbXML to EnergyPlus translator for supporting building life cycle performance analysis

Yan D, O'Brien W, Hong T, Feng X, Burak Gunay H, Tahmasebi F et al (2015) Occupant behavior modeling for building performance simulation: current state and future challenges. Energy Build 107:264–278. https://doi.org/10.1016/j.enbuild.2015.08.032

Yang Z, Zhang J, Kintner-Meyer MCW, Lu X, Choi D, Lemmon JP et al (2011) Electrochemical energy storage for green grid. Chem Rev 111(5):3577–3613. https://doi.org/10.1021/cr100290v

Yang C, Chen Y, Li N, Sun Y, Wu R (2020) EnergyPlus and Python co-simulation model to support machine learning-based control of ground-source heat pump system. In: Wang Z, Zhu Y, Wang F, Wang P, Shen C, Liu J (eds) Proceedings of the 11th international symposium on heating, ventilation and air conditioning (ISHVAC 2019). Springer Singapore, Singapore, pp 759–66

Yu S, Cui Y, Xu X, Feng G (2015a) Impact of civil envelope on energy consumption based on EnergyPlus. Procedia Eng 121:1528–1534. https://doi.org/10.1016/j.proeng.2015.09.130

Yu W, Li B, Jia H, Zhang M, Wang D (2015b) Application of multi-objective genetic algorithm to optimize energy efficiency and thermal comfort in building design. Energy Build 88:135–143. https://doi.org/10.1016/j.enbuild.2014.11.063

Zandi H, Kuruganti T, Vineyard EA, Fugate D (2018) Home energy management systems: an overview

Zhang W, Tan S, Lei Y, Wang S (2014) Life cycle assessment of a single-family residential building in Canada: a case study. Build Simul 7(4):429–438. https://doi.org/10.1007/s12273-013-0159-y

Zhang Z, Ding T, Zhou Q, Sun Y, Qu M, Zeng Z et al (2021) A review of technologies and applications on versatile energy storage systems. Renew Sustain Energy Rev 148:111263. https://doi.org/10.1016/j.rser.2021.111263

Zhou B, Li W, Chan KW, Cao Y, Kuang Y, Liu X et al (2016) Smart home energy management systems: concept, configurations, and scheduling strategies. Renew Sustain Energy Rev 61:30–40. https://doi.org/10.1016/j.rser.2016.03.047

Zhu Z (2014) Assessment of the technical potential for multifunction building zero-carbon renovation with EnergyPlus. Int J f Low-Carbon Technol 9(3):178–188. https://doi.org/10.1093/ijlct/cts066

Chapter 29
Advancing Calibration Technologies: A New Approach for Environmental Monitoring System Transmitter Calibration in Regulated Industries

Mirna Osama and Ahmed Y. Shash

Abstract The research highlights the importance of temperature and humidity in product quality control, emphasizing the need for accurate environmental monitoring systems. The existing calibration method for wall-mounted sensors was found to have limitations, including high uncertainty. In response, a new advanced device has been developed to address these shortcomings. This device offers enhanced precision and versatility in calibrating various irregularly shaped wall-mounted sensors, revolutionizing environmental monitoring technology. The updated device represents a significant advancement in ensuring efficient and reliable evaluation of temperature and humidity levels in industrial settings, promising improved product quality control across diverse industries.

29.1 Introduction

The advancement of technology in environmental monitoring systems (EMS) continues to be a focal point in ensuring product quality and operational efficiency across industries. Building upon the foundation laid by previous research on the "Development of a Site Calibration Set for Prefixed Wall Transmitters of EMS in Regulated Industries" (Osama and Shash 2024), this new study aims to enhance the calibration process further by introducing innovative improvements to the existing device.

Temperature and humidity monitoring remain critical factors influencing product quality in various industrial settings, from production lines to sterile environments. The accuracy of these measurements is paramount, particularly in regulated sectors

M. Osama · A. Y. Shash (✉)
Faculty of Engineering and Materials Science, German University in Cairo, Cairo, Egypt
e-mail: ahmed.shash@cu.edu.eg; ahmed.shash@guc.edu.eg

A. Y. Shash
Mechanical Design and Production Dept, Faculty of Engineering, Cairo University, Giza, Egypt

H. Altenbach et al. (eds.), *Lectures Notes on Advanced Structured Materials 3*,
Advanced Structured Materials 226, https://doi.org/10.1007/978-3-031-77403-4_29

like food, medical, and pharmaceutical industries where adherence to stringent standards such as current good manufacturing practices (CGMP) is imperative.

The proposed research seeks to refine the calibration methodology for wall-mounted sensors within EMS by leveraging advancements in technology. By addressing the limitations identified in the previous calibration method, this study aims to streamline the calibration process, reduce time, and cost inefficiencies, and enhance the accuracy and reliability of calibrating odd-shaped wall-mounted sensors.

Drawing inspiration from the initial research's focus on developing a portable micro-environmental chamber for calibration (Osama and Shash 2024), this recent investigation will explore additional enhancements to optimize temperature and humidity control within the chamber. The goal is to create a more efficient and precise calibration procedure that not only saves time and resources but also ensures valid traceability and compliance with regulatory requirements. In addition to eliminating most of the negatives that were discovered in the mentioned research in which the first prototype of the micro-environmental chamber was developed.

This updated research endeavors to push the boundaries of environmental monitoring technology by refining the calibration set for wall-mounted sensors, contributing to improved product quality control and operational efficiency in regulated industries.

Calibrating transmitters that are connected to an environmental monitoring system is a critical process to ensure accurate and reliable data collection. This process involves comparing the output of a transmitter with a known standard and making necessary adjustments to align the transmitter's output with the standard. Transmitter calibration is essential for maintaining the accuracy of measurements in environmental monitoring systems. Over time, transmitters can drift from their calibrated state due to numerous factors such as environmental conditions, mechanical wear, and electrical disturbances. Regular calibration ensures that transmitters provide accurate readings, which is crucial for making informed decisions based on environmental data (Lake et al. 2003).

Regular calibration procedures for temperature and humidity transmitters:

1. Preparation: Ensure the transmitter is professionally installed, with adequate ventilation and correct wiring (Mahan and Yeager 2019).
2. Calibration: Use a reference device to calibrate the transmitter. This may involve adjusting the offset and span settings to ensure accurate readings (Mahan and Yeager 2019).
3. Configuration: Configure the transmitter using its programming interface, setting the output type, range, and measurement units as needed (Mahan and Yeager 2019).

29.2 Methodology

The first micro-environmental chamber prototype that was developed in the former research as shown in Fig. 29.1, had several boxes that the air had to pass by to ensure the desired temperature and humidity.

Such an approach has proven to have major setbacks which were:

- The need for various air blowers to ensure consistent air flow through the entire system.
- The need for a strong peltier.
- The need for a strong heater.
- The need for intense humidification
- The constant loss of energy.
- The stability of the temperature and humidity for a long amount of time was almost impossible to achieve.

Avoiding these drawbacks a second prototype was developed in this study. The second prototype has only two boxes as shown in Fig. 29.2, where one is the control unit containing the pre-programmed arduino microcontroller as shown in Fig. 29.3, display screen, and control buttons as shown in Fig. 29.4; The second box is the micro-environmental chamber which contains an air blower, a cooling peltier, a heater plate, a humidifier, a built-in sensor, and heat insulation lining all over as shown in Fig. 29.4.

The calibration process using the updated version of the micro-environmental chamber is conducted by the following steps:

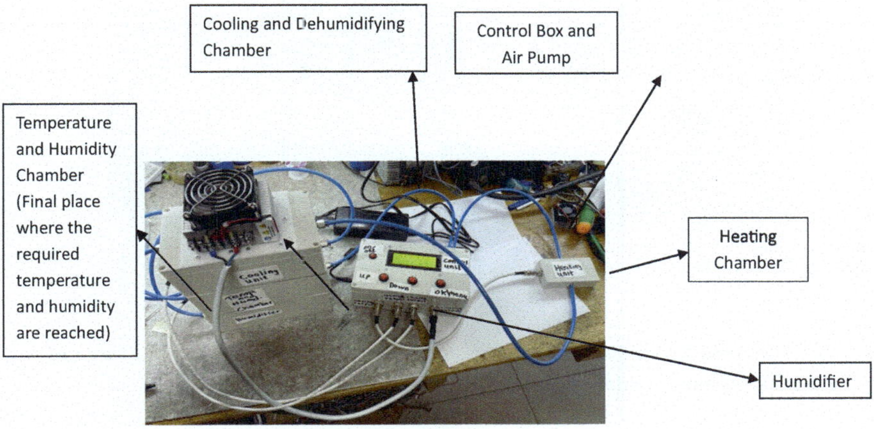

Fig. 29.1 First prototype of the micro-environmental chamber

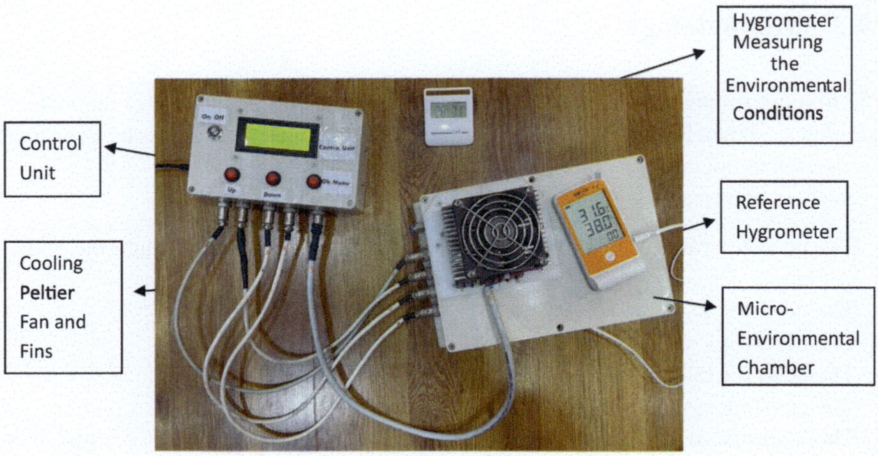

Fig. 29.2 The second prototype of the micro-environmental chamber

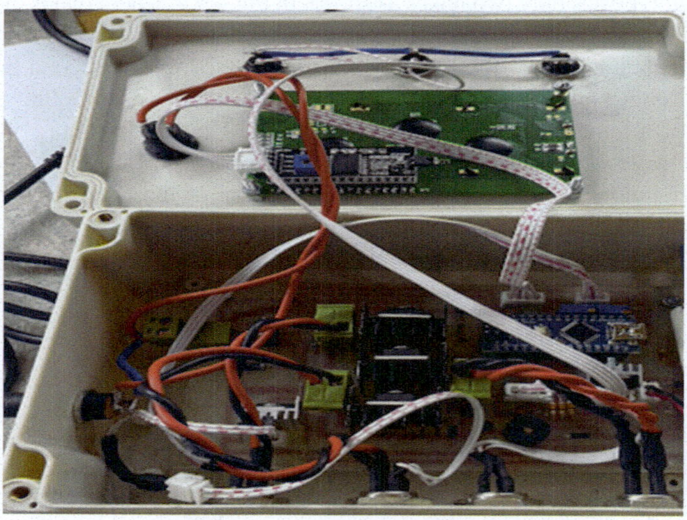

Fig. 29.3 Pre-programmed arduino microcontroller

(1) The micro-environmental chamber covers the entire wall-mounted sensor (unit under test) as it is attached to the wall.
(2) A hygrometer reference with an outer sensor is attached to the chamber from the inside to ensure the proper calibration process as shown in Fig. 29.5.
(3) Using the chamber's controller both temperature and humidity are set.
(4) Results are recorded for both the reference and the unit under test.

According to the psychrometric chart invented by Willis Carrier, the temperature is first changed then the humidity is changed (Teitelbaum et al. 2023). In air

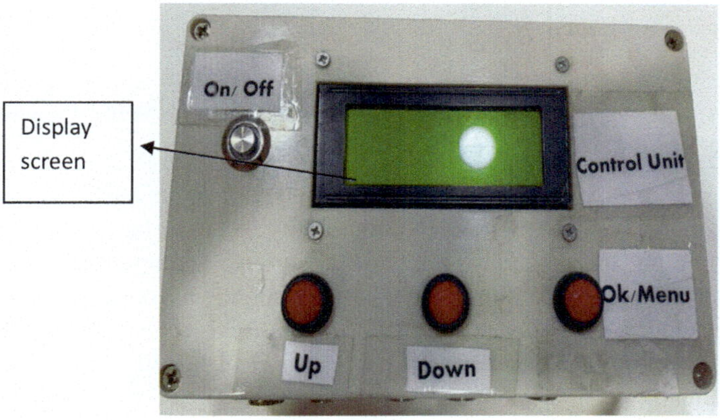

Fig. 29.4 Calibration set control unit

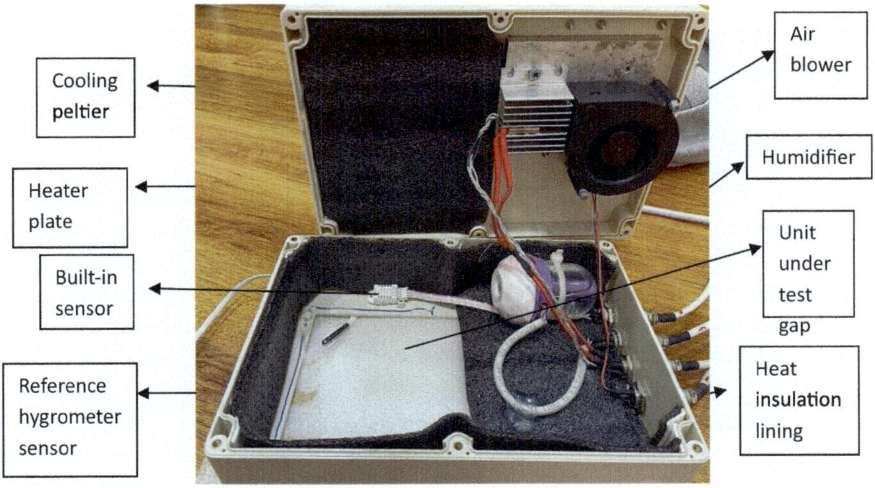

Fig. 29.5 Micro-environmental chamber

conditioning and environmental control, the temperature is adjusted first (horizontal movement on a psychrometric chart) because it directly affects the dry bulb temperature. Then, humidity is adjusted (adding or removing moisture) to achieve the desired conditions (Teitelbaum et al. 2023).

To comprehensively evaluate the performance of the micro-environmental chamber across a range of conditions, evaluating the conditions at both higher and lower temperatures, as well as at a wider range of humidity levels ranging from 35 to 90% RH. This will help ensure that the chamber can maintain stable conditions across the full spectrum of environments it might be used in. Here are the test points:

High Temperature Range

25 °C: Test for stability at humidity levels ranging from 40 to 60% RH. This range is typical for comfortable indoor environments and would be a useful test for applications requiring simulating such conditions.

40 °C: High humidity range: Aim for a RH of 90–95%. This range is often used in constant-humidity tests for evaluating the durability and performance of materials and electronic components under high humidity conditions. Testing at this high humidity level can simulate conditions that might be encountered in tropical environments or during the summer in temperate climates.

Mid-range humidity: Consider testing at an RH of 60–70%. This range is not explicitly mentioned in the sources but represents a midpoint that could be relevant for simulating less extreme, yet still humid, conditions. It could be particularly useful for applications related to storage or operational environments for various products.

Mid-Range Humidity Levels at Existing Temperatures

Testing mid-range humidity levels is a crucial step.

At **20 °C**, test for stability at 55 and 60% RH.

At **15 °C**, test stability at 70 and 75% RH.

Extreme Conditions

High humidity: To evaluate the chamber's ability to reach and maintain exceedingly high humidity levels, such as 95% RH, at a relevant temperature, such as **20 °C** or **25 °C**.

It is important that the relative humidity is the amount of moisture in the air compared to what the air can hold at that temperature. As temperature increases, the air can hold more moisture, so the same absolute amount of water vapor results in a lower RH.

By expanding the testing to include these points one will gain a complete understanding of the capabilities and limitations of the micro-environmental chamber across a broader range of conditions. This will be invaluable for calibrating the chamber for specific applications or making any necessary adjustments to improve its performance.

29.3 Results

The main aim of the results is to determine the temperature and humidity accuracy of the whole developed calibration set.

The accuracy of temperature readings is evaluated across a range of humidities (from low to high) by taking measurements at temperatures between 15 and 45 °C.

Table 29.1 shows the obtained temperature accuracy for each specific temperature.

Table 29.1 Temperature accuracy corresponding to different temperatures

Temperature °C	Temperature accuracy °C
15	±0.6
20	±0.2
25	±0.7
30	±0.5
40	±0.4
45	±0.9

The accuracy for humidity at 25 and 40 °C is also evaluated by maintaining a constant temperature and changing the humidities as time passes and allowing stability time (to ensure temperature and humidity stabilization). Tables 29.2 and 29.3 show the obtained results for humidity accuracy at 25 and 40 °C.

Table 29.2 Humidity accuracy at 40 °C

Relative humidity %	Humidity accuracy
35	±2.3
40	±2.1
50	±2.3
60	±2.6
70	±2.6
80	±3.0
90	±3.0

Table 29.3 Humidity accuracy at 25 °C

Relative humidity %	Humidity accuracy
35	±1.2
40	±1.3
50	±2.3
60	±1.0
70	±2.5
80	±2.8
90	±2.7

29.4 Conclusion

The second prototype is way more enhanced, compact and easily transported to different sites; it allows stability of temperature and humidity as it only has an individual chamber where all temperature and humidity controllers are located therefore no energy is lost as air is moved from one chamber to the other like the first prototype.

The accuracy of the second prototype was discovered to be ±0.9 °C (maximum accuracy reached) for the temperature which is considered an acceptable accuracy but since it is a calibration device, it was aimed to be 0.25 °C; such high accuracy is required to be able to calibrate sensors with lower accuracy levels.

In addition to the temperature accuracy, the humidity accuracy was found to be ±3.0% which is a low accuracy as it was aimed to be ±2.0% to be considered a proper calibration device.

The temperature and humidity accuracy can be enhanced in the upcoming third prototype by installing a more advanced heater, cooling peltier, humidifier, and higher accuracy sensor model. However, the main goal of the study was to overcome the drawbacks of the first prototype and develop a higher accuracy calibration set which was successfully established.

References

Lake BJ, Nuss SE, Carlson DE, Weisberg RH (2003) Calibration procedures and instrumental accuracy estimates of ATLAS air temperature and relative humidity measurements. J Atmos Ocean Technol 20(6):733–750. https://doi.org/10.1175/1520-0426(2003)20%3c733:CPAIAE%3e2.0.CO;2

Mahan JR, Yeager JR (2019) Humidity sensor calibration using saturated salt solutions and microwave transmission measurements. IEEE Trans Instrum Meas 68(2):636–644. https://doi.org/10.1109/TIM.2018.2853878

Osama M, Shash AY (2024) Development of site calibration set for prefixed wall transmitters of environmental monitoring systems in regulated industries. In: Altenbach H, Hitzler L, Johlitz M, Merkel M, Öchsner A (eds) Lectures notes on advanced structured materials 2. Advanced structured materials, vol 203. Springer, Cham. https://doi.org/10.1007/978-3-031-49043-9_24

Teitelbaum E, Miller C, Meggers F (2023) Highway to the comfort zone: history of the psychrometric chart. Buildings 13(3):797. https://doi.org/10.3390/buildings13030797

The manufacturer's authorised representative in the EU is Springer Nature Customer Service Centre GmbH, Europaplatz 3, 69115 Heidelberg, Germany. If you have any concerns regarding our products, please contact ProductSafety@springernature.com

Printed and bound by CPI Group (UK) Ltd, Croydon, CR0 4YY
23/09/2026
02221471-0006